Kinetics and Mechanism

Kinetics and Mechanism

THIRD EDITION

John W. Moore
Eastern Michigan University

Ralph G. Pearson
University of California, Santa Barbara

A WILEY-INTERSCIENCE PUBLICATION

JOHN WILEY & SONS

New York / Chichester / Brisbane / Toronto / Singapore

Library of Congress Cataloging in Publication Data:

Moore, John W.
    Kinetics and Mechanism.

    Revision of: 2nd ed./Arthur A. Frost, Ralph G.
Pearson. 1961.
    "A Wiley-Interscience publication."
    Includes bibliographies and index.
    1. Chemical reaction, Rate of. I. Pearson, Ralph G.
II. Frost, Arthur Atwater, 1909. Kinetics and
mechanism. III. Title.

QD502.M66 1981   541.3'94   81-981
ISBN 0-471-03558-0   AACR2

Printed in the United States of America

10 9 8 7 6 5 4 3 2

PREFACE TO THE THIRD EDITION

It is now 20 years since the second edition of *Kinetics and Mechanism* was written. Obviously, a revision was more than called for. There was also a strong possibility that an up-to-date version would differ so much from the original that a new title and a new literary identity should be created. Indeed the third edition appears quite different from the first and second. A full 80% of the material is new. Nevertheless, enough remains of the old to make its heritage obvious, and we felt that calling it *Kinetics and Mechanism*, the third, was in order.

There is also the spirit of the parent volumes, which we hope remains. That is, the attempt to emphasize the role of kinetics in unraveling reaction mechanisms. Of course in the last 20 years the subject of reaction mechanisms, in organic, inorganic, and biochemistry, has been very highly developed. Many excellent treatises have been written. As a result we did not feel it important to give as many examples of reactions whose mechanisms have been elucidated as in earlier editions.

Instead, we have concentrated more on recent developments in both the experimental and theoretical aspects of kinetics of which there have been many. The routine study of very rapid reactions and the use of computers to solve complex rate equations may be particularly mentioned on the experimental side. Plug and stirred flow, relaxation methods, molecular beam studies, laser applications, and other modern techniques for fast kinetics are described at appropriate points in the text when example reactions that have been studied by these methods are introduced. Chapters 2 and 3 have been expanded to include computerized analysis of experimental data, as well as exercises that require students to apply this technique. Theoretical advances are reflected in an expanded treatment of collision theory, an introduction to the calculation of potential energy surfaces and their use in trajectory studies, and considerable attention to the theory of unimolecular reactions. The concept of control of reaction rates by orbital topology is the major theoretical advance that applies most directly to an understanding of reaction mechanisms.

While reaction mechanism remains the primary goal of most kinetics research, considerable progress has been made in the ability to predict rates of reaction. In addition to *ab initio* theory we include empirical and semiempirical methods for estimating activation energies and reaction rates. We also describe how rate constants obtained from experiments, predicted by theory, or otherwise estimated can be combined in a computer simulation of a complex mechanism to obtain concentration-time behavior of all species involved.

The most important development in gas phase kinetics, and in the chemical physicist's view of kinetics, has been in the study of state-to-state transitions. At this time such research provides information that is too detailed to use in many studies of reaction mechanism. We nevertheless devote considerable attention to state-to-state chemistry, because it represents an important step toward our goal of a complete understanding of chemical reactions.

We hope this volume will make a small contribution towards such understanding.

We acknowledge the aid of a number of institutions and individuals. The libraries of Eastern Michigan University and of the University of Michigan, Ann Arbor, were essential to our efforts. The following persons supplied preprints or reprints that would otherwise have been unavailable: T. Bartfai, G.-M. Côme, David Edelson, V. Gold, Darrel G. Hopper, P. J. Kuntz, Bengt Mannervik, Michael L. Michelsen, William M. Moore, David T. Pratt, B. S. Rabinovitch, Warren L. Reynolds, Robert F. Sawyer, Henry F. Schaefer III, T. S. Sørensen, I. Ugi, and Jacques Weber. Dr. Kenneth W. Hicks read and constructively criticized significant portions of the manuscript. Much of JWM's contribution to this revision was accomplished during a sabbatical leave granted by Eastern Michigan University and supported by the National Science Foundation. Finally, and most importantly, we thank Elizabeth A. Moore. She typed the entire manuscript, did extensive literature searching and other bibliographic work, checked all references for accuracy, and did much of the work of indexing this volume. Without these indispensable efforts this revision would not have been possible.

RALPH G. PEARSON
JOHN W. MOORE

Santa Barbara, California
Ypsilanti, Michigan
November 1980

PREFACE TO THE SECOND EDITION

This revision was made necessary by the substantial advances in chemical kinetics made during the last eight or nine years. The most important gains have been made in the areas of elementary reactions in the gaseous phase and the study of very rapid chemical reactions. The new edition attempts to do justice to these topics at an introductory level.

Since A.A.F. has succumbed to the siren call of quantum mechanics, the revision is almost entirely the work of R.G.P., who must be blamed for its shortcomings.

Thanks are due to Dr. E. W. Schlag for reading much of the revised manuscript and for many helpful suggestions.

RALPH G. PEARSON

Evanston, Illinois
January, 1961

PREFACE TO THE FIRST EDITION

When we started to write this book we were particularly struck by the fact that the existing textbooks on kinetics treated reaction mechanisms in a rather perfunctory style. There are, of course, excellent books on mechanisms particularly of organic reactions in which some mention of the use and value of kinetics is made. However, there seemed to be no work which showed enough of the intimate relationship between kinetics and mechanism to enable the student to understand exactly how much detail of reaction mechanism can be found from reaction kinetics and to understand what the limitations of the kinetic method of studying mechanism are.

A study of the recent literature will show that the great majority of the work on reaction velocities now being done is primarily concerned with trying to find out exactly in what manner the reactions are proceeding. Thus, while the theories of kinetics, mathematical and experimental details, and the calculation of energetics are all of great importance and, we hope, have not been neglected in this book, a neglect of mechanism would be to ignore the most important application of kinetics. Consequently we have included a great deal of rather detailed stereochemical discussion of the reaction steps.

We have not tried to include a catalogue of all the chief kinds of reactions that may be encountered, but we have tried to select a number of varied and fairly typical examples. Even in this we were regretfully forced to forego discussing a number of topics which might properly be included in a course in reaction kinetics, such as heterogeneous reactions and photochemistry, for example.

We hope that the absence of several such topics will be compensated for by the added material on mechanism and that we are presenting a work which will be useful as a textbook for courses in kinetics on the graduate level, and as a reference book for those interested in the study of mechanisms of chemical reactions.

We should like to take this opportunity to thank the following persons who contributed in one way or another to the writing and completion of this book: Professors R. L. Burwell, Jr., L. Carroll King, Ronald P. Bell, Louis P. Hammett, Richard E. Powell, Frank H. Seubold, Frank J. Stubbs, Lars Melander; Misses Elaine Strand, Marianne Fält, Mrs. Lenore Pearson, and Mrs. Faye Frost.

ARTHUR A. FROST
RALPH G. PEARSON

Evanston, Illinois
November, 1952

CONTENTS

Kinetics and Mechanism

ONE

INTRODUCTION

Kinetics is part of the science of motion. In physics the science of motion is termed dynamics and is subdivided into kinematics, which treats the motion of bodies, and kinetics, which deals with the effect of forces on motion. In chemistry no such distinction is made. Kinetics deals with the rate of chemical reaction, with all factors that influence the rate of reaction, and with the explanation of the rate in terms of the reaction mechanism. In its most detailed form a reaction mechanism describes, as a function of time, the relative positions of all microscopic particles whose motion is necessary for the reaction to occur. Since it is concerned with both the motions and the forces among these particles, chemical kinetics might very well be called chemical dynamics.

Chemical kinetics with its emphasis on dynamics may be contrasted with the static viewpoint of chemical thermodynamics. Equilibrium thermodynamics[†] is interested only in the initial and final states of a system; the mechanism whereby the system is converted from one state to another and the time required are of no importance. Time is not one of the thermodynamic variables. The most important subject in thermodynamics is the state of equilibrium, and, consequently, thermodynamics is the more powerful tool for investigating the conditions at equilibrium. Kinetics is concerned fundamentally with the details of the process whereby a system gets from one state to another and with the time required for the transition. Equilibrium can also be treated in principle on the basis of kinetics as that situation in which the rates of the forward and reverse reactions are equal. The converse is not true; a reaction rate cannot be understood on the basis of equilibrium thermodynamics alone. Therefore chemical kinetics may be considered a more fundamental science than chemical thermodynamics. Unfortunately the complexities are such that the theory of chemical kinetics is difficult to apply with accuracy. As a result, we find that thermodynamics can tell with precision the extent of reaction, but only kinetics can tell (perhaps crudely) the rate of the reaction.

[†] Thermodynamics can be applied to nonequilibrium systems. See, for example, references 1 and 2 at the end of this chapter.

1

Theoretical methods for predicting reaction rates are based on a conclusion originally drawn by Arrhenius[3] : the most commonly observed relationship between rate of reaction and temperature implies an energy barrier to reaction. The kinetic-molecular theory can tell us the number of molecular collisions per unit volume per unit time as well as the fraction of those collisions whose energy exceeds some threshold energy. It does not, however, speak to the issue of how large the threshold energy must be, nor to the effectiveness of a collision as a function of relative molecular orientation. These issues can be addressed by using quantum mechanics[4] to evaluate the electronic and nuclear-repulsion energies of the collection of atoms involved in a reaction as a function of suitable bond distances and angles. Motion of the atomic nuclei, whether or not it results in the desired reaction, thus corresponds to movement on a multidimensional "potential" energy surface.[5, 6] Holes or valleys (energy minima) on such a hypersurface correspond to stable molecular species—reactants, products, and intermediates. As a reaction occurs the coordinates of the collection of nuclei change in such a way that the energy rises and a mountain pass must be surmounted. The saddle point or col at the top of the pass corresponds to the *activated complex* or *transition state*. When allowance is made for the zero-point vibrational energies of reactant(s) and activated complex, the height of the col above the valley floor is the energy barrier to reaction.

A variety of paths or trajectories are available to a collection of nuclei passing over an energy barrier, and many of them may be used, depending on initial velocities and momenta of the nuclei. Large, high-speed digital computers have made possible accurate quantum-mechanical calculations of energy surfaces for reactions involving small numbers of nuclei and electrons. Once the energy surface is known, *molecular dynamics* calculations, in which the (usually classical) trajectories of collections of nuclei are followed over time, can be carried out. By starting with weighted random initial conditions, repeating the procedure a large number of times, and averaging the results, a computer can arrive at reaction rates and evaluate the parameters on which those rates depend.[7]

Such lengthy calculations can be avoided if we concentrate on the minimum-energy path from ground-state reactants to activated complex. This path is called the *reaction coordinate*. If it is assumed that reactants and activated complexes are in equilibrium, statistical mechanics may be used to calculate the concentration of activated complexes. This concentration can be combined with the speed at which activated complexes pass along the reaction coordinate toward products, enabling the reaction rate to be evaluated. This is the basis of the transition-state theory or theory of absolute reaction rates developed principally by Eyring.[8] In cases where statistical mechanics is difficult to apply from first principles, a "thermodynamic" approach may be taken. The increase in thermodynamic parameters such as enthalpy, entropy, and Gibbs free energy on going from reactants to activated complex may be obtained from rate data and used to estimate properties of the activated complex.

REACTION MECHANISMS

From the chemist's molecular viewpoint, the most interesting aspect of the study of reaction rates is the insight it provides into the mechanism of a reaction. *The dependence of rate of reaction on concentrations of reactants, temperature, and other factors is the most general method for weeding out unsuitable reaction mechanisms.* Here the term mechanism is used in the classical sense (developed chiefly by physical chemists) to mean all the individual collisional or other elementary processes involving molecules (atoms, radicals, and ions included) that take place simultaneously or consecutively to produce the observed overall reaction. For example, the rate of the reaction

$$H_2(g) + Br_2(g) = 2 HBr(g) \qquad 498\,K < T < 574\,K$$

was found by Bodenstein and Lind[9] to be proportional to the concentration of H_2 and to the square root of the concentration of Br_2. Furthermore, the rate was inhibited by the increasing concentration of HBr as the reaction proceeded. These observations are not consistent with a mechanism involving bimolecular collisions of a single molecule of each kind. The currently accepted mechanism is considerably more complicated, involving the dissociation of bromine molecules into atoms followed by reactions between atoms and molecules:

$$Br_2 \rightleftharpoons 2\,Br$$

$$Br + H_2 \rightarrow HBr + H \qquad slow$$

$$H + Br_2 \rightarrow HBr + Br \qquad fast$$

$$H + HBr \rightarrow H_2 + Br \qquad fast$$

It should be clear from this example that the mechanism cannot be predicted from the overall stoichiometry. This point is emphatically reiterated in the case of gas-phase formation of water from its elements:

$$2 H_2 + O_2 \rightarrow 2 H_2O$$

The reaction certainly does not involve simultaneous, trimolecular collisions of $2 H_2$ with O_2. As many as 40 elementary steps have been suggested for the mechanism, and about 15 steps are needed to account for the slow reaction under simplified conditions.[10]

It cannot be overemphasized that a reaction mechanism is essentially a theory that has been devised to explain currently known experimental facts, such as the overall stoichiometry and the dependence of reaction rate on concentrations, temperature, or other variables. In general such experimental facts can be interpreted in several ways; that is, there are several mechanisms consistent with the data. Further experimentation may eliminate some of these, but even if only one mechanism remains that is in agreement with all the known facts, there is no assurance that it is unique or that new experiments will not add evidence discrediting it. A case in point is the reaction of $H_2(g)$ with $I_2(g)$. Because its rate is directly proportional to the concentrations of H_2 and I_2, this reaction was thought for over 60 years to occur in a single step involving bimolecular collisions of H_2 and I_2. However, a reinterpretation of the variation of reaction rate with temperature suggested that a mechanism involving atom–molecule collisions (like the mechanism of the $H_2 + Br_2$ reaction) could account for part of the reaction above 633 K.[11] Furthermore, the supposedly bimolecular $H_2 + I_2$ step has been shown more recently[12] to probably involve individual I atoms rather than I_2 molecules. Thus one must be careful not to place absolute faith in any mechanism (or any other theory, for that matter).

As with other theories, new information can modify a reaction mechanism, rather than eliminate it completely. An example is provided by the nitration of benzene and other aromatic compounds (ArH):

$$ArH + HNO_3 \rightarrow ArNO_2 + H_2O$$

Until recently this represented an example of a mechanism that was considered as firmly established as a mechanism could be.[13] The steps may be written as

$$H^+ + HNO_3 \xrightleftharpoons{\text{fast}} H_2ONO_2^+ \rightarrow H_2O + NO_2^+$$

$$NO_2^+ + ArH \rightarrow HArNO_2^+ \xrightarrow{\text{fast}} ArNO_2 + H^+$$

Depending on circumstances, the formation of NO_2^+ or the reaction of NO_2^+ with ArH may be the slowest, or rate-determining, step.

It now appears that the initial reaction of NO_2^+ with the aromatic molecule is an electron transfer to form a pair of free radicals, which then collapse to the intermediate $HArNO_2^+$.[14]

$$NO_2^+ + ArH \rightarrow NO_2 \cdot + ArH^+ \cdot \rightarrow HArNO_2^+$$

The evidence for the modified mechanism consists of two parts. The first is the experimental demonstration that energetically an electron transfer was favored, except for benzene itself, and perhaps toluene. The second is a review of old data on intramolecular selectivity (*ortho* versus *para* substitution) and the demonstration that it could be better explained by the radical pair intermediate.

Many students find it difficult to accept the fact that a theory is fundamentally different from an experimental fact, and that it can never be taken as completely established. Are there not some theories, such as the atomic theory, that are so in accord with countless observations that they can be considered as fact? The following discussion may be helpful.[15]

A theory is a mental model constructed to explain a set of observations on a subject. Like any model, it has some, but not all, of the properties of the real subject. Our view of the atom is like that of an astronomer on Jupiter who observes human beings on Earth with a superpowerful telescope. He sees us as black dots scurrying about engaged in various forms of activity. If our Jovian astronomer is as clever as we are, he will eventually develop a model of humans that has great detail. He may deduce that we are bipedal, that we communicate by sound waves, even that we have hands with an opposed thumb. Still it is clear that we, at least, would never mistake his model for a human being, if we met it face to face.

Despite these difficulties, there are theories that are so reasonable and so in accord with all experience, that we accept them as essentially true. The same applies to reaction mechanisms. The justification for this becomes apparent when it is observed that a mechanism can successfully predict reaction products or the optimum conditions for running a reaction. That is, a theory is tested by the pragmatic criterion: Does it work? Does it make useful predictions about experiments as yet undone? When it no longer works, of course, we must abandon it.

In most cases, especially those involving molecules that contain numerous atoms and have more complicated structures, it is desirable to infer more than just the collection of elementary processes that constitutes the classical reaction mechanism mentioned above. Organic, and more recently, inorganic chemists have therefore broadened the concept of reaction mechanism to include a *detailed stereochemical picture* of each step in the overall reaction. This concept implies a knowledge not only of the composition of the activated complex in terms of the various atoms and molecules of reactants, but also of the geometry of the activated complex in terms of interatomic distances and angles. For example, it is generally accepted that the conversion of hypochlorite ion to chlorate ion occurs in two steps:

$$ClO^- + ClO^- \rightarrow ClO_2^- + Cl^- \quad \text{slow}$$
$$ClO_2^- + ClO^- \rightarrow ClO_3^- + Cl^- \quad \text{fast}$$

The formation of chlorite ion is the slower, rate-limiting step, and the formation of chlorate ion is rapid.[16] This can be deduced from the fact that the rate of formation of chlorate ion is proportional to the square of the hypochlorite concentration. Furthermore, separate experiments show that the reaction of chlorite ion with hypochlorite ion is fast. These facts would suffice to fix a probable mechanism according to the older definition. The new approach adds to the mechanism a picture such as (1.1) for the transfer of an oxygen atom between ions.

$$\text{Cl—O}^- + \text{Cl—O}^- \rightarrow \begin{bmatrix} \text{Cl} \cdots \text{O—Cl} \\ | \\ \text{O} \end{bmatrix}^{2-} \rightarrow \begin{bmatrix} \text{Cl—O} \\ | \\ \text{O} \end{bmatrix}^- + \text{Cl}^- \tag{1.1}$$

This stereochemical representation is guessed at from chemical intuition and experience. It enables the reaction to be classified as a member of a large class of similar reactions, a nucleophilic displacement of one base (Cl^-) by another (ClO^-). This classification encourages us to focus attention on the making and breaking of bonds between atoms and to attempt to understand the influence of minor alterations in molecular structure. In the case under discussion, for example, we might consider the effect of protonating the reactant ions by changing pH or the effect of changing the halogen from chlorine to, say bromine. We can better understand any one reaction, in other words, by drawing on a large body of information on similar reactions.[17] The stereochemical picture also suggests the possibility that the mechanism may be more complex in that the oxygen atom that leaves one hypochlorite ion may not be the same as the oxygen that appears on the other. Thus the solvent might be involved in the oxygen transfer,[18] possibly as shown in (1.2). In this particular example the products are the same whether the

$$2ClO^- + H_2O \rightarrow \begin{bmatrix} O - Cl \\ \vdots \\ O - H \\ | \quad \vdots \\ H \cdots O - Cl \end{bmatrix}^{2-} \rightarrow \begin{bmatrix} O - Cl \\ | \\ O \end{bmatrix}^- + Cl^- + H_2O \quad (1.2)$$

solvent is involved or not, and so whether (1.1) or (1.2) is more accurate may be of theoretical significance only. When we reflect, however, that hydrolysis of an organic halide may lead to an alcohol with an inversion of original configuration, retention of configuration, or a mixture of the two, then the necessity for a detailed and pictorial representation of a mechanism becomes more apparent.

Quite recently physical chemists have begun to develop techniques by which an even more detailed concept of mechanism can be elucidated. Because there is a distribution of energy states for reactants, products, and activated complexes, each step of the classical mechanism can be thought of as an average over a large number of distinct but related subelementary processes. In the most detailed formulation each subelementary process could be completely defined in terms of molecular variables such as momenta characterizing translational energy states and quantum numbers characterizing vibrational, rotational, electronic, or other energy states that are internal to a molecule. Although ultimate detail has not yet been attained, steps are being taken in that direction. For example, when atomic chlorine reacts with hydrogen iodide in the gas phase at very low pressures,

$$Cl + HI \rightarrow HCl + I \qquad \Delta E \simeq -133\,kJ\,mol^{-1}$$

the average time between collisions of product HCl molecules and other molecules is long enough that the HCl molecules lose energy by emission of infrared radiation rather than by collision. This infrared chemiluminescence can be used to determine the distribution among the vibrational levels of HCl of the energy released by the reaction.[19] Such experiments show that product HCl molecules most commonly have vibrational quantum number $v = 3$, and that the $v = 4$ and $v = 2$ levels are next most highly populated. To achieve the Boltzman distribution of vibrational level populations that is observed when the reaction is run at more normal pressures requires that physical, inelastic-collisional processes occur following the elementary chemical reaction. Furthermore, the elementary chemical reaction actually consists of several processes, each leading to a different vibrational level of the product HCl molecule. The ultimately detailed mechanism, then, indicates for each step not only the stereochemistry in terms of the average position of each atomic nucleus, but also the motions of the nucleus about that average position. Rotational motions of the collection of nuclei and changes in electron density distribution corresponding to different electronic states are also within the purview of this most recently developed concept of mechanism.

METHODS FOR ELUCIDATING REACTION MECHANISMS

Whereas studies of reaction rates are very useful in determining the individual steps of the classical reaction mechanism, such studies are of limited value in giving stereochemical or more intimate molecular details. Therefore it is usually necessary to refer to other methods of obtaining information about mechanisms. Many of these are incomplete in themselves and are best used in combination with kinetic evidence. Some of them are mentioned now, with further and more complete illustrations given later in the text.

Reaction Stoichiometry

The most important circumstantial evidence as to reaction mechanism is the identity of the products formed. This seems so obvious that it is difficult to believe that rate studies have frequently been reported in which the exact nature of the products was unknown. Such, however, is the case. As an example of the type of reasoning involved after the products have been identified, consider the alkaline hydrolysis of a simple ester such as ethyl acetate. Formation of ethanol and acetate ion indicates reaction of a hydroxide ion with an ester molecule, breaking one C—O bond:

$$CH_3-\overset{\overset{\displaystyle O}{\|}}{C}-OC_2H_5 + OH^- \rightarrow CH_3-\overset{\overset{\displaystyle O}{\|}}{C}-O^- + C_2H_5OH$$

That the reaction occurs by attack of OH^- on the ester is supported by the fact that the rate of formation of acetate ion is proportional both to the concentration of ester and to the concentration of hydroxide ion.[20]

While discussing stoichiometry it should be mentioned that it is equally important to know what the reactants are. Sometimes it is possible for a reactant to be transformed by a rapid reaction before a rate study begins. An example is hydrolysis of bis-2,4-pentanediononickel(II) in acidic, aqueous solution.[21] Below pH = 3, one 2,4-pentanedione ligand[†] is removed from the nickel(II) ion within the time of mixing the solutions, and the reaction actually studied is the second step of a two-step process. Thus it is important that the identity of both reactants and products be carefully ascertained before postulating a mechanism.

Stereochemical Evidence

Considerable information as to the intimate details of a mechanism can be gained by comparing the stereochemistry of reactants with that of products. In the ester hydrolysis reaction above, for example, if an optically active alcohol, RR'CHOH, is used instead of ethanol in forming the ester, the original alcohol can be recovered without racemization or inversion after hydrolysis.[22] This suggests strongly that in both the synthesis of the ester and its hydrolysis the C—O bond that is formed or broken is the one connecting the carbonyl carbon to oxygen, and that the C—O

[†] The anion of acetylacetone, $C_5H_7O_2^-$.

bond involving the alcohol carbon remains intact. So far as reaction mechanism is intended to include stereochemical detail, methods for determining structures of reactants and products are investigative tools whose importance equals that of kinetic studies.

Use of Isotopes

The conclusion that the bond between carbonyl carbon and oxygen is the one broken can be confirmed by the use of hydroxide ion enriched with ^{18}O. The heavy oxygen is then found in the anion of the acid rather than in the alcohol.[23] The question of solvent participation in the transfer of oxygen between two hypochlorite ions could be approached in a similar manner[18] using $H_2^{18}O$. To rule out solvent participation in this case it would be necessary to show that under the experimental conditions oxygen exchange between hypochlorite ion and the solvent[24] was not as rapid as formation of chlorite ion.

Detection of Intermediates

One of the best tests of a multistep mechanism is illustrated above in the case of the hypochlorite ion disproportionation. The intermediate ClO_2^- was prepared independently and was shown to react rapidly enough that the proposed mechanism was not contradicted. Often, however, proposed intermediates are extremely reactive, short-lived, and difficult to study independently. This is almost invariably true in the case of free atoms or radicals, for example. In such cases evidence of an intermediate's existence can be sought using spectroscopic or other physical techniques, substances can be added to trap the intermediate by means of known, rapid reactions that form stable products, or the suspected intermediate may be generated *in situ* by a well-characterized reaction and its incremental effect observed. Examples of these techniques include the use of electron spin resonance (ESR) spectroscopy to observe and often identify free radicals, detection of simple inorganic ions in solution by means of specific ion electrodes, trapping of radicals by means of NO, I_2, or other scavengers, and generation of atoms by photochemical dissociation. The last-mentioned technique was instrumental in elucidating the mechanism of the $H_2 + I_2$ reaction mentioned earlier.[12]

Study of Closely Related Systems

It is often valid to assume that minor changes in molecular structures of reactants or in the environment in which reactant molecules collide will not change the sequence of steps by which a reaction occurs. Thus additional information as to mechanism can be obtained by studying the effect on rate of reaction when substituents on reactants, ionic strength of solution, or the solvent is varied. For example, in the alkaline hydrolysis of esters mentioned earlier, if a series of ethyl benzoates is studied kinetically it is found that substituents (such as the nitro group) that withdraw electrons from the carbonyl group increase the rate of hydrolysis, while electron-repelling substituents (such as methoxyl) decrease the rate.[25] The interpretation is that increased positive charge on the carbonyl carbon facilitates addition of the negative hydroxide ion to form an intermediate containing tetrahedral carbon. The effect of changing the dielectric constant of the solvent[26] or the ionic strength can also be used in conjunction with modern theories as strong evidence that the reaction is between a negative ion and a neutral, but polar, molecule.

Although the study of closely related systems can provide much valuable information, one must always beware of generalizing beyond the known experimental facts. This is certainly true in the case of alkaline hydrolysis of esters, where four different types of mechanisms have been observed.[27] Clearly one cannot make too large a change of substituent or surroundings without having alternative reaction pathways become dominant.

Micro Techniques

To obtain data to support the ultimately detailed concept of mechanism mentioned earlier, it is necessary to study individual subelementary processes where the energy states of reactants and products are carefully controlled and measured. A variety of such micro techniques are available,[28] of which molecular beam scattering is the most generally applicable. In a typical experiment reactant molecules are formed into two collimated beams which then intersect in a scattering region of ambient pressure below 10^{-6} torr. The angular and velocity distributions of scattered reactants and products can be measured, and it is often possible to vary the velocity and even the orientation of reactants in the crossed beams. Techniques by which the distribution of internal vibrational and rotational states of reactants and products may be selected or determined are also available. Such methods have tremendous potential for increasing the level of detail at which we can understand those atomic and energetic rearrangements that occur during a gas-phase reaction. At the same time the micro techniques can provide confirmation of concepts previously supported by more circumstantial evidence. For example, reactions of oriented molecules show definite steric effects whose magnitudes are roughly proportional to van der Waals radii,[29] and molecular beam experiments have helped to confirm the conclusion that the $H_2 + I_2$ reaction does not occur in one step.[30]

Agreement with Theory

Although precise prediction of rates of reaction is still quite difficult, one can nevertheless make strong arguments against otherwise plausible mechanisms on the basis of their contradiction of accepted theory. For example, theory can assign a maximum value that a particular rate can have. Mechanisms requiring a significantly larger value can be ruled out. Thermodynamic data on reactants and products can also be used to assign maximum or minimum values of rates. Quite often an elementary process can be assigned a high activation energy on the basis of the symmetry or topology of molecular orbitals for reactants and products.[31] This method, too, has been applied to the $H_2 + I_2$ reaction.[32] Another general rule derived from theory is that elementary processes are usually unimolecular or bimolecular, since collisions of higher molecularity occur much less often (Chapter 4). Other generalities that contribute to a kineticist's intuition as to what constitutes a reasonable mechanism are described later in the text. It is true, of course, that any rule based on theory cannot be applied with absolute confidence, but a mechanism including steps that contradict theory will require strong experimental support before it will gain acceptance.

REFERENCES

1. I. Prigogine, *Introduction to Thermodynamics of Irreversible Processes*, Wiley, New York, 1967.

2. W. Yourgrau, A. van der Merwe, and G. Raw, *Treatise on Irreversible and Statistical Thermophysics*, Macmillan, New York, 1966.

3. S. Arrhenius, *Z. Phys. Chem. (Leipz.)*, **4**, 226 (1889).

4. H. Eyring, J. Walter, and G. Kimball, *Quantum Chemistry*, Wiley, New York, 1944, Chap. 16.

5. A. Marcelin, *Ann. Phys.*, **3**, 158 (1915).

6. S. Glasstone, K. J. Laidler, and H. Eyring, *The Theory of Rate Processes*, McGraw-Hill, New York, 1941, Chap. 3.

7. F. T. Wall, L. A. Hiller, and J. Mazur, *J. Chem. Phys.*, **29**, 255 (1958); *ibid.*, **35**, 1284 (1961).

8. H. Eyring, *J. Chem. Phys.*, **3**, 107 (1935).

9. M. Bodenstein and S. C. Lind, *Z. Phys. Chem.*, **57**, 168 (1906).

10. R. R. Baldwin and R. R. Walker, *Essays Chem.*, **3**, 1 (1972).

11. S. W. Benson and R. Srinivasan, *J. Chem. Phys.*, **23**, 200 (1955).

12. J. H. Sullivan, *J. Chem. Phys.*, **46**, 73 (1967).

13. C. K. Ingold, *Structure and Mechanism in Organic Chemistry*, 2nd ed., Cornell University Press, Ithaca, New York, 1969, Chap. 6.

14. C. L. Perrin, *J. Am. Chem. Soc.*, **99**, 5516 (1977).

15. M. Walker, *The Nature of Scientific Thought*, Prentice-Hall, Englewood Cliffs, N.J, 1963.

16. F. Foerster and P. Dolch, *Z. Elektrochem.*, **23**, 137 (1917).

17. H. Taube, *Rec. Chem. Prog.*, **17**, 25 (1956); J. O. Edwards, *J. Chem. Educ.*, **31**, 270 (1954).

18. J. Halperin and H. Taube, *J. Am. Chem. Soc.*, **72**, 3319 (1950); T. C. Hoering, F. T. Ishimori, and H. O. McDonald, *ibid.*, **80**, 3876, (1958).

19. D. H. Maylotte, J. C. Polanyi, and K. B. Woodall, *J. Chem. Phys.*, **57**, 1547 (1972).

20. G. M. Blackburn and H. L. H. Dodds, *J. Chem. Soc., Perkin Trans., II*, **1974**, 377.

21. J. W. Moore and R. G. Pearson, *Inorg. Chem.*, **5**, 1523 (1966).

22. B. Holmberg, *Ber.*, **45**, 2997 (1912).

23. M. Polanyi and A. L. Szabo, *Trans. Faraday Soc.*, **30**, 508 (1934).

24. M. Anbar and H. Taube, *J. Am. Chem. Soc.*, **80**, 1073 (1958).

25. K. Kindler, *Ann. Chem.*, **450**, 1 (1926); **452**, 90 (1927); **464**, 278 (1928); R. W. Taft, M. S. Newman, and F. H. Verhoek, *J. Am. Chem. Soc.*, **72**, 4511 (1950); L. B. Jones and T. M. Sloane, *Tetrahedron Letters*, **1966**, 831.

26. A. J. Parker, *Chem. Rev.,* **69**, 1 (1969); M. Balakrishnan, G. V. Rao, and N. Venkatasubramanian, *J. Chem. Soc., Perkin Trans. II,* **1974**, 1093.

27. C. K. Ingold, *Structure and Mechanism in Organic Chemistry,* 2nd ed., Cornell University Press, Ithaca, New York, 1969, p. 1131; R. F. Pratt and T. C. Bruice, *J. Am. Chem. Soc.,* **92**, 5956 (1970).

28. R. D. Levine and R. B. Bernstein, *Molecular Reaction Dynamics,* Oxford University Press, New York, 1974, Chap. 6.

29. G. Marcelin and P. R. Brooks, *J. Am. Chem. Soc.,* 97, 1710 (1975).

30. S. B. Jaffe and J. B. Anderson, *J. Chem. Phys.,* **51**, 1057 (1969); R. N. Porter, D. L. Thompson, L. B. Simms, and L. M. Raff, *J. Am. Chem. Soc.,* **92**, 3208 (1970).

31. R. B. Woodward and R. Hoffman, *J. Am. Chem. Soc.,* 87, 395 (1965); R. B. Woodward and R. Hoffman, *The Conservation of Orbital Symmetry,* Verlag Chemie, Weinbeim/ Bergstrasse, 1970; R. G. Pearson, *Symmetry Rules for Chemical Reactions,* Wiley-Interscience, New York, 1976.

32. R. Hoffman, *J. Chem. Phys.,* **49**, 3739 (1968).

EMPIRICAL TREATMENT OF REACTION RATES

KINDS OF SYSTEMS

A closed system is one in which no matter is gained or lost, as in a typical reaction in a liquid phase in a flask or in a gas-phase reaction taking place in a closed vessel. An open system, or flow system, involves gain or loss of matter and is exemplified by reactions of a flowing gas in a heated tube or at a solid catalyst (as in an automobile's catalytic converter), by a flame, and by living organisms, which exchange nutrients and metabolic products with the surroundings. Precise rate measurements, as well as their theoretical interpretation, are more convenient in the case of closed systems, and in what follows it is to be assumed that closed systems are always being considered unless an open system is explicitly indicated. The theory of the open system may be obtained by generalization of the theory of the closed system.

Both closed and open systems may be either homogeneous or heterogeneous, but most open systems are heterogeneous and may have concentration gradients in the reaction zone that make such reactions quite different from typical homogeneous reactions. Heterogeneous closed systems are of particular interest in connection with surface catalysis, but, because of their more fundamental value to chemical kinetic theory, homogeneous closed systems are the principal subject of this treatise. It should be pointed out that a homogeneous system does not necessarily imply a reaction whose occurrence is homogeneously distributed. The walls of a container, for example, often influence rates, and experiments must be designed to check for such effects. Conversely, a heterogeneous system may be such that a reaction occurs homogeneously in one of the system's phases. Examples are provided by many gas–liquid systems, where, provided that dissolution and diffusion of the gas are sufficiently rapid, the reaction has the same rate everywhere in the liquid phase.

Isothermal systems are convenient from a theoretical standpoint because temperature can be considered an independent variable. However, it must not be forgotten that reaction rates and rate constants are almost invariably temperature dependent. Nonisothermal or approximately adiabatic systems are of practical importance in connection with rapid exothermic reactions such as flames and explosions, where the heat of reaction cannot be conducted away rapidly enough.

DEFINITION OF REACTION RATE[†]

A general equation for any chemical reaction can be written as

$$0 = \sum_{B} \nu_B B \tag{2.1}$$

where B is the chemical symbol for a molecule, atom, ion, or radical, and ν_B is the stoichiometric number (positive for a product and negative for a reactant) for species B. The stoichiometric numbers must satisfy the usual conditions of conservation of atoms and conservation of charge for a balanced chemical equation. The extent of reaction ξ is defined by

$$n_B = n_{B,0} + \nu_B \xi \tag{2.2}$$

where n_B is the amount of substance B and $n_{B,0}$ is a chosen amount of B (such as that at the beginning of the reaction) that fixes the zero of ξ. An alternative form of this definition is

$$dn_B = \nu_B d\xi \tag{2.3}$$

The rate of reaction $\dot{\xi}$ is defined as the time rate of increase of the extent of reaction

$$\dot{\xi} = \frac{d\xi}{dt} = \nu_B^{-1} \frac{dn_B}{dt} \tag{2.4}$$

A derivative is used because the rate almost invariably changes as time goes on. So long as (2.1) represents a single elementary process, the rate defined in (2.4) is the same regardless of which species B is observed. If the reaction mechanism involves more than one step, however, it is possible for one reactant to disappear at a rate quite different from that at which other reactants are consumed or at which products are formed. This can occur, for example, if one reactant is involved in two or more elementary processes, only one of which includes another reactant or a particular product. It can also occur if significant concentrations of intermediate species build up during the reaction. Observation of rates of disappearance of reactants or rates of formation of products significantly different from those predicted by (2.4) is good evidence that a reaction mechanism involves several steps or that parallel side reactions are taking place.

[†] The conventions and terminology developed here are those recommended by IUPAC. Slightly different conventions are recommended by the Committee on Data for Science and Technology of the International Council of Scientific Unions. See reference 1.

As an example of the general definition stated above consider the reaction

$$N_2 + 3H_2 = 2NH_3$$

The chemical equation can be rewritten

$$0 = 2NH_3 - N_2 - 3H_2$$

and so

$$\dot{\xi} = \frac{1}{2} \frac{dn_{NH_3}}{dt} = -\frac{dn_{N_2}}{dt} = -\frac{1}{3} \frac{dn_{H_2}}{dt}$$

Note that the signs of the ν_B ensure that the rate of reaction will be positive. It is also important to realize that the rates of increase in amounts of substances involved in the reaction are not necessarily the same as the rate of reaction. Because 2 mol of NH_3 is produced for every 1 mol of N_2 that reacts,

$$\frac{dn_{NH_3}}{dt} = -2 \frac{dn_{N_2}}{dt} = 2\dot{\xi}$$

and the amount of NH_3 increases twice as rapidly as the amount of N_2 decreases. If reaction rates are not reported in terms of $\dot{\xi}$, it is necessary to specify for which substance (e.g., N_2, H_2, or NH_3 in this example) the rate of increase in amount of substance has been measured.

The definition of rate of reaction in terms of ξ, a variable that involves amount of substance, is completely general. It is independent of the choice of B, of the type of system (open or closed, homogeneous or heterogeneous), and of properties (such as the volume of the system) that might vary with time. However, for a homogeneous reaction occurring in a volume V that is constant over time, it is convenient and conventional to make the rate an intensive property (independent of the volume of the system) by specifying $\dot{\xi}/V$, the rate of reaction per unit volume, or ν_B, the rate of increase of the concentration (c_B or $[B]$) of species B.

$$\nu_B = \frac{dc_B}{dt} = \frac{d[B]}{dt}$$

Here again it is important to indicate for which substance the rate of increase of concentration has been measured, since this rate depends on the stoichiometric numbers. In the case of the reaction of N_2 with H_2 at constant volume, for example, the rate of reaction per unit volume is

$$\frac{\dot{\xi}}{V} = \frac{1}{2} \frac{d[NH_3]}{dt} = -\frac{d[N_2]}{dt} = -\frac{1}{3} \frac{d[H_2]}{dt} = \frac{1}{2} \nu_{NH_3} = -\nu_{N_2} = -\frac{1}{3} \nu_{H_2}$$

Since this book is primarily concerned with homogeneous reactions, the rate of increase of concentration ν_B and the rate of reaction per unit volume $\dot{\xi}/V$ are used almost exclusively. The quantity $\dot{\xi}/V$ has dimensions of concentration/time and is commonly expressed in $mol\,dm^{-3}\,s^{-1}$ ($mol\,liter^{-1}\,s^{-1}$), abbreviated $M\,s^{-1}$. For gas-phase reactions other concentration units are often convenient.

For a homogeneous reaction at constant volume it is quite useful to define x, the extent of reaction in terms of concentration, as $x = \xi/V$. All concentrations can be expressed in terms of their initial values and the reaction variable x by means of equations analogous to (2.2) or (2.3), and for a general reaction (2.1)

$$\frac{\dot{\xi}}{V} = \frac{dx}{dt} = \nu_B^{-1} \frac{d[B]}{dt} \tag{2.5}$$

The value and convenience of the reaction variable x becomes evident as examples in this and the next chapter are studied.

The definitions presented above are valid for macroscopic systems. They include the effects of all microscopic processes that change the amount or concentration of a given species, and so, according to these definitions, the rate of reaction is zero for a system at equilibrium. A different kind of rate is referred to in the familiar statement that "at equilibrium the rate of the forward reaction equals the rate of the reverse reaction." The latter rate is defined in terms of a hypothetical change in amount or concentration that would occur if a single process took place *by itself*, even though this is not possible in the real reaction mixture. Usually this distinction in usage of the word rate is implicit in the context of specific chemical kinetic discussions.[†]

EFFECT OF CONCENTRATION ON REACTION RATE – EMPIRICAL RATE EXPRESSIONS

An important step in any kinetic study is to determine which components of the reaction system are kinetically active, that is, for which substances does a change in concentration alter the rate per unit volume. The most commonly encountered situation is that some or all of the reactants are kinetically active. If the concentration of a product affects the rate per unit volume, the effect is called either autoinhibition or autocatalysis. If a substance that is neither a reactant nor a product affects the rate, that substance is called an inhibitor, retarder, sensitizer, or catalyst, depending on the nature of the effect. The functional relationship between rate of increase in concentration of one species and the concentrations of all kinetically active species is called a rate expression. In general it is not possible to predict the rate expression for a given reaction by just knowing the stoichiometric equation. Although the reactions of bromine with iodine and hydrogen follow similar stoichiometric equations, the rate expressions are of quite different form:

for

$$Br_2 + I_2 = 2BrI \qquad \frac{d[BrI]}{dt} = k[Br_2][I_2]$$

for

$$Br_2 + H_2 = 2HBr \qquad \frac{d[HBr]}{dt} = \frac{k[H_2][Br_2]^{1/2}}{1 + k'[HBr]/[Br_2]}$$

The rate expression for the gas-phase reaction of Br_2 with H_2 can be accounted for by the complicated series of steps mentioned in Chapter 1. The simple rate

---

[†] For a more detailed exposition of this point see reference 21.

expression for the reaction of Br_2 with I_2 has been observed in carbon tetrachloride solution.[2] While this seems to suggest a simple mechanism, orbital topology considerations strongly imply a complex mechanism for this reaction also (see Chapter 1, p. 9).

ORDER OF REACTION AND MOLECULARITY

Rate expressions that are of the form of a product of powers of concentrations such as

$$\frac{-dc_B}{dt} = kc_A^a c_B^b c_D^d \ldots$$

are easier to handle mathematically than expressions of a more complex type, such as that given above for the hydrogen–bromine reaction. For this restricted type of rate expression, and for this type only, there are defined the concepts of *overall order of reaction* and *order with respect to each kinetically active component*. In the case of the rate expression given immediately above, for example, the reaction may be described as of order a with respect to component A, of order b with respect to component B, . . . , and of overall order $(a + b + d + \ldots)$. The bromine–iodine reaction is second-order overall, and the order with respect to bromine or with respect to iodine separately is one. For the bromine–hydrogen reaction the concept of order does not apply, because the rate expression is not of the restricted form required for this concept.

If conditions for a reaction to which the concept of order applies are such that the concentrations of one or more of the kinetically active components are constant or nearly constant during a "run," these constant concentrations (raised to appropriate powers) may be included in the constant k. In this case the reaction is said to be *pseudo-nth order* or *kinetically of the nth order* where n is the sum of the exponents of those concentrations that change during the run. This is the situation for catalytic reactions with the catalyst concentration remaining constant during the run, or where there is a buffering action that keeps a certain concentration such as that of the hydrogen ion nearly constant, or where one reactant is present in large excess over another so that during the run there is only a small percentage change in the concentration of the former. (To ensure that conditions are pseudo-first order, at least a forty-fold excess is required.) Take, for example, the inversion of sucrose[†] catalyzed by strong acids. The rate is given by

$$\frac{-d[S]}{dt} = k[S][H_2O][H^+]$$

where S denotes sucrose. The reaction is third order. However, since H^+ is a catalyst and its concentration remains constant during a run and also since $[H_2O]$ is

---

[†] This reaction is of historical interest as it is considered to be the first one studied kinetically. See reference 3.

essentially constant when water is the solvent, the reaction is pseudo-first order. The rate expression can be rewritten as

$$\frac{-d\,[S]}{dt} = k'[S]$$

where the constant concentrations of water and hydrogen ion have been incorporated into the pseudo-first-order rate constant $k' = k\,[H_2O]\,[H^+]$. If some inert solvent were used so that water was present only as a reactant and its concentration would change over time, the reaction would become pseudo-second order.

The exponents in a rate expression are often integers, but occasionally they may be fractional or even negative, depending on the complexity of the reaction. For overall order 1, 2, or 3, there is some confusion in the older literature with the terms unimolecular, bimolecular, and ter- or trimolecular. These terms describe the *molecularity* of an elementary process, that is, the number of particles involved in a simple collisional reaction process. Molecularity is a theoretical concept, whereas order is empirical. Order and molecularity are generally different numerically. However, as is shown later, a bimolecular reaction is usually second order and a trimolecular reaction third order, but the reverse of these statements is less often true.

The rate constant, or the specific reaction rate, is denoted by k. It is seen from the general rate expression that the rate constant has the dimensions

$$(\text{concentration})^{(1-a-b-d-\cdots)}(\text{time})^{-1}$$

In base units of the International System concentration would be expressed in $\text{mol}\,\text{m}^{-3}$ and time in s. However, it is more common to express concentrations in $\text{mol}\,\text{dm}^{-3}$ (mol/liter) or $\text{mol}\,\text{cm}^{-3}$. For a first-order reaction a rate constant is expressed typically in s^{-1}, the value being independent of the concentration unit. For a second-order reaction typical units would be $\text{mol}^{-1}\,\text{dm}^3\,\text{s}^{-1}$ ($\text{mol}^{-1}\,\text{liter}\,\text{s}^{-1}$), which is usually abbreviated to $M^{-1}\,s^{-1}$ in the literature. For gas-phase reactions at low pressures it is convenient to use the number of molecules per cubic centimeter instead of the concentration, and a second order rate constant is often reported in $\text{cm}^3\,\text{s}^{-1}$. It is also common to find rate constants reported in terms of properties such as gas pressure that are proportional to concentration. However, it is preferable to convert such units to concentration units when rate constants are published.[1]

INTEGRATED FORMS OF SIMPLE RATE EXPRESSIONS

A rate of reaction is not usually obtained directly from experiment. Instead one or more concentrations, or one or more physical properties that can be related to concentration, are usually measured as a function of time in an isothermal system. A number of kinetic runs of this type, carried out under varying conditions, provide the data required to deduce the rate expression and perhaps the variation of the rate constant with temperature.

A typical result for a single run would be as shown in Fig. 2.1, where the concentration c of a reactant starts at zero time at some initial value c_0 and decreases more

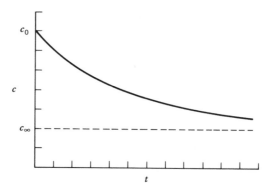

Figure 2.1.   Concentration versus time for a typical reaction.

and more slowly, approaching zero or some equilibrium value c_∞ asymptotically. The rate at any time would be the negative of the slope of the curve expressed in the appropriate units. In some cases the experimental data are more conveniently handled if the rate is determined approximately by graphical or numerical methods. This is especially true when concentrations can be obtained continuously (or nearly so) as a function of time and when a computer is available to automate the process.

More commonly, however, the concentration data are compared directly with an integrated form of the rate expression. Good agreement indicates that an appropriate rate expression has been found. It also permits evaluation of the rate constant and hence calculation of the rate. In the following sections of this chapter a number of typical rate expressions are integrated and discussed, beginning with the simplest example where only one concentration factor is involved. Chapter 3 describes a number of additional techniques by which rate expressions and rate constants may be obtained from experimental data.

nth-ORDER REACTION OF A SINGLE COMPONENT WITH STOICHIOMETRIC EQUATION $A = \ldots$

For this case the form of the rate expression is

$$\frac{-dc}{dt} = kc^n \tag{2.6}$$

It may be readily integrated after first multiplying by dt and dividing by c^n to get the variables separated. The limits of integration are taken as $c = c_0$ at $t = 0$ and $c = c$ at $t = t$, respectively.

$$-\frac{dc}{c^n} = kdt$$

$$-\int_{c_0}^{c} \frac{dc}{c^n} = k \int_{0}^{t} dt \tag{2.7}$$

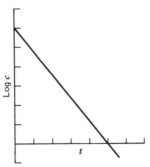

Figure 2.2. Linear plot for a first-order reaction.

For $n = 1$ *first order* the result is

$$\ln \frac{c_0}{c} = kt \tag{2.8}$$

or

$$\ln c = -kt + \ln c_0 \tag{2.9}$$

or

$$c = c_0 e^{-kt} \tag{2.10}$$

To test data for a first order reaction the logarithmic form is preferable since a plot of $\ln c$ or $\log c$ (base 10) versus t should be linear with a slope of $-k$ or $-k/2.303$, respectively (see Fig. 2.2). The rate constant k may be evaluated from this slope. Alternatively, the equation can be checked numerically by solving (2.8) for k,

$$k = \frac{1}{t} \ln \frac{c_0}{c}$$

and substituting successive values of c and t as obtained experimentally. If the values of k so calculated show no significant trend during the course of the reaction run, the conclusion is that the reaction is first order.

For $n = 2$ *second order* the integration yields

$$\frac{1}{c} = kt + \frac{1}{c_0} \tag{2.11}$$

According to this a plot of $1/c$ versus t should be linear, and with a positive slope equal to k (see Fig. 2.3). Or, again, k can be calculated, if the initial concentration is known, for each successive c, t pair and its constancy verified.

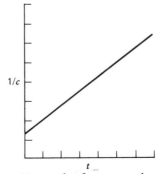

Figure 2.3. Linear plot for a second-order reaction.

For $n = 3$ *third order* the result of integration (2.6) is

$$\frac{1}{2}\left(\frac{1}{c^2} - \frac{1}{c_0^2}\right) = kt \tag{2.12}$$

In general, for $n \neq 1$ integration gives

$$\left(\frac{1}{n-1}\right)\left(\frac{1}{c^{n-1}} - \frac{1}{c_0^{n-1}}\right) = kt \tag{2.13}$$

This is valid for fractional n's as well as integers.

It is sometimes more convenient to use as the dependent variable x, the extent of increase in concentration defined earlier. For the reaction of a single component, $A = \ldots$, the stoichiometric coefficient $\nu_A = -1$ and from (2.5) $dx/dt = -d[A]/dt = -dc/dt$. Using a to represent the initial concentration of reactant A (earlier c_0), we have $c = c_0 + \nu_A x = a - x$, and (2.6) and (2.8) become

$$\frac{dx}{dt} = k(a-x)^n \tag{2.14}$$

$$\ln\left(\frac{a}{a-x}\right) = kt \qquad \text{first order} \tag{2.15}$$

Also (2.11) becomes

$$\frac{1}{a-x} - \frac{1}{a} = \frac{x}{a(a-x)} = kt \qquad \text{second order} \tag{2.16}$$

This method is particularly useful in more complicated situations discussed shortly, where concentrations of several reactants, and perhaps products, appear in the rate expression. In such cases all concentrations can be expressed in terms of one variable x.

PLOTS USING DIMENSIONLESS PARAMETERS

A more general way of presenting these relationships is in terms of dimensionless variables α and τ:

$$\alpha = \frac{c}{c_0} \qquad \text{relative concentration} \tag{2.17}$$

$$\tau = kc_0^{n-1}t \qquad \text{time parameter}$$

By substitution in (2.9) and (2.13) there results for *first order*

$$\ln \alpha = -\tau \tag{2.18}$$

and *nth order* ($n \neq 1$)

$$\alpha^{1-n} - 1 = (n-1)\tau \tag{2.19}$$

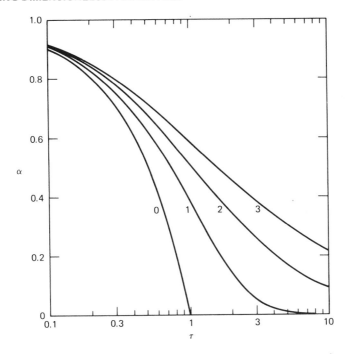

Figure 2.4. Fraction remaining α as a function of time parameter τ for zeroth-, first-, second-, and third-order reactions. (Plot according to R. E. Powell, private communication.)

These equations no longer contain c_0 or k explicitly. For any given order there is a unique relationship between α and τ. Powell[4] has shown the usefulness of a plot of α versus $\log \tau$ for various orders as diagrammed in Fig. 2.4. Suppose that for a given reaction of this type experimental values of α are plotted versus $\log t$ (τ is not experimental because of the at-first-unknown values of k and n relating τ to t). This experimental curve should match the form of the theoretical curve for a reaction of the same order except that because $\log \tau = \log t + \log kc_0^{n-1}$ the experimental curve is shifted along the $\log t$ axis by $-\log kc_0^{n-1}$. Therefore the Powell plot gives a unique form for α versus $\log t$ for each order.

Wilkinson[5] has described a related method. Equation (2.19) can be put into the form

$$(1-p)^{1-n} = 1 + (n-1)Kt \qquad (2.20)$$

where $p = 1 - \alpha$, the fraction reacted, and $K = kc_0^{n-1}$. If $(1-p)^{1-n}$ is expanded by the binomial theorem and terms higher than the second degree in p are discarded, (2.20) becomes

$$\frac{t}{p} = \frac{nt}{2} + \frac{1}{K} \qquad (2.21)$$

A plot of t/p against t should be a straight line with slope $n/2$ and intercept $1/K$. Because (2.21) is only approximate, actual plots might be expected to be curves with a limiting slope of $n/2$. Over the range up to $p = 0.4$ the lines for actual data are remarkably straight with apparent slopes rather close to $n/2$. At least it is easy to distinguish the order of the reaction to the nearest half integer. After the order is obtained, an approximate rate constant can be calculated using the initial concentration and the K obtained from the intercept.

SECOND-ORDER REACTION. FIRST ORDER WITH RESPECT TO EACH REACTANT A AND B WITH STOICHIOMETRIC EQUATION $A + B = \cdots$

$$\frac{-dA}{dt} = kAB \tag{2.22}$$

A and B are here used for the concentrations of the corresponding substances. From the stoichiometric equation $dA = dB$, and integration gives

$$A_0 - A = B_0 - B \tag{2.23}$$

and

$$B = B_0 - A_0 + A$$

where A_0 and B_0 are initial concentrations. Multiplying (2.22) by dt/AB and substituting for B by (2.23) gives an equation in the two variables A and t, and with these variables separated:

$$\frac{-dA}{A(B_0 - A_0 + A)} = kdt \tag{2.24}$$

Now use the method of partial fractions[†] to write the left side as a sum of two simpler terms. Let

$$\frac{1}{A(B_0 - A_0 + A)} \equiv \frac{p}{A} + \frac{q}{B_0 - A_0 + A} \tag{2.25}$$

where p and q are constants and the identity symbol means equality of the two sides of the equation for all values of A. The p and q are evaluated by using a common denominator and equating coefficients of like powers of A in the numerators. Equating the numerators gives

$$1 \equiv p(B_0 - A_0 + A) + qA$$

Equating constant terms gives

$$1 = p(B_0 - A_0) \tag{2.26}$$

and equating coefficients of A leads to

$$0 = p + q \tag{2.27}$$

[†] Shillady has described a general theorem for carrying out this type of integration. See reference 6.

Solving (2.26) and (2.27) simultaneously for p and q yields

$$p = \frac{1}{B_0 - A_0} \qquad q = -\frac{1}{B_0 - A_0}$$

Equation (2.24) then becomes

$$-\frac{dA}{(B_0 - A_0)A} + \frac{dA}{(B_0 - A_0)(B_0 - A_0 + A)} = k\,dt \qquad (2.28)$$

Integration of each term is simple, and the result is

$$\frac{1}{B_0 - A_0}\ln\frac{A_0}{A} + \frac{1}{B_0 - A_0}\ln\frac{(B_0 - A_0 + A)}{B_0} = kt$$

which reduces to

$$\frac{1}{B_0 - A_0}\ln\frac{A_0(B_0 - A_0 + A)}{B_0 A} = kt$$

or

$$\frac{1}{B_0 - A_0}\ln\frac{A_0 B}{B_0 A} = kt \qquad (2.29)$$

Experimental data may be plotted linearly either by plotting the left side of the equation against t, or with less calculation by plotting just $\log (B/A)$ against t. Or data may be tested without plotting by solving the equation for k, substituting successive observed values of A, B, and t, noticing the constancy of k.

In terms of the variable x representing the decrease in concentration of a reactant in a given time, (2.22) becomes

$$\frac{dx}{dt} = k(a - x)(b - x) \qquad (2.30)$$

where a and b are now used to represent the initial concentrations, and the integrated form, which is equivalent to (2.29), is

$$\frac{1}{b - a}\ln\frac{a(b - x)}{b(a - x)} = kt$$

or

$$\frac{1}{a - b}\ln\frac{b(a - x)}{a(b - x)} = kt \qquad (2.31)$$

Grant and Hinshelwood have applied this equation to the data given in Table 2.1 for the reaction of KOH with ethyl bromide in ethyl alcohol. The expected reaction is that of the ethoxide ion formed by acid–base interchange:

$$OH^- + C_2H_5OH \rightleftharpoons H_2O + C_2H_5O^- \qquad \text{fast}$$

$$C_2H_5O^- + C_2H_5Br = C_2H_5OC_2H_5 + Br^- \qquad \text{slow}$$

Grant and Hinshelwood removed aliquots at various times and back titrated the remaining alkalinity with standard acid. A second-order rate constant was calculated for each sample using (2.31). Figure 2.5 shows a plot of $\ln [(b - x)/(a - x)]$

Table 2.1. KOH and C_2H_5Br in ethanol at 332.90 K

t/min	Acid titer[a]	$k/10^{-3}\ M^{-1}\ s^{-1}$ [b]
0	10.00	—
12	8.59	4.39
24	7.33	4.48
41	6.09	4.33
63	4.71	4.32
84	3.74	4.29
108	2.90	4.28
131	2.25	4.33
148	1.93	4.27
171	1.45	4.39
218	0.96	4.28

Source. Grant and Hinshelwood.[7]
[a] In ml of $0.100\ M$ acid/aliquot.
[b] Initial concentrations: KOH, $0.0125\ M$; C_2H_5Br, $0.050\ M$. Average $k = 4.34 \times 10^{-3}$ $M^{-1}\ s^{-1}$.

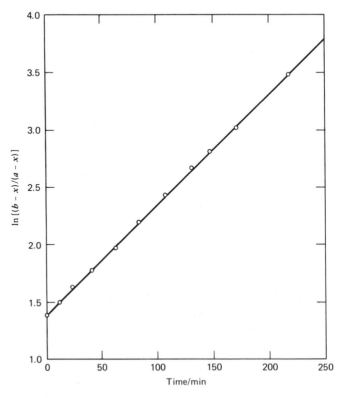

Figure 2.5. Second-order plot for the reaction of KOH with C_2H_5Br in ethanol. (Data from Grant and Hinshelwood[7]).

against t for these data. The straight line shows second-order behavior, and a mean rate constant $k = 4.26 \times 10^{-3} M^{-1} s^{-1}$ is obtained from the slope.

Two side reactions complicate the study of this reaction. Depending on the dryness of the solvent, some OH^- reacts with C_2H_5Br:

$$OH^- + C_2H_5Br = C_2H_5OH + Br^-$$

Also, some ethylene is formed by base-induced elimination of HBr:

$$C_2H_5O^- + C_2H_5Br = C_2H_5OH + C_2H_4 + Br^-$$

Fortunately each side reaction consumes 1 mol of base per mole of alkyl bromide reacted. Consequently the rate constant derived from the data is accurate for the disappearance of ethyl bromide. To get the rates of formation of the products it would be necessary to determine each of their concentrations individually at different times. The data of Table 2.1 go back to 1933, and with the direct chemical methods of analysis available at that time such a determination would have been quite difficult. Today indirect physical methods of the kind described in Chapter 3 are almost always used. Gas chromatography, for example, would greatly simplify determination of product concentrations in this reaction.

The left sides of (2.29) and (2.31) become indeterminate if the initial concentrations are made equal. But this situation of equal initial concentrations is equivalent mathematically to a second-order reaction of one reactant with (2.11) or (2.16) as the solution. This equation may also be obtained from (2.29) or (2.31) by an expansion of the logarithmic term followed by a limiting process.

If the initial concentrations are not quite equal, exact equality being difficult to attain experimentally, another formula may be necessary since the argument of the logarithm in (2.29) or (2.31) is close to unity. Here again an expansion of the logarithm is useful. Following Wideqvist,[8] let $a = d + s$ and $b = d - s$, where d is the mean initial concentration and $2s$ is the excess of a over b. Substitution in (2.31) results in

$$\ln\left(1 + \frac{s}{d-x}\right) - \ln\left(1 - \frac{s}{d-x}\right) + \ln\left(1 - \frac{s}{d}\right) - \ln\left(1 + \frac{s}{d}\right) = 2skt$$

Each logarithm is of the form $\ln(1+y)$, where $y < 1$ at least in the early stages of the reaction. Expansion and collection of terms results in

$$\frac{1}{d-x} - \frac{1}{d} + \frac{s^2}{3}\left[\frac{1}{(d-x)^3} - \frac{1}{d^3}\right] + \ldots = kt \qquad (2.32)$$

When $s = 0$ this reduces to

$$\frac{1}{d-x} - \frac{1}{d} = kt \qquad (2.33)$$

which is the equivalent of (2.16). Since terms in s to the first degree are missing in (2.32) owing to the use of the mean concentration d, it is apparent that (2.33) may also be useful when the concentrations are only slightly unequal, that is, when $s \ll d$. In such a case a simple plot of $1/(d-x)$ against time is linear with the slope equal to k.

SECOND-ORDER AUTOCATALYTIC REACTION

Let the stoichiometric equation be $A = B + \cdots$, and the rate expression

$$\frac{-dA}{dt} = kAB$$

In this case $A_0 - A = B - B_0$ so that $B = A_0 + B_0 - A$. Integration, again by partial fractions, yields

$$\frac{1}{A_0 + B_0} \ln \frac{A_0 B}{B_0 A} = kt \tag{2.34}$$

Examples of such kinetics are found in the acid-catalyzed hydrolyses of esters and similar compounds, and in various biochemical processes. In the hydrolysis of simple esters, such as those of carboxylic acids, the observed rate expression is more complicated. Both the molecular acid and hydrogen ion from its dissociation may act as catalysts. Also, it may be necessary to account for an uncatalyzed reaction. Indeed, if there were not some process to start the reaction, B_0 and the rate would remain at zero forever. Figure 2.6 shows a typical plot of B as a function of t. Solution of (2.34) gives

$$B = \frac{A_0 + B_0}{1 + (A_0/B_0)e^{-k(A_0+B_0)t}} \tag{2.35}$$

The curve is seen to be S-shaped and is typical of autocatalytic reactions. It is also characteristic of many growth processes, including the growth of populations.[9]

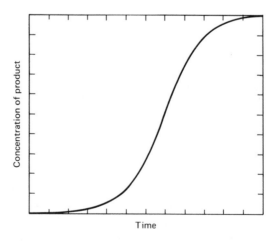

Figure 2.6. Product concentration as a function of time for a second-order auto-catalytic reaction.

THIRD-ORDER REACTION WITH THREE REACTANTS

For the stoichiometric equation $A + B + C = \ldots$ the rate expression is

$$\frac{-dA}{dt} = kABC \tag{2.36}$$

but

$$A_0 - A = B_0 - B = C_0 - C$$

and

$$B = B_0 - A_0 + A \qquad C = C_0 - A_0 + A$$

Then

$$\frac{-dA}{dt} = kA(B_0 - A_0 + A)(C_0 - A_0 + A) \tag{2.37}$$

Integration by partial fraction results in

$$\frac{1}{(B_0 - A_0)(C_0 - A_0)} \ln \frac{A_0}{A} + \frac{1}{(A_0 - B_0)(C_0 - B_0)} \ln \frac{B_0}{B}$$

$$+ \frac{1}{(A_0 - C_0)(B_0 - C_0)} \ln \frac{C_0}{C} = kt$$

or

$$\frac{1}{(A_0 - B_0)(B_0 - C_0)(C_0 - A_0)} \ln \left(\frac{A}{A_0}\right)^{(B_0 - C_0)} \left(\frac{B}{B_0}\right)^{(C_0 - A_0)} \left(\frac{C}{C_0}\right)^{(A_0 - B_0)} = kt$$

$$\tag{2.38}$$

THIRD-ORDER REACTION WITH TWO REACTANTS

Suppose that $2A + B = \cdots$ and

$$\frac{-dA}{dt} = kA^2 B$$

Here $A_0 - A = 2(B_0 - B)$ and $B = B_0 - A_0/2 + A/2$ and

$$\frac{-dA}{dt} = kA^2 \left(B_0 - \frac{A_0}{2} + \frac{A}{2}\right)$$

The partial fractions in this case are of the form

$$\frac{1}{A^2(B_0 - A_0/2 + A/2)} \equiv \frac{p}{A^2} + \frac{q}{A} + \frac{r}{B_0 - A_0/2 + A/2}$$

with p, q, and r constants. Integration yields

$$\frac{2}{(2B_0 - A_0)}\left(\frac{1}{A} - \frac{1}{A_0}\right) + \frac{2}{(2B_0 - A_0)^2}\ln\frac{B_0 A}{A_0 B} = kt \tag{2.39}$$

This equation applies to such reactions as those between NO and O_2, Cl_2, or Br_2.

A different situation exists for a third-order reaction with two reactants if the stoichiometric equation is $A + B = \cdots$ and

$$\frac{-dA}{dt} = kA^2 B$$

In this case $A_0 - A = B_0 - B$ and

$$\frac{-dA}{dt} = kA^2(B_0 - A_0 + A)$$

The result of integration is

$$\frac{1}{B_0 - A_0}\left(\frac{1}{A} - \frac{1}{A_0}\right) + \frac{1}{(B_0 - A_0)^2}\ln\frac{B_0 A}{A_0 B} = kt \tag{2.40}$$

An example of this is given in Table 2.2. Triphenyl methyl (trityl) chloride was

Table 2.2.   Reaction of $0.106\,M$ trityl chloride with $0.054\,M$ methanol in dry benzene solution in the presence of pyridine at 298 K.

Run[a]	t/min	x/M	x_{cor}/M	$k_2/M^{-1}\,s^{-1}$	$k_3/M^{-2}\,s^{-1}$
50	20	0.0010	—	—	—
47	22	0.0003	—	—	—
46	22	0.0010	—	—	—
47	168	0.0067	0.0091	0.0107	0.224
46	174	0.0086	0.0110	0.0127	0.278
47	418	0.0157	0.0181	0.0101	0.234
50	426	0.0165	0.0189	0.0105	0.248
46	444	0.0183	0.0207	0.0115	0.278
50	1,150	0.0294	0.0318	0.0089	0.272
47	1,440	0.0310	0.0334	0.0077	0.252
46	1,510	0.0321	0.0345	0.0080	0.264
50	1,660	0.0330	0.0354	0.0077	0.263
47	2,890	0.0394	0.0418	0.0066	0.296
46	2.900	0.0390	0.0414	0.0064	0.281
50	3,120	0.0392	0.0416	0.0060	0.269
47	193,000	0.0490	0.0514	—	—
				Average	0.263

Source. Swain.[10] Reprinted with permission from *J. Am. Chem. Soc.*, **70**, 1119 (1948). Copyright 1948 American Chemical Society.

[a] Run 46, $0.064\,M$ pyridine; run 50, $0.108\,M$ pyridine; run 47, $0.215\,M$ pyridine. $b/a = 1.963$.

reacted with methanol in dry benzene solution as follows:

$$(C_6H_5)_3CCl + CH_3OH = (C_6H_5)_3COCH_3 + HCl$$

Pyridine was added to remove the HCl and so prevent the reverse reaction. The reaction was followed by taking advantage of the fact that pyridine hydrochloride is only slightly soluble in benzene and precipitates out as the reaction proceeds. After a given time, each sample was filtered, unreacted trityl chloride was hydrolyzed with water, and the resulting hydrochloric acid was titrated with standard sodium hydroxide. A correction was made for the solubility of pyridine hydrochloride. The stoichiometry might lead one to expect a second-order reaction, but it was suspected of being third order, that is, first order with respect to the trityl chloride and second order with respect to the methanol. Both second-order constants k_2 and third-order constants k_3 were calculated for the same runs. The data and calculated results are in Table 2.2. It is evident that k_3 is a reasonable constant where k_2 is not. The values of k_3 were calculated from the reaction variable form of (2.40).

$$k = \frac{1}{t(b-a)}\left[\frac{x}{a(a-x)} + \frac{2.303}{(b-a)}\log\frac{b(a-x)}{a(b-x)}\right] \qquad (2.41)$$

where a is the initial concentration of methanol, b is the initial concentration of trityl chloride, and x is the concentration of methanol or trityl chloride that reacts in time t.

Mechanisms involving equilibria, reverse reactions, side reactions, and consecutive reactions are often met in practice. These situations are discussed in Chapter 8, "Complex Reactions". Capellos and Bielski[11] give integrated forms for most of the rate expressions that are likely to be encountered, and Margerison[22] presents a thorough discussion of the treatment of experimental kinetic data.

APPLICATIONS OF COMPUTERS TO REACTION RATE PROBLEMS

A large number of data are usually obtained in any kinetic study, and, because of the difficulty of determining concentrations that are changing over time, individual data points are often not highly accurate. For both of these reasons it is often advantageous to employ computers and sophisticated numerical methods to obtain rate parameters and estimate their accuracies. The most obvious applications of computers in chemical kinetics involve automations of procedures that formerly were carried out by hand. Integrated rate expressions such as (2.9), (2.11), (2.12), (2.15), (2.16), (2.29), (2.31), (2.33), (2.34), (2.38), (2.39), and (2.40) are all of the form

$$f(c_j, c_{j,0}) = kt$$

Provided that the initial concentrations $c_{j,0}$ of all kinetically active species are known and the concentrations c_j of those same species can be obtained as a function of time, a plot of $f(c_j, c_{j,0})$ versus t should be linear and have a slope equal to k. The computer can be programmed to perform a least-squares analysis of the data and obtain the best value of k, together with the standard deviation in that value. In cases such as (2.9), (2.11), and (2.12) the single initial concentration may also be obtained from the intercept of a plot, provided an appropriate rearrangement of the equation is performed.

Wiberg[12] has described a subroutine LEASQ1 that performs the least-squares analysis and has applied it to first- and second-order reactions. DeTar[13] has written a general least-squares program LSG that adjusts any or all of three parameters to minimize the scalar error in a general variable. This least-squares routine is also the basis for LSKIN1, a highly flexible program that calculates the rate constant and (if desired) the zero-time and infinity-time values of any variable that is directly or inversely proportional to concentration. (The use of such experimental variables is described in detail in the next chapter.) Input to LSKIN1 consists of preliminary values of the parameters followed by a series of experimental values of concentration variable and time together with appropriate weighting factors. The use of weighting factors can compensate for variations in the accuracy of the data over the course of the experiment. Roseveare[14] has shown, for example, that for a simple first-order reaction (no side reactions) the most accurate points are to be found at two-thirds of complete reaction. Weighting factors should not be too different, however, since they can result in some of the data being essentially ignored when parameters are fitted. DeTar has also published a program, LSKIN2,[15] that fits a second-order rate expression to experimental concentration—time data. An annotated bibliography of computer programs that treat experimental data for a number of cases other than those listed above is available,[16] and Wiberg[17] has reviewed the use of computers in kinetics in detail. Pattengill and Sands[23] present a good discussion of the linear least-squares method and the statistical significance of the parameters obtained.

It cannot be overemphasized that using a computer to handle the mechanical process of data reduction should not replace careful examination of the data by the person doing the experiments. Unless it is very ingeniously programmed to do so, a computer will not apply "common sense" checks on the data it is given or on its results. The method of least squares can be used to adjust parameters to any data,[†] whether or not the data are consistent with the assumptions or model that suggested the parameters in the first place. Thus one must be very careful not to ascribe to the computer supernatural powers of accuracy and/or infallibility. Instead every effort should be made to produce computer output in forms that indicate just how well the data have been fitted. One means for doing this is to compare variables calculated from the rate parameters to those obtained experimentally. Differences between experimental and calculated values may show regularities in magnitude, sign, or both that suggest systematic deviation from the assumed rate expression. These differences should be printed along with other criteria, such as correlation coefficient and standard deviations in the fitted parameters, that indicate how well the rate expression fits the experimental data.

Most computing centers now have available digital plotters, and it is strongly recommended that plots corresponding to Figs. 2.2, 2.3, 2.5, and so on also form part of the output from the computer program. Experimental data can be shown as individual points and the curve predicted by the rate expression can be drawn

---

[†] See reference 15, pp. 25 and 88–95.

through them to reveal systematic or periodic deviations. The rapidity with which calculations and plots can be handled by computer also permits alternative graphical displays of data. In the case of a first-order reaction, for example, both the linear graph of (2.9) and the corresponding exponential form (2.10) may be displayed without excessive programming effort. Each plot is potentially capable of revealing a different aspect of the experimental data, should a good fit to the rate expression not be obtained.

Application of the method of least squares is not confined to situations in which an equation may be rearranged into the form $y = mx + b$ so as to give a linear graph.[18,19] All that is required is that the equation may be written as

$$F(v_1, v_2, \ldots, v_i, p_1, p_2, \ldots, p_i) = 0$$

where the v_i and the p_i represent variables and parameters, respectively, and that all partial derivatives $\partial F/\partial p_1$, $\partial F/\partial p_2$, ... be obtainable. If variable weighting is to be used, $\partial F/\partial v_1$, $\partial F/\partial v_2$, ... must also be available. Applying this method to (2.10), for example, we have

$$0 = F(c, t, c_0, k) = c_0 e^{-kt} - c$$

$$\frac{\partial F}{\partial c} = -1 \qquad \frac{\partial F}{\partial c_0} = e^{-kt}$$

$$\frac{\partial F}{\partial t} = -c_0 k e^{-kt} \qquad \frac{\partial F}{\partial k} = -c_0 t e^{-kt}$$

A computer program is given in the Appendix that carries out the nonlinear least-squares fit of (2.10) to experimental data. The advantage of this technique over application of the method of least squares to (2.9) is that (2.10) involves no transformation (by taking logarithms in this case) of the concentration data obtained from experimental measurements. Such a transformation may force an inappropriate weighting on the data, resulting in poor values of the parameters.

EFFECT OF TEMPERATURE ON REACTION RATE

Observed rates or rate constants as a function of temperature T may be of various forms as indicated in Fig. 2.7. Curve a is most typical and is discussed in detail; b represents an explosion where the sudden rise in rate occurs at the ignition

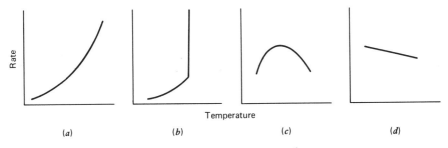

Figure 2.7. Various forms for the dependence of reaction rate on temperature.

temperature; c is observed, for example, in catalytic hydrogenations and in enzyme reactions; and d is observed in the reaction of NO with O_2.

Case a may be called an Arrhenius temperature dependence. Cases $b-d$ are sometimes referred to as anti-Arrhenius. They are usually indicative of a multistep mechanism or a mechanism that changes as temperature increases. Case a corresponds to the common statement that a reaction rate increases by a factor of 2 or 3 for each $10°C$ rise in temperature. It is usually found experimentally that a plot of log k versus $1/T$ is nearly linear with negative slope. This result is equivalent to the Arrhenius equation

$$\frac{d \ln k}{dT} = \frac{E_a}{RT^2} \tag{2.42}$$

If E_a, the Arrhenius activation energy, is a constant with respect to temperature, integration results in

$$\ln k = \frac{-E_a}{RT} + \text{const} \quad \text{or} \quad k = Ae^{-E_a/RT} \tag{2.43}$$

A plot of k versus T according to (2.43) would have the appearance of Fig. 2.8 rather than case a of Fig. 2.7, so that k would approach the constant value A asymptotically. For most reactions the accessible temperature range includes only the lower, rising part of the curve, but there are some reactions involving free atoms or radicals with very small or zero activation energy such that the upper portion of the curve is approached. However, under such conditions the factor A is also found to be a function of temperature and a more accurate equation must be used. Such an equation is

$$k = BT^m e^{-E_b/RT} \tag{2.44}$$

or

$$\ln k = \frac{-E_b}{RT} + m \ln T + \ln B \qquad B = \text{constant} \tag{2.45}$$

Equations (2.44) and (2.45) also have a theoretical justification, the exponent m having a particular value depending on the kind of theory used and the nature of the reaction considered (see Chapters 4 and 5).

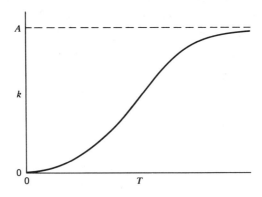

Figure 2.8.   Arrhenius temperature dependence of a rate constant.

Treating experimental data by (2.43) and (2.44) results in different activation energies. It is of importance to know the relationship between E_a and E_b. Differentiate (2.45) with respect to T, assuming E_b constant, and set this equal to (2.42):

$$\frac{E_b}{RT^2} + \frac{m}{T} = \frac{E_a}{RT^2} \quad \text{and} \quad E_a - E_b = mRT \tag{2.46}$$

Since $R = 8.314 \ \text{J K}^{-1} \ \text{mol}^{-1}$, the difference between the two activation energies is typically a few kilojoules per mole.

If (2.45) is correct with m not zero, a plot of log k versus $1/T$ shows a slight curvature. There are cases, however, where nonlinearity is much more pronounced. The curve may sometimes be resolved into two parts, each of which approaches linearity. This can result if there are two competing reactions with different activation energies and is often observed where the same reaction can occur both homogeneously and heterogeneously. The homogeneous reaction usually has the higher activation energy and so it is favored at high temperatures, whereas the heterogeneous reaction predominates at lower temperatures. Such a case is shown in Fig. 2.9.

The activation energy E_a and the constant A may be obtained from experimental values of k and T by applying to (2.43) the same least-squares methods discussed in the preceding section. Similarly B, E_b, and perhaps m may be treated as adjustable parameters in (2.44). In either case the values obtained will have large standard deviations, since the experimentally accessible range of temperatures and values of k is usually small. Nevertheless, all parameters should be reported to four or five significant figures with error limits in terms of standard deviations indicated.[20] For example, for (2.37) one might obtain $A = (16.47 \pm 1) \times 10^{10} M^{-1} s^{-1}$ and $E_a = (164.13 \pm 10) \text{kJ mol}^{-1}$. The error limits serve to define the accuracy of the values, should it be desired to compare them with theoretical predictions. The large number of significant digits is needed so that (2.43) or (2.44) can be used to

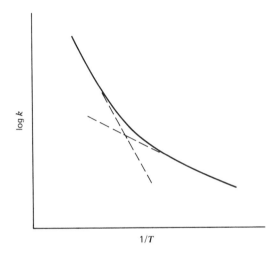

Figure 2.9.   Transition from homogeneous to heterogeneous reaction.

compute the k values adequately. In this latter case A and E_a (or B, m, and E_b) provide an excellent way of averaging and summarizing all experimental data.

It is clear from (2.43) that the frequency factor A must have the same units as the rate constant, and therefore that the value of A depends on the units in which k was expressed prior to making the Arrhenius plot. These units, however, may be disguised by taking a logarithm, or even worse, by feeding numbers into a computer. Almost invariably the units of k and A are $M^{-1} s^{-1}$ for a second-order reaction and s^{-1} for a first order process, but it is good practice to specify them explicitly and avoid ambiguity. In the case of (2.42) and the logarithmic form of (2.43), units can be specified simply by writing $\ln (k/M^{-1} s^{-1})$, for example, ensuring that the argument of the logarithm is a pure number. The potential for confusion is even greater in the case of (2.44), where units for B depend on the choice of m as well as on the units of k.

PROBLEMS

2.1. For the formation of hydrogen iodide from hydrogen and iodine a graph of the reciprocal of the concentration of hydrogen iodide in units of moles per 22.4 liters versus time in minutes has a slope of 3.58. Calculate the rate constant for disappearance of hydrogen in units of liters $mol^{-1} s^{-1}$, $m^3 mol^{-1} s^{-1}$, $dm^3 mol^{-1} s^{-1}$, and $cm^3 s^{-1}$.

2.2. Liebhafsky and Mohammad, *J. Am. Chem. Soc.*, **55**, 3977 (1933) have determined the rate expression and rate constants for the reaction of iodide ions with hydrogen peroxide. Use their data to calculate the initial rate of disappearance of I^- from solutions of pH $= 3, 4, 5, 6$, and 7. Assume the initial concentrations are $[H_2 O_2] = 1.00 \times 10^{-2} M$ and $[I^-] = 5.00 \times 10^{-3} M$. Express the calculated rates in units of $M s^{-1}$. Compare your result at pH $= 7$ with the value of $5.75 \times 10^{-7} M s^{-1}$ obtained from data reported by Surfleet and Wyatt, *J. Chem. Soc. A*, **1967**, 1564.

2.3. Consider a reaction with the stoichiometric equation

$$A + 2B \rightarrow C + D$$

that is first order in each of the two reactants. Set up the rate expression in terms of a reaction variable x and integrate. How could this integrated result have been obtained directly from (2.31)?

2.4. The oxidation of certain metals is found to obey the parabolic equation

$$y^2 = k_1 t + k_2$$

where y is the thickness of the oxide film at time t. What order could be ascribed to this reaction? How could this be interpreted? [See W. J. Moore, *J. Chem. Phys.*, **18**, 231 (1950).]

2.5. The decomposition of nitrogen dioxide is a second-order reaction with rate constants k as follows:

T/K	592	603.2	627	651.5	656
$k/cm^3 \, mol^{-1} \, s^{-1}$	522	755	1700	4020	5030

Determine the Arrhenius activation energy E_a, and also E_b of the equation

$$k = BT^{1/2} e^{-E_b/RT}$$

Compare the two results with (2.46).

2.6. Show that the Powell plot could be generalized so that a unique curve results for each order of reaction when log c is plotted against log t. By shifting in two dimensions, without rotation, all curves for the same order of reaction should be superimposable. This would presumably make it possible to determine the order of reaction without knowing the initial concentration. Test the method on the data of Table 3.2. Is the method practicable?

2.7. The following data were obtained for the concentration c of a reactant during the course of a reaction presumed to be first order:

$c/10^{-3} \, M$	4.09	3.36	2.73	1.83	1.12	0.45	0.27	0.17	0.10	0.06
t/s	10.0	20.0	30.0	50.0	75.0	100.0	125.0	150.0	175.0	200.00

Analyze the data to obtain c_0 and k by the following means:

(a) Use the first-order least squares computer program of Wiberg (reference 17, p. 746) to fit c_0 and k to (2.9).

(b) Use the computer program of the Appendix to fit c_0 and k to (2.10).

(c) Plot ln c versus t for the experimental data and draw the straight line calculated from (2.9) and the parameters of part **a** on the same axes.

(d) Plot c versus t for the experimental data and draw the exponential curve calculated from (2.10) and the parameters of part **b** on the same axes.

(e) Comment on the validity of the assumption that the reaction is first order and/or on the quality of the data.

2.8. Use the computer program of the Appendix to obtain E_a and A from the following data for the reaction

$$2HI(g) \rightarrow H_2(g) + I_2(g)$$

T/K	556	629	666	716	781
$k/M^{-1} \, s^{-1}$	3.52×10^{-7}	3.02×10^{-5}	2.19×10^{-4}	2.50×10^{-3}	3.95×10^{-2}

2.9. Use data from the literature to test the Wilkinson method for finding the order of a reaction.

2.10. Show that for a first-order reaction with stoichiometric equation $2A = \ldots$ the rate constant is one-half of the slope of a plot of ln (c_A) versus t. Generalize your result for a first-order reaction with stoichiometric equation $\nu A = \ldots$. Find two examples in the literature where the result of your derivation has not been taken into account when a rate constant is reported.

REFERENCES

1. M. L. McGlashan and M. A. Paul, *Manual of Symbols and Terminology for Physico-chemical Quantities and Units,* Butterworths, London, 1973, pp. 31–32; M. L. McGlashan, *Physicochemical Quantities and Units,* 2nd ed., Royal Institute of Chemistry, Monographs for Teachers, No. 15, London, 1971, pp. 67–69; M. L. McGlashan, "Internationally Recommended Names and Symbols for Physicochemical Quantities and Units, *Ann. Rev. Phys. Chem.,* **24,** 51 (1973); "The Presentation of Chemical Kinetics Data in the Primary Literature," CODATA Bulletin No. 13, Paris, December 1974.

2. P. R. Walton and R. M. Noyes, *J. Am. Chem. Soc.,* **88,** 4324 (1966).

3. L. Wilhelmy, *Ann. Phys. Chem. (Poggendorf),* **81,** 413, 499 (1850).

4. R. E. Powell, private communication.

5. R. W. Wilkinson, private communication.

6. D. D. Shillady, *J. Chem. Educ.,* **49,** 347 (1972); J. R. Dias, *J. Chem. Educ,* **49,** 851, (1972).

7. G. H. Grant and C. N. Hinshelwood, *J. Chem. Soc.,* **1933,** 258.

8. S. Wideqvist, *Ark. Kemi,* **2,** 303 (1950).

9. H. T. Davis, *Theory of Econometrics,* Principia Press, Bloomington, Indiana, 1941, Chap. 11; C. J. Krebs, *Ecology,* Harper and Row, New York, 1972, Chap. 11; D. H. Meadows, D. L. Meadows, J. Randers, and W. W. Behrens, III, *The Limits to Growth,* Universe Books, New York, 1972, p. 91.

10. C. G. Swain, *J. Am. Chem. Soc.,* **70,** 1119 (1948).

11. C. Capellos and B. H. J. Bielski, *Kinetic Systems,* Wiley-Interscience, New York, 1972.

12. K. B. Wiberg, *Computer Programming for Chemists,* Benjamin, New York, 1965; for updated versions of these programs see reference 17.

13. D. F. DeTar, *Computer Programs for Chemistry,* Vol. 1, Benjamin, New York, 1968, p. 117.

14. W. E. Roseveare, *J. Am. Chem. Soc.,* **53,** 1651 (1931).

15. D. F. DeTar, *Computer Programs for Chemistry,* Vol. 4, Academic, New York, 1972, p. 1.

16. J. L. Hogg, *J. Chem. Educ.,* **51,** 109 (1974).

17. K. Wiberg, "Use of Computers," in E. S. Lewis, Ed., *Investigation of Rates and Mechanisms of Reactions,* Part I, in Techniques in Chemistry, A. Weissberger, Ed., Vol. 6, 3rd ed., Wiley-Interscience, New York, 1974.

18. W. E. Deming, *Statistical Adjustment of Data,* Wiley, New York, 1943, p. 148.

19. D. F. DeTar, *Computer Programs for Chemistry,* Vol. 4, Academic, New York, 1972, p. 71.

20. D. F. DeTar, *J. Chem. Educ.,* **44,** 759 (1967).

21. V. Gold, *Nouv. J. Chim.,* **3,** 69 (1979).

22. D. Margerison, "The Treatment of Experimental Data," in *Comprehensive Chemical Kinetics,* Vol. 1, C. H. Bamford and C. F. H. Tipper, Eds., Elsevier, Amsterdam, 1969.

23. M. D. Pattengill, and D. E. Sands, *J. Chem. Educ.,* **56,** 244 (1979).

THREE

EXPERIMENTAL METHODS AND TREATMENT OF DATA

The determination of reaction rates by conventional methods reduces to a study of concentrations as a function of time. In these cases a problem in quantitative analysis is always present, and one or more of the countless analytical procedures that have been devised may be needed in a particular kinetics problem. It may be mentioned also that kinetic studies can be made without any measurement of concentration and without any reference to laboratory time, for example, by nuclear magnetic resonance and electron paramagnetic resonance methods.[1]

In general, analytical procedures may be divided into two broad categories, chemical and physical. Chemical analysis implies a direct determination of one of the reactants or products by volumetric or gravimetric procedures, the former being preferred because of their rapidity. An important restriction on any chemical method is that it must be rapid compared with the reaction being studied. Or if the method is relatively slow, the reaction must be stopped or frozen by some sudden change, such as lowering the temperature, removal of a catalyst, addition of an inhibitor, and removal of a reactant. Chemical methods of analysis have the advantage of giving an absolute value of the concentration.

On the other hand, physical methods of analysis are usually much more convenient than chemical methods. A physical method measures some physical property of the reaction mixture that changes substantially as the reaction proceeds. Common among physical methods are pressure measurements in gaseous reactions; dilatometry, or measurement of volume change; optical methods such as polarimetry, refractometry, colorimetry, fluorimetry, and spectrophotometry; electrical methods such as conductivity, potentiometry, and polarography; and mass spectrometry, nuclear magnetic resonance (NMR) and electron spin resonance (ESR) spectrometry, and gas–liquid partition chromatography (GC). Theoretically any property that changes sufficiently could be used to follow the course of a reaction. Thermal conductivities, solidification temperatures, viscosities (for polymerization reactions), coagulating power toward colloids, and heats of reaction are among the more unusual properties that have been utilized.

In general, a physical method of analysis has the advantage of being rapid so that more experimental points are available in a given time. Measurements can frequently be made in the reaction vessel so that sampling with its attendant errors is eliminated. The system is usually not destroyed by the method nor even perceptively disturbed. Often it becomes possible to make automatic and continuous recordings of the changes in property. Also, several different physical properties can be monitored simultaneously, each elucidating a different aspect of the reaction. Physical methods, however, often have the limitation of not giving absolute values of concentration directly. Furthermore, errors due to side reactions may be enormously magnified. For example, in spectrophotometric studies, small amounts of highly colored impurities or by-products obscure the desired quantities. For a complete study of any given reaction, more than one method of analysis should be used. It is especially desirable that the stoichiometry of the reaction be verified to be sure that the reaction being studied is one in which the products are known with certainty.

CORRELATION OF PHYSICAL PROPERTIES WITH CONCENTRATIONS

One requirement of any physical measurement as a criterion of extent of reaction is that the property being measured differ appreciably from reactants to products. Another requirement is that the property vary in some simple manner with the concentrations of reactants and products. The most common and useful relationship is when the physical property is a linear function of the concentration. Such a relationship exists, for example, between concentration and electrical conductance, absorbance, rotation of polarized light, and pressure of gases. In dilute solutions, many physical properties, such as the specific volume, refractive index, vapor pressure, and fluidity, become linear functions of the concentration. In practice, of course, many of these linear relationships break down if applied over too wide a range of concentration, the reasons being not only deviations from ideal behavior, but also nonlinearities in the mathematical forms of the ideal laws relating the properties to concentration.

General equations can be derived to relate a measured physical quantity with the reaction variable $x = \xi/V$ defined in Chapter 2, provided that the physical quantity is a linear function of concentration. Using the convention established in Chapter 2 (2.1), any reaction can be written as

$$0 = \sum_{B} \nu_B B \qquad (2.1)$$

Suppose that substances A, B, and D react completely to form products represented by Z, and that A is the reactant present in limiting amount. The equation for this reaction can be written

$$-\nu_A A - \nu_B B - \nu_D D = \nu_Z Z \qquad (3.1)$$

where ν_A, ν_B, and ν_D are negative since A, B, and D are reactants. The concentrations of A, B, D, and Z are given in terms of x by

$$c_A = a + \nu_A x \qquad c_D = d + \nu_D x$$
$$c_B = b + \nu_B x \qquad c_Z = \nu_Z x \tag{3.2}$$

where a, b, and d represent initial concentrations and it has been assumed that no product was present initially. Upon completion of the reaction

$$c_A = 0 = a + \nu_A x \qquad x = \frac{-a}{\nu_A}$$

Let λ be the value of the physical property at any time t. Then

$$\lambda = \lambda_M + \lambda_A + \lambda_B + \lambda_D + \lambda_Z \tag{3.3}$$

where λ_M is the contribution of the medium and the others vary with concentration as, for example,

$$\lambda_A = k_A c_A \tag{3.4}$$

k_A being a proportionality constant. Combining (3.2) with (3.3) and (3.4) yields

$$\lambda = \lambda_M + k_A(a + \nu_A x) + k_B(b + \nu_B x) + k_D(d + \nu_D x) + k_Z \nu_Z x \tag{3.5}$$

Substituting the initial value $x = 0$ gives

$$\lambda_0 = \lambda_M + k_A a + k_B b + k_D d \tag{3.6}$$

and the final value $x = -a/\nu_A$ yields

$$\lambda_\infty = \lambda_M + k_B\left(b - \frac{\nu_B a}{\nu_A}\right) + k_D\left(d - \frac{\nu_D a}{\nu_A}\right) - \frac{k_Z \nu_Z a}{\nu_A} \tag{3.7}$$

Subtracting (3.6) from (3.7) leads to

$$\lambda_\infty - \lambda_0 = -k_A a - \frac{k_B \nu_B a}{\nu_A} - \frac{k_D \nu_D a}{\nu_A} - \frac{k_Z \nu_Z a}{\nu_A} \tag{3.8}$$

and subtracting (3.6) from (3.5) gives

$$\lambda - \lambda_0 = k_A \nu_A x + k_B \nu_B x + k_D \nu_D x + k_Z \nu_Z x \tag{3.9}$$

and

$$\lambda_\infty - \lambda = -k_A(a + \nu_A x) - k_B \nu_B\left(\frac{a}{\nu_A} + x\right) - k_D \nu_D\left(\frac{a}{\nu_A} + x\right) - k_Z \nu_Z\left(\frac{a}{\nu_A} + x\right) \tag{3.10}$$

If we set

$$\Delta k = k_A \nu_A + k_B \nu_B + k_D \nu_D + k_Z \nu_Z$$

then we may write

$$\lambda - \lambda_0 = x\,\Delta k \qquad \lambda_\infty - \lambda_0 = -\left(\frac{a}{\nu_A}\right)\Delta k \tag{3.11}$$

and

$$\lambda_\infty - \lambda = -\left(\frac{a}{\nu_A} + x\right)\Delta k \tag{3.12}$$

From these we obtain the kinetically useful relationships

$$-\frac{\nu_A x}{a} = \frac{\lambda - \lambda_0}{\lambda_\infty - \lambda_0} \tag{3.13}$$

$$\frac{a}{a + \nu_A x} = \frac{\lambda_\infty - \lambda_0}{\lambda_\infty - \lambda} \tag{3.14}$$

It is also possible to express $(b + \nu_B x)$ and $(d + \nu_D x)$ in terms of the measured physical property. The net result is of the form

$$\frac{b}{b + \nu_B x} = \frac{(b/a)(\lambda_\infty - \lambda_0)}{(b/a)(\lambda_\infty - \lambda_0) - (\nu_B/\nu_A)(\lambda - \lambda_0)} \tag{3.15}$$

Considerable simplification can be obtained by using equivalent concentrations of reactants, so that $b/a = \nu_B/\nu_A$, and so on.

Reactions that do not go to completion can be handled also if the equilibrium constant is known independently. We illustrate the use of (3.13) and (3.14) with several examples taken from the literature for reactions both in solution and in the gas phase.

REACTIONS IN THE GAS PHASE

Lack of space prevents an extended discussion of apparatus and experimental methods. Fortunately, a number of excellent references are available to the reader in various treatises.[2-13]

Kinetic studies of gas-phase reactions have been aided enormously by the development of gas–liquid partition chromatography. Virtually all compounds can be determined quantitatively by GC, isomeric forms are readily distinguished, and the method is sufficiently sensitive that very small samples (whose removal does not appreciably alter concentrations) may be taken from closed reaction vessels. The chief limitation of GC is that the time required for analysis cannot be reduced below the millisecond range. This precludes application of GC to reactive intermediates. Also, in specific cases, one must beware of heterogeneous reactions that may occur on the GC column; indeed, gas chromatographic columns may be used as chemical reactors for the collection of kinetic data.[14]

Table 3.1 gives some of the results obtained by Cocks and Egger[15] for thermal isomerization of N-propylidenecyclopropylamine. Nuclear magnetic resonance and mass spectrometry were used to establish that at temperatures below 700 K the sole product was 5-ethyl-1-pyrroline:

$$\tag{3.16}$$

Each kinetic run was carried out in a 1-liter cylindrical glass vessel. The reaction was quenched by condensing the contents of the vessel into a liquid-nitrogen trap, and the liquid products were injected into a gas chromatograph. Because reactant and

Table 3.1. Thermal isomerization of N-propylidenecyclopropyl amine at 573 K[a]

Time/min	Pressure/torr	Fraction isomerized	$k/10^{-5}\ s^{-1}$
20	55.2	0.0783	6.79
30	15.4	0.1113	6.56
60	18.1	0.2104	6.56
100	18.3	0.3313	6.71
210	17.5	0.5784	6.85
			Average $k = 6.69 \pm 0.13$

Source. Data from Cocks and Egger.[15]
[a] Temperatures varied from 572.9 to 573.3 K.

product are isomeric the fraction isomerized could be obtained by direct comparison of chromatographic peak areas. No changes in the analytical results were observed when the GC column was varied. Rate constants obtained in a reaction vessel whose surface-to-volume ratio had been increased by a factor of 13 were essentially the same as those obtained under normal conditions, suggesting that any heterogeneous component of the reaction is small.

Rate constants were calculated from the equation

$$k = \frac{1}{t} \ln \frac{c_0}{c} = -\frac{1}{t} \ln \frac{a-x}{a} = -\frac{1}{t} \ln \left(1 - \frac{x}{a}\right) = -\frac{2.303}{t} \log (1 - f)$$

where f is the fraction isomerized. The first-order rate constants obtained do not vary significantly with extent of reaction or with pressure.

Manometric methods are also generally applicable to the study of gas-phase reactions. This may involve direct measurement of pressure in systems where there is a change in the total amount of substance such as the decomposition of phosgene

$$COCl_2 = CO + Cl_2 \tag{3.17}$$

or where a reaction product is removed continuously by absorption or condensation. For example, in the reaction

$$H_2 + Cl_2 = 2HCl \tag{3.18}$$

the acid may be removed by absorption in water. Or the pressure may be read intermittently after removing a product, so that in (3.18) chlorine and hydrogen chloride may be condensed by liquid nitrogen and the residual pressure of the hydrogen measured.

Table 3.2 gives some manometric results obtained by Takezaki and Takeuchi[16] for thermal decomposition of dimethyl peroxide. This reaction has also been studied by infrared spectrophotometry[17] and by gas chromatography.[18] The stoichiometry is clearly established as

$$2CH_3OOCH_3 = 3CH_3OH + CO \tag{3.19}$$

and so the pressure should increase by 100% of the initial partial pressure of dimethyl-peroxide as the reaction proceeds. Experimentally it was found that the pressure increased by 93%.

Table 3.2.   Decomposition of dimethyl peroxide at 439.8 K^a

Time/s^b	Total pressure/torr	$(P_\infty - P)$/torr	$\ln [(P_\infty - P)/\text{torr}]$
15	427.12	7.84	2.06
90	428.27	6.69	1.90
240	429.83	5.13	1.64
390	431.23	3.73	1.32
570	432.48	2.48	0.91
780	433.18	1.78	0.58
990	433.67	1.29	0.25
2910	434.96	0	
4590	434.96	0	

Source. Data from Takezaki and Takeuchi.[16]

^{a} Partial pressures of methanol and dimethyl peroxide were 400 and 23.0 torr, respectively, upon initiation of reaction.

^{b} Measured from time of initiation plus 210 s.

However, (3.19) occurs by a complex chain reaction mechanism, which makes it difficult to interpret the kinetic results. The initiating reaction is dissociation into $CH_3O\cdot$ radicals, which then react further.

$$CH_3OOCH_3 \xrightarrow{k} 2CH_3O\cdot \qquad (3.20)$$

If excess methanol is added initially, this can trap the methoxy radicals and supress the chain reaction. Ethylene glycol is then formed

$$CH_3O\cdot + CH_3OH \rightarrow CH_3OH + \cdot CH_2OH$$

$$2\cdot CH_2OH \rightarrow (CH_2OH)_2$$

In the presence of methanol, the stoichiometry of (3.19) was changed to

$$2CH_3OOCH_3 \rightarrow 1.4CH_3OH + 0.4CO + 0.2CH_2O + 1(CH_2OH)_2 \qquad (3.21)$$

After a short induction period, this stoichiometry remained constant; that is, the relative amounts of CH_3OH, CO, CH_2O, and glycol remained constant with time as the peroxide decomposed. This was verified, of course, by a complete analysis of the products at various times in preliminary experiments.[16]

Under these circumstances, the change in pressure can still be used to follow the reaction. The expected pressure increase is now 50% of the initial pressure of the peroxide, according to (3.21). Also for the experiment reported in Table 3.2, the initial pressure is mainly due to added methanol. Nevertheless, we set $x/a = (P - P_0)/(P_\infty - P_0)$ and $a/(a - x) = (P_\infty - P_0)/(P_\infty - P)$. If (3.21) is a first-order reaction, $kt = \ln [a/(a - x)]$, which also can be expressed as

$$\ln (P_\infty - P) = -kt + \ln (P_\infty - P_0) \qquad (3.22)$$

Figure 3.1 shows the natural logarithm of $P_\infty - P$ plotted against the time. The linearity confirms first-order kinetics, and the negative of the slope gives a rate constant $k = 1.9 \times 10^{-3}$ s^{-1}. The interpretation of this k depends on the assumed mechanism. Takezaki and Takeuchi considered it to be the rate constant for (3.20).

Equation (3.22) clearly shows that for a first-order reaction the initial concentration need not be known; the value of P_0 does not necessarily enter into the calculations. Actually in Takezaki and Takeuchi's experiment time was not measured from the instant of mixing and heating the reagents because the reaction is not precisely first order until steady-state concentrations of certain intermediates are achieved. Instead, the times reported in Table 3.2 were measured beginning at 210 s after initiation of the reaction. Since only P_∞ and P are required for the first-order plot, such a shift along the time axis is perfectly legitimate, even though the value of P_0 is not available and must be obtained by extrapolation of the graph.

The value of P_∞, on the other hand, is very important. A change of a few percent in the final value of a physical property will cause a similar alteration in the value of the rate constant. Moreover, the effect of such a change increases with the extent of reaction, resulting in curvature of a plot like Fig. 3.1. In many cases a theoretical value of λ_∞ can be calculated ($P_\infty = 2P_0$ for (3.19) in the absence of methanol). It is also possible to treat λ_∞ as an adjustable parameter when a computer is used for data processing.[19] Nevertheless, it is generally preferable to use an experimental value of the equilibrium reading unless there is reason to believe that some anomaly or error exists in the experimental value. For example, a slow secondary reaction of a product might give a false experimental value of λ_∞. This would necessitate use of a theoretical or estimated value. The fact that a computer can estimate λ_∞ (and λ_0 as well), however, does not mean that correct values can be obtained from incomplete data. Data must be collected over a range from 10 to 90% of completion if accurate estimates are to be obtained.[20] It is

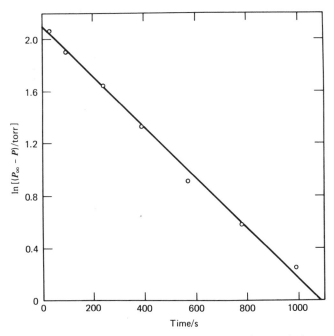

Figure 3.1. First-order plot for the decomposition of dimethyl peroxide (Data from Takezaki and Takeuchi[16]).

especially dangerous to adjust λ_∞ when attempting to determine the order of a reaction, because second-order data can be fitted very nicely to a first-order equation when k, λ_0, and λ_∞ are all varied. It is much to be preferred if only the rate constant is adjusted to fit the data, at least until the order of the reaction has been ascertained.

As an example of a complex gaseous reaction that requires a correction to each measured value of the pressure, the results of Kistiakowsky and Lacher[21] on the condensation of acrolein and 1,3-butadiene for form tetrahydrobenzaldehyde may be cited. This Diels-Alder reaction proceeds essentially to completion in the

$$(3.23)$$

temperature range 155–300°C. The reaction is second order and follows the rate expression

$$\frac{dx}{dt} = k\,[\text{acrolein}]\,[\text{butadiene}] \tag{3.24}$$

It is complicated by a simultaneous second-order polymerization reaction of the butadiene:

$$2C_4H_6 = C_8H_{12} \tag{3.25}$$

This reaction can be studied separately (by omitting acrolein) and is found to be about one-tenth as fast as the main reaction. A correction was made for the side reaction even though the error is small. Table 3.3 gives the values of the total pressure of the system at various times, starting with a known initial pressure of acrolein and of butadiene. From the total drop in pressure for each successive time interval the pressure change for acrolein and for butadiene was calculated. First the

Table 3.3.   Condensation of acrolein and butadiene at 564.4 K

Time/s	$P_{\text{total}}/$ torr	$-\Delta P/$ torr	$-\Delta P_{\text{dim}}/$ torr	$P_{\text{acrolein}}/$ torr	$P_{\text{butadiene}}/$ torr	$k/10^{-7}\,\text{torr}^{-1}\,\text{s}^{-1}$
0	658.2			418.2	240.0	
63	652.1	6.1	0.2	412.3	233.7	9.6
181	641.4	10.7	0.3	401.9	222.7	9.5
384	624.1	17.3	0.5	385.1	204.9	9.9
542	612.2	11.9	0.3	373.5	192.7	9.7
745	598.1	14.1	0.3	359.7	178.3	10.0
925	587.1	11.0	0.3	349.0	167.0	9.7
1145	574.9	12.2	0.3	337.1	154.5	9.8
1374	564.1	10.8	0.3	326.6	143.4	9.3
1627	552.8	11.3	0.2	315.5	131.9	9.9
1988	539.4	13.4	0.3	302.4	118.2	9.4
					Average	9.7

Source. Data from Kistiakowsky and Lacher.[21]

change in pressure due to dimerization of butadiene was calculated approximately from $\Delta P_{\text{dim}} = k'(P_{\text{butadiene}})^2 \Delta t$, where $P_{\text{butadiene}}$ is the partial pressure of butadiene at the start of the time interval and k' is the specific rate constant for dimerization (previously evaluated). Then ΔP_{dim} was subtracted from ΔP_{total} to give the pressure drop due to the Diels-Alder reaction. From the stoichiometry of (3.23) and (3.25) it then follows that

$$\Delta P_{\text{acrolein}} = \Delta P_{\text{total}} - \Delta P_{\text{dim}} \tag{3.26}$$

$$\Delta P_{\text{butadiene}} = \Delta P_{\text{total}} + \Delta P_{\text{dim}} \tag{3.27}$$

so that the new pressure of each at the end of the time interval could be found. This enabled the next two columns of Table 3.3 to be filled out. The second-order rate constant given in the last column was evaluated by using the unintegrated form (3.24):

$$\frac{-\Delta P_{\text{acrolein}}}{\Delta t} = k(P_{\text{acrolein}})(P_{\text{butadiene}}) \tag{3.28}$$

where the pressures are the average partial pressures of each component during the time interval Δt. The constancy of k (average 9.7×10^{-7} torr^{-1} s^{-1}) is evidence that the reaction is second order. Another way of treating the data is to use the integrated form of (3.24), which, as is shown in (2.29), becomes

$$kt = \frac{1}{(P^0_{\text{acrolein}} - P^0_{\text{butadiene}})} \ln \frac{(P_{\text{acrolein}})}{(P_{\text{butadiene}})} + \text{constant} \tag{3.29}$$

Figure 3.2 shows the result of plotting $\log[(P_{\text{acrolein}})/(P_{\text{butadiene}})]$ against time.

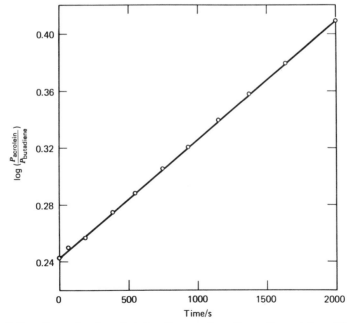

Figure 3.2. Second-order plot for the condensation of acrolein with butadiene (Data from Kistiakowsky and Lacher[21]).

The slope multiplied by 2.303 and divided by $(418.2 - 280.0)$ torr gives k as 10.9×10^{-7} torr^{-1} s^{-1}. Part of the difference between 10.9 and 9.7 is due to the dimerization reaction, which is included in the rate calculated by (3.29), and the remainder is due to the difference in using integrated and unintegrated forms of the rate equation.

REACTIONS AT CONSTANT PRESSURE

It must be emphasized that all rate expressions given up to this point are based on the implicit assumption that the volume of the reacting system remains constant. However, this assumption is not always valid. When a gas-phase reaction involving a change in the amount of substance is carried out at constant pressure, for example, there is a significant change in volume. The effect of this change on the reaction rate must be taken into account.

Consider the first-order gas-phase reaction

$$A \to \nu B \tag{3.30}$$

The first-order rate expression may be rewritten as follows to include volume explicitly:

$$\frac{-dc_A}{dt} = \frac{-(1/V)dn_A}{dt} = \frac{kn_A}{V} = kc_A \tag{3.31}$$

Integrating 3.31 in terms of n_A yields

$$\frac{n_A}{n_{A,0}} = e^{-kt} \tag{3.32}$$

Assuming ideal gas behavior and that the initial system consists of pure A, the volume of the system can be expressed as a function of the extent of reaction ξ:

$$V = V_0[1 + (\nu - 1)\xi] = V_0\left[1 + \frac{(\nu - 1)(n_{A,0} - n_A)}{n_{A,0}}\right]$$

$$= V_0\left[\nu + \frac{(1 - \nu)n_A}{n_{A,0}}\right] \tag{3.33}$$

Combining 3.32 and 3.33, we have

$$\frac{V}{V_0} = \nu + (1 - \nu)e^{-kt} \tag{3.34}$$

and, assuming $\nu > 1$,

$$\ln\left(\nu - \frac{V}{V_0}\right) = -kt + \ln(\nu - 1) \tag{3.35}$$

Thus k may be obtained from a plot of $\ln(\nu - V/V_0)$ versus t. If it is more convenient to measure concentration than volume, another useful relationship may be obtained by dividing (3.34) by (3.32)

$$\frac{c_{A,0}}{c_A} = \nu e^{kt} + (1 - \nu) \tag{3.36}$$

and

$$\ln\left(\frac{c_{A,0}}{c_A} - 1 + \nu\right) = \ln\left[\frac{a}{(a-x)} - 1 + \nu\right] = kt + \ln \nu \qquad (3.37)$$

This result differs appreciably from the usual equation, unless, of course, $\nu = 1$, in which case (3.37) becomes identical to (2.9).

Special forms of the rate equation for systems in which the volume varies have been reported by a number of workers.[22-24] This situation is most commonly encountered in the case of flow reactors,[25, 26] where pressure is constant or nearly so. Flow reactors are discussed in some detail later in this chapter.

REACTIONS IN SOLUTION

Methods for the study of reactions in solution have been described by a number of authors,[13, 27-30] and details are not given here. Instead we describe briefly a number of methods, giving examples of several that are very widely used.

If a gas is consumed or produced during a reaction in solution, manometric methods of the type described for gas-phase reactions are valuable. However, one must be certain that the gas is essentially insoluble in the reaction medium and that the rate of diffusion to the liquid surface does not affect the observed reaction rate.[31, 32] Gas–liquid partition chromatography can also be applied to liquid-phase reactions that involve volatile solutes, the reaction solution usually being injected directly into the gas chromatograph. Spectral methods, especially ultraviolet-visible spectrophotometry, but also infrared and fluorescence spectrophotometry, polarimetry, and NMR spectrometry, are widely used. If a range of wavelengths can be scanned repetitively during the course of a reaction, evidence regarding the presence and identity of intermediate species may also be obtained. Dilatometry, the measurement of small volume changes that accompany most reactions in solution, is in principle broadly applicable, but it requires very precise temperature control in order that thermal expansion not obscure the effect of a chemical reaction.[33] For reactions that involve a change in the total concentration or charge of ions, very precise rate determinations can be achieved by monitoring the resistance of a solution. Many reactions consume or produce hydrogen ions, and so pH measurements are widely applicable. The speed of response of the glass electrode limits of the range of rates for which a pH meter can be used, but acid–base equilibria involving indicators are established rapidly and spectrophotometric methods can be used to measure rapid changes in hydrogen ion concentration. For specific reactions in solution a variety of other methods, including chemical analysis, may be useful, and the references mentioned above should be consulted for detailed descriptions of their application.

The demetalation of $\alpha,\beta,\gamma,\delta$-tetra-($p$-sulfophenyl)porphineiron(III) in sulfuric acid–ethanol–water media has been reported by Reynolds et al.[34] and provides a good example of the application of ultraviolet-visible spectrophotometry to the study of reactions in solution. Successive scans of the spectrum of the reaction mixture are shown in Fig. 3.3. Since the initial and final spectra agree closely

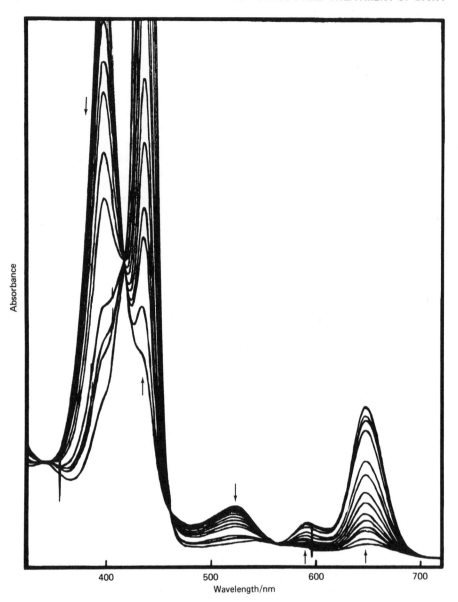

Figure 3.3.　Successive scans of the visible spectrum during demetalation of α, β, γ, δ-tetra-(p-sulfophenyl)porphineiron(III) in sulfuric acid [(Reprinted with permission from W. L. Reynolds, J. Schufman, F. Chan, and R. C. Brasted), *Int. J. Chem. Kinet.*, **9**, 777 (1977).]

with independently determined spectra of $Fe(TPPS)^{3-}$ and $H_4 TPPS^{2-}$, respectively [TPPS = completely deprotonated $\alpha,\beta,\gamma,\delta$-tetra($p$-sulfophenyl)porphine], Reynolds et al. concluded that the overall reaction was

$$Fe(TPPS)^{3-} + 4H^+ = Fe^{3+} + H_4 TPPS^{2-}$$

The absorbance change as a function of time was followed at 437.5 nm, a wavelength at which H_4TPPS^{2-} has an absorbance maximum and at which there is a large difference between spectra of reactant and product. Since concentrations of all other species were of the order of 10^3 times that of $Fe(TPPS)^{3-}$, the reaction conditions were pseudo-first order. Therefore $kt = 2.303 \log [(A_\infty - A_0)/(A_\infty - A)]$, where A represents the absorbance at 437.5 nm, and

$$\log (A_\infty - A) = \frac{-kt}{2.303} + \log (A_\infty - A_0)$$

Accordingly, rate constants were obtained as $-2.303 \times$ slope from a plot of log $(A_\infty - A)$ versus t. Such a plot is shown in Fig. 3.4. The pseudo-first-order rate constant is $1.30 \times 10^{-3} s^{-1}$.

Figure 3.3 exhibits three well-defined isosbestic points, that is, wavelengths at which the absorbance does not change with time during the course of the reaction. It is generally accepted that occurrence of an isosbestic point in a closed system requires that the changes in the concentrations of the various components be linearly related,[35] although situations are known where changes in molar absorption coefficients with temperature or solvent composition may produce isosbestic points.[36] In kinetic studies this latter complication is unlikely, and the presence of

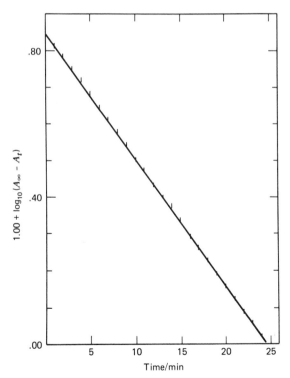

Figure 3.4. Plot of log $(A_\infty - A)$ against t for demetalation of $\alpha, \beta, \gamma, \delta$-tetra-($p$-sulfophenyl)porphineiron(III) in sulfuric acid [(Reprinted with permission from W. L. Reynolds, J. Schufman, F. Chan, and R. C. Brasted), *Int. J. Chem. Kinet.*, **9**, 777 (1977).]

an isosbestic point may be taken as evidence that a single set of reactants gives a single set of products in constant proportions. Such a reaction may be described by a single extent-of-reaction parameter ξ, or by $x = \xi/V$ as in the case of (3.1). For the general reaction (3.1) we obtained the relationships $\lambda - \lambda_0 = x\,\Delta k$ and $\lambda_\infty - \lambda_0 = (a/\nu_A)\Delta k$ (3.11) above. Since a and ν_A are nonzero, $\Delta k = 0$ under any conditions in which $\lambda_\infty = \lambda_0$. Consequently $\lambda = \lambda_0$ throughout the reaction, and an isosbestic point will occur.

An illustration of the implications of the absence of isosbestic points is provided by the work of Childers et al.[37] on the aquation of tetraaquoethylenediamine-chromium(III) cation. Repeated scans of the spectrum of the reaction mixture revealed isosbestic points at 402, 434, and 566 nm during the early stages of the reaction. Eventually, however, the first and last "isosbestic points" shifted toward longer wavelengths, revealing that they were not true isosbestic points. This was taken as evidence for a two-stage reaction:

$$Cr(en)(OH_2)_4^{3+} + H_2O + H^+ = Cr(enH)(OH_2)_5^{4+} \qquad (3.38a)$$

$$Cr(enH)(OH_2)_5^{4+} + H_2O = Cr(OH_2)_6^{3+} + enH^+ \qquad (3.38b)$$

Here en represents $H_2NCH_2CH_2NH_2$, a bidentate ligand, and enH^+ represents $H_2NCH_2CH_2NH_3^+$, a unidentate ligand. Childers et al. were able to prepare the proposed intermediate and measure its spectrum, thus confirming the intersection points observed early in the reaction. The isosbestic point at 434 nm remained throughout the reaction because at this wavelength reactant, intermediate, and product all had the same molar absorption coefficient. Such a coincidence can result in an isosbestic point even though concentrations are not linearly related.

In some cases spectral scans cross at such a small angle that isosbestic points are difficult to identify unambiguously. Special techniques involving linear combinations of the spectral curves can be used to reveal genuine isosbestic points when this happens.[38] Another useful technique for stepwise reactions such as (3.38) is to observe the first step at a wavelength that is an isosbestic point for the second. In this way an A_∞ value is obtained that is not affected by the subsequent reaction.

The application of resistance measurements to kinetic studies is illustrated by the data of McGuire[39] in Table 3.4 for the reaction of pyridine with phenacyl bromide to form a quaternary ammonium salt, phenacylpyridinium bromide, in methanol solution. Neutral molecules are forming ions, and so the resistance decreases sharply as the reaction (3.39) goes on. Since resistance R is inversely

$$C_6H_5-\overset{\overset{\displaystyle O}{\|}}{C}-CH_2Br + C_5H_5N = C_6H_5-\overset{\overset{\displaystyle O}{\|}}{C}-CH_2-\overset{+}{N}C_5H_5 + Br^- \qquad (3.39)$$

proportional to conductance, which in turn is proprotional to concentration, (3.8) and (3.9) become

$$-\frac{\nu_A x}{a} = \frac{1/R - 1/R_0}{1/R_\infty - 1/R_0} = \frac{(R_0 - R)R_\infty}{R(R_0 - R_\infty)} \qquad (3.40)$$

and

$$\frac{a}{a + \nu_A x} = \frac{1/R_\infty - 1/R_0}{1/R_\infty - 1/R} = \frac{(R_\infty - R_0)R}{R_0(R_\infty - R)} \qquad (3.41)$$

Table 3.4. Phenacyl bromide and pyridine in methyl alcohol at 308 K[a]

Time/min	Resistance/Ω	$-R/(R_\infty - R)$
7	45,000	1.019
28	11,620	1.074
53	9,200	1.096
68	7,490	1.120
84	6,310	1.145
99	5,537	1.170
110	5,100	1.186
127	4,560	1.213
153	3,958	1.253
203	3,220	1.330
368	2,182	1.580
∞	801	

Source. Data from McGuire.[39]
[a] Value of k from slope $7.4 \times 10^{-4}\ M^{-1}\ s^{-1}$.

Reaction (3.39) is second order and both reactants were present initially at equal concentrations so (2.16) applies. Substitution of (3.41) into (2.16) gives

$$kt = \frac{1}{2}\left(\frac{a}{a-x} - 1\right) = \frac{1}{2}\left[\frac{(R_\infty - R_0)R}{R_0(R_\infty - R)} - 1\right]$$

or

$$\frac{R}{R_\infty - R} = \frac{aR_0kt}{R_\infty - R_0} + \frac{R_0}{R_\infty - R_0} \tag{3.42}$$

Figure 3.5 shows that $R/(R_\infty - R)$ is linear with time, as required by (3.42). Dividing the slope of this plot by the initial concentration, $a = 0.0385\ M$, and by the intercept, $R_0/(R_\infty - R_0)$, gives a rate constant $k = 7.41 \times 10^{-4}\ M^{-1}\ s^{-1}$.

In addition to the obvious case where there is no change in the number and/or charge of ions, there are several situations where resistance measurements are inappropriate for kinetic studies. If a large concentration of ions is present throughout a reaction (to maintain approximately constant ionic strength or pH, for example) relative changes in resistance are small and difficult to measure precisely. In solvents of low dielectric constant and solvating ability the relationship between resistance and concentration is complicated by ionic association. Even under favorable circumstances highly accurate results require correction for the fact that concentration is not exactly proportional to conductance.[40]

The use of pH measurements in the study of reaction rates is illustrated by the work of Hay and Cropp[41] on the base hydrolysis of *cis*-chloroaminebis(ethylene-diamine)cobalt(III) complexes. The reaction is

$$cis\text{-}[Co\ en_2(NH_2R)Cl]^{2+} + OH^- = [Co\ en_2(NH_2R)OH]^{2+} + Cl^-$$

It is second order overall — first order in complex ion and first order in hydroxide ion. The reaction was followed in an automatic titrator by recording as a function

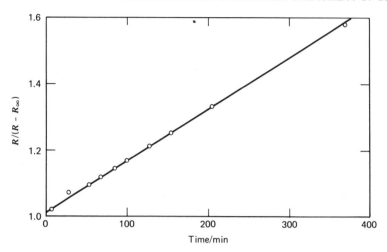

Figure 3.5. Reaction of phenacyl bromide and pyridine as second order with equal initial concentrations (Data from McGuire[39]).

of time the volume V of base required to maintain constant pH. Since the hydroxide ion concentration did not change, the reaction met the conditions described in Chapter 2 for pseudo-first-order kinetics. Plots of log $(V_\infty - V)$ versus time were found to be linear as shown in Fig. 3.6 for the case where $R = CH_3$. These plots yield pseudo-first-order rate constants k_{obs} of 2.67×10^{-4} s^{-1} at pH = 9.20, 4.23×10^{-4} s^{-1} at pH = 9.40, and 6.55×10^{-4} s^{-1} at pH = 9.60. Hydroxide ion concentrations were calculated from pH values using estimated activity coefficients, and second-order rate constants $k = k_{obs}/[OH^-]$ of $12.9\,M^{-1}$ s^{-1}, $12.9\,M^{-1}$ s^{-1}, and $12.6\,M^{-1}$ s^{-1}, respectively, were obtained. The pH-stat method is generally applicable to reactions that consume or produce hydrogen or hydroxide ions, but one must be careful that the stoichiometry of the reaction remains the same over the entire pH range studied. For example, although base-catalyzed hydrolysis of *cis*-[Co en$_2$(NH$_2$R)Cl]$^{2+}$ complexes occurs at pH values as low as 3–4, the pH-stat method is only applicable above pH = 7. This is a consequence of rapid protonation of the product hydroxo complex, changing the stoichiometry to

$$\textit{cis-}[Co\ en_2(NH_2R)Cl]^{2+} + H_2O = [Co\ en_2(NH_2R)(H_2O)]^{3+} + Cl^-$$

below pH = 6–7.

Although physical measurements that involve a linear relationship with concentration are by far the most common, other methods may also be used provided suitable calibrations are performed or theoretical equations of sufficient validity are available. A good example is provided by the work of Rolston and Yates,[42] who took advantage of the logarithmic relationship between concentration and electromotive force measurements. The reaction being studied was bromination of substituted styrenes in acetic acid:

The rate depends on the concentrations of bromine and styrene, and the reaction is accelerated in the presence of bromide ions. The sensitivity of their potentiometric method allowed Rolston and Yates to use initial bromine concentrations as small as 10^{-4} M and still follow the reaction to 95% completion. Concentrations of styrene and bromide ion were sufficiently large that the reaction was pseudo-first order in bromine. Therefore the pseudo-first-order rate constant k_{obs} could be obtained as the negative of the slope of a plot of ln $[Br_2]$ versus t, that is

$$\frac{d \ln [Br_2]}{dt} = -k_{obs} \tag{3.43}$$

Applying the Nernst equation under conditions of constant bromide ion concentration, Rolston and Yates obtained

$$E_{obs} = \text{constant} + \frac{RT}{2F} \ln [Br_2] \tag{3.44}$$

where the constant includes the standard potential of the Pt$|$Br$_2$, Br$^-$ half cell referred to the modified calomel electrode, an unknown but constant liquid junction potential, and a potential due to the constant bromide ion concentration. Equation (3.44) readily leads to the relationship

$$\frac{dE_{obs}}{dt} = \frac{RT}{2F} \frac{d \ln [Br_2]}{dt} = -\frac{RT}{2F} k_{obs}$$

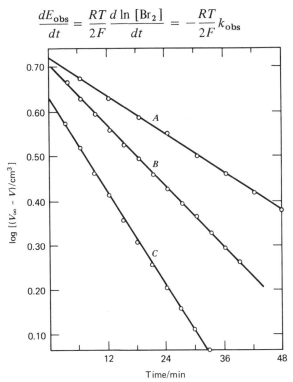

Figure 3.6. Plot of log $(V_\infty - V)$ against t for base-catalyzed hydrolysis of cis-[Co en$_2$(NH$_2$CH$_3$)Cl]$^{2+}$. V = volume of base required to maintain constant pH of A, 9.20; B, 9.40; and C, 9.60. [Reprinted with permission from R. W. Hay and P. L. Cropp, *J. Chem. Soc. (A)*, **1969**, 42.]

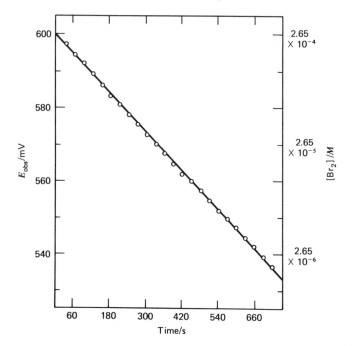

Figure 3.7.   Plot of E_{obs} (and $[Br_2]$) against t for the bromide-ion-catalyzed addition of bromine to 3-chlorostyrene [Reprinted with permission from J. H. Rolston and K. Yates, *J. Am. Chem. Soc.*, **91**, 1483 (1969).]

and so k_{obs} can be determined from the slope of a plot of E_{obs} as a function of time. Figure 3.7 shows that such a plot is linear over an approximately hundredfold decrease in bromine concentration. Instead of the theoretical value of $RT/2F = 0.0128$ V calculated from the gas constant, temperature, and Faraday constant, Rolston and Yates used the factor 0.0135 V (determined from the slope of an experimentally derived plot of E_{obs} versus ln $[Br_2]$) to calculate k_{obs} values from plots such as that of Fig. 3.7. Thus in acetic acid calibration gives results significantly different from the theoretical factor, although the latter is perfectly adequate in aqueous systems.

FLOW METHODS

Flow methods are the best way to follow fast reactions in which the reagents cannot be previously mixed. (Relaxation methods, in which a system is perturbed and its rapid return to equilibrium is followed, are discussed in Chapter 8.) All fast flow methods are based on the pioneering work of Hartridge and Roughton.[43] These methods may be used to study reactive intermediates that are too unstable to be isolated as well as to follow rapid changes involving stable reactants.[44] Flow reactors also are widely used for industrial processes because of the ease of handling large quantities of reactants in a limited reactor space and because of the possibility of continuous operation. This is particularly advantageous for reactions involving solid catalysts, where emptying and filling operations may change the activity of the catalyst.

For the measurement of reaction rates, mixtures of reactants of known initial concentration flow through a region of known volume at a known, usually constant, rate. The mixture is analyzed after it has passed through the reaction space or at fixed points in the space. Complete mixing of reactants before they enter the zone of reaction is assumed. Specially designed mixing chambers just upstream of the reaction space are usually employed.

If, after flow enters the reaction space, partial mixing occurs between different parts of the flow stream, such as would be caused by diffusion and convection, the results are difficult to interpret.[45] There are two limiting cases, however, at least one of which can usually be approached in a particular system and each of which can be treated mathematically in a straightforward way. Plug flow corresponds to flow down a tubular reactor with essentially no longitudinal mixing. Stirred flow corresponds to complete longitudinal mixing, so that composition is uniform throughout the reaction volume.

To derive the basic equations for a plug-flow reactor, consider a tubular reactor space of constant cross-sectional area A as shown in Fig. 3.8 with a steady flow u of reaction mixture expressed as volume per unit time. Select a small cylindrical volume element dV such that the concentration of a component i entering the unit is c_i and the concentration leaving the unit is $c_i + dc_i$. Within the volume unit the component is changing in concentration owing to chemical reaction with a rate equal to r_i. This rate is of the form of the usual chemical rate equation and is a function of the rate constants of all reactions involving the component i and of the various concentrations in the volume unit. The time rate of change in the amount of i in the volume unit is

$$\frac{dn_i}{dt} = r_i \, dV - u \, dc_i \tag{3.45}$$

Eventually a steady state is reached where the concentration of each component within each volume element becomes constant. The composition from one unit to the next is different, however. In this steady state the condition is that

$$r_i \, dV = u \, dc_i \tag{3.46}$$

This may be integrated to give

$$\frac{V}{u} = \int_{c_0}^{c} \frac{dc_i}{r_i} \tag{3.47}$$

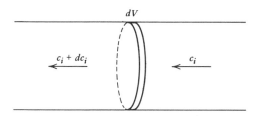

Figure 3.8. Cylindrical volume element in a flow system.

where c_0 is the concentration of the ith component entering the reaction space and V is the total volume of the reactor up to the point where the concentration is c. For a tubular reactor $V = Al$, where l is the distance from the entrance of the reactor to the point in question.

To use (3.47) one must have a definite expression for the rate r. Consider the case of a first-order reaction

$$A \xrightarrow{k_1} B \tag{3.48}$$

where c is the concentration of substance A.

$$r = \frac{dc}{dt} = -k_1 c \tag{3.49}$$

Putting (3.49) into (3.47) and integrating gives

$$k_1 = \left(\frac{u}{V}\right) \ln\left(\frac{c_0}{c}\right) \tag{3.50}$$

The resemblance of this equation to that for a first-order reaction in a closed system,

$$k_1 = \frac{1}{t} \ln\left(\frac{c_0}{c}\right) \tag{3.51}$$

is obvious. The result can be generalized in that the integrated equation for a reaction of any order in a plug-flow system is the same as for a closed system except that V/u replaces the time variable. Since for a tubular reactor $V/u = lA/u$ and A and u are constants, the distance down the tube replaces time as a variable. Special cases have been discussed by Harris.[46]

It should be noted that the derivation just given assumes constancy of the volume element dV occupied by an element of reaction mixture passing down the tube. Volume changes, which can be taken into account by mathematical methods described earlier in this chapter, may result from the existence of a significant pressure difference along the tube or from a change in amount of gaseous substance due to the reaction itself. The latter effect can be mitigated by diluting reactants with a carrier gas or by making observations at very small extents of reaction.

The case of complete mixing in a stirred-flow reactor has been discussed thoroughly by Denbigh.[47] Since composition is uniform as a result of efficient stirring, the volume element dV may be replaced by the total volume V in (3.45) and dc_i may be replaced by $c - c_0$. Making the substitutions and dividing by V transforms (3.45) into

$$\frac{dc}{dt} = r - \frac{u}{V}(c - c_0) \tag{3.52}$$

which is the equation for the approach to the steady state in a constant-volume system. At the steady state $dc/dt = 0$, the composition in the reactor becomes constant, and

$$r = \frac{u}{V}(c - c_0) \tag{3.53}$$

A similar equation can, of course, be derived for each component in the system if the proper expression for r is used. If u and V are known, determination of concentrations c and c_0 gives r, the value of the rate for the particular constant conditions in the reactor. By changing the initial concentration c_0 or the flow u, the rate can be found for other conditions. This permits explicit determination of the form of the rate expression, without integration.

Consider again the first-order reaction (3.48). In a stirred-flow reactor with B initially absent, (3.53) becomes

$$k_1(a - x) = \frac{u}{V}x \qquad (3.54)$$

so that knowledge of a, the initial concentration, and x, the steady-state concentration of product, enables the rate constant to be evaluated. Changing the initial concentration and the flow would simply result in a change in x so that $ux/V(a - x)$ remains constant. For related equations generalized to a system in which volume is not constant, see Boudart.[48]

Extension of (3.54) to more complex rate equations can readily be seen. An important advantage of the stirred-flow reactor is that complicated rate expressions can be handled since integration is avoided.[49] Another important advantage of the stirred-flow method for reactions in solution is that constant conditions of solvent composition, ionic strength, catalyst concentration, and so forth are readily maintained. The method can also be used to build up the concentration of a transient intermediate to an optimum value that can be maintained long enough for the intermediate to be detected and its concentration measured.

Stirred-flow reactors have several disadvantages. The method depends on knowledge of concentrations of all the reactants, intermediates, and products for a complete solution. In a simple case analysis for one component might suffice, but as complexity of the reaction system increases, so does the number of separate analyses that are required. The method is rather slow. Large volumes of solution go to waste while the steady state is being established, and each experiment gives essentially only one point toward determining the rate equation. Hammett and coworkers[50] give valuable details on apparatus and operation. Johnson and Edwards[51] discuss using a number of stirred-flow reactors in series.

Although the stirred-flow method is more commonly applied to reactions in solution, Mulcahy and Williams[52] have shown that this technique is suitable for gas-phase studies. They studied pyrolysis of di-t-butyl peroxide (3.55) at pressures

$$(CH_3)_3COOC(CH_3)_3 \rightarrow 2(CH_3)_2CO + C_2H_6 \qquad (3.55)$$

between 2 and 30 torr and over the temperature range 430–550 K. Their reactor was an 80-mm spherical flask into which reactants were introduced by means of a 14-mm spherical perforated diffuser at the center. The entering gases were mixed rapidly by vigorous convection and diffusion. Residence times in the reactor ranged from 0.1 to 100 s. Temperature was maintained within ± 2.5 K by an electric furnace that completely surrounded the reactor. Experiments were conducted with pure peroxide and in the presence of CO_2 carrier gas. Flow rate of peroxide was determined from loss in weight of a thermostated reservoir. Ethane and carbon dioxide were collected at 77 K after separation of unreacted peroxide and acetone product. Quantities of C_2H_6 and CO_2 were determined by fractionation into vessels of known volume provided with manometers. Flow rates and reactor pressure were regulated by capillary orifices that could be changed readily. Since (3.55) involves a threefold increase in number of molecules, volume changes had to be taken into account when calculating rate constants. The equation used was

$$k = \frac{v_P RT}{\nu PV} \left[\frac{\nu(v_0 + v_c) + (\nu - 1)v_P}{\nu v_0 - v_P} \right] \tag{3.56}$$

where v_0, v_c, and v_P are the rates of flow of reactant, carrier gas, and product in amount of substance per unit time, and ν is the number of product molecules formed per reactant molecule destroyed. For decomposition of di-t-butyl peroxide $\nu = 3$. The rate of flow of product out of the reactor v_P was calculated as three times the rate of accumulation of ethane. Results obtained from (3.56) were in excellent agreement with previous studies of the same reaction by static and plug-flow methods, indicating that under the conditions of Mulcahy and Williams' experiments there was essentially complete mixing in the reactor.

As is the case for stirred-flow reactors, flow systems without mixing can be used to maintain unstable reactants and intermediates at steady concentrations so that they may be detected and measured. However, the most important advantage of unstirred-flow methods is that they may be applied to very rapid reactions. Since the time coordinate is replaced by V/u, using a reactor of minimum volume and a high rate of flow can result in equivalent reaction times as small as 1 ms. Thus reactions with half-lives of less than 10 ms may be measured if a suitable method of analysis can be found. Methods that have been used include spectrophotometry, conductance and EMF measurements, temperature change due to heat of reaction, and quenching of the reaction mixture as it exits the reactor.[43, 44, 53]

Since different lengths of a tubular reactor represent different values of V, a series of measurements may be made to obtain in one experiment a number of concentration/equivalent-time data. It is also possible to vary the equivalent time by changing u. In Chance's accelerated-flow method[54] the flow rate is increased steadily while at the same time a photomultiplier output measuring solution absorbance is fed to an oscilloscope. This provides a complete concentration–time record but requires a much smaller volume of reactant solutions than does the continuous plug-flow method. Small solution volume is also an advantage of the stopped-flow method,[55] which in reality is a static reactor connected to a rapid-mixing device. Reactants are injected through a special mixing chamber with multiple entry ports that are arranged tangentially. The mixed solutions then pass into the reactor. After a flow time sufficient to reach a steady state (usually a few milliseconds) the flow is suddenly stopped and the concentration is followed by spectrophotometric or other methods in the static reactor. The stopped-flow method permits mixing of two solutions in less than 1 ms. Currently it is the most popular flow method, especially for enzyme-catalyzed reactions where cost of reagents is a major consideration. Instruments are available commercially, and computer-automated devices have been described.[56]

As an example, Fig. 3.9 shows some data obtained in a continuous-flow study[57] of the moderately rapid reaction.

$$H_2CO_3 \rightarrow H_2O + CO_2 \tag{3.57}$$

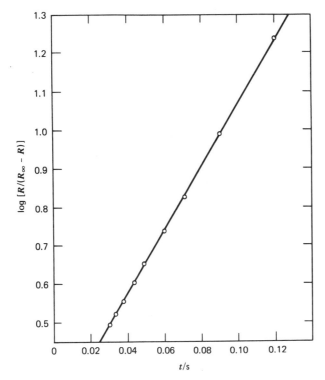

Figure 3.9. Decomposition of carbonic acid at 22°C (Data from Pearson, Meeker, and Basolo[57]).

The experimental procedure involved mixing aqueous solutions of sodium bicarbonate and hydrochloric acid (about $0.005M$) in equivalent amounts. Measurements of electrical conductivity were then made as a function of time, as calculated from the flow rate and distance to the point of observation. The reaction is first order with a rate constant of $20.9\,\mathrm{s}^{-1}$. The resistance increases with time as the partly ionized carbonic acid is converted to unionized carbon dioxide.

DETERMINATION OF THE RATE EXPRESSION

Assuming that concentration–time data have been obtained for several kinetic runs at the same temperature, the problem next arises of obtaining a rate expression that can account for the data. The rate expression provides an important clue to the mechanism of the reaction as well as summarizing in its parameters all the experiments that have been done.

No single approach to the problem of determining the form of the rate expression and evaluating the parameters involved is convenient in all cases. The most common procedure is to make an intelligent guess at the rate expression. Such a guess might be based on the stoichiometry of the reaction or on some assumption concerning the mechanism. The assumed rate expression is then integrated when possible to give a relationship between concentration (or some physical property) and time, and that relationship is tested against the experimental data by numerical or graphical methods. This trial-and-error. procedure is repeated until, as in the examples given in Chapter 2 and the earlier portions of this chapter, the integrated form of the assumed rate expression closely reproduces the data.

Occasionally it may happen that a rate expression cannot be integrated analytically. In such a case numerical integration can be carried out by means of a digital computer, and the rate expression can still be tested as described above. Even though integration can be done analytically, numerical integration or computer-based data reduction are still valuable in the case of all but the simplest rate expressions. Unfortunately, however, no single computer-based system is available that, without becoming cumbersome, can handle the multitude of combinations of rate expressions, physical methods of analysis, and experimental conditions that may be encountered in kinetic studies. Instead the kineticist must select intelligently among a variety of conventional and computerized methods, each of which is restricted to a specified set of conditions.

FRACTIONAL-LIFE PERIOD METHODS

In the event that the rate expression is of the form

$$\frac{dx}{dt} = k(a-x)^n \tag{3.58}$$

the half-life (or other fractional-life) method can be used. The half-life period is the time required for one-half of a given reactant to be consumed. Notice that this is not one-half of the time required to complete the reaction, which is presumably infinite except for a zeroth-order reaction. Substituting $(a-x) = a/2$ in the integrated forms of (3.58) gives for $n = 1$

$$t_{1/2} = \frac{\ln 2}{k} \tag{3.59}$$

and for $n \neq 1$

$$t_{1/2} = \frac{(2^{n-1} - 1)}{k(n-1)a^{n-1}} \tag{3.60}$$

The dependence of the half-life period on the initial concentration is of particular importance. It is evident that for all values of n

$$t_{1/2} = \frac{f(n, k)}{a^{n-1}} \tag{3.61}$$

where f is some function of n and k and therefore constant for a given reaction at constant temperature. This relationship includes the usually special first-order case, which has the unusual feature of a $t_{1/2}$ independent of a. Putting (3.61) in logarithmic form yields

$$\log t_{1/2} = \log f - (n-1) \log a \tag{3.62}$$

A log–log plot of $t_{1/2}$ versus a should be linear with a slope of $1 - n$.

By applying (3.62) to any two pairs of $(t_{1/2}, a)$ data, say $t_{1/2}$ and a, and $t'_{1/2}$ and

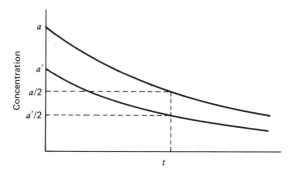

Figure 3.10. Demonstration of the independence of half-life on concentration for a first-order reaction.

a', and subtracting one equation from the other, the log f term drops out. Solving for n gives the Noyes equation (3.63). With this equation the order of reaction can

$$n = 1 + \frac{\log t'_{1/2} - \log t_{1/2}}{\log a - \log a'} \tag{3.63}$$

be calculated directly from data on two runs at sufficiently different initial concentration, as in Figure 3.10.

Equations (3.59) and (3.63) can be generalized to apply to t_y, the time for the fraction reacted to equal y, when the concentration has dropped from a to $a(1 - y)$. Equation (3.62) becomes

$$\log t_y = \log f - (n - 1) \log a \tag{3.64}$$

and, in general, a log–log plot of t_y versus a should be linear. Equation (3.63) becomes (3.65).

$$n = 1 + \frac{\log t'_y - \log t_y}{\log a - \log a'} \tag{3.65}$$

So far the discussion of fractional-life periods has depended on the experimental observation of two or more runs with different initial concentrations. But it is evident that two or more successive time intervals in a *single run* may be used in a similar manner, the concentration resulting at the end of one time interval being

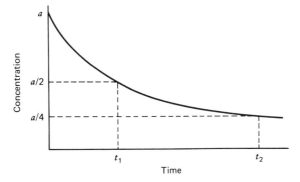

Figure 3.11. Two successive half-life periods in a second-order reaction; $t_2 = 3t_1$.

considered the initial value for a new time interval. Figure 3.11 shows such an experiment. If t_1 corresponds to the concentration $a(1-y)$, then t_2 corresponds to the concentration $a(1-y)^2$. Also $t'_y = t_2 - t_1$ so that (3.65) becomes (3.66).

$$n = 1 + \frac{\log [(t_2/t_1) - 1]}{\log [1/(1-y)]} \qquad (3.66)$$

For the case shown in Fig. 3.11, $t_2/t_1 = 3$ and $y = \frac{1}{2}$, so that

$$n = 1 + \frac{\log 2}{\log 2} = 2$$

Because rate data may be more accurate at the beginning of a reaction, a value of y less than one-half should be used. On the other hand, y should not be too small since as y decreases the log ratio in (3.66) tends toward an indeterminate form, and thus slight experimental errors are greatly amplified in calculating n. A suitable value of y might be 0.2, in which case t_1 and t_2 would be the times for the concentration of reactant to have fallen to 0.80 and 0.64 of its original value.

It should be remembered that this treatment is valid only if the rate expression is given by (3.58). If this is not true, equations such as (3.65) and (3.66) will still give a value of n, but one that will change with the concentration. Whenever the order by this method turns out to be fractional it may be expected that further investigation will show that a more complex rate expression is needed. The Powell plot of Chapter 2 may be considered a generalization of these fractional life methods.

Suppose that the rate expression is now given by

$$\frac{dx}{dt} = k [A]^{n_a} [B]^{n_b} [C]^{n_c} \qquad (3.67)$$

where $[A]$, $[B]$, and $[C]$ are the concentrations of substances A, B, and C, respectively. There are certain conditions that, if satisfied, make for simplification in the determination of the order.

Suppose that there is effectively constant concentration of all components except one, which for convenience is taken as component A, so that we have a pseudo-n_ath-order reaction. Then during a reaction run $[B]$ and $[C]$ and their powers may be lumped in with the rate constant k to form a new effective (or pseudo-n_ath-order) rate constant k', which is given by

$$k' = k [B]^{n_b} [C]^{n_c} \qquad (3.68)$$

so that

$$\frac{dx}{dt} = k'[A]^{n_a} \qquad (3.69)$$

This is of the same form as (3.58) and all the equations (3.58)–(3.66) apply to this case, but with the reinterpretation of the symbols, in particular, k becomes k' the effective rate constant, and n becomes n_a the pseudo-order of the reaction for the given conditions, which also is the order with respect to A.

In principle, it is possible in this way to get the order with respect to each reactant (and product, if it affects the rate) by making each one successively the least concentrated component.

A second method of simplifying the solution of (3.67) is to choose the initial concentration of each reactant so that all concentrations throughout the run are in constant proportion to each other. This will be true if equivalent amounts of the reactants are used. For example, if in the reaction

$$2NO + O_2 = 2NO_2$$

the concentration or partial pressure of NO is twice as great as that of O_2, then neither reactant is in excess. For a general reaction such as (3.1) under these conditions, the reactant concentrations are in constant proportion during the run:

$$[B] = \left(\frac{\nu_B}{\nu_A}\right)[A] \qquad [C] = \left(\frac{\nu_C}{\nu_A}\right)[A] \qquad \text{etc.}$$

where ν_A, ν_B, and ν_C are stoichiometric coefficients in the equation of the reaction. Then (3.67) becomes

$$\frac{dx}{dt} = k\left(\frac{\nu_B}{\nu_A}\right)^{n_b}\left(\frac{\nu_C}{\nu_A}\right)^{n_c}[A]^{n_a+n_b+n_c} \tag{3.70}$$

or

$$\frac{dx}{dt} = k''(a-x)^n$$

where the new constant $k'' = k(\nu_B/\nu_A)^{n_b}(\nu_C/\nu_A)^{n_c}$ and $n = n_a + n_b + n_c$ is the total order of the reaction. Equation (3.70) is again of the same form as (3.58) and the preceding discussion again applies. A different constant, k'', now results, and the total order of the reaction with respect to all reactants rather than to just one is obtained. There is an important restriction in using this method. No product of the reaction must influence the rate, for, if it did, its concentration could not be proportional to the concentration of a reactant during a run.

METHODS THAT INVOLVE NUMERICAL INTEGRATION

Suppose that we have an irreversible reaction

$$A + 2B + 3C = \text{products}$$

and we assume a rate expression of the form of (3.67). If a, b, and c represent initial concentrations of A, B, and C, then

$$\frac{dx}{(a-x)^{n_a}(b-2x)^{n_b}(c-3x)^{n_c}} = k\,dt \tag{3.71}$$

The left-hand side of (3.71) can be integrated by the method of partial fractions, but the resulting integrated form is so complicated that it is difficult to test against experimental data by graphing, especially when n_a, n_b, and n_c are not known. Skinner[58] has suggested the following computer-based procedure for handling this and similar situations. Integrate the right-hand side of (3.71), replace dx by Δx, a small but finite change in extent of reaction, and solve for k:

$$k_i = \left(\frac{\Delta x}{t_i}\right)\sum_{x=0}^{x=x_i}(a-x)^{-n_a}(b-2x)^{-n_b}(c-3x)^{-n_c} \tag{3.72}$$

Here x_i, t_i is the ith experimentally determined concentration–time data point. For a particular choice of orders n_a, n_b, and n_c, a computer can readily perform the summation in (3.72) for each data point, giving a rate constant for each pair of experimental values. Check the accuracy of the numerical integration by halving Δx, repeating the calculation, and looking for a significant change in any of the k_i values. (Better methods for numerical integration than the crude one described here are discussed in Chapter 8.)

Next, program the computer to vary n_a, n_b, and n_c systematically over a reasonable range. For example, each order might be allowed to vary from 0 to 2.0 in increments of 0.2. This gives $11^3 = 1331$ combinations, each of which involves a calculation of the sort described in the preceding paragraph. For each calculation determine the average of the k_i values and the relative standard deviation in that average. The best n_a, n_b, and n_c correspond to the minimum relative standard deviation, that is, the set of rate constants that is most nearly constant. To improve the accuracy of n_a, n_b, and n_c, program the computer to vary each order in increments of, say 0.02 about the best values chosen in the first round of calculations. To save computer time, use one of the other methods discussed in this chapter to obtain initial estimates of n_a, n_b, and n_c, and reduce the range over which the orders need to be varied. Or, when possible, carry out the integration analytically, thereby eliminating the need for a summation during calculation of each of the k_i. When using this method, obtain data from enough different kinetic runs so that the initial concentration of each kinetically active species is varied.

Lampugnani et al.[59] have described a related numerical integration method that can be used only if the rate expression is of the form of (3.58), (3.69), or (3.70). They evaluate statistically the standard deviation in the reaction order. If the order differs from an integer by more than three standard deviations, then the rate expression must contain more than a single term, each term involving a different integer power of the concentration being studied. The reaction of thiocyanate with iron(III), for example, is reversible,

$$\text{Fe}^{3+}(\text{aq}) + \text{SCN}^-(\text{aq}) = \text{Fe}(\text{SCN})^{2+}(\text{aq})$$

and if iron(III) is greatly in excess the rate expression is

$$\frac{d\,[\text{Fe}(\text{SCN})^{2+}]}{dt} = k_1\,[\text{Fe}^{3+}]\,[\text{SCN}^-] - k_{-1}\,[\text{Fe}(\text{SCN})^{2+}]$$

$$= k'\,[\text{SCN}^-] - k_{-1}\,[\text{Fe}(\text{SCN})^{2+}]$$

As we show in Chapter 8, for such a reversible reaction the approach to equilibrium is first order, the usual first-order plot is linear, and a rate constant equal to $k' + k_{-1}$ is obtained from the slope. Only if a number of experiments are done with different initial concentrations so that different equilibrium concentrations (or λ_∞ values) are obtained will it be discovered that the rate expression contains a term that is zeroth order in thiocyanate as well as one that is first order. The method of Lampugnani et al., on the other hand, yields orders that lie between zero and one and are clearly distinguishable from unity for any run in which the reaction does not go to completion. Provided that precise data can be obtained for 90% of the reaction, a single kinetic run suffices to indicate that the rate expression is not simple first order.

Numerical integration may also be of value in calculating rate constants, once the rate expression is known. If analytical methods are available that permit simultaneous and nearly continuous monitoring of the concentrations of several species in the reaction mixture, the method of concentration–time integrals[60, 61] can be used. Consider, for example, a third-order reaction $2A + B = $ products whose rate expression is

$$\frac{dx}{dt} = k[A]^2[B] \tag{3.73}$$

The integrated rate expression was given as (2.39). It is so cumbersome that treatment of experimental data is difficult. However (3.73) can be rearranged to

$$\frac{dx}{(b-x)} = k[A]^2 dt$$

and the left-hand side can be integrated readily

$$\ln\frac{b}{b-x} = k\int_0^t [A]^2 dt$$

If, in addition to the usual measurement of x as a function of t, $[A]$ can be determined nearly continuously during the reaction, then the integral on the right-hand side can be evaluated by summation for each time at which x was determined. A plot of $\ln[b/(b-x)]$ versus the concentration–time integral at successive times is then a straight line of slope k. Provided that $[A]$ can be measured over sufficiently small intervals Δt, the method of concentration–time integrals permits considerable simplification of the integrated rate expression.

INITIAL RATE AS A FUNCTION OF INITIAL CONCENTRATION

The most obvious use of (3.67) is to make a direct comparison between rate and concentration. This can be done approximately by taking as a measure of the derivative dx/dt the corresponding ratio of finite increments, $\Delta x/\Delta t$. For reasonable accuracy the fraction reacting should be no more than, say, 0.1. Then, by making a run at each of two different initial concentrations of any one component, say, B, the other concentrations remaining constant, the data will enable the determination of the order of reaction with respect to that component. Let the two rates and corresponding initial concentrations be $(dx/dt)_1$, $(dx/dt)_2$, and $[B]_1$, $[B]_2$. Then

$$\left(\frac{dx}{dt}\right)_1 = (k[A]^{n_a}[C]^{n_c})[B]_1^{n_b}$$

$$\left(\frac{dx}{dt}\right)_2 = (k[A]^{n_a}[C]^{n_c})[B]_2^{n_b}$$

Dividing, taking the logarithm, and solving for n_b yields

$$n_b = \frac{\log(dx/dt)_1 - \log(dx/dt)_2}{\log[B]_1 - \log[B]_2} \tag{3.74}$$

This equation is due to van't Hoff.

Varying successively the initial concentration of each component will give the order with respect to each component, following which the rate constant k may be evaluated approximately from any one run. Equation (3.74) is also applicable to the simpler case of one variable concentration, but the methods discussed above are more accurate and convenient.

An example of the use of (3.74), as well as of (3.65), is found in the work of Klute and Walters[62] on the thermal gas-phase decomposition of tetrahydrofuran. This reaction is complex, yielding a variety of products. The pressure–time data do not correspond to a simple order of reaction. Accordingly, a number of runs were made, and the maximum rates, $(\Delta P/\Delta t)_{max}$, and the half-lives were determined for different starting pressures. A plot of $\log(\Delta P/\Delta t)_{max}$ against $\log P_0$ gave a straight line of slope equal to 1.5, and $\log t_{1/2}$ plotted against $\log P_0$ gave a straight line with a slope of -0.55. This indicated that the reaction was of the 1.5 order.

The method of initial rates has been applied to another complex reaction by Sharma and Schubert.[63] The catalytic decomposition of H_2O_2 by a copperimidazole complex was followed by determing the initial rate of O_2 evolution. The rate expression is of the form

$$\frac{dx}{dt} = k_1 [Cu(Im)_2^{2+}] [HOO^-] + k_2 [Cu(Im)_2(OH)^+] [HOO^-]$$

$$= k_1 \beta_1 [Cu^{2+}] [Im]^2 [HOO^-] + k_2 \beta_2 [Cu^{2+}] [Im]^2 [OH^-] [HOO^-] \tag{3.75}$$

where Im represents imidazole and β_1 and β_2 are formation constants for the complexes $Cu(Im)_2^{2+}$ and $Cu(Im)_2(OH)^+$. Dividing both sides of (3.75) by $[Im]^2 \cdot [HOO^-]$ gives

$$\frac{dx/dt}{[Im]^2 [HOO^-]} = (k_1 \beta_1 + k_2 \beta_2 [OH^-])[Cu^{2+}] \tag{3.76}$$

A plot of the quantity on the left of (3.76) versus $[Cu^{2+}]$ at pH $= 7.95$ gave a straight line with zero intercept over a tenfold variation in $[Cu^{2+}]$, confirming that both terms in the rate expression are first order in Cu^{2+}. Similar plots were made to determine the order with respect to imidazole and peroxide. To obtain k_1 and k_2 Sharma and Schubert rearranged (3.75) to

$$\frac{dx/dt}{[Cu(Im)_2^{2+}] [HOO^-]} = k_1 + \frac{k_2 [Cu(Im)_2(OH)^+]}{[Cu(Im)_2^{2+}]} \tag{3.77}$$

A plot of the left-hand side of (3.77) versus $[Cu(Im)_2(OH)^+]/[Cu(Im)_2^{2+}]$ over the pH range from 6.4 to 8.4 was linear, although uncertainties in β_1 and β_2 caused some scatter in the points. Values of $k_1 = 530 M^{-1} s^{-1}$ and $k_2 = 120 M^{-1} s^{-1}$ were obtained from the intercept and slope.

The method of initial rates provides considerable simplification in the case of reversible reactions or in cases where side reactions occur as the reaction proceeds, because all observations are made at or near $t = 0$. By the same token, however, initial rate measurements often do not detect any contribution to the rate expression by the products of a reaction. Also, if rate constants are to be calculated from $\Delta x/\Delta t$, a sensitive method of analysis must be used. This permits a small enough Δt so that reactant concentrations do not decrease significantly from their initial values.

Hall, Quickenden, and Watts[64] have described an alternative method for obtaining accurate initial rates without the requirement of very small Δt. For a first-order process the reaction variable is given by

$$x = a(1 - e^{-kt})$$

and expanding e^{-kt} as a power series gives

$$x = a\left(kt - \frac{(kt)^2}{2!} + \frac{(kt)^3}{3!} - \cdots\right) \tag{3.78}$$

Similar series expansions are possible for other orders, and x can therefore be expressed as a polynomial in t

$$x = bt + ct^2 + dt^3 + \cdots \tag{3.79}$$

where $b, c, d, \ldots$ are constants. When $t = 0$, the slope $dx/dt = b$, and it can be obtained by fitting the parameters $b, c, d, \ldots$ in (3.79) to the observed concentration–time data. For less than 10% of reaction terms involving t^3 and higher powers are negligible, and dividing both sides of (3.79) by t gives

$$\frac{x}{t} = b + ct$$

Under these circumstances b is the intercept of a plot of x/t against t.

RATE EXPRESSION DETERMINED DIRECTLY FROM RATE

In most cases an unintegrated rate expression is simpler mathematically than the integrated form. The chief reason for using integrated forms is that concentration–time data are much easier to measure than are concentration–rate data. However, it is possible to obtain rates from concentration–time data or to measure rates directly. Malmstadt, Delaney, and Cordos,[65] for example, have described integrated circuits and automated instruments for analytical methods based on direct measurement of reaction rates. If the rate is plotted against concentration, the simplicity of the unintegrated rate expression often permits discovery of the relationship between rate and concentration. Such a relationship can be verified by a suitable plot. In the case of a second-order reaction of two components with no product initially present, for example,

$$A + B = C \qquad \frac{dx}{dt} = kAB$$

$$x = A_0 - A = B_0 - B = C$$

The following relationships can then be derived:

$$\frac{dx/dt}{A} = kB = kA - k(A_0 - B_0) \tag{3.80}$$

$$\frac{dx/dt}{B} = kA = kB - k(B_0 - A_0) \tag{3.81}$$

$$\frac{dx}{dt} = k(A_0 - C)(B_0 - C) = kC^2 - kC(A_0 + B_0) + kA_0B_0 \tag{3.82}$$

A plot of $(dx/dt)/A$ versus A, then, should be linear and k can be determined from the slope.

If a continuous or nearly continuous plot of concentration versus time can be obtained experimentally, the normal (and hence the tangent) can be obtained by using a mirror, prism, or glass rod, all of which show a smooth continuation of the curve when oriented normal to it. More convenient than this mechanical method is a digital computer program that performs numerical differentiation to obtain dx/dt. Wen[66] has described such a program and applied it to the data of Kistiakowsky and Lacher shown in Table 3.3. Rate constants of 10.3×10^{-7}, 11.3×10^{-7}, and 9.6×10^{-7} torr^{-1} s^{-1} were obtained using (3.80), (3.81), and (3.82). These are in good agreement with the value of 9.7×10^{-7} torr^{-1} s^{-1} reported by Kistiakowsky and Lacher, and the value of 10.9×10^{-7} torr^{-1} s^{-1} calculated earlier from the integrated rate expression. Wen's method has the added advantage that initial concentrations of both reactants need not be known exactly. He has shown that when the integrated rate expression is used for a second-order reaction, considerable error can be introduced as a result of error in the initial concentrations, even if the concentration–time data are highly accurate. This is especially true if the rate constant is calculated during the first 30% or so of reaction. Less error is introduced when the rate expression itself is used to calculate k, provided that dx/dt can be determined accurately.

Krause[67] has reported a variation of the method of dimensionless parameters (Chapter 2) that makes use of the rate directly. Using the general rate expression $-d\alpha/dt = k\alpha^n c_0^{n-1}$ and the fact that $\tau = kc_0^{n-1}t = -\int_1^\alpha d\alpha/\alpha^n$, where $\alpha = c/c_0$ is the relative concentration of reactant, Krause shows that for $n = 1$

and for $n \neq 1$

$$-t\frac{d\alpha}{dt} = \alpha \ln \alpha \tag{3.83}$$

$$-t\frac{d\alpha}{dt} = \frac{\alpha(1 - \alpha^{n-1})}{n-1} \tag{3.84}$$

The rate, whether measured directly or obtained from concentration–time data, may be used to calculate $d\alpha/dt$. Then an experimental plot of $-t\,d\alpha/dt$ versus α can be compared with theoretical curves obtained from (3.83) and (3.84).

EVALUATION OF RATE CONSTANTS

Once the form of the rate expression has been determined, the next step in treating experimental data is to evaluate the rate constant or constants. Complex reactions, which involve more than one rate constant, require special techniques, each case usually being sufficiently different so that no general method applies. In Chapter 8 a number of examples of complex reactions are discussed in detail.

For simple first-, second-, or third-order reactions where there is only a single rate constant and the integrated form of the rate expression is readily available, four principal methods are in common use for obtaining rate constants from concentration–time data:

1.   Calculation of a rate constant for each experimental point, using the integrated rate expression. In the case of a first-order reaction, for example,

$$k = \frac{1}{t} \ln \frac{a}{a-x} \tag{3.85}$$

The rate constants obtained in this way are then averaged.

2.   Calculation of a rate constant for each adjacent pair of points. Again using a first-order reaction as an example,

$$k = \frac{1}{t-t'} \ln \left(\frac{a-x'}{a-x} \right) \tag{3.86}$$

where t, x and t', x' are adjacent data points.

3.   The graphical methods, using the slope of an appropriate plot to obtain the rate constant. For a first-order reaction $\ln (a-x)$ is plotted versus t and

$$k = -\text{slope} = \frac{-[\Delta \ln (a-x)]}{\Delta t} \tag{3.87}$$

4.   Least-squares regression analysis applied to the data x, t. For a first-order reaction this involves evaluation of the parameters a and k in (3.85) or, preferably, in

$$a - x = a\,e^{-kt} \tag{3.88}$$

When (3.88) is used the relative errors and hence the weights of the $a-x$ data are not transformed by taking logarithms.

Roseveare[68] has analyzed the problem of determining rate constants. He recommends against method 1 because it places very heavy emphasis on the initial time and concentration. These enter into every calculation of k while the rest of the x, t values are each used but once. Method 1 is also very sensitive to errors in x values near the beginning of the reaction. Roseveare's analysis also shows that when method 2 is applied using equal intervals $t-t'$, it is equivalent to calculating the rate constant from (3.86) using only the first and last experimental points. Since these are often the least accurate, method 2 is not recommended either, unless Roseveare's special but laborious method for weighting the data is adopted. The same arguments against methods 1 and 2 hold true for second- and third-order reactions.

The graphical method has the advantage of readily showing trends and deviations from straight-line behavior. Isolated points that are obviously in error can be ignored in drawing the best straight line through the data. The initial concentration, or the initial reading if some physical property is measured, may be known, in which case it furnishes an extra point on the graph. Otherwise it may be calculated from the intercept. The best straight line is readily found by moving a transparent straightedge until it appears to fit the data with minimum deviation.

Least-squares regression analysis requires a good deal of calculation, but this can easily be done by a computer. If a digital plotter is used to produce appropriate graphs, computerized regression analysis has all the advantages of the graphical method. Furthermore, the least-squares method can easily handle data that are too precise for plotting on a reasonable scale, weighting of data is quite simple to do, and numerical estimates of how well the parameters fit the data can be calculated with no extra effort. Methods of weighting data have been discussed by several authors,[19, 69, 70] and Wiberg[71] has described several measures of the goodness of fit of a least-squares regression analysis. In cases where a physical property is measured, least-squares analysis can include the infinity value of the physical property as an adjustable parameter. In the first-order case, for example, (3.88) becomes

$$\lambda_\infty - \lambda = (\lambda_\infty - \lambda_0)e^{-kt} \tag{3.89}$$

and λ_∞, as well as k and λ_0, may be varied to fit the data. However, the caveats of an earlier section of this chapter regarding adjustment of λ_∞ must not be ignored. For second-order and higher-order reactions the number of parameters is so large that regression analysis on all parameters in an equation analogous to (3.89) is of little value.

The accepted rate constant should never be based on a single kinetic run, even if a large number of points are taken. A minimum of three runs should be made, each with different initial concentrations, and the rate constant should be calculated as a weighted average[13] of the separate constants found for each run. Wagner, Czerlinski, and Pring[72] have described a general procedure for optimizing the initial conditions for separate runs as well as the times at which data points are taken within a run so as to obtain rate constants of maximum accuracy. Their procedure involves maximizing the curvature of the hypersurface representing the sum of the squares of the deviations of theoretical from experimental concentration–time data. The maximization is done with respect to each rate constant, ensuring that the least-squares minimum is as sharp as possible. Even for simple reactions the procedure of Wagner, Czerlinski, and Pring requires extensive calculations and must be implemented on a computer.

An alternative to averaging rate constants from separate runs is to write $F(c) = kt$, where $F(c)$ is some function of the concentration. That is, for a first-order reaction $F(c) = \ln [a/(a - x)]$. Then a plot of $F(c)$ for several runs against time should give a single straight line with an intercept at the origin and a slope equal to the rate constant. This procedure is helpful when only a few determinations can be made in each run, and when the points tend to scatter.

METHODS WHERE THE FINAL READING IS UNKNOWN

In addition to the treatment of λ_∞ as an adjustable parameter in least-squares regression analysis, several methods are available to handle the situation where the

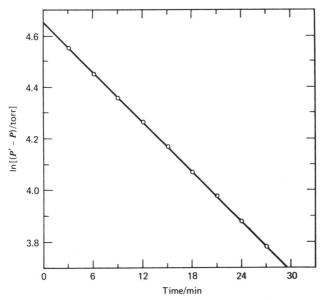

Figure 3.12. Guggenheim plot for thermal decomposition of chlorocyclohexane, (Data from Swinbourne[74]).

initial concentrations of reactants or the infinity reading of a physical property is not obtainable. Such a situation arises when a reaction is rapid enough so that measurements cannot be made at zero time or a side reaction or equilibrium or instrumental drift causes the final reading of a physical property to be incorrect.

For a first-order reaction the Guggenheim method [73] is widely used. Measurements $\lambda_1, \lambda_2, \lambda_3, \ldots$ of a physical property are made at times $t_1, t_2, t_3, \ldots$, and a second series of measurements $\lambda_1', \lambda_2', \lambda_3', \ldots$ are made at times $t_1 + \Delta, t_2 + \Delta, t_3 + \Delta, \ldots$, where Δ is a constant increment. For accurate results Δ should be at least one and preferably two or three times the half-life of the reaction. Substituting λ_1 and λ_1' into (3.89) gives

$$\lambda_\infty - \lambda_1 = (\lambda_\infty - \lambda_0)e^{-kt_1} \tag{3.90}$$

$$\lambda_\infty - \lambda_1' = (\lambda_\infty - \lambda_0)e^{-k(t_1 + \Delta)} \tag{3.91}$$

Subtracting (3.91) from (3.90) yields

$$\lambda_1' - \lambda_1 = (\lambda_\infty - \lambda_0)e^{-kt_1}(1 - e^{-k\Delta})$$

or

$$\ln(\lambda_1' - \lambda_1) = -kt_1 + \ln[(\lambda_\infty - \lambda_0)(1 - e^{-k\Delta})] = -kt_1 + \text{constant} \tag{3.92}$$

Since equations similar to (3.90) and (3.91) can be written for t_2 and $t_2 + \Delta$, and so on, the subscript 1 can be dropped, making (3.92) general for any λ and λ'.

The Guggenheim method can be applied to the data of Swinbourne [74] on thermal decomposition of chlorocyclohexane in the gas phase (Table 3.5). Swinbourne followed the reaction

$$C_6H_{11}Cl(g) = C_6H_{10}(g) + HCl(g)$$

Table 3.5. Thermal decomposition of chlorocyclohexane at 641 K

Time/min	Pressure/torr	P'/torr[a]	$(P' - P)$/torr[a]
3	237.2	332.1	94.9
6	255.3	341.1	85.8
9	271.3	349.3	78.0
12	285.8	356.9	71.1
15	299.0	363.7	64.7
18	311.2	369.9	58.7
21	322.2	375.5	53.3
24	332.1	380.5	48.4
27	341.1	384.9	43.8
30	349.3		
33	356.9		
36	363.7		
39	369.9		
42	375.5		
45	380.5		
48	384.9		

Source. Data from Swinbourne.[74]
[a] P' is total pressure at $t + 21$ min.

by measuring the total pressure P. Figure 3.12 is a plot of $\ln (P' - P)$ against t. The slope gives a rate constant $k = 5.31 \times 10^{-4}$ s^{-1}. An alternative but closely related method has been proposed independently by Kezdy, Jaz, and Bruylants[75] and by Swinbourne.[74] Dividing (3.90) by (3.91) gives

$$\frac{\lambda_\infty - \lambda_1}{\lambda_\infty - \lambda_1'} = e^{-kt_1} e^{k(t_1 + \Delta)}$$

or

$$\lambda_1 = e^{k\Delta} \lambda_1' + \lambda_\infty (1 - e^{k\Delta}) \qquad (3.93)$$

Thus a plot of λ against λ' should be linear and $k = [\ln (\text{slope})]/\Delta$. Further, at very long times $\lambda = \lambda' = \lambda_\infty$; therefore λ_∞ is the point at which the graph of (3.93) crosses the line $\lambda = \lambda'$, and λ_∞ is readily obtained. Figure 3.13 shows a Kezdy-Swinbourne plot for the data of Table 3.5 with $\Delta = 21$ min. From the slope $k = 5.38 \times 10^{-4}$ s^{-1} and from the intersection with $P = P'$, $\lambda_\infty = 430$ torr.

The techniques described just above should only be applied to data for reactions that have been found by an independent method to be first order. Reversible and concurrent first-order reactions give apparent rate constants by the Guggenheim method. When data are taken through only the first two half-lives, even a second-order reaction yields an acceptably linear plot, although calculating λ_∞ reveals the incorrect order because an unreasonable value is obtained. Also, since the $\lambda_1', \lambda_2', \lambda_3', \ldots$ data are used to find λ_∞, only half the data, namely, $\lambda_1, \lambda_2, \lambda_3,$ $\ldots$, are effectively applied to determining k. These data often include no more than the first half-life or so.

Tobey[76] has described a way of treating the data from a second-order reaction

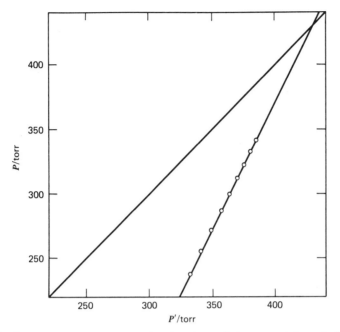

Figure 3.13. Kezdy-Swinbourne plot for thermal decomposition of chlorocyclo-hexane (Data from Swinbourne[74]).

with equivalent concentrations so that k is obtained directly in concentration units. When the concentrations of reactants are equivalent the integrated rate expression is (2.16), which yields upon multiplication of both sides by the initial concentration a and substitution of (3.13) and (3.14) with $\nu_A = -1$,

$$\frac{xa}{a(a-x)} = \frac{\lambda - \lambda_0}{\lambda_\infty - \lambda_0} \times \frac{\lambda_\infty - \lambda_0}{\lambda_\infty - \lambda} = akt$$

Rearranging, we have

$$\lambda = \lambda_0 - akt(\lambda - \lambda_\infty) \qquad (3.94)$$

and for a subsequent measurement λ' at $t' = t + \Delta$,

$$\lambda' = \lambda_0 - akt'(\lambda' - \lambda_\infty) \qquad (3.95)$$

Subtracting (3.95) from (3.94) gives

$$\lambda - \lambda' = ak(t'\lambda' - t\lambda) - ak\lambda_\infty \Delta \qquad (3.96)$$

Consequently a plot of $\lambda - \lambda'$ against $t'\lambda' - t\lambda$ for a series of pairs of points separated by a constant interval Δ should be linear and have a slope of ak. The value of a must be known to obtain k in proper units.

Shank[77] has given a method for a second-order reaction with unequal concentration, as well as for certain other cases. The integrated rate equation must be of the form

$$\ln [f(A) + a] = bt + c \qquad (3.97)$$

where $f(A)$ is a function of a single concentration and a, b, and c are constants. If two series of measurements of the concentration of A are made at times t and $t + \Delta$, (3.97) becomes

$$\ln \left[\frac{f(A_2) + a}{f(A_1) + a} \right] = b\Delta$$

$$f(A_2) = f(A_1)e^{b\Delta} + a(e^{b\Delta} - 1) \tag{3.98}$$

For a second-order reaction, this becomes

$$\frac{1}{[A_2]} = \frac{\exp(B_0 - A_0)k\Delta}{[A_1]} + \frac{\exp(B_0 - A_0)k\Delta - 1}{B_0 - A_0} \tag{3.99}$$

A plot of $1/[A_2]$ versus $1/[A_1]$ is linear. From the slope and intercept, both k and $(B_0 - A_0)$ may be found.

Notice that the actual concentrations of A must be known at various times. However, neither B_0 nor A_0 need be known. For a reaction that goes to completion, this is equivalent in some respects to not knowing λ_∞.

PROBLEMS

3.1. The rate of iodination of nitroethane in the presence of pyridine according to the equation

$$C_2H_5NO_2 + C_5H_5N + I_2 \rightarrow C_2H_4INO_2 + C_5H_5NH^+ + I^-$$

was followed by measuring the change in electrical conductivity. The accompanying data were obtained at 25°C in a water–alcohol solvent. Original concentrations: nitroethane and pyridine, each 0.1 M; iodine, 0.0045 M.

Time/min	Resistance/Ω
0	2503
5	2295
10	2125
15	1980
20	1850
25	1738
30	1639
∞	1470

Determine the apparent order of the reaction and the apparent rate constant in suitable units. Assuming that the reaction is first order in nitroethane and first order in pyridine, convert to the correct second-order constant.

3.2. The thermal decomposition of dimethyl ether in the gas phase has been studied by measuring the increase in pressure.

$$(CH_3)_2O \rightarrow CH_4 + H_2 + CO$$

Some measurements made at $504°C$ and an initial pressure of 312 torr of ether are as follows:

Time/s	390	777	1195	3155	∞
Pressure increase/torr	96	176	250	467	619

Calculate the rate constant after determining the order of the reaction. [C. N. Hinshelwood and P. J. Askey, *Proc. Roy. Soc. (Lond.)*, **A115**, 215 (1927).]

3.3. The substance 3,3'-dicarbazyl phenyl methyl chloride exists in an alkaline solution partly as a negative ion and partly as a neutral anhydro base. The two forms have different colors, the negative ion being green and absorbing strongly at 730 nm and the anhydro base being red and absorbing strongly at 500 nm. One or both forms combine with water or methanol to form colorless carbinols or ethers. The accompanying data were collected using a spectrophotometer and a water–acetone solvent at $25°C$. What conclusions can be drawn concerning the rate of disappearance of each of the colored forms? What conclusions can be drawn concerning whether both forms are reacting to form carbinol? Original concentrations: NaOH, $2.03 \times 10^{-2} M$; 3-3'-dicarbazyl phenyl methyl chloride, $\sim 10^{-5} M$.

500 nm		730 nm	
Time/min	$A - A_\infty^a$	Time/min	$A - A_\infty$
3.7	0.320	2.7	0.562
5.7	0.246	4.7	0.423
7.7	0.188	6.7	0.333
9.7	0.143	8.7	0.243
11.7	0.109	10.7	0.191
13.7	0.084	12.7	0.148
15.7	0.064	14.7	0.111
17.7	0.049	16.7	0.087
19.5	0.036	18.7	0.065
23.7	0.023	21.7	0.045

^{a} A refers to absorbance and A_∞ to the absorbance for complete reaction. The latter value is essentially zero. [G. E. K. Branch and B. M. Tolbert, *J. Am. Chem. Soc.*, **69**, 523 (1947).]

3.4. The cleavage of diacetone alcohol by alkali to form acetone obeys the equation

$$CH_3\overset{O}{\overset{\|}{C}}CH_2\overset{OH}{\underset{\underset{CH_3}{|}}{C}}CH_3 = 2CH_3\overset{O}{\overset{\|}{C}}CH_3$$

It can be followed with a dilatometer since there is a substantial increase in volume in concentrated solutions as the reaction proceeds. The given data below were collected by Åkerlof at $25°C$. Calculate the rate constant after determining the order of the reaction. Original concentrations: KOH in water, $2M$; diacetone alcohol, 5% by volume. [G. Åkerlof, *J. Am. Chem. Soc.*, **49**, 2955 (1927).]

Time/s	Cathetometer reading
0.0	8.0
24.4	20.0
35.0	24.0
48.0	28.0
64.8	32.0
75.8	34.0
89.4	36.0
106.6	38.0
133.4	40.0
183.6	42.0
∞	43.3

3.5. The decomposition of nitrogen pentoxide in the gas phase can be followed manometrically but is complicated by the reversible dissociation of the product nitrogen tetroxide:

$$2N_2O_5 \rightarrow 2N_2O_4 + O_2$$

$$N_2O_4 \rightleftharpoons 2NO_2$$

The latter reaction rapidly reaches equilibrium, and the equilibrium constant can be independently measured [E. and L. Natanson, *Wied. Ann.*, **24**, 454 (1885)]. Its value at 25°C is 97.5 torr. From the accompanying data of F. Daniels and E. H. Johnston [*J. Am. Chem. Soc.*, **43**, 53 (1921)] calculate the rate constant for the decomposition of nitrogen pentoxide. Temperature, 25°C.

Time/min	Pressure/torr
0	268.7
20	293.0
40	302.2
60	311.0
80	318.9
100	325.9
120	332.3
140	338.8
160	344.4
∞	473.0

Use may be made of the following list, which shows the fraction of nitrogen tetroxide dissociated as a function of the pressure of oxygen in the system, assuming the decomposition of the pentoxide to be the only source of both substances.

Pressure O_2/torr	Fraction dissociated
5	0.761
25	0.496
50	0.386
100	0.292
150	0.247

3.6. Ammonium nitrite breaks down to give water and nitrogen and the rate can be followed by measuring the volume of nitrogen evolved. An aqueous solution containing ammonium ions and nitrite ions also contains ammonia and nitrous acid molecules formed by hydrolysis (nitrous acid is weak, $K_a = 6.3 \times 10^{-4}$). Also such ions as H_2ONO^+ and NH_2^- are possible. From the following data, collected by J. H. Dusenberry and R. E. Powell [*J. Am. Chem. Soc.*, **73**, 3266 (1951); see also A. T. Austin et al., *J. Am. Chem. Soc.*, **74**, 555 (1952)], decide the order with respect to each of the possible reactants (some combinations may be kinetically indistinguishable). Temperature is $30°C$ for all reactions.

$$NH_4^+ + NO_2^- \rightarrow N_2 + 2H_2O$$

Total nitrous acid/M^a	Total ammonia/M	Initial rate $\times 10^8/M\,s^{-1}$
	pH constant at 2.85	
0.009 04	0.395	128
0.008 96	0.197	64
0.009 16	0.098	34.9
0.009 24	0.049	16.6
	pH constant at 2.93	
0.0940	0.186	643
0.0507	0.196	338
0.0249	0.196	156
0.0100	0.198	65
0.0049	0.198	32.6
0.0024	0.198	17.5

^{a} Total nitrous acid means summation of all forms in solution.

Total nitrous acid constant at $0.047\,M$ and total ammonia constant at $0.197\,M$:

pH	0.43	0.96	1.81	2.95	3.68	4.11	4.33	5.00	6.18
Initial rate $\times 10^8/M\,s^{-1}$	517	513	371	305	140	52	31	9.2	0.6

3.7. A gaseous reaction takes place between A and B. A is in great excess. The half-life, $t_{1/2}$, as a function of initial pressures at $50°C$ is as follows:

p_A/torr	500	125	250	250
p_B/torr	10	15	10	20
$t_{1/2}$/min	80	213	160	80

(a) Show that the rate expression is

$$\text{rate} = kp_A p_B^2$$

(b) Evaluate the rate constant for concentration units of mol liter^{-1} and time in seconds.

3.8.　Hydrolysis of phenyl-*m*-nitrobenzoate follows the equation

S. W. Tobey [*J. Chem. Educ.*, **39**, 473 (1962)] measured the resistance of the solution as a function of time in 50% aqueous dioxane at 25.00°C. The initial concentration of OH⁻ was 6.275×10^{-3} M and the initial concentration of phenyl-*m*-nitrobenzoate was 3.138×10^{-3} M. The reaction is first order in each reactant. Use the data in the list below to calculate the second-order rate constant by Tobey's method using (3.96).

Time/s	Resistance/Ω	Time/s	Resistance/Ω
25	734	130	960
30	753	140	970
40	786	150	980
50	815	160	990
60	841	170	998
70	864	180	1006
80	884	190	1014
90	902	200	1021
100	919	230	1040
125	954	260	1055

3.9.　The gas-phase decomposition of di-*t*-butyl peroxide gives essentially acetone and ethane. Raley, Rust, and Vaughn [*J. Am. Chem. Soc.*, **70**, 88 (1948)] obtained

$$(CH_3)_3 COO(CH_3)_3 = 2(CH_3)_2 CO + C_2 H_6$$

the values given below for the total pressure as a function of time at 427.8 K. The reaction is first order and the initial pressure of peroxide is 169 torr. (The additional pressure is due to N_2 used to force the peroxide into the reaction vessel.)

Time/min	Pressure/torr	Time/min	Pressure/torr
0	173.5	12	244.4
2	187.3	14	254.5
3	193.4	15	259.2
5	205.3	17	268.7
6	211.3	18	273.9
8	222.9	20	282.0
9	228.6	21	286.8
11	239.8	∞	491.8

(a)　Obtain *x* as a function of *t* from the pressure data given, use the method of Hall, Quickenden, and Watts to calculate the initial rate, and calculate the first-order rate constant.

(b) Compare your result from part **a** with that obtained from the usual first-order plot of the data.

(c) Calculate $\Delta x/\Delta t$ for $t = 2$ min and from it obtain the rate constant. Compare this with your results from parts **a** and **b**.

3.10. Modify the computer program of the Appendix so that it will fit four parameters to data obtained for two experimental variables. Use the modified program and the data of problem **9** to obtain the four parameters in (3.80) when the equation is truncated after the t^4 term. Use the initial rate b to calculate the rate constant for decomposition of di-t-butyl peroxide and compare it with the results of problem **9**.

3.11. Write the integrated rate expressions for a first-order reaction and a second-order reaction of a single reactant in terms of a physical property λ that is directly proportional to concentration. Rearrange each equation to the form $0 = f(\lambda, t, k, \lambda_0, \lambda_\infty)$ and obtain the partial derivative of f with respect to each variable and parameter.

(a) Use the computer program of the Appendix to fit k and P_∞ to the data of problem **9** (including P_0) for both a first- and second-order reaction. Compare your results with the experimental values.

(b) Modify the computer program of the Appendix so that it will fit three parameters to data obtained for two experimental variables. Use the modified program to fit k, λ_0, and λ_∞ to the data of problem **9** for both the first-order and the second-order equations. Compare your results with the experimental values.

(c) Comment on the advisability of using regression analysis to determine rate constants and reaction orders when a reaction has been followed to less than 50% of completion.

3.12. Derive (3.56).

REFERENCES

1. T. J. Swift, "Nuclear Magnetic Resonance," in *Investigation of Rates and Mechanism of Reactions*, Part II, G. G. Hammes, Ed., Techniques of Chemistry, Vol. 6, A. Weissberger, Ed., Wiley-Interscience, New York, 1974; K. S. Chen and N. Hirota, "Electron Paramagnetic Resonance," *ibid.*

2. A. Farkas and H. W. Melville, *Experimental Methods in Gas Reactions,* Macmillan, London, 1939; and H. W. Melville and B. G. Gowenlock, *Experimental Methods in Gas Kinetics,* 2nd ed., Macmillan/St. Martins, New York, 1964.

3. E. W. R. Steacie, *Atomic and Free Radical Reactions,* 2nd ed., Reinhold, New York, 1954.

4. C. N. Hinshelwood, *The Kinetics of Chemical Change,* Oxford University Press, Oxford, 1941.

5. W. A. Noyes and P. A. Leighton, *Photochemistry of Gases,* Reinhold, New York, 1941; Dover, New York, 1966.

6. A. F. Trotman-Dickenson, *Gas Kinetics,* Academic, New York, 1955.

7.   Z. G. Szabo, *Advances in the Kinetics of Homogeneous Gas Reactions*, Methuen, London, 1964.

8.   A. Maccoll, "Homogeneous Gas-Phase Reactions," in *Investigation of Rates and Mechanisms of Reactions*, Part I, E. S. Lewis, Ed., Techniques of Chemistry, Vol. 6, 3rd ed., A. Weissberger, Ed., Wiley-Interscience, New York, 1974.

9.   P. G. Ashmore, Ed., *Reaction Kinetics*, Vol. 1, Specialist Periodical Reports, The Chemical Society, London, 1975; and P. G. Ashmore and R. G. Donovan, Eds., *Gas Kinetics and Energy Transfer*, Vol. 2, Specialist Periodical Reports, The Chemical Society, London, 1976.

10. G. L. Pratt, *Gas Kinetics*, Wiley, New York, 1969.

11. M. F. R. Mulcahy, *Gas Kinetics*, Wiley/Halsted, New York, 1973.

12. V. N. Kondratiev, *Chemical Kinetics of Gas Reactions*, Pergamon, Oxford, 1964.

13. C. H. Bamford and C. F. H. Tipper, Eds., *Comprehensive Chemical Kinetics*, Vol. 1, Elsevier, Amsterdam, 1969.

14. D. W. Bassett and H. W. Habgood, *J. Phys. Chem.*, **64**, 769 (1960); E. Gil-Av and Y. Herzberg-Minzly, *Proc. Chem. Soc.*, **1961**, 316; S. H. Langer and J. E. Patton, *J. Phys. Chem.*, **76**, 2159 (1972); S. H. Langer and T. D. Griffith, *J. Phys. Chem.*, **82**, 1327 (1978).

15. A. T. Cocks and K. W. Egger, *Int. J. Chem. Kinet.*, **4**, 169–174 (1972).

16. Y. Takezaki and C. Takeuchi, *J. Chem. Phys.*, **22** (9), 1527 (1954).

17. P. L. Hanst and J. G. Calvert, *J. Phys. Chem.*, **63**, 104 (1959).

18. L. Batt and R. D. McCulloch, *Int. J. Chem. Kinet.*, **8**, 491 (1976); J. R. Barker, S. W. Benson, and D. M. Golden, *Int. J. Chem. Kinet.*, **9**, 31 (1977).

19. P. Moore, *J. Chem. Soc., Faraday Trans. I*, **68**, 1890 (1972).

20. D. F. DeTar, Ed., *Computer Programs for Chemistry*, Vol. 4, Academic, New York, 1972, p. 25.

21. G. B. Kistiakowsky and J. R. Lacher, *J. Am. Chem. Soc.*, **58**, 123 (1936).

22. A. F. Benton, *J. Am. Chem. Soc.*, **53**, 2984 (1931).

23. M. S. Peters and E. J. Skorpinski, *J. Chem. Educ.*, **42**, 329 (1965).

24. S. G. Carragaratna, *J. Chem. Educ.*, **50**, 200 (1973).

25. G. M. Harris, *J. Phys. Chem.*, **51**, 505 (1947).

26. O. A. Hougen and K. M. Watson, *Chemical Process Principles*, Part III, Wiley, New York, 1947, pp. 834 ff.

27. J. Reilly and W. N. Rae, *Physicochemical Methods*, Vol. 6, Van Nostrand, New York, 1954.

28. J. F. Bunnett, "Kinetics in Solution," in *Investigation of Rates and Mechanisms of Reactions*, Part I, E. S. Lewis, Ed., Techniques of Chemistry, Vol. 6, 3rd ed., A. Weissberger, Ed., Wiley-Interscience, New York, 1974.

29. A. G. Sykes, *Kinetics of Inorganic Reactions*, Pergamon, New York, 1966, Chap. 2.

30. R. G. Wilkins, *The Study of Kinetics and Mechanism of Reactions of Transition Metal Complexes*, Allyn and Bacon, Boston, 1974, Chap. 3.

31. G. Astarita, *Mass Transfer with Chemical Reaction*, Elsevier, Amsterdam, 1967.

32. P. V. Danckwerts, *Gas–Liquid Reactions*, McGraw-Hill, New York, 1970.

33. L. K. J. Tong and A. R. Olson, *J. Am. Chem. Soc.*, **65**, 1704 (1943).

34. W. L. Reynolds, J. Schufman, F. Chan, and R. C. Brasted, Jr., *Int. J. Chem. Kinet.,* **9,** 777 (1977).

35. M. D. Cohen and E. Fischer, *J. Chem. Soc.,* **1962,** 3044.

36. J. Brynestad and G. Pedro Smith, *J. Phys. Chem.,* **72,** 296 (1968).

37. R. F. Childers, Jr., K. G. VanderZyl, Jr., D. A. House, R. G. Hughes, and C. S. Garner, *Inorg. Chem.,* **7,** 749 (1968).

38. T. Nowicka-Jankowska, *J. Inorg. Nucl. Chem.,* **33,** 2043 (1971).

39. W. J. McGuire, M. S. Thesis, Northwestern University, Evanston, Illinois, 1949.

40. B. L. Murr, Jr. and V. J. Shiner, Jr., *J. Am. Chem. Soc.,* **84,** 4672 (1962).

41. R. W. Hay and P. L. Cropp, *J. Chem. Soc. (A),* **1969,** 42.

42. J. H. Rolston and K. Yates, *J. Am. Chem. Soc.,* **91,** 1483 (1969).

43. H. Hartridge and F. J. W. Roughton, *Proc. R. Soc. (Lond.),* **A104,** 376 (1923).

44. B. Chance, "Rapid Flow Methods," in *Investigation of Rates and Mechanisms of Reactions,* Part II, G. G. Hammes, Ed., Techniques of Chemistry, Vol. 6, A. Weissberger, Ed., Wiley-Interscience, New York, 1974.

45. I. Förster and K. H. Geib, *Ann. Phys.,* **20,** 250 (1934); H. M. Hurlburt, *Ind. Eng. Chem.,* **36,** 1012 (1944).

46. G. M. Harris, *J. Phys. Colloid Chem.,* **51,** 505 (1947).

47. K. G. Denbigh, *Trans. Faraday Soc.,* **40,** 352 (1944); B. Stead, F. M. Page, and K. G. Denbigh, *Disc. Faraday Soc.,* **2,** 263, (1947).

48. M. Boudart, *Kinetics of Chemical Processes,* Prentice-Hall, Englewood Cliffs, NJ, 1968, Chap. 1.

49. H. H. Young and L. P. Hammett, *J. Am. Chem. Soc.,* **72,** 280 (1950); J. E. Taylor, *J. Chem. Educ.,* **46,** 742 (1969).

50. J. Saldick and L. P. Hammett, *J. Am. Chem. Soc.,* **72,** 283 (1950); M. J. Rand and L. P. Hammett, *J. Am. Chem. Soc.,* **72,** 287 (1950).

51. J. D. Johnson and L. J. Edwards, *Trans. Faraday Soc.,* **45,** 286 (1949).

52. M. F. R. Mulcahy and D. J. Williams, *Aust. J. Chem.,* **14,** 534 (1961).

53. R. N. J. Saal, *Rec. Trav. Chim.,* **47,** 73 (1928); V. K. LaMer and C. L. Read, *J. Am. Chem. Soc.,* **52,** 3098 (1930); P. D. Bartlett, F. E. Condon, and A. Schneider, *J. Am. Chem. Soc.,* **66,** 1531 (1944).

54. B. Chance, *J. Franklin Inst.,* **229,** 455, 737 (1940); *J. Biol. Chem.,* **179,** 1299, 1311, 1331, 1341 (1949); **180,** 865 (1949).

55. B. Chance, *Rev. Sci. Instr.,* **22,** 619 (1951); Q. H. Gibson, *Disc. Faraday Soc.,* **17,** 137 (1954).

56. B. G. Willis, J. A. Bittikofer, H. L. Pardue, and D. W. Margerum, *Anal. Chem.,* **42,** 1340 (1970).

57. R. G. Pearson, R. E. Meeker, and F. Basolo, *J. Am. Chem. Soc.,* **78,** 709 (1956).

58. G. B. Skinner, *Introduction to Chemical Kinetics,* Academic, New York, 1974, pp. 21–26.

59. L. Lampugnani, L. Meites, P. Papoff, and T. Rotunno, *Ann. Chim.,* **65,** 257 (1975).

60. D. French, *J. Am. Chem. Soc.,* **72,** 4806 (1950).

61. B. Saville, *J. Phys. Chem.,* **75,** 2215 (1971).

62. C. H. Klute and W. D. Walters, *J. Am. Chem. Soc.,* **68,** 506, (1946).

63. V. S. Sharma and J. Schubert, *J. Am. Chem. Soc.*, **91**, 6291 (1969).

64. K. J. Hall, T. I. Quickenden, and D. W. Watts, *J. Chem. Educ.*, **53**, 493 (1976).

65. H. V. Malmstadt, C. J. Delaney, and E. A. Cordos, *Anal. Chem.*, **44**, 79A (1972).

66. W. Y. Wen, *Int. J. Chem. Kinet.*, **5**, 621 (1973).

67. E. Krause, Report ARL 64-177, Aerospace Research Laboratories, Wright-Patterson Air Force Base, Ohio, 1964.

68. W. E. Roseveare, *J. Am. Chem. Soc.*, **53**, 1651 (1931).

69. C. J. Collins, *Adv. Phys. Org. Chem.*, **2**, 62 (1964).

70. D. F. DeTar, Ed., *Computer Programs for Chemistry*, Vol. 4, Academic, New York, 1972, pp. 92–94.

71. K. Wiberg, "Use of Computers," in *Investigation of Rates and Mechanisms of Reactions*, Part I, E. S. Lewis, Ed., Techniques of Chemistry, Vol. 6, 3rd ed., A. Weissberger, Ed., Wiley-Interscience, New York, 1974.

72. M. Wagner, G. Czerlinski, and M. Pring, *Comput. Biol. Med.*, **5**, 105 (1975).

73. E. A. Guggenheim, *Philos. Mag.*, **2**, 538 (1926).

74. E. S. Swinbourne, *Aust. J. Chem.*, **11**, 314 (1958).

75. F. J. Kezdy, J. Jaz, and A. Bruylants, *Bull. Soc. Chim. Belg.*, **67**, 687 (1958).

76. S. W. Tobey, *J. Chem. Educ.*, **39**, 473 (1962); see also J. H. Espenson, *J. Chem. Educ.*, **57**, 160 (1980).

77. N. E. Shank, *Int. J. Chem. Kinet.*, **5**, 577 (1973); see also J. M. Sturtevant, *J. Am. Chem. Soc.*, **59**, 699 (1937).

FOUR

ELEMENTARY PROCESSES: MOLECULAR COLLISIONS

The occurrence of any chemical reaction involves collision processes. This is obvious in the case of a bimolecular elementary process, in which reaction occurs at a collision between two molecules, or a trimolecular process, in which reaction occurs at a collision among three molecules. (In the gas phase, collisions involving four or more molecules are so improbable that they are of no importance in chemical kinetics. No known gas-phase reaction is suspected of having a molecularity greater than 3. In the liquid phase, molecules are always in close proximity so that such multiple collisions where the solvent participates are more likely.) The role of collisions is less obvious in the case of a unimolecular process, in which a given molecule decomposes spontaneously, until it is realized that the supply of molecules that have sufficient energy to react is maintained by means of energy transfer during collisions. (For the time being we ignore photochemical processes, which involve absorption or emission of photons. These are treated in Chapter 10.)

We begin this chapter by calculating. the number of bimolecular collisions per unit volume per unit time in a gas-phase sample, assuming that molecules can be approximated by rigid spheres undergoing elastic collisions. Comparison of this result with experimentally measured rate constants reveals that in most cases only a fraction of all collisions result in reaction. This is primarily due to the activation energy barrier to reaction, but even when only those collisions that are sufficiently energetic are considered, there are still many reactions whose rates cannot be predicted accurately. Thus it becomes necessary to examine molecular collisions in greater detail, both from a theoretical and an experimental perspective. Finally we turn to the somehwat more complicated cases of unimolecular and trimolecular reactions.

THE DISTRIBUTION LAW OF A COMPONENT OF VELOCITY

Assuming that gas-phase molecules move independently of each other except for the short duration of each collision, there is a definite distribution of molecular velocities that can be calculated from the kinetic theory of gases.[†] According to the Boltzmann distribution of energy, the probability that a molecule is in a level of energy ϵ_i and statistical weight g_i is proportional to $g_i e^{-\epsilon_i/kT}$, where k is the Boltzmann constant and T is the thermodynamic temperature. The probability may be represented by dN/N_0, the fractional number of molecules having the designated energy.

Consider first the distribution of a given component of velocity, say $\dot{x}$. The energy is translational with $\epsilon_i = 1/2 m \dot{x}^2$, where m is the mass of a molecule. For a continuous range of energy as in this case, the statistical weight is the volume in phase space in units of h^f for f degrees of freedom. Here

$$g_i = \frac{m \, d\dot{x} \, dx}{h}$$

where h is Plank's constant. Therefore, for the fractional number of molecules with x component of velocity between $\dot{x}$ and $\dot{x} + d\dot{x}$ and x coordinate between x and $x + dx$

$$\frac{dN}{N_0} = A \frac{m \, d\dot{x} \, dx}{h} e^{-m\dot{x}^2/2kT}$$

A is a proportionality constant that may be determined by integrating with respect to $\dot{x}$ and x over their complete ranges $-\infty$ to $+\infty$ and 0 to a, respectively, where a is the length of the corresponding edge of the (rectangular) container.

$$\int \frac{dN}{N_0} = 1 = \frac{Aam}{h} \int_{-\infty}^{+\infty} e^{-m\dot{x}^2/2kT} \, d\dot{x}$$

$$= \left(\frac{Aam}{h}\right) \left(\frac{2\pi kT}{m}\right)^{1/2}$$

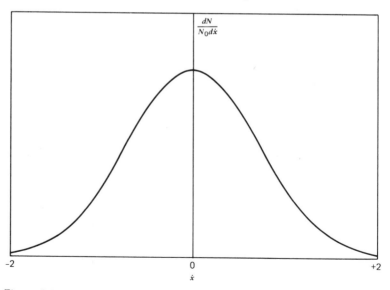

Figure 4.1. Distribution of a Cartesian component of molecular velocity.

[†] For general treatments see references 1–4.

Therefore,

$$A = \left(\frac{h}{a}\right)\left[\frac{1}{(2\pi mkT)^{1/2}}\right]$$

and

$$\frac{dN}{N_0} = \left(\frac{m}{2\pi kT}\right)^{1/2} e^{-m\dot{x}^2/2kT} \, d\dot{x} \qquad (4.1)$$

This distribution, which has been integrated to include all positions x, has a maximum at $\dot{x} = 0$ and falls off as shown in Fig. 4.1 for positive or negative $\dot{x}$.

For all three components in specified ranges $\dot{x}$ to $\dot{x} + d\dot{x}$, $\dot{y}$ to $\dot{y} + d\dot{y}$, and $\dot{z}$ to $\dot{z} + d\dot{z}$, the probabilities multiply. Therefore

$$\frac{dN}{N_0} = \left(\frac{m}{2\pi kT}\right)^{3/2} e^{-mc^2/2kT} \, d\dot{x} \, d\dot{y} \, d\dot{z} \qquad (4.2)$$

where $c^2 = \dot{x}^2 + \dot{y}^2 + \dot{z}^2$ is the square of the magnitude of the velocity vector.

DISTRIBUTION OF MAGNITUDE OF VELOCITY. MAXWELL DISTRIBUTION LAW

The distribution of magnitude of velocity without regard to direction is often desired. The Cartesian coordinates of velocity of (4.2) may be replaced by spherical polar coordinate velocity variables c, θ, φ, where c is the magnitude of velocity and θ and φ are colatitudinal and longitudinal angles giving the direction of velocity. The volume element $d\dot{x} \, d\dot{y} \, d\dot{z}$ becomes $c^2 \sin \theta \, dc \, d\theta \, d\varphi$, and integrating over θ from 0 to π and φ from 0 to 2π gives

$$\int_0^\pi \int_0^{2\pi} \sin \theta d\theta \, d\varphi = 4\pi$$

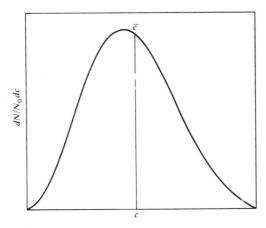

Figure 4.2. Distribution of magnitude of molecular velocity c for motion in three dimensions. Average velocity $\bar{c}$ is also shown.

Therefore the distribution without regard to direction is

$$\frac{dN}{N_0} = 4\pi \left(\frac{m}{2\pi kT}\right)^{3/2} c^2 e^{-mc^2/2kT} \, dc \tag{4.3}$$

This is the common form of the Maxwell distribution law and has the graph shown in Fig. 4.2. The maximum shifts to higher c values as T increases or as m decreases.

The average speed $\bar{c}$ of molecules in a sample of gas can be obtained by multiplying (4.3) by c and integrating from 0 to ∞

$$\bar{c} = \left(\frac{8kT}{\pi m}\right)^{1/2} \tag{4.4}$$

COLLISION NUMBER

The collision number is defined as the number of bimolecular collisions per unit time per unit volume. If there are two types of molecules A and B and they are considered to be rigid spheres with diameters d_A and d_B, then a bimolecular collision may be defined as the situation where there is contact of the surfaces of the two spheres. Suppose that there are N_A/V molecules of A per unit volume and N_B/V molecules of B per unit volume. To calculate the collision number of a single A molecule, consider such a molecule moving in an arbitrary direction with a *mean relative velocity* $\bar{r}$ relative to a molecule of type B. If the center of molecule B is within a distance $d_{AB} = (d_A + d_B)/2$ of the line of flight of the center of molecule A during the passage of A, a collision results (see Fig. 4.3). The total number of collisions per unit time of molecule A with molecules of type B can then be estimated from the volume swept out by a sphere of radius d_{AB} multiplied by the number of type B molecules per unit volume. The volume swept out per unit time is $\pi d_{AB}^2 \bar{r}$ and the number of collisions for a single A molecule is $\pi d_{AB}^2 \bar{r} N_B/V$. To obtain the collision number Z_{AB} we multiply this by the number of A molecules per unit volume.

$$Z_{AB} = \frac{\pi d_{AB}^2 \bar{r} N_A N_B}{V^2} \tag{4.5}$$

The mean relative velocity $\bar{r}$ is slightly greater than the mean speed $\bar{c}$ of a single kind of molecule. It can be calculated by using the product of two distribution laws (4.2), one for each kind of molecule. The fraction of A molecules whose

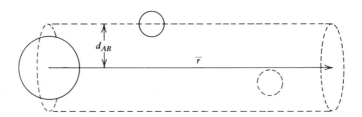

Figure 4.3.   Cylinder space swept out by a molecule in unit time. Molecules whose centers are within the cylinder would undergo collision.

velocity components are between $\dot{x}_A$ and $\dot{x}_A + d\dot{x}_A$, $\dot{y}_A$ and $\dot{y}_A + d\dot{y}_A$, and $\dot{z}_A$ and $\dot{z}_A + d\dot{z}_A$ is given by (4.2) as

$$\frac{dN_A}{N_A} = (m_A/2\pi kT)^{3/2} e^{-m_A c_A^2/2kT} d\dot{x}_A d\dot{y}_A d\dot{z}_A \tag{4.6}$$

Multiplying by the similar distribution for B molecules yields

$$\frac{dN_A dN_B}{N_A N_B} = \frac{(m_A m_B)^{3/2}}{(2\pi kT)^3} e^{-(m_A c_A^2 + m_B c_B^2)/2kT} d\dot{x}_A d\dot{y}_A d\dot{z}_A d\dot{x}_B d\dot{y}_B d\dot{z}_B \tag{4.7}$$

Now transform (4.7) to the new velocity coordinates u, v, w, r, θ, φ, where the first three are Cartesian components of the velocity of the center of mass of the two molecules and the latter three are spherical coordinates for the relative velocity of the two molecules.

$$u = \frac{m_A \dot{x}_A + m_B \dot{x}_B}{m_A + m_B}$$

$$v = \frac{m_A \dot{y}_A + m_B \dot{y}_B}{m_A + m_B}$$

$$w = \frac{m_A \dot{z}_A + m_B \dot{z}_B}{m_A + m_B}$$

$$r = [(\dot{x}_B - \dot{x}_A)^2 + (\dot{y}_B - \dot{y}_A)^2 + (\dot{z}_B - \dot{z}_A)^2]^{1/2}$$

$$\theta = \cos^{-1}\left(\frac{\dot{z}_B - \dot{z}_A}{r}\right)$$

$$\varphi = \tan^{-1}\left(\frac{\dot{y}_B - \dot{y}_A}{\dot{x}_B - \dot{x}_A}\right) \tag{4.8}$$

The volume element in (4.7) becomes

$$d\dot{x}_A d\dot{y}_A d\dot{z}_A d\dot{x}_B d\dot{y}_B d\dot{z}_B = r^2 \sin\theta \, dr d\theta d\varphi du \, dv \, dw \tag{4.9}$$

Introducing the reduced mass,

$$\mu = \frac{m_A m_B}{m_A + m_B}$$

and integrating over the complete range of all variables except r gives $dN_{AB}/N_A N_B$, the fraction of pairs of molecules with the magnitude of relative velocity in the range r to $r + dr$.

$$\frac{dN_{AB}}{N_A N_B} = \left[\left(\frac{m_A m_B}{2\pi kT}\right)^{3/2}\left(\frac{\mu}{2\pi kT}\right)^{3/2}\right] \int_0^\pi \int_0^{2\pi} \int_{-\infty}^\infty \int_{-\infty}^\infty \int_{-\infty}^\infty$$

$$\exp\left(\frac{-[(m_A + m_B)(u^2 + v^2 + w^2) + \mu r^2]}{2kT}\right) r^2 \cos\theta \, dr d\theta d\varphi du dv dw$$

$$= 4\pi \left(\frac{\mu}{2\pi kT}\right)^{3/2} \exp\left(-\mu r^2/2kT\right) r^2 \, dr \tag{4.10}$$

Note that integration over u, v, w simply cancels the factor $(m_A m_B/2\pi kT)^{3/2}$ and integration over θ, φ yields the factor 4π. To obtain the mean relative velocity $\bar{r}$ multiply the fraction of collisions given by (4.10) by r and integrate

$$\bar{r} = 4\pi \left(\frac{\mu}{2\pi kT}\right)^{3/2} \int_0^\infty r^3 e^{-\mu r^2/2kT} dr \tag{4.11}$$

This yields

$$\bar{r} = \left(\frac{8kT}{\pi\mu}\right)^{1/2} \tag{4.12}$$

and substituting into (4.5) gives

$$Z_{AB} = (\pi d_{AB}^2)\left(\frac{8kT}{\pi\mu}\right)^{1/2} \frac{N_A N_B}{V^2} \tag{4.13}$$

It is often useful to know the collision number for a sample of a single gas where all molecules are alike. This can be derived from (4.13) by dropping the subscripts A and B and applying a factor of $1/2$ so as not to count each collision twice. Since $m_A = m_B = m$,

$$\mu = \frac{m}{2} \quad \text{and} \quad \bar{r} = \sqrt{2}\,\bar{c}$$

Therefore (4.13) becomes

$$Z = \frac{1}{2}\left(\frac{16kT}{\pi m}\right)^{1/2} \frac{\pi d^2 N^2}{V^2} = 2\left(\frac{kT}{\pi m}\right)^{1/2} \frac{\pi d^2 N^2}{V^2} \tag{4.14}$$

For a typical gas at standard conditions Z is approximately 10^{28} collisions per cubic centimeter per second.

The average distance traveled by a single molecule between collisions is called the mean free path. It is the quotient of the average molecular speed $\bar{c}$ divided by the collision number for a single A molecule $Z_{AB}/(N_A/V)$. In the case where there are N/V like molecules per unit volume, the mean free path is

$$l = \frac{\bar{c}}{ZV/N} = \frac{(8kT/\pi m)^{1/2}}{(16kT/\pi m)^{1/2} \pi d^2 N/V} = \frac{1}{\sqrt{2}\pi d^2 (N/V)} \tag{4.15}$$

For standard conditions of temperature and pressure with typical gas molecules $l \cong 10^{-7}$ m and is inversely proportional to pressure at constant temperature. In gas reactions at low pressures the mean free path can become of the same order of magnitude as the size of the container and wall effects may complicate the reaction.

The mean free path enters into the formula for gas viscosity η,

$$\eta = \tfrac{1}{3}\bar{c}\rho l \tag{4.16}$$

where ρ is the density. Measurement of viscosity, therefore, enables calculation of molecular diameters through (4.15) and (4.16). Molecular diameters can also be obtained from data on heat conductivity, diffusivity, and deviations from ideal gas behavior.

RATE OF A BIMOLECULAR REACTION

If a reaction occurred at every collision between given reacting molecules, the rate would be much greater than is usually observed. For example, when the collision number for a single A molecule $Z_{AB}/(N_A/V)$ is calculated from (4.13) for typical molecules under standard conditions, the result is on the order of 10^{10} collisions per molecule per second. Thus the half-life period of the reaction under standard initial conditions would be of the order of 10^{-10} s — smaller by a factor of 10^{-15} to 10^{-20} than results usually obtained. Furthermore, the temperature dependence of the collision number, $T^{1/2}$, does not agree with the exponential temperature dependence of the Arrhenius equation. However, the collision-number formula (4.13) does predict correctly that a bimolecular reaction will be second order.

Let us assume that only a small fraction q of all collisions is effective, q being a function of temperature and of the nature of the reacting molecules. Then the number of effective collisions per unit volume per unit time (that is, the rate of disappearance of A or of B molecules per unit volume) is

$$-\frac{1}{V}\frac{dN_A}{dt} = -\frac{1}{V}\frac{dN_B}{dt} = qZ_{AB} = q\left(\frac{8kT}{\pi\mu}\right)^{1/2}\frac{\pi d_{AB}^2 N_A N_B}{V^2} \qquad (4.17)$$

For like molecules, if two molecules react at each collision,

$$-\frac{1}{V}\frac{dN}{dt} = 2qZ = 4q\left(\frac{kT}{\pi m}\right)^{1/2}\frac{\pi d^2 N^2}{V^2} \qquad (4.18)$$

The second-order rate constant predicted by hard-sphere collision theory can be obtained by comparing (4.17) with the macroscopic rate expression

$$-\frac{d[A]}{dt} = -\frac{1}{V}\frac{dn_A}{dt} = -\frac{1}{LV}\frac{dN_A}{dt} = k[A][B] = \frac{k}{V^2}n_A n_B$$

$$= \frac{k}{L^2 V^2}N_A N_B \qquad (4.19)$$

In (4.19) we use the fact that $n = N/L$, where n is the amount of substance and L is the Avogadro constant. Multiplying both sides of (4.17) by $1/L$, dividing by (4.19), solving for the rate constant k, and, to avoid confusion, substituting R/L for the Boltzmann constant (R is the gas constant) yields

$$k = L\left(\frac{8RT}{\pi L\mu}\right)^{1/2}q(\pi d_{AB}^2) = L\left[\left(\frac{8RT}{\pi}\right)\left(\frac{1}{M_A}+\frac{1}{M_B}\right)\right]^{1/2}q(\pi d_{AB}^2) \qquad (4.20)$$

Similarly, for like molecules

$$k = 2L\left(\frac{RT}{\pi M}\right)^{1/2}q(\pi d^2) \qquad (4.21)$$

In (4.20) and (4.21), M_A, M_B, and M represent molar masses.

Careful examination of (4.20) and (4.21) reveals that the right-hand side of each contains three factors: the Avogadro constant to convert from molecular to molar scale; the mean relative velocity $\bar{r} = (8RT/\pi L\mu)^{1/2}$; and an effective cross-sectional area, $q(\pi d^2)$. The mean relative velocity was evaluated by integration of the product of the velocity distributions for the two types of colliding molecules, but all that the theory tells us about the effective cross-sectional area is that it usually is much smaller for reactive collisions than for the simple collisions that distribute energy among the gas molecules. Under these circumstances it is reasonable to introduce the *reactive cross section* σ_R, a molecular-scale parameter that includes all factors that determine the effectiveness of collisions. Then for the general case of any distributions $f_A(c_A)$, $f_B(c_B)$ of molecular speeds c_A, c_B, the rate constant at a particular temperature can be calculated as

$$k = L \int \int \sigma_R r f_A f_B dc_A dc_B \tag{4.22}$$

Or, assuming Boltzmann distributions of energy and carrying out the same integration over center-of-mass and angular coordinates that led to (4.10),

$$k = 4\pi L \left(\frac{L\mu}{2\pi RT}\right)^{3/2} \int_0^\infty r^3 e^{-\mu r^2 L/2RT} \sigma_R \, dr \tag{4.23}$$

[As in (4.20) and (4.21), the Boltzmann constant has been replaced by R/L.] Note that in general σ_R is a function of relative speed r and must remain within the integral. Often it is more convenient to obtain σ_R as a function of energy rather than relative speed. Equation (4.23) can be transformed by recognizing that the kinetic energy of collision $E = \mu r^2/2$ and $dE = \mu r dr$. Thus

$$k = L \left[\frac{8}{\pi\mu(RT/L)^3}\right]^{1/2} \int_0^\infty E e^{-LE/RT} \sigma_R \, dE \tag{4.24}$$

According to (4.24), once the energy dependence of the reactive cross section σ_R is known for a given type of bimolecular collision, the second-order rate constant can be calculated.

ENERGY DEPENDENCE OF EFFECTIVE CROSS SECTION

The energy requirement is by far the most important of the factors that determine the effectiveness of a collision. Comparison of bond lengths in a normal molecule with internuclear distances between atoms in different molecules undergoing collision implies that excess energy is required for a collision to result in reaction. In general van der Waals radii are significantly larger than covalent radii, and so collisions must be sufficiently energetic to cause a compression of the reacting molecules so that the atoms to be bonded approach more closely their normal distance in the product molecules. In the case of an endothermic reaction there is an additional energy requirement, since, for successful reaction, sufficient energy must be supplied in the collision to put the product molecules in at least their lowest energy levels.

A complete theory of collisional reaction rate must include a detailed treatment of the forces and motions involving all the atoms of the colliding molecules. This is a difficult quantum-mechanical problem that we bypass for the time being by assuming that for reaction the energy must exceed a minimum value or threshold. energy E_0. E_0 can be determined empirically from the activation energy E_a rather than by a quantum-mechanical calculation.

The simplest starting assumption about the energy dependence of σ_R is

$$\sigma_R = 0 \qquad E < E_0$$

$$\sigma_R = \pi d_{AB}^2 \qquad E \geqslant E_0$$

That is, all collisions for which the relative energy equals or exceeds the threshold energy are effective, and all collisions whose relative energy is less than the threshold energy are ineffective. The collision diameter is the one obtained from the kinetic theory of gases. Thus, from (4.24)

$$k = L \left[\frac{8}{\pi\mu(RT/L)^3} \right]^{1/2} \int_{E_0}^{\infty} E e^{-LE/RT} \pi d_{AB}^2 \, dE$$

and

$$k = L \left(\frac{8RT}{\pi L\mu} \right)^{1/2} (\pi d_{AB}^2) \left(\frac{1 + LE_0}{RT} \right) e^{-LE_0/RT} \qquad (4.25)$$

However, since the repulsive force resisting compression of spherically symmetric molecules is exerted along the line connecting their centers, it seems more reasonable to assume that only the component of the relative kinetic energy of approach of two molecules that is directed along their line of centers at the moment of collision can be effective.[5] Kinetic energy associated with motion perpendicular to the line of centers corresponds to a sideswiping or a rotation and would not be expected to be effective in reaction. For a given relative kinetic energy, then, the effective energy depends on the *impact parameter b* defined in Fig. 4.4. The impact

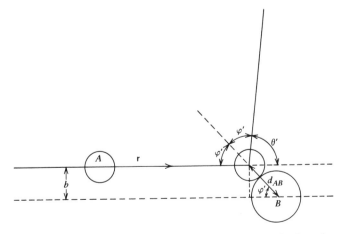

Figure 4.4. Collision of two hard spheres whose relative velocity of approach is r. The angle between the velocity vector and the line of centers at the instant of collision is φ', the scattering angle is θ', the impact parameter is b, and the distance of closest approach is d_{AB}.

parameter is the distance between two lines, each parallel to the relative velocity vector and passing through the center of one of the molecules prior to their collision. From Fig. 4.4 the component of relative velocity along the line of centers upon collision is $r \cos \varphi'$, and thus the desired component of kinetic energy is $\mu r^2 (\cos^2 \varphi')/2$. Also from the geometry of Fig. 4.4, $\sin \varphi' = b/d_{AB}$, and so

$$\frac{\mu r^2 (\cos^2 \varphi')}{2} = \frac{\mu r^2 (1 - \sin^2 \varphi')}{2} = \frac{\mu r^2 (1 - b^2/d_{AB}^2)}{2}$$

Since the total relative kinetic energy is $E = \mu r^2/2$, the line-of-centers component is $E(1 - b^2/d_{AB}^2)$. This must exceed E_0 for the collision to result in reaction. For a given value of $E \geqslant E_0$ there is some impact parameter b_0 at which the line-of-centers component of kinetic energy exactly equals E_0. Thus

$$E\left(1 - \frac{b_0^2}{d_{AB}^2}\right) = E_0$$

and

$$b_0^2 = d_{AB}^2 \left(1 - \frac{E_0}{E}\right)$$

For that particular value of E, then, all collisions having impact parameter $b \leqslant b_0$ are effective and the reactive cross section is πb_0^2. Thus

$$\sigma_R = 0 \qquad E < E_0$$

$$\sigma_R = \pi b_0^2 = \pi d_{AB}^2 \left(1 - \frac{E_0}{E}\right) \qquad E \geqslant E_0$$

Substituting into (4.24) gives

$$k = L \left| \frac{8}{\pi \mu (RT/L)^3} \right|^{1/2} \int_{E_0}^{\infty} E e^{-LE/RT} \pi d_{AB}^2 \left(1 - \frac{E_0}{E}\right) dE$$

and integration yields

$$k = L \left(\frac{8RT}{\pi L \mu}\right)^{1/2} (\pi d_{AB}^2) e^{-LE_0/RT} \tag{4.26}$$

Equation (4.26) is usually referred to as "the" collision-theory rate constant expression, although the term line-of-centers rate constant might be more appropriate since other modifications of the collision theory exist [(4.25), for example]. If E_0 is set equal to zero in (4.26), so that all collisions are effective, (4.26) is identical to (4.20) with $q = 1$. Thus (4.26) provides the surprisingly simple result that the fraction of all collisions having the necessary line-of-centers component of energy is given by $e^{-E_0/kT}$ or $e^{-LE_0/RT}$, where LE_0 is the required energy per mole of collisions.

The relationship between E_0 and the activation energy can be obtained using the definition of E_a given in (2.42):

$$\frac{d \ln k}{dT} = \frac{E_a}{RT^2} \tag{2.42}$$

Taking the required derivative of (4.26) yields

$$\frac{d \ln k}{dT} = \frac{1}{2T} + \frac{LE_0}{RT^2} = \frac{E_a}{RT^2}$$

and

$$E_0 = \left(\frac{1}{L}\right)\left(E_a - \frac{RT}{2}\right) \tag{4.27}$$

Substituting (4.27) back into (4.26) gives

$$k = L\left(\frac{8RT}{\pi L \mu}\right)^{1/2} (\pi d_{AB}^2)e^{-E_a/RT}e^{1/2}$$

and thus the Arrhenius A factor from (2.43) is given by

$$A = L\left(\frac{8RT}{\pi L \mu}\right)^{1/2} (\pi d_{AB}^2)e^{1/2} \tag{4.28}$$

All parameters in (4.28) can be obtained from nonkinetic experiments, and so calculated A factors can be compared with those obtained from kinetic studies.

OTHER FACTORS THAT AFFECT THE REACTIVE CROSS SECTION

1. *Orientation of Molecules — Steric Factor p.* For any molecular species more complicated than a free atom, it is apparent that reaction is not to be expected at a collision unless the molecules are so oriented relative to each other that the groups reacting or the bonds to be shifted are relatively close. A steric factor p is considered to represent the fraction of collisions that have the proper orientation for the colliding molecules. It is expected to range from a value near unity down to 0.1 or 0.01, depending on the complexity of the molecules. The efficiency factor introduced earlier may now be written as

$$q = pe^{-E_0/kT}$$

and the factor p is introduced on the right-hand side of expressions such as (4.25), (4.26), and (4.28). Unfortunately, there appears to be no way to predict the value of p. Hence p can only be determined empirically by comparing calculated and experimental values of the frequency factor A. When determined this way p is not only a steric factor, but also includes all other errors of simplification inherent in the collision theory. Some of these are listed in the remainder of this section.

2. *Restrictions on Bimolecular Association Reactions.* Consider a reaction of the type $A + B \rightarrow AB$ or $2A \rightarrow A_2$. To form a stable molecule in such a reaction there is always an evolution of energy. If the product molecule has several vibrational degrees of freedom, the energy to be evolved in forming the new bond or bonds

may be distributed among the other degrees of freedom, thus stabilizing the molecule until the extra energy is removed by later collisions with other molecules. In the association of atoms to form diatomic molecules, however, there is only the one vibrational degree of freedom, and the energy remains to cause the molecule to dissociate again in one-half a period of vibration ($\sim 10^{-13}$ s). It might be supposed that the molecule could be stabilized by emission of radiation, but this process is forbidden for the formation of symmetric diatomic molecules and too slow for the formation of others. Association reactions of atoms and of some simple radicals like OH do not take place as bimolecular reactions, but require trimolecular collisions where the third molecule may be of any kind and serves to remove excess energy in such forms as translation, vibration, or possibly rotation.

3. *Quantum Considerations.* There is no reason to assume, as we do up to this point, that the reactive cross section is independent of the internal (rotational, vibrational, or electronic) energy states of colliding molecules, or that all collisions are elastic with no change in the total translational kinetic energy (*adiabatic* collisions). Internal energy can contribute to overcoming an activation energy barrier, and it may happen that certain internal energy states are so much more reactive than others as to constitute the only viable reaction pathway. Consequently rates of energy transfer can be an important factor determining overall reaction rates. Several detailed descriptions of energy transfer are available.[6-8]

Energy transfer is usually defined by specifying the source of the energy followed by the form to which it is converted. The abbreviations V (vibrational), R (rotational), and T (translational) are commonly used. Thus the inelastic collision process

$$HF(v = 0) + Ar \rightarrow HF(v = 1) + Ar$$

is an example of T–V energy transfer. Translational energy of HF and Ar is required to raise the vibrational quantum number v of HF. Techniques for studying rates of energy transfer include[6] measurements of absorption and dispersion of sound, shock wave studies, the optic-acoustic effect, quenching of vibrational fluorescence, measurements of transport phenomena such as thermal conductivity, molecular beam methods, and microwave double resonance. Typical second-order rate constants for various energy transfer processes at 298 K are as follows:[9] R–R, $10^8 - 10^9 \, M^{-1} \, s^{-1}$; R–T, $10^6 - 10^7 \, M^{-1} \, s^{-1}$; V–V, $10^6 - 10^7 \, M^{-1} \, s^{-1}$; V–R, $10^4 - 10^5 \, M^{-1} \, s^{-1}$; V–T, $10^3 - 10^4 \, M^{-1} \, s^{-1}$. In general the efficiency of energy transfer is greater the more closely matched are the differences between energy levels from and to which energy is transferred. Thus V–T energy transfer, especially for diatomic molecules having large vibrational spacing, is least efficient. Also, R–R and V–V energy transfer is most likely under near-resonant conditions, in which there is little change in translational energy.

Eliason and Hirschfelder[10] have carried out a general collision-theory treatment of the rate of a bimolecular reaction. They define a separate reaction cross section for each definite internal quantum state of reactants and products as σ_{ij}^{kl}, where i indexes vibrational, rotational, and possibly electronic states of one reactant and j of the other, and k and l index the states of the respective products after collision. The total reactive cross section, then, must be obtained by summation over all combinations of i, j with k, l. Each term in the sum is the product of the occupation numbers of the specified states times the reactive cross section for that combination of states. It is not necessary to assume equilibrium Boltzmann distributions of translational, rotational, or vibrational energy, but if this assumption is made, (4.24) becomes

$$k = L\left[\frac{8}{\pi\mu(RT/L)^3}\right]^{1/2} q_A^{-1} q_B^{-1} \sum_{ij}\sum_{kl} e^{-L(E_i + E_j)/RT} \int_0^\infty \sigma_{ij}^{kl} e^{-LE/RT} E \, dE \qquad (4.29)$$

where $q_A = \sum_i e^{-E_i/kT}$ and $q_B = \sum_j e^{-E_j/kT}$ are partition functions for the internal quantum states of molecules A and B, and i and j index those quantum states.

Inclusion of internal energy states can be simplified if such states are approximated by a collection of s harmonic oscillators all of whose energies lie well below kT. Thus the oscillators may be treated by classical methods. The rate constant can be calculated as the hard-sphere collision frequency times the probability that the total energy in the s harmonic oscillators exceeds E_0.[11] The fraction of time during which the energy of the s oscillators exceeds E_0 is

$$[(s-1)!]^{-1}\left(\frac{E_0}{kT}\right)^{s-1} e^{-E_0/kT}$$

so long as $E_0 \gg (s-1)kT$. When this factor is incorporated into (4.24) and the integration is performed, the line-of-centers rate constant expression becomes

$$k = L\left(\frac{8RT}{\pi L\mu}\right)^{1/2} (\pi d_{AB}^2)\left[\frac{1}{(s-1)!}\right]\left(\frac{LE_0}{RT}\right)^{s-1} e^{-LE_0/RT} \qquad (4.30)$$

This is known as the *activation-in-many-degrees-of-freedom* rate constant. Since $LE_0 \gg RT$ and $(LE_0/RT)^{s-1} > (s-1)!$, including the energy of internal degrees of freedom increases the rate constant. Equation (4.30) can be applied to rotational as well as vibrational energy by recalling that a pair of rotational degrees of freedom makes a contribution to the classical energy that is equivalent to a single vibrational degree of freedom.

Because electronic energy levels are much more widely spaced than vibrational, rotational, and translational levels, collisions are usually inefficient with respect to electronic energy transfer. If the ground electronic state of the products differs from that of the reactants, such reactions can be expected to be slow. Examples of this are situations in which ground-state products have different multiplicities or different symmetries from ground-state reactants.[12] As long as no change in electronic state occurs during collision, products must be formed in an excited state — an energetically unfavorable process. However, only a small number of reactions, some of which are important in flames, explosions, and photochemistry, are included in this category. An electronic restriction that is much more widely applicable is related to the symmetry or topology of molecular orbitals before, during, and after a collision. This topic is taken up in some detail in Chapter 5.

4. *Inadequacy of the Hard-Sphere Model.* Molecules are not hard spheres — most are not even spherically symmetric — and the treatment of molecular collisions becomes much more complex when this fact is taken into account. In general it is

necessary to consider the energy as a function of the positions of all nuclei and all electrons at each stage of the collision process. Electronic motions, however, are rapid compared to the time scale of a collision, and the Born–Oppenheimer approximation can be extended to treat the nuclei as moving in the average potential field of the electrons. Quantum mechanics can then be applied to compute the energy as a function of each possible set of relative nuclear positions. This approach generates a multidimensional "potential" energy surface over which the collection of atomic nuclei move during the collision, and we develop it more completely in Chapter 5.

A much simpler approach is to assume (as we already do) that one molecule of a colliding pair is at rest while the other moves at some relative velocity in the potential field of the first molecule. If that potential field is spherically symmetric, the situation can be diagrammed as shown in Fig. 4.5, where the cylindrical symmetry about the line $b = 0$ is obvious. A typical spherically symmetric potential that has been found to give a reasonable representation of the interaction of real molecules is the Lennard-Jones or 6–12 potential

$$U = 4\epsilon \left[\left(\frac{d_0}{d} \right)^{12} - \left(\frac{d_0}{d} \right)^6 \right] \tag{4.31}$$

where U is the potential energy, d is the internuclear separation, and ϵ and d_0 are

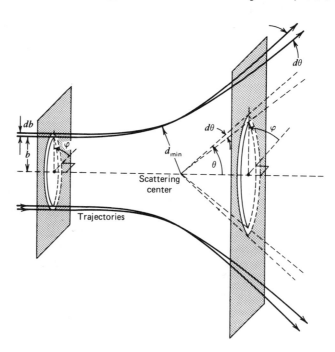

Figure 4.5.   Deflection of a particle trajectory by a scattering center. The relationship of the impact parameter b and the scattering angle θ is shown. [Reprinted with permission from E. F. Greene and A. Kuppermann, *J. Chem. Educ.*, **45** (6), 361 (1968).]

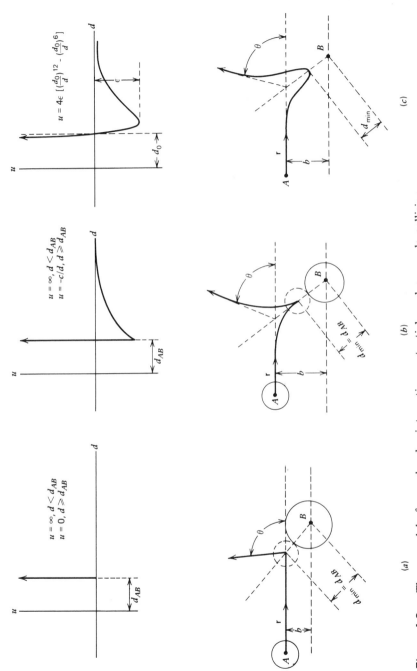

Figure 4.6. Three models for molecular interaction potentials and a sample collision trajectory based on (a) hard spheres; (b) hard spheres with attraction; and (c) Lennard-Jones 6–12 potential.

parameters that determine the depth of the potential well and the smallest internuclear separation at which $U = 0$ (see Fig. 4.6). Figure 4.6 also shows the hard-
sphere potential and one based on hard-sphere molecules with attractive forces. A
typical collision trajectory for molecules governed by each of the potential
functions is also illustrated. Note that when the potential function includes an
attractive term, collisions can occur even though the impact parameter b exceeds
the sum of the radii of the molecules.

Calculations based on the Lennard-Jones potential function are rather involved,
and important points regarding the influence of the form of the potential on the
cross section for collision can be made using the model of hard spheres with attraction, so we confine our analysis to this less realistic but simpler case. If we define
a collision as involving contact of the hard spheres, ignoring deflections of
trajectories where the moving molecule passes the stationary one without contact,
then the cross section for collision σ depends on the relative kinetic energy as
shown in (4.32).[13]

$$ \sigma = (\pi d_{AB}^2)\left[1 - \frac{U(d_{AB})}{E}\right] \tag{4.32} $$

Here $U(d_{AB}) = -C/d_{AB}$, where C is a constant that depends on the nature of the
colliding molecules. $U(d_{AB})$ is the potential energy when the two hard spheres are
just in contact. Note that since $U(d_{AB})$ is a negative quantity, the effect of the
second term in brackets in (4.32) is to increase the cross section for collision over
the simple hard-sphere value of πd_{AB}^2. Only when $E \gg |U(d_{AB})|$ does the hard-
sphere-with-attraction model make the same prediction for the number of collisions
as the simple hard-sphere model. Now the reactive cross section σ_R can still be set
to zero if the relative line-of-centers energy is less than the threshold value E_0,
but when $E > E_0$, σ_R can be identified with σ of (4.32). If A and B are oppositely
charged ions, $|U(d_{AB})|$ is quite large and the reactive cross section may exceed
πd_{AB}^2, the cross section obtained by measurement of gas viscosity or transport
properties, by a significant amount. This explains why some ion-recombination
reactions have steric factors that exceed unity.

EXPERIMENTAL DETERMINATION OF REACTIVE CROSS SECTIONS

Since calculation of reactive cross sections from first principles is difficult and time-
consuming in all but the simplest cases, the alternative of using empirically derived
values is attractive. However, rate studies of thermal reactions (those in which all
energy levels of all reactants are accurately described by Boltzmann distribution
laws) provide information about the integral of σ_R over a wide range of energies.
The distribution of molecular energies can be changed by changing the temperature,
but this provides a very dull tool for obtaining the desired information about σ_R
as a function of energy or relative molecular speed. The situation is akin to the
problem of measuring an absorption spectrum using a blackbody source but no
monochromator — extraction of a molar absorptivity at a given wavelength from

the change in total absorbance as a function of color temperature would be well-nigh impossible. To obtain σ_R as a function of energy, then, requires experiments in which molecular speeds are restricted to a narrow range and/or internal energy states can be specifically selected. The latter requirement is being achieved in the field of state-to-state chemistry,[14] in which non-Boltzmann distributions over internal energy states are induced in reactant molecules and detected in product molecules. Techniques for achieving non-Boltzmann distributions of translational or internal energy and studying their effect on cross sections or rate constants may be divided into two categories: bulk methods and beam methods.

Bulk Methods[†]

Here a mixture of gases in a typical reaction vessel is perturbed so as to introduce a reactant whose energy lies within a limited range. A typical example is the introduction of translationally "hot" atoms by short-wavelength photolysis of an appropriate molecule.

The first direct measurement of the threshold energy for a reaction involving the formation and breaking of chemical bonds was carried out in this way.[16] At wavelengths between 366 and 303 nm, photolysis of DI produces D and I atoms in their ground electronic states. Therefore the photon energy in excess of the D—I bond-dissociation energy (298.9 kJ mol^{-1}) appears as translational energy of the product atoms, with 98.5% of it in the D atoms. These hot D atoms (indicated by D*) were produced in a room-temperature mixture of DI and H$_2$, where, until they became thermalized, they could react with H$_2$:

$$D^* + H_2 \rightarrow DH + H \tag{4.33}$$

The D atoms (both hot and thermal) that escape this reaction react with DI to form D$_2$:

$$D^* + DI \rightarrow D_2 + I \tag{4.34}$$

$$D + DI \rightarrow D_2 + I \tag{4.35}$$

Also, H atoms produced by reaction (4.33) can react with DI, forming additional DH:

$$H + DI \rightarrow HD + I \tag{4.36}$$

The ratio [D$_2$]/[DH] was determined by mass spectrometry, and, to eliminate the effect of reaction (4.34), results obtained for several different ratios of [DI]/[H$_2$] were extrapolated to [DI]/[H$_2$] = 0. After a correction for the effect of 2–4% HI impurity in the DI, the nonzero intercept was taken as a measure of the relative yields of the hot reaction (4.33) and the thermal reaction (4.35). Experiments were carried out at 366, 334, 313, and 303 nm, corresponding to translational energies of 27.5, 58.3, 82.0, and 94.4 kJ mol^{-1}, for D atoms. Since the masses of D and H$_2$ are nearly equal, the relative collision energy would be half the translational energy of D if the H$_2$ and DI molecules were motionless initially. This value is increased by about 2.9 kJ mol^{-1} when initial thermal motion is included,[‡] yielding

[†] For recent reviews see reference 15.

[‡] The distribution of relative energies may be obtained as indicated in reference 17.

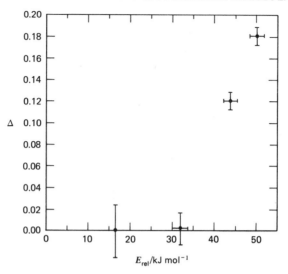

Figure 4.7.   Effect of average relative initial kinetic energy E_{rel} of reactants of the reaction $D + H_2 \rightarrow DH + H$. The fraction of D atoms undergoing reaction, corrected for the effect of minor DI impurity, is represented by Δ. Error limits on the observables are shown. (Data from Kupperman and White.[16])

average relative energies of collision of 16.6, 32.0, 43.9, and 50.1 kJ mol^{-1}. Experimental results at these energies are shown in Fig. 4.7, from which a threshold energy of 31.8 ± 1.9 kJ mol^{-1} was obtained for reaction (4.33).

A similar bulk study by Gann, Ollison, and Dubrin[18] has yielded information on the variation of reactive cross section with energy for the reaction

$$H + n\text{--}C_4 D_{10} \rightarrow HD + sec\text{--}C_4 D_9 \tag{4.37}$$

These authors note several factors that complicate the interpretation of data obtained from experiments on bulk systems. Even at low temperatures and with a line photolytic source there is considerable spread of the initial relative kinetic energy about the average values calculated in the preceding paragraph. If a hot H atom fails to react on its first collision with the target molecule it may still retain more than the threshold kinetic energy, and even after it falls below the threshold it may become reactivated as a result of subsequent collisions or suffer a collision where its partner is energetic enough that the relative energy exceeds the threshold. The latter "thermal" reactions become especially important when the initial kinetic energy of the H atom is near the threshold value, and their relative importance depends on the (unknown) cross section for scavenging of thermal H atoms by HX to form H_2. Thus unless a source of photolysis light is available at just the wavelength needed to provide initial relative kinetic energies equal to the threshold value, extrapolation of results from other wavelengths may not be very accurate.

Information about energy transfer in nonreactive collisions and its effect on the distribution of relative kinetic energies can be obtained by measuring the effect of diluting the system with an unreactive gas whose collision cross section with H is known or can be calculated from a known intermolecular potential function. In

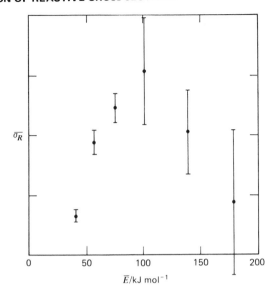

$\bar{\sigma}_R$

$\bar{E}$/kJ mol^{-1}

Figure 4.8.   Average reactive cross section $\bar{\sigma}_R$ versus laboratory energy $\bar{E}$ for the reaction $H + C_4 D_{10} \rightarrow HD + C_4 D_9$. H–Xe interaction potential $\propto r^{-7}$ (see text). (Data from Gann, Ollison, and Dubrin.[18])

the case of (4.37), Gann et al. used xenon as the diluent and found that their results were not very sensitive to the exact form of the H–Xe potential. The plot of average reactive cross section versus laboratory energy shown in Fig. 4.8 for a potential proportional to r^{-7} is nearly identical to one based on a potential proportional to r^{-4}, a range that includes all potentials suggested for the H–Xe interaction. The most interesting feature of Fig. 4.8 is that the reactive cross section rises to a maximum at an energy not too far above the threshold and then decreases at higher energies. Because the average reactive cross sections in Fig. 4.8 were obtained as the differential increment to the total yield and that increment becomes smaller at higher energies, accurate results could not be obtained above $200 \, kJ \, mol^{-1}$. Consequently Fig. 4.8 could not be extended to higher energies to see whether additional maxima exist. The rise in σ_R as a function of energy lies between the step function cross section $\sigma_R = 0, E < E_0$; $\sigma_R = \pi d_{AB}^2, E \geqslant E_0$ and the line-of-centers cross section $\sigma_R = 0, E < E_0$; $\sigma_R = \pi d_{AB}^2(1 - E_0/E), E \geqslant E_0$. The falloff in σ_R at higher energies is not predicted by either of these simple models.

Vibrational energy of a reactant molecule can also influence the reactive cross section.[19] The data[20] in Fig. 4.9 for the reaction

$$H_2^+ + He \rightarrow HeH^+ + H \qquad (4.38)$$

show this effect clearly since the abcissa is the total energy (translational + internal). Thus differences among curves for different vibrational quantum numbers v can be attributed to dynamic rather than energetic factors. In this study photoionization of H_2 with photons of 0.014-nm bandwidth was employed to generate H_2^+ in specific vibrational states from $v = 0$ to $v = 5$. The kinetic energy of the ions was

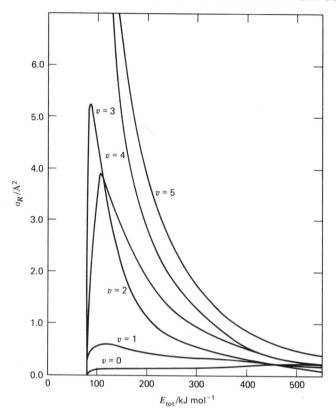

Figure 4.9.   Reactive cross section σ_R for $H_2^+ + He \rightarrow HeH^+ + H$ as a function of total energy $E_{tot}(= E_{trans} + E_{internal})$ for different vibrational states of H_2^+ ($v = 0$ to $v = 5$). [Reprinted with permission of Weizmann Science Press of Israel from *Isr. J. Chem.* **9**, 615 (1971), based on data from W. A. Chupta, J. Berkowitz, and M. E. Russell.[20]]

varied between 0 and 96 kJ mol^{-1} by modifying the repeller voltage in the ionization chamber of the mass spectrometer used for product analysis. This reaction is one of several[21] that have been shown to exhibit a vibrational threshold — for reactant states above $v = 3$ there is no translational threshold and cross sections are quite large.

Beam Methods[†]

The most detailed information regarding reaction cross sections can be obtained from experiments involving crossed molecular beams. At sufficiently low pressures the mean free path (4.15) for gas molecules becomes so large that the molecules obey the laws of geometric optics and can be collimated into beams. A schematic diagram of a typical molecular beam apparatus is shown in Fig. 4.10. Beams from

[†] For recent reviews see references 22–25.

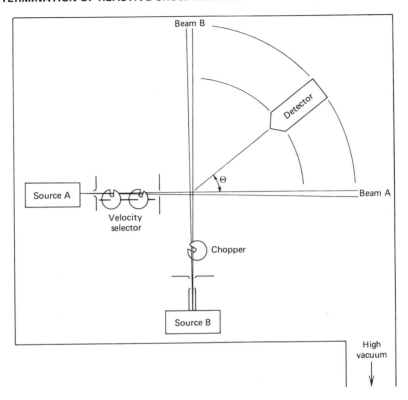

Figure 4.10.   Schematic diagram of apparatus for experiments on crossed molecular beams. Rotation of the detector about the point of intersection of the beams provides for measurements at various laboratory scattering angles Θ.

two sources intersect at a 90° angle, and reaction products are observed by means of a detector that can be moved in a circular path centered on the intersection point and lying in the plane defined by the two beams. With such an apparatus the total reactive cross section can be obtained by measuring the flux of product particles as a function of angle and integrating over all possible angles in three dimensions. Refinements of the crossed-beam technique permit the following additional types of experiments (in order of increasing detail): determination of total reaction cross section as a function of collision energy; determination of angular and/or velocity distribution of products; internal energy state selection of reactants and/or determination for products; alignment of reactants and/or alignment analysis of products (to determine steric factors). However, the greater the level of detail to be extracted from an experiment the smaller the fraction of all involved particles that reaches the detector.

A true molecular beam must be sufficiently dilute that molecular collisions do not occur within it. Otherwise the molecules would be diverted from their straight-line paths and the beam would diverge. Therefore detector sensitivity is a major factor influencing what kinds of experiments can be done and how much information can be extracted from a reaction system. Most of the earliest beam studies of chemical reactions involved alkali metals because atoms of these metals can be detected by surface ionization.[26] When an alkali metal atom strikes a hot wire, the atom loses an electron and a positive ion leaves the wire. The positive ion current provides a sensitive means for detecting alkali atoms, whether they are free or combined in molecular species. Mass spectrometry[27] is not limited to a particular type of product molecule and therefore is widely used. If one of the reactant beams is periodically interrupted by a chopper, the time of flight of product molecules to the mass spectrometer provides information about the velocity distribution of products. Other detection methods include infrared luminescence,[28] which is especially valuable for analysis of product-state distributions, and laser-induced fluorescence,[29] in which a tunable laser is swept in wavelength, inducing transitions of product molecules to excited states that can be detected by the wavelengths of their fluorescence emission. Short-lived radioisotopes may also be used in one of the reagent beams, and their presence in the product beam can be detected by counting of radioactive disintegrations.[30] This technique provides an increase in sensitivity of detection of several orders of magnitude if a suitable isotope is available.

In some cases, such as that of the alkali metals, molecular beams can be formed by merely heating the reactant substance in an oven. Beams of H, D, or halogen atoms can be produced by thermal dissociation of H_2, D_2, or halogen molecules. Beams of O or N atoms require a radiofrequency discharge to dissociate the molecules. For molecular species that are permanent gases supersonic expansion through a pinhole nozzle from a source at about 100 torr into a vacuum provides a beam of much more uniform direction and much narrower velocity distribution than the Maxwell-Boltzmann distribution that results from the thermal sources listed above. Nozzle expansion has a further advantage. Since collisions in a nozzle bring all molecules to nearly the same velocity, "seeding" the beam with a large excess of a lighter nonreactive gas, such as He, increases the translational energies of the reactive gas molecules in the ratio of the molar masses. In the case of O_2 seeded with He, for example, an eightfold increase is attained. Velocity selection in one or both reactant beams can be accomplished by means of slotted rotating disks whose orientation or speed can be varied.[31] Internal states of the reactants can be selected by laser pumping of vibrational states,[32] or, for polar molecules, by using static electric fields to deflect or focus molecules in particular rotational states or orientations.[33]

Teloy and Gerlich[34] have measured the total (integrated) cross section for the reaction

$$Ne^+ + CO \rightarrow Ne + C^+ + O$$

as a function of kinetic energy of reactants. In their apparatus neon ions produced by electron impact are stored for about 1 ms by means of an inhomogeneous radiofrequency field, greatly narrowing their kinetic energy distribution. The ions are then passed through a mass and velocity filter, down a flight tube where they are contained by rf "walls," and into a scattering chamber that contains CO. Because of the very narrow energy distribution of Ne^+, thermal motion of CO is by far the most important factor causing a range of collision energies. After passing through the reaction chamber, reactant and product ions continue along the flight tube, guided by rf fields, and effuse into a mass spectrometer. Because all product ions are directed along the flight tube to the detector, this apparatus performs the

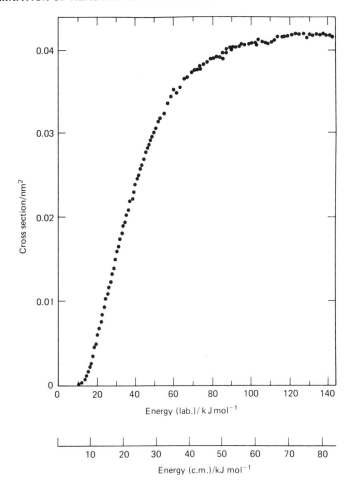

Figure 4.11. Cross section as a function of energy for the reaction $Ne^+ + CO \rightarrow Ne + C^+ + O$. [Reprinted with permission from E. Teloy and D. Gerlich, *Chem. Phys.*, **4**, 417 (1974).]

integration of differential cross sections that would be necessary in the case of the device shown in Fig. 4.10. Figure 4.11 shows the results of many experiments of the type just described as a function of kinetic energy of the reactant beam. The reaction exhibits a sharp threshold, presumably because there is total dissociation. A closely spaced series of thresholds corresponding to different vibrational and rotational states of the products is not possible because all products are atomic.

More detailed information about the collision process can be obtained if the flux of product particles is measured as a function of the laboratory scattering angle Θ defined in Fig. 4.10. However, the collision can be better understood if it is described in terms of a coordinate system that is centered on and moves with the center of mass (also called the centroid) of the reactant and product particles. Such a coordinate system emphasizes the relative velocities of the colliding particles and

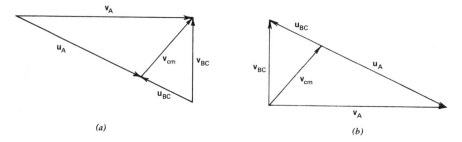

Figure 4.12. Vector diagram for collision of atom A with molecules BC. (*a*) Center-of-mass velocity vectors for the case in which the paths of A and BC are at right angles. (*b*) A Newton diagram of the same collision. All velocity vectors in each coordinate system have been shifted to a common origin.

permits a description of the collision in terms of the angles θ and φ defined in Fig. 4.5. Transformation from laboratory to center-of-mass coordinates is described in detail in several references.[35] Our procedure here is to develop a convenient diagrammatic representation of the collision process.

Consider a beam of atoms A that intersects a beam of molecules BC at right angles in the apparatus of Fig. 4.10. Assume that the reaction

$$A + BC \rightarrow B + AC$$

occurs as a result of some of the collisions. The velocity of each particle can be represented by a vector, and we shall adopt the convention that velocities in the center-of-mass coordinate system are represented by the letter $\mathbf{u}$, while velocities in the laboratory coordinate system are represented by $\mathbf{v}$. Then the approach and collision of reactants A and BC can be diagrammed as shown in Fig. 4.12*a*. In the laboratory coordinate system the vectors $\mathbf{v_A}$ and $\mathbf{v_{BC}}$ are at right angles, but to an observer traveling along with the center of mass the reactants appear to travel directly toward one another, as indicated by $\mathbf{u_A}$ and $\mathbf{u_{BC}}$. It is convenient and conventional to shift the velocity vectors of Fig. 4.12*a* so that all vectors in the same coordinate system have a common origin. A diagram in which this has been done is called a Newton diagram. An example is given in Fig. 4.12*b*. In both the Newton diagram and Fig. 4.12*a* the hypotenuse of the vector triangle represents the relative velocity of approach $\mathbf{r}$ of the two particles. That is,

$$\mathbf{r} = \mathbf{u_A} - \mathbf{u_{BC}} = \mathbf{v_A} - \mathbf{v_{BC}} \tag{4.39}$$

where the minus sign is needed to take account of the direction of the vectors. Also, the velocity vector of the center of mass $\mathbf{v_{cm}}$ is readily obtained from the velocities and masses of the reactants as

$$\mathbf{v_{cm}} = \frac{m_A \mathbf{v_A} + m_{BC} \mathbf{v_{BC}}}{m_A + m_{BC}} \tag{4.40}$$

Equations (4.39) and (4.40) can be combined to yield

$$\mathbf{u_A} = \mathbf{v_A} - \mathbf{v_{cm}} = \frac{m_{BC} \mathbf{r}}{m_A + m_{BC}} \tag{4.41a}$$

and

$$u_{BC} = v_{BC} - v_{cm} = \frac{m_A r}{m_A + m_{BC}} \qquad (4.41b)$$

entirely determining the vectors in each coordinate system, provided that v_A and v_{BC} are known.

Conservation of linear momentum requires that the velocity vector of the center of mass remain constant in magnitude and direction after the collision unless some external force is acting on the system of particles. This also requires that the kinetic energy $\frac{1}{2}(m_A + m_{BC})v_{cm}^2$ of the system remain unchanged. However, the translational energy E_t relative to the center of mass can change, as can the internal energy E_i of the particles. If the collision results in a chemical reaction, then it is also necessary to account for the difference ΔE_0 between dissociation energies of products and reactants, measured from zero-point energy levels. Conservation of energy requires that

$$E_t + E_i = E_t' + E_i' + \Delta E_0 \qquad (4.42)$$

where the primes indicate conditions after collision. For an elastic collision $\Delta E_0 = 0$ and $E_i = E_i'$, so that the relative translational energy and hence the magnitude of the relative velocity vector must remain the same. However, the direction of the relative velocity vector can change, as diagrammed in Fig. 4.13. Once the collision occurs the particles are no longer constrained to travel in the plane defined by the original beams. Assuming a uniform flux of both A and BC within their respective beams and random orientation of the axis of BC, all values of the azimuthal angle φ of Fig. 4.5 are equally probable, and scattering at a particular center-of-mass angle θ is equally likely along all rays of a cone whose axis corresponds to the original relative velocity vector. Such a cone intersects the plane of the molecular beam apparatus in which the detector moves at four different laboratory angles Θ; two of these correspond to paths of one of the product particles and two to paths of the other. By convention the center-of-mass scattering angle θ for each

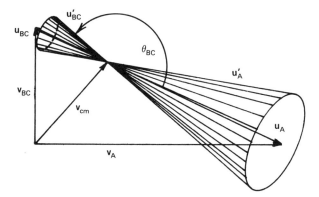

Figure 4.13. Elastic scattering. The angle θ_{BC} between the relative velocity vector u_A before collision and the vector u_{BC}' after collision is the center-of-mass scattering angle for species BC.

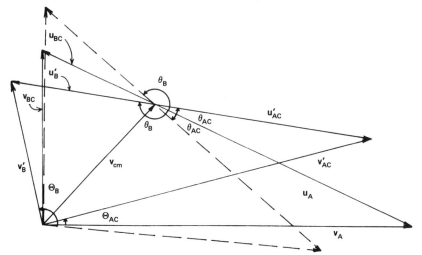

Figure 4.14.   Velocity-vector diagram showing the relationship between laboratory velocities (v and v') and center-of-mass velocities (u and u') in the plane of the crossed molecular beams diagrammed in Fig. 4.10. Dashed lines denote the third and fourth intersections of the cone of the product paths (see Fig. 4.13) with the plane of the molecular beam detector.

product particle is measured from the center-of-mass velocity vector of the atomic reactant A. Consequently the sum $\theta_A + \theta_{BC}$ of the scattering angles must be $\pi/2$. Similarly, the laboratory scattering angle Θ for any product is measured from the laboratory velocity vector of the atomic reactant.

When a collision results in chemical reaction, the Newton diagram shown in Fig. 4.14 applies. The relative product velocity vector $r' = u'_B - u'_{AC}$ is no longer equal in length to the relative reactant velocity vector, since internal energy changes and/or reaction endo- or exoergicity may affect the energy balance of (4.42). The cone-shaped distribution of product paths shown in Fig. 4.13 still remains, however, and for each center-of-mass scattering angle θ, two laboratory scattering angles Θ are observed. Conservation of momentum can be used to partition the relative product velocities in the center-of-mass coordinate system,

$$m_{AC}u'_{AC} + m_B u'_B = 0 \qquad (4.43)$$

requiring that u'_{AC} and u'_B be opposite in sign and have magnitudes inversely proportional to the corresponding masses. Hence only one of the two products need be observed — the velocity of the other can be calculated from (4.43). The observed (laboratory) velocity vector v'_{AC} for species AC is seen from Fig. 4.14 to be the sum of the velocity v_{cm} of the center of mass (which is unchanged by the collision and can be calculated using equation 4.40) plus the relative velocity u'_{AC}. Thus u'_{AC} can be calculated from

$$u'_{AC} = v'_{AC} - v_{cm}$$

provided that both the magnitude and direction of v'_{AC} can be determined experimentally. This requires analysis of the speed of the product particle as well as measurement of the angle Θ_{AC}. As shown in Fig. 4.14 the same scattering angle

θ_{AC} corresponds to two different values of Θ_{AC}. The speed of AC particles differs at these two angles. Conversely, different scattering angles θ_{AC} can result in observation of product flux at the same laboratory angle Θ_{AC}, but the different scattering angles can be identified because they result in different speeds for the AC molecules. In addition to permitting completion of the Newton diagram for a collision, determination of the speeds of product particles has a second advantage. Since the average internal energies of reactants are usually available, velocity analysis of the products provides the last piece of information necessary to extract E_i', the internal energy of the products, from the energy balance equation (4.42).[†]

When both the speed and angle of scattering of products can be measured, results are usually reported as a contour diagram in a polar coordinate system. The contour lines represent loci of equal product flux, the product speed u' is represented by distance from the origin, and the center-of-mass scattering angle θ is represented by the polar angle in the diagram. An example is shown in Fig. 4.15 for the reaction[23]

$$H + Cl_2 \rightarrow HCl + Cl$$

In this case θ is measured from the velocity vector of H (that is, H approaches from

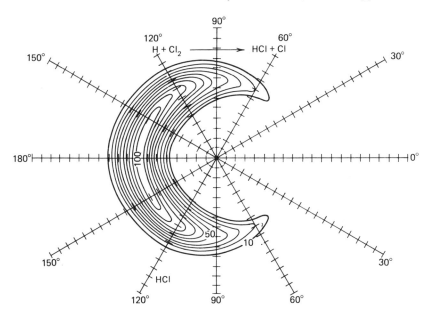

Figure 4.15. Contour diagram showing product flux velocity-angle distribution of HCl produced by reaction of H with Cl_2. Angles are measured from the relative velocity vector of H. Conservation of momentum requires that the contours representing flux of Cl atoms be nearly symmetric (by inversion) with those for HCl molecules, so the Cl contours have not been shown. Each tick mark along a radial line represents an increment in velocity of 200 m s^{-1}. [Reprinted with permission from D. R. Herschbach, *Pure Appl. Chem.*, **47**, 61 (1976).]

[†] A good discussion of the advantages of product velocity analysis is given by Levine and Bernstein in reference 24, pp. 191–203.

the left and Cl_2 from the right). The diagram is symmetric from top to bottom because there is no dependence of scattering on the azimuthal angle φ of Fig. 4.5.

It is clear from Fig. 4.15 that the product angular distribution is anisotropic. HCl recoils backward with respect to the incoming H atom. Conservation of momentum requires that Cl recoil in the opposite direction, although this is not shown explicitly. Any highly anisotropic velocity-angle distribution such as this implies that the colliding species do not remain in contact long enough for rotational motion of the collision complex to cause them to "forget" the original direction of relative approach. Thus the lifetime of the complex must be less than about 10^{-13} s, a typical period for rotation. Any reaction that takes place on such a short time scale is referred to as *impulsive* or *direct*. The backward recoil of the HCl in this specific case suggests that the preferred orientation of the atoms at the onset of reaction is collinear H—Cl—Cl. The large recoil velocity of HCl (about $1600 \, m \, s^{-1}$ in the region of maximum flux) indicates that about half of the $189 \, kJ \, mol^{-1}$ difference in bond energies between HCl and Cl_2 is released as translational energy of the HCl and Cl product particles. The remainder appears as vibrotational excitation of HCl and produces infrared chemiluminescence.[36] Such a reaction is said to have a *rebound* mechanism. The cross section for a rebound process is usually small because a nearly head-on collision is required. That is, only collisions having small impact parameters result in reaction.

Other reactions yield quite different velocity—angle distributions of products. Figure 4.16 shows results obtained by Gillen, Rulis, and Bernstein[37] for the reaction

$$K + I_2 \rightarrow KI + I$$

In this case most of the product KI molecules are scattered in the same direction as the relative velocity of K prior to collision. Such a process is referred to as a *stripping* reaction, and it has a relatively large cross section ($> 1 \, nm^2$). The most probable recoil velocity is small relative to the maximum possible value and so most of the product KI molecules must be in highly excited vibrotational states. This is in accord with the so-called harpoon mechanism,[38] which can be written as

$$K + I\text{—}I \rightarrow K^+ + I^-\text{—}I \rightarrow K^+\text{—}I^- + I$$

As the alkali metal atom approaches the halogen molecule a distance is reached at which an electron transfers from K to I_2. The force between the charged species is strongly attractive and the I atom separates from the K^+—I^- molecule. The cross section for such a process can be quite large because the transferred electron serves as a "harpoon" by which the alkali metal hooks the halogen atom and pulls it in with Coulombic attraction.

Mechanisms intermediate between the rebound and stripping cases also exist. For example, scattering of DI from the reaction

$$D + I_2 \rightarrow DI + I$$

occurs principally at a center of mass angle $\theta \cong 105°$.[39] Similar results are obtained

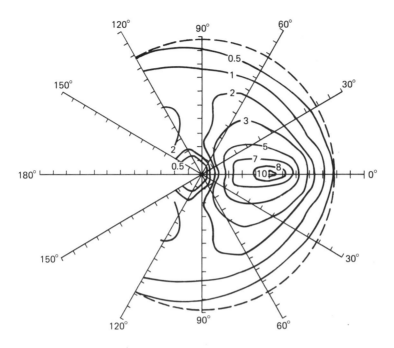

Figure 4.16. Product flux velocity-angle distribution for KI produced by the reaction $K + I_2 \rightarrow KI + I$. The dashed line corresponds to the highest recoil velocity permitted by conservation of energy (that is, to KI molecules in their lowest internal energy state). Angles are measured from the relative velocity vector of K. Tick marks on radial lines are at intervals of 10 m s^{-1}. [Reprinted with permission from K. T. Gillen, A. M. Rulis, and R. B. Bernstein, *J. Chem. Phys.*, **54**, 2831 (1971).]

with the interhalogens IBr and ICl. Such results can be interpreted in terms of a bent geometry of the DI_2 transition state. Since the D atom is so much lighter than I_2, the two heavy atoms should not move much during the D-atom attack. The reaction is very much like a photodissociation, with D serving as the photon. (A similar situation obtains in the case of the $H + Cl_2$ reaction discussed earlier. A contour map for photodissociation of Cl_2 is nearly identical to Fig. 4.15.)[23] The product DI and I should therefore separate along the direction of the breaking I–I bond, and the angular distribution of HI product should reflect the orientation of the I_2 molecule at the instant of collision. Thus in reactive collisions the D atom must have approached from a direction nearly perpendicular to the I-I bond.

By contrast with direct reactions such as those discussed above, there are situations where a reactive collision lasts long enough for the collision complex to undergo vibrations and rotations. In such an *indirect* process the product velocity–angle distribution gives information about the unimolecular dissociation of a *persistent complex*, and the contour map is symmetric with respect to frontward versus backward scattering as shown in Fig. 4.17.[23] In this example, the reaction

$$Cl + CH_2 = CHBr \rightarrow CH_2 = CHCl + Br$$

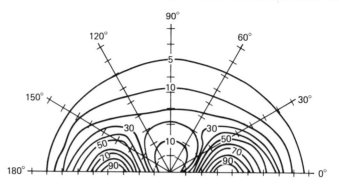

Figure 4.17. Product flux velocity-angle distribution for CH_2CHCl produced by the reaction $Cl + CH_2CHBr \rightarrow CH_2CHCl + Br$. Angles are measured from the relative velocity vector of Cl, and each tick mark represents a velocity increment of $100\ m\,s^{-1}$. Since the contour diagram must be symmetric from top to bottom, only the top half is shown. In this case, left-to-right symmetry implies a persistent complex. [Reprinted with permission from D. R. Herschbach, *Pure Appl. Chem.*, **47**, 61 (1976).]

the persistent complex is the chlorobromoethyl radical,

$$\begin{array}{c} Cl \\ | \\ H-C-C \end{array} \begin{array}{c} H \\ \diagup \\ \diagdown \\ Br \end{array}$$

a known, stable species. This radical is vibrationally excited by about $125\ kJ\,mol^{-1}$, and because no subsequent collisions occur during the beam experiment, the vibrational energy remains trapped and becomes randomized prior to dissociation of the complex. The situation is analogous to that of an activated molecule in a unimolecular reaction, and the product energy distribution can be predicted using the theory of unimolecular reactions developed later in this chapter.

The reason that the contours in Fig. 4.17 are not circles centered on the origin (corresponding to a completely random angular distribution) arises from the rotational motion of the collision complex.[23] To simplify the explanation, suppose that all atoms of the complex lie on the line defined by the bond that breaks when the complex dissociates. Then any rotational motion of the complex must be around an axis perpendicular to the relative velocity vectors of the approaching reactants (see Fig. 4.18a). However, all axes in the plane perpendicular to the direction of approach of the reactants are equally probable, and so any effect of rotation must be averaged over all orientations of the rotation axis. The effect of rotation about an axis perpendicular to the line of centers upon dissociation is a centrifugal one — product particles are released with extra translational energy as a result of the rotational motion. Moreover, products are uniformly distributed over a circle in the plane perpendicular to the rotation axis. The overall product distribution corresponds to the intersections of circles of latitude and longitude on a globe whose north–south axis is the original direction of approach of the reactants (Fig.

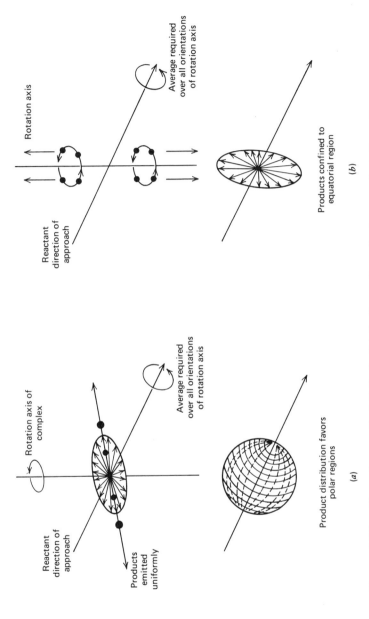

Figure 4.18. Dependence of statistical distribution of products on orientation of rotation axis of persistent collision complex. (*a*) Rotation axis perpendicular to line of separation of products. (*b*) Rotation axis parallel to line of separation (spinning component of rotation).

4.18*a*). Consequently the product intensity is higher in the polar than in the equatorial regions, and it is possible to account for the maxima at $0°$ and $180°$ in Fig. 4.17.

If all atoms of the collision complex do not lie on the line of dissociation, a spinning component of rotation that makes no centrifugal contribution is also possible. Figure 4.18*b* shows that such a spinning rotation alone leads to a heavy concentration of products in the plane perpendicular to the line of approach of reactants. For a multiatom complex such as the chlorobromoethyl radical both centrifugal and spinning components of rotation are present, but the experimental results of Fig. 4.17 imply that the former is more important.

The product flux velocity–angle distribution for the reaction of chlorine atoms with allyl bromide,

$$Cl + CH_2{=}CHCH_2\,Br \rightarrow ClCH_2\,CH{=}CH_2 + Br$$

is shown in Fig. 4.19. This differs from the contour diagram for the reaction of vinyl bromide in that here the frontward peak is roughly three times as intense as the backward peak. This has been interpreted in terms of an "osculating complex" model,[40] and the data indicate that dissociation occurs (on average) after half a rotation (about 10^{-12} s). This behavior suggests a difference in mechanisms of the vinyl bromide and allyl bromide reactions with atomic chlorine.[41] A reasonable hypothesis is that the vinylic reaction proceeds by atom migration:

$$Cl + {=}\underset{Br}{\diagup} \;\rightarrow\; \overset{Cl}{\diagdown}\underset{Br}{\diagup}\cdot \;\rightarrow\; \overset{Cl}{\triangle}\underset{Br}{} \;\rightarrow\; {=}\diagup^{Cl} + Br$$

while the allylic reaction involves bond migration:

$$Cl + \diagup\!\diagup\!\diagdown^{Br} \;\rightarrow\; Cl\diagdown\!\diagup\!\diagdown Br \;\rightarrow\; Cl\diagdown\!\diagup\!\diagup\!\diagup + Br$$

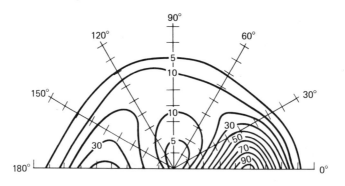

Figure 4.19. Product flux velocity-angle distribution for allyl chloride produced by the reaction $Cl + CH_2\,CHCH_2\,Br \rightarrow ClCH_2\,CHCH_2 + Br$. Angles are measured from the relative velocity vector of Cl, and each tick mark represents a velocity increment of $100\,\mathrm{m\,s^{-1}}$. [Reprinted with permission from D. R. Herschbach, *Pure Appl. Chem.*, **47**, 61 (1976).]

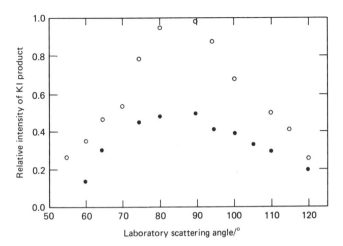

Figure 4.20. Laboratory angular distribution of KI produced by reaction of K with oriented CH_3I molecules. (○) K approaching I end of CH_3I. (●) K approaching CH_3 end of CH_3I. (Data from Brooks.[43])

Presumably the Cl atom migration should be much slower than the electron migration, accounting for the statistical distribution of products in the vinylic reaction. Many other applications of beam studies to the elucidation of reaction mechanisms have been reviewed by Hershbach.[23]

Molecular beam experiments also provide a means for measuring the steric factor p directly. Specific orientations of the rotation axis of a symmetric top molecule, such as CH_3I, $CHCl_3$, CF_3I, or PH_3, can be selected by passing a molecular beam along the axis of a six-pole electric field.[42,43] The molecules can then be oriented by a weak dipolar field so that the distribution of orientations that remains results in one or the other end of the rotation axis being closer to the incoming beam containing the second reactant. When this is done for the reaction

$$K + CH_3I \rightarrow CH_3 + KI$$

the results shown in Fig. 4.20 are obtained.[44] Clearly the orientation in which K approaches the I end of CH_3I is more reactive than the one in which steric hindrance by CH_3 prevents K from colliding directly with I. The steric factor obtained from these results is $p = 0.5$, and experimental results for the reactions of K with methyl iodide and t-butyl iodide can be accounted for by a model based on van der Waals radii.[43,45] However, the angular distribution of product is less easily explained. The fact that maxima occur in Fig. 4.20 at about the same laboratory angle for both orientations of CH_3I shows that KI is always scattered back towards the direction from which the alkali atom approached, whether or not K preferentially strikes the I end of the CH_3I molecule. No completely satisfactory model for this process is currently available.[43,46]

Exchange of internal (vibrational and rotational) energy with translational energy requires either a collision or a unimolecular dissociation. Because there are very few collisions at the low pressures under which molecular beam experiments are carried out, it is possible to prepare reactants with non-Boltzmann distribution of internal energy. Product molecules can also be analyzed to determine nascent distributions of internal energy produced immediately after a reactive collision, because such distributions cannot be thermalized rapidly by further collisions. An excellent example of the opportunities this aspect of beam studies affords for further elucidation of mechanistic details is the study of the reaction

$$Ba + HF \rightarrow BaF + H$$

by Pruett and Zare.[47] They permitted a collimated beam of Ba atoms to enter a reaction chamber containing HF at a pressure of 10^{-4} torr. The product BaF was observed by means of laser-excited fluorescence.[48] A tunable dye laser whose output was flat in the vicinity of 500 nm was focused on the beam of Ba in the reaction chamber, exciting fluorescence in the $C^2\Pi_{1/2} - X^2\Sigma^+$ band system. Since the laser-excited BaF molecules undergo fluorescence emission much more rapidly than they undergo changes in vibrational energy, peak heights in the observed fluorescence spectrum can be related to the populations of nascent vibrational states of BaF. These populations correspond to rotationally and translationally averaged reaction rates for transformation of HF in its ground vibrational state (which was the only one appreciably populated under the conditions of the experiment) to BaF in each of the observed vibrational states. Pruett and Zare's results are shown in Fig. 4.21.

The effect of vibrational excitation of the reactant HF was determined by employing a second laser to excite about 1% of the HF molecules to the $v = 1$ state. The difference in fluorescence intensity between spectra measured when this second laser was on and when it was off yielded the reaction rates shown in Fig. 4.21 for

$$Ba + HF(v = 1) \rightarrow BaF(v = 3{-}12) + H$$

The distribution of product vibrational energy is very dependent on the vibrational excitation of HF, with 57% of the excess reactant vibration appearing as product vibration. However, this fraction is unusually low. In other systems excess reactant vibrational energy is entirely retained as product vibrational energy.[49] Speculation about the reason for this unusually low retention of vibrational energy requires a consideration of possible trajectories of the reaction system across a potential energy hypersurface of the type discussed in Chapter 5.

ARRHENIUS ACTIVATION ENERGY, FREQUENCY FACTOR, AND EXCITATION FUNCTION

The variation of the reactive cross section with energy is referred to as the excitation function. Experiments described in the preceding section have demonstrated that excitation functions for different reactions can vary significantly, usually rising from zero at some threshold energy E_0 to a maximum at higher energy, and then often decreasing at even greater energies. Equation (4.27) has already demonstrated that, in the case of the hard-sphere line-of-centers excitation function, the Arrhenius activation energy E_a is closely related to but not the same as the molar threshold energy LE_0. We now consider the relationship between Arrhenius

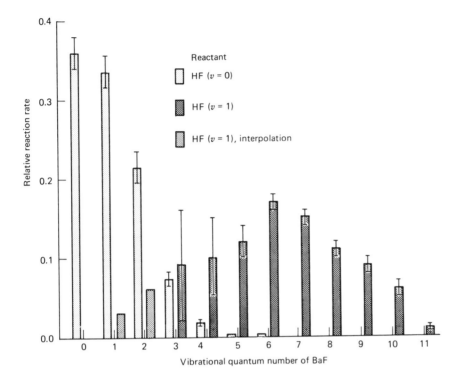

Figure 4.21. Bar graph showing state-to-state reaction rates for Ba + HF ($v = 0, 1$) → BaF ($v = 0, 1, 2, \ldots, 11$) + H. Error bars indicate uncertainties in the relative rates. Values for production of BaF ($v = 1, 2$) from HF ($v = 1$) are estimated by linear interpolation between the experimental value for BaF ($v = 3$) and an assumed value of zero for BaF ($v = 0$). (Data from Pruett and Zare.[47])

activation energy, frequency factor, and excitation function with greater generality, following the treatment of Truhlar.[50, 51]

The activation energy is obtained experimentally from the slope of a plot of ln k versus $1/T$, and is therefore defined according to (2.42):

$$\frac{d \ln k}{dT} = \frac{E_a}{RT^2} \tag{2.42}$$

Assuming Boltzmann distributions of translational energy for both colliding particles, but leaving completely general the energy dependence of the reactive cross section σ_R, leads to (4.24) in the manner described earlier in this chapter.

$$k = L \left[\frac{8}{\pi\mu(RT/L)^3} \right]^{1/2} \int_0^\infty E e^{-LE/RT} \sigma_R \, dE \tag{4.24}$$

Here σ_R is, of course, an average over all participating internal energy states. Taking natural logarithms of both sides of (4.24), differentiating with respect to T, and substituting in (2.42) yields

$$E_a = -\frac{3}{2}RT + L\frac{\int_0^\infty E^2 e^{-LE/RT}\sigma_R\,dE}{\int_0^\infty Ee^{-LE/RT}\sigma_R\,dE} + RT^2\frac{\int_0^\infty Ee^{-LE/RT}(d\sigma_R/dT)\,dE}{\int_0^\infty Ee^{-LE/RT}\sigma_R\,dE}$$

$$(4.44)$$

In the special case where σ_R is independent of temperature (that is, where the cross sections of different internal energy states are all the same) the last term in (4.44) drops out. Then

$$E_a = L\frac{\int_0^\infty E^2 e^{-LE/RT}\sigma_R\,dE}{\int_0^\infty Ee^{-LE/RT}\sigma_R\,dE} - \frac{3}{2}RT$$

a result originally derived by Tolman in a slightly different way.[52] The term $3/2RT$ is the average molar energy of all collisions, whether or not reaction occurs, and the quotient of integrals is the average molar energy of all collisions that result in reaction. That is, in the absence of effects due to internal energy, E_a is the difference between the average energies of two Maxwell-Boltzmann distributions — one over reactive collisions and the other over all collisions. Both distributions and the excitation function σ_R for the hard-sphere line-of-centers case are shown in Fig. 4.22.

The energy distribution of reactive collisions is the product of the Maxwell-Boltzmann energy distribution for all collisions times the excitation function, and therefore E_a varies depending on the form of σ_R versus E, even though the temperature and threshold energy remain the same. LeRoy[53] has derived analytic expressions relating E_a and σ_R for several general forms of σ_R as a function of E. These results are summarized in Table 4.1. The forms of σ_R labeled class I and class II both increase from zero at $E = E_0$ to a maximum value and then decrease, the rate of increase depending on n and the rate of decrease on m. The parameters m and n can be adjusted to fit the class I function to experimental observations for reactions involving neutral and some ionic species. Also, when $m = n = 0$, the class I function reduces to a step function corresponding to hard-sphere collisions without regard to whether velocity is directed along the line-of-centers. Class II functions can also be fitted to the same experimental data as class I, and the primary difference is that class II includes the line-of-centers excitation function when $m = 0$ and $n = 1$. Another special case of class II, $m = 0$ and $n = 1/2$, yields the result that the molar threshold energy exactly equals the Arrhenius activation energy.

Assuming that E_a is independent of temperature, (2.42) can be rewritten as

$$k = Ae^{-E_a/RT} \qquad (2.43)$$

This can be combined with (4.24) to yield a general form for the frequency factor

$$A = L\left[\frac{8}{\pi\mu(RT/L)^3}\right]^{1/2}\int_0^\infty Ee^{(E_a-LE)/RT}\sigma_R\,dE \qquad (4.45)$$

This expression has no simple conceptual interpretation, but it does show that the

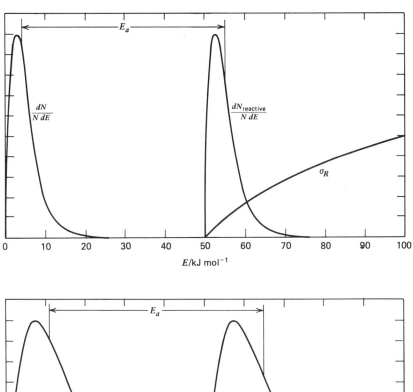

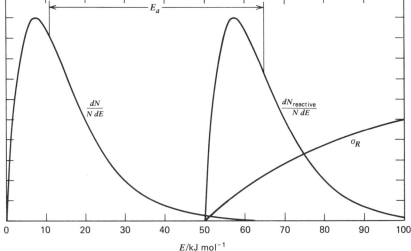

Figure 4.22. Relationship between threshold energy LE_0 and Arrhenius activation energy E_a. Maxwell-Boltzmann distribution of all collisions (dN/NdE) and of reactive collisions (dN_{reactive}/NdE) are shown at temperatures of: (a) 300 K; (b) 900 K. The threshold energy $LE_0 = 50\,\text{kJ mol}^{-1}$ is the same in both parts of the figure, as is the form of the reactive cross section $\sigma_R = (E - E_0)/E$, which is also plotted. Note that E_a is temperature dependent: at 300 K $E_a = 51.25\,\text{kJ mol}^{-1}$, while at 900 K $E_a = 53.74\,\text{kJ mol}^{-1}$. All functions are arbitrarily normalized.

frequency factor depends on the magnitude of σ_R, as well as on the functional form of the variation of σ_R with E.

When different internal energy states make different contributions to the rate of reaction, the third term in (4.44) must also be considered. Truhlar[50] has shown that

Table 4.1 Arrhenius activation energy for several forms of reactive cross section as a function of energy

Class	Form of σ_R for $E \geqslant E_0$ [a]	Expression for E_a
I	$\sigma_R = C(E - E_0)^n e^{-m(E-E_0)}$ $m \geqslant 0, \quad n \geqslant 0$	$E_a = LE_0 + RT(n + 1/2) - \dfrac{m(n+2)(RT)^2}{1 + mRT} - \dfrac{LE_0}{(n+1) + (1+mRT)LE_0/RT}$
II	$\sigma_R = C\dfrac{(E - E_0)^n}{E} e^{-m(E-E_0)}$ $m \geqslant 0, \quad n \geqslant 0$	$E_a = LE_0 + RT(n - 1/2) - \dfrac{m(n+1)(RT)^2}{1 + mRT}$
III[b]	$\sigma_R = CE^n$	$E_a = RT(n + 1/2) + LE_0 \dfrac{(LE_0/RT)^{n+1} e^{-LE_0/RT}}{\Gamma(n+2) - \gamma(n+2)LE_0/RT}$

Source. Data from LeRoy.[53]

[a] $\sigma_R = 0$ for $E < E_0$ in all cases; C represents a constant of proportionality.

[b] $\Gamma(x)$ and $\gamma(x,y)$ are the gamma function and the incomplete gamma function; see M. Abramowitz and I. A. Stegun, Eds., *Handbook of Mathematical Functions*, Dover, New York, 1965, pp. 260–267.

this term corresponds to the difference between the average molar internal energy of reacting pairs of colliding molecules and that of all collision pairs of reagents. That is,

$$E_a = \langle LE_{\text{trans}} + LE_{\text{internal}} \rangle_{\text{reactive collisions}} - \langle LE_{\text{trans}} \rangle_{\text{all collisions}}$$
$$- \langle LE_{\text{internal}} \rangle_{\text{all collisions}}$$

where values enclosed in angle brackets are averages over Boltzmann distributions. This result assumes that both the relative translational energies and the internal energies of reactant molecules correspond to the equilibrium Boltzmann distributions. As long as the reactive cross section is an order of magnitude or more smaller than the cross section for all collisions, which is the case for the majority of reactions, the condition of equilibrium distribution should be met.[54] In a few cases, for example, dissociation of diatomic molecules in shock tubes[55] and unimolecular reactions in the low-pressure limit, the more energetic molecules may be removed faster than they can be supplied by energy-exchange collisions. In such cases the observed steady-state rate constant is less than expected, and Tolman's interpretation of the Arrhenius activation energy does not apply.

UNIMOLECULAR REACTIONS

Since a collision involves at least two molecules, it would appear that a unimolecular reaction must be one that takes place without collision, perhaps as a spontaneous disruption or transformation of the reacting molecule. Typical examples are dissociations such as

$$C_4H_8 \text{ (cyclobutane)} \rightarrow 2C_2H_4$$

and isomerizations such as

$$CH_3NC \rightarrow CH_3CN$$

Let the probability of such a spontaneous reaction in time dt be $k\,dt$, where k is a constant characteristic of the molecule and the temperature. For a large number of reacting molecules the fraction that reacts in time dt can be set equal to the probability $k\,dt$, where $-dN$ is the number reacting

$$\frac{-dN}{N} = k\,dt \qquad (4.46)$$

or

$$\frac{-dN}{dt} = kN$$

Dividing both sides by the volume and the Avogadro constant yields

$$\frac{-dc}{dt} = kc \qquad (4.47)$$

where c is the concentration. This is the familiar first-order rate equation with rate constant k. It shows that a unimolecular reaction is expected to be first order and that the rate constant can be interpreted as a probability of reaction per unit time. Simple radioactive disintegrations follow this same law with constants characteristic of the particular disintegrations but independent of temperature.

The Lindemann Mechanism

In first-order chemical reactions suspected of being unimolecular, the temperature is of great importance, the rate constant following an exponential Arrhenius expression. In unimolecular reactions energy must be supplied to activate the molecule, whereas in a radioactive disintegration the energy is already on hand in the nucleus. The manner in which the molecules become activated was long a mystery, because activation by collision was at first ruled out. It was thought (incorrectly) that because collision processes are bimolecular the overall reaction would also have to be bimolecular and hence second order. In 1919 Perrin[56] proposed that molecules could be activated by means of absorption of infrared radiation emitted by the walls of the container, but it was soon demonstrated that the density of radiation was insufficient to account for the reaction rates and that unimolecular reactions did not occur in molecular beams where there were no collisions.[57] All current theories of unimolecular reactions are based on the idea that such reactions involve a multistep mechanism. This idea was published almost simultaneously by Lindemann[58] and Christiansen,[59] and it is referred to as the Lindemann mechanism.

For a reaction whose stoichiometry is $A = B + C$ the Lindemann mechanism is given below,

1. *Activation by Collision:* $\quad A + A \xrightarrow{\ k_1\ } A^* + A$ $\hspace{4cm}$ (4.48a)

where A^* is a molecule that has sufficient energy for reaction. The rate constant for this bimolecular process is k_1.

2. *Deactivation by Collision:*

$$A^* + A \xrightarrow{\ k_2\ } A + A \hspace{4cm} (4.48b)$$

This is the reverse of the activation process and is expected to occur at the first collision of A^* after it is formed, if it has not already reacted in the meantime. Because the rate of the activation processes is limited by the energy requirement, $k_2 \gg k_1$.

3. *Spontaneous Reaction of an Activated Molecule:*

$$A^* \xrightarrow{\ k_3\ } B + C \hspace{4cm} (4.48c)$$

This is a "true" unimolecular process in the same sense as a radioactive disintegration, and k_3 is a first-order rate constant.

On the basis of the Lindemann mechanism the following expressions for the rates of formation of A^* and B apply:

$$\frac{d[A^*]}{dt} = k_1 [A]^2 - k_2 [A^*][A] - k_3 [A^*] \hspace{3cm} (4.49)$$

$$\frac{d[B]}{dt} = k_3 [A^*] \hspace{4cm} (4.50)$$

It is difficult to solve these equations accurately, but an excellent simplifying approximation is to assume that there is a *steady state*[†] where the concentration of A^*, which is always very small, does not change with time. That is,

$$\frac{d[A^*]}{dt} = 0$$

This gives from (4.49)

$$k_1[A]^2 = (k_2[A] + k_3)[A^*] \tag{4.51}$$

Solving this for $[A^*]$ and substituting into (4.50) yields

$$\frac{d[B]}{dt} = \frac{k_3 k_1[A]^2}{k_2[A] + k_3} \tag{4.52}$$

Most techniques for measuring concentrations in the gaseous reaction mixture determine $[A] + [A^*]$, but this is nearly identical to $[A]$ because $[A^*]$ is very small.

Equation (4.52) shows that the expected rate is neither first nor second order with respect to A; however, there are two limiting cases where it does become so. At high pressure where $[A]$ is large so that $k_2[A] \gg k_3$, (4.52) simplifies to

$$\frac{d[B]}{dt} = \left(\frac{k_3 k_1}{k_2}\right)[A] \tag{4.53}$$

Under this condition a first-order rate is expected, despite the fact that bimolecular processes occur in activation and deactivation. At low pressure where $[A]$ is small so that $k_2[A] \ll k_3$, (4.52) simplifies to

$$\frac{d[B]}{dt} = k_1[A]^2 \tag{4.54}$$

so that the reaction becomes second order. Here the rate of reaction is the rate of activation. Deactivation is unimportant because there is enough time between collisions for essentially all A^* to decompose to $B + C$ before undergoing a deactivating collision.

It is necessary to generalize this mechanism to allow for activation and deactivation by collision with molecules other than reactant A. The symbol M is used to designate any such molecule — a reactant, a product, or an added nonreactive gas molecule. Assuming that the efficiency of the various molecules in causing activation or deactivation is not very different,[‡] all such effects can be designated by the mechanism:

$$A + M \rightarrow A^* + M \qquad k_1$$

$$A^* + M \rightarrow A + M \qquad k_2$$

$$A^* \rightarrow B + C \qquad k_3$$

[†] See Chapter 8 for further discussion of the steady-state approximation.

[‡] This depends on the efficiency of energy transfer between reactant A and the various other molecules. See reference 60 for a review.

Now applying the steady-state approximation gives

$$k_1 [M] [A] = (k_2 [M] + k_3)[A^*] \tag{4.55}$$

and (4.52) becomes

$$\frac{d[B]}{dt} = \frac{k_3 k_1 [M] [A]}{k_2 [M] + k_3} \tag{4.56}$$

This has high- and low-pressure limiting expressions similar to (4.53) and (4.54). If [M] is approximately constant during a run, as when the product compensates for disappearing reactant as activator and deactivator, or when a foreign gas is present, (4.56) is that of a pseudo-first-order reaction. Define k as

$$k = \frac{k_3 k_1 [M]}{k_2 [M] + k_3} \tag{4.57}$$

so that

$$\frac{d[B]}{dt} = k[A] \tag{4.58}$$

Here k is the observed first-order rate constant. It also has limiting values for high and low pressure:

High pressure

$$k_2 [M] \gg k_3$$

$$k = k_\infty = \frac{k_3 k_1}{k_2} \tag{4.59}$$

Low pressure

$$k_2 [M] \ll k_3$$

$$k = k_0 = k_1 [M] \tag{4.60}$$

Thus k has a value that decreases if the pressure is lowered sufficiently. For the purpose of verifying this theory, (4.57) can be inverted to give

$$\frac{1}{k} = \frac{k_2}{k_3 k_1} + \frac{1}{k_1 [M]} \tag{4.61}$$

A plot of $1/k$ versus $1/[M]$ should yield a straight line.

For a multistep mechanism such as this it is often possible to identify a single elementary process whose rate governs the overall reaction rate. Sometimes such a *rate-limiting step* is the slowest one in a sequence. In the high-pressure limit, for example, the rate-limiting step is the third, unimolecular elementary process. This can be shown by applying the inequality $k_2 [M] \gg k_3$ to (4.55), yielding

$$k_1 [M] [A] = k_2 [M] [A^*] \tag{4.62}$$

Thus the rate of activation equals the rate of deactivation. Furthermore,

$$k_2 [M] [A^*] \gg k_3 [A^*]$$

showing that the rate of the third step is much less than for the first two and, consequently, limits the overall rate. The equality (4.62) between rates of activation and deactivation implies that the concentration of activated molecules A^* is that to be expected from an equilibrium Boltzmann distribution of energy.

In the low-pressure limit, on the other hand, collisions are so few and far between that the concentration of activated molecules is significantly lower than would be predicted on the basis of an equilibrium distribution. This is because activated molecules are consumed by the third unimolecular step as rapidly as they are produced. At low pressure $k_2 [M] \ll k_3$ and (4.55) becomes

$$k_1 [M] [A] = k_3 [A^*] \tag{4.63}$$

Also,

$$k_2 [M] [A^*] \ll k_3 [A^*]$$

so that the slowest elementary process is deactivation. In this case the overall rate is limited by the rate of activation (step 1), despite the fact that this step is the *fastest* elementary process in the mechanism (along with step 3). The rate of the spontaneous unimolecular reaction is limited by the low concentration of A^* supplied by step 1, not by the rate constant k_3, which is much larger than $k_1 [M]$, the pseudo-first-order rate constant for step 1. This example illustrates the ambiguity inherent in defining the rate-limiting step as the slowest step in a sequential mechanism.[61] If elementary processes of different molecularities are involved, the rate-limiting step may be different at different concentrations. Furthermore, at intermediate concentrations it may be that no single step is rate limiting. This is true of the Lindemann mechanism in the *fall-off region*, where the overall reaction is mixed first and second order.

A complete theory of unimolecular reaction rates should predict suitable values for the constants k_1, k_2, and k_3. Both k_1 and k_2 are bimolecular rate constants and might be expected to be given satisfactorily by (4.20), (4.21), (4.25), or (4.26). In particular, k_2 for deactivation can be calculated directly from the collision number as $Z_{MA^*}/L[M][A]$,[†] because no threshold energy is required in the process and no important steric effect is expected. Such a calculation is referred to as the strong-collision assumption — every collision is assumed to be successful in deactivating an activated molecule. To obtain the rate constant k_1

[†] Solving (4.19) for k and using (4.17) gives

$$k = - \frac{(1/LV)(dN_A/dt)}{[A][B]} = \frac{(1/L)qZ_{AB}}{[A][B]}$$

and since $q = pe^{-E_0/kt}$, setting $p = 1$ and $E_0 = 0$ gives

$$k = \frac{Z_{AB}}{L[A][B]}$$

from experimental data let $[M]_{1/2}$ be the concentration where $k = k_\infty/2$. From (4.57) and (4.59)

$$k = \frac{k_1 k_3/k_2}{1 + k_3/k_2[M]} = \frac{k_\infty}{1 + k_3/k_2[M]} \tag{4.64}$$

so that

$$[M]_{1/2} = \frac{k_3}{k_2} = \frac{k_1 k_3}{k_1 k_2} = \frac{k_\infty}{k_1} \tag{4.65}$$

Measurements in the high-pressure limit determine k_∞, and $[M]_{1/2}$ is available from experiments in the fall-off region. Values of k_1 calculated from (4.65) are several orders of magnitude larger than those obtained from (4.26). Also, when k is measured over the full range of pressures and $1/k$ plotted against $1/[M]$, there are significant deviations from the straight line predicted by (4.61). The Lindemann mechanism appears to be reasonable qualitatively in that it accounts for the falloff in the pseudo-first-order rate constant k at low pressures, but improved means for predicting the rate constants k_1, k_2, and k_3 are needed.

The Hinshelwood Modification

Hinshelwood[62] used a more realistic model, which included the contribution of internal degrees of freedom to overcoming the energy barrier, in calculating the rate constant k_1 for activation. The activation-in-many-degrees-of-freedom rate constant is derived earlier in this chapter (4.30), where it can be seen that including internal energy increases the rate constant over the line-of-centers value by a factor of $(E_0/kT)^{s-1}/(s-1)!$, where s is the number of vibrational degrees of freedom over which energy may be distributed. Thus for polyatomic molecules s may be as large as $3n - 5$, n being the number of atoms, and $(E_0/kT)^{s-1}/(s-1)!$ may be 10^5 or larger. Good agreement with $[M]_{1/2} = k_\infty/k_1$ is usually obtained when s is about half the actual number of modes of vibration of the molecule under consideration.

The Hinshelwood theory does not attempt to calculate k_3. Instead values are obtained empirically from k_∞. Equation (4.30) can be rewritten as

$$k_1 = \frac{(Z_{MA}/L[M][A])(E_0/kT)^{s-1}e^{-E_0/kT}}{(s-1)!} \tag{4.66}$$

Assuming that $k_2 = Z_{MA*}/L[M][A^*]$ and making the further reasonable assumption that $Z_{MA}/L[M][A] = Z_{MA*}/L[M][A^*]$, that is, that the collision numbers are directly proportional to concentrations, we have

$$\frac{k_1}{k_2} = \frac{(E_0/kT)^{s-1}e^{-E_0/kT}}{(s-1)!} \tag{4.67}$$

Then, solving (4.59) for k_3,

$$k_3 = \frac{k_\infty}{(k_1/k_2)} = k_\infty(s-1)!\left(\frac{E_0}{kT}\right)^{1-s}e^{E_0/kT} \tag{4.68}$$

Calculations of k_3 using typical parameters yield the following results:[63]

s	1	5	10
k_3/s^{-1}	10^{13}	$10^{9.5}$	$10^{6.5}$

The value $k_3 = 10^{13}$ s^{-1} that would be obtained from the line-of-centers rate constant expression ($s = 1$) leads to an anomaly.[64] In the high-pressure limit where no falloff of k is observed it is necessary that $k_2[M] \gg k_3$. However, because k_2 is limited by the collision number, it can be calculated that, for a gas under ordinary conditions of temperature and pressure, $k_2[M]$ is no greater than 10^{10} s^{-1}. Thus $k_3 \gtrsim 10^8$ s^{-1}, that is, $s \lesssim 6$, would violate the inequality $k_2[M] \gg k_3$ under such conditions. The calculated values of k_3 indicate that the more degrees of freedom available to store energy, the greater the lifetime of the activated molecule. A diatomic molecule, with only one degree of freedom, is predicted to dissociate in 10^{-13} s, that is, during the course of a single vibration.

The Lindemann-Hinshelwood theory provides considerable improvement in predicting the pressure (that is, [M]) of the fall-off region, provided that the parameter s is chosen appropriately. However, this theory does not solve the problem of nonlinearity in a plot of $1/k$ versus $1/[M]$. The reason for this curvature is that k_3 is a function of the energy of the activated molecule, whereas k_3 has been assumed so far to be a true constant.

The Rice-Ramsperger-Kassel (RRK) Theory

Rice and Ramsperger[65] and Kassel[66] extended the Lindemann-Hinshelwood theory almost simultaneously and in very similar ways. Both Rice-Ramsperger and Kassel concentrated on the energy dependence of the rate constant k_3 for the unimolecular step. Rice and Ramsperger considered that in order for a true unimolecular process to occur the critical energy E_0 must concentrate in a specific part of the molecule corresponding to one squared term in the energy expression (such as $1/2\ mu^2$ for kinetic energy or $1/2\ kx^2$ for potential energy of an oscillator). They used classical statistical mechanics to calculate the probability that such a concentration of energy would occur. Kassel assumed that the energy had to be concentrated into a single oscillator (that is, two squared terms, one potential and one kinetic energy). Kassel also developed a quantum theory, but here we consider the classical version. Both the Rice-Ramsperger and Kassel theories assume that the total energy E is rapidly redistributed among the various vibrational or other degrees of freedom of the molecule.

For a system of s classical oscillators with total energy E, the probability P that a particular oscillator will have energy $\geq E_0$ is[66]

$$P(E \geq E_0 \text{ in one oscillator}) = \left(\frac{E - E_0}{E}\right)^{s-1}$$

and the rate constant k_3 should be proportional to this probability

$$k_3 = A \left(\frac{E - E_0}{E} \right)^{s-1} \qquad E \geqslant E_0$$

$$k_3 = 0 \qquad\qquad\qquad E < E_0 \qquad (4.69)$$

The observed rate constant at high pressure can be calculated from (4.59), but since k_3 is a function of energy and k_1 must be evaluated by the integration of (4.24), the appropriate expression is

$$k_\infty = \int_{E_0}^\infty A \left(\frac{E - E_0}{E} \right)^{s-1} \frac{(E/kT)^{s-1}}{(s-1)!} \frac{e^{-E/kT}}{kT} \, dE$$

which integrates to

$$k_\infty = A e^{-E_0/kT} \qquad (4.70)$$

Typical frequency factors for unimolecular reactions in the high-pressure limit are of the order of 10^{13} s^{-1}, and it is not unreasonable that the proportionality constant A of (4.69) should be approximately a vibrational frequency. Once the critical energy E_0 is available to an appropriate vibrational mode, reaction can be expected during a single vibration. The overall variation of k with [M] can be obtained from (4.57) by an integration similar to that by which (4.70) is derived. The result is

$$k = \frac{A e^{-E_0/kT}}{(s-1)!} \int_{E_0}^\infty \frac{[(E - E_0)/kT]^{s-1} e^{(E-E_0)/kT}}{kT \{ 1 + (A/k_2 [M])[(E - E_0)/E]^{s-1} \}} \, dE$$

This integral can be evaluated numerically for various choices of s. As in the case of Hinshelwood's modification of the Lindemann theory, agreement with experiment requires that s be considerably less than the number of vibrational degrees of freedom of the molecule—usually one half to two-thirds that number given a reasonable choice of the hard-sphere collision diameter. However, the RRK treatment accounts much better for the curvature of plots of k versus [M] in the falloff region because the contribution to the overall rate from molecules with different excitation energies changes with the pressure. At high pressures, low-energy activated molecules (k_3 small) contribute little to the rate because they are deactivated by collision before decomposing. At low pressures the low-energy molecules contribute proportionately more because of the longer interval before deactivation.

The Slater Theory

Beginning in 1939 Slater[67] attempted to relate the rates of unimolecular reactions to existing knowledge of molecular vibrations. Slater treats each activated molecule of the Lindemann-Hinshelwood theory as a collection of simple, uncoupled harmonic oscillators. When a chosen coordinate (some combination of bond lengths and angles) reaches a critical value, reaction occurs. A "specific dissociation probability," which is the frequency at which the critical value of the chosen co-ordinate is exceeded, replaces the rate constant k_3. Application of the theory requires a vibrational analysis of the molecule so that the contribution of each normal mode to the critical coordiante can be determined. The threshold energy E_0 required by an activated molecule turns out to be E_a (high), the observed Arrhenius activation energy in the high-pressure limit. Thus, other than the choice of a critical coordinate for reaction, no arbitrary parameters are necessary.

At high pressures Slater's theory predicts

$$k = A e^{-E_0/kT} \qquad (4.71)$$

which is identical in form to the RRK result (4.70), but now A is a weighted root-mean-square average of frequencies of normal modes that contribute to extension of the critical coordinate. At low pressures the result is

$$k = k_{\text{bim}} [\text{M}] \qquad (4.72)$$

where k_{bim} is a second-order rate constant. However, k_{bim} is not k_1, the second-order constant for collisional activation of the Lindemann mechanism. Instead

$$k_{\text{bim}} = k_2 \left(\frac{4\pi E_0}{kT} \right)^{1/2 (s-1)} e^{-E_0/kT} \left(\prod_{k=1}^{s} \mu_k \right) \qquad (4.73)$$

where k_2 is the rate constant for collisional deactivation, s is the number of con-tributing vibrational modes, and μ_k is the weighting factor for the contribution of the kth mode to the critical coordinate. From (4.71) the observed Arrhenius activation energy at high pressure is seen to be LE_0, but the activation energy decreases at low pressures, becoming

$$E_a(\text{low}) = LE_0 - \left(\frac{n}{2} - 1 \right) RT \qquad (4.74)$$

Because it requires vibrational analysis of the reacting molecule, the complete Slater treatment has been applied in only a few cases. Also, the restriction to simple harmonic oscillators, which does not permit energy flow between different modes, is unrealistic. Despite considerable success in reproducing experimental observations over a wide range of pressures, Slater's theory has now been largely supplanted by an extension of the RRK approach.

The Rice-Ramsperger-Kassel-Marcus (RRKM) Theory

This model, devised by Rice and Marcus[68] and developed by Marcus,[69] makes use of the terminology of transition-state theory, and therefore its description is postponed until Chapter 5, where the transition-state theory is developed. Several recent books and articles describe RRKM theory and its inherent assumptions in detail.[70-72] A computer program for automated RRKM calculations is available.[73]

TRIMOLECULAR REACTIONS

The hard-sphere collision model embodies a fundamental difficulty when trimolecular collisions are considered. If a collision is defined as simultaneous contact of the spherical surfaces of the three molecules, the chance of having such a collision is zero. Two spheres would be in contact only for an instant, and there is only an infinitesimal probability that a third molecule would make contact with the other two at that instant. To have a finite number of trimolecular collisions, it is necessary to consider either that the spheres are not rigid or that a collision consists of the approach of rigid spheres to within some arbitrary distance of each other. This latter technique was used by Tolman.[74] The result is that for collision between molecules A, B, and C such that A and C are to be within a distance δ of B, the collision number is

$$Z_{ABC} = 8(2kT/\pi)^{1/2}(\pi d_{AB}^2)(\pi d_{BC}^2)\delta\,[(1/\sqrt{\mu_{AB}}) + (1/\sqrt{\mu_{BC}})]\,[N_A N_B N_C/V^3]$$

$$(4.75)$$

The symbols have the same or corresponding significance to those used previously. Nothing can be said about δ except that it should be of the order of 1 Å. Typical values of Z_{ABC} are 1000 times or more smaller than those for bimolecular collisions.

The alternative of spherical molecules interacting in accord with some intermolecular potential $U(d)$ where d is the distance between the centers of the spheres leads to "sticky" collisions whose lifetimes are of the order of the period of a molecular vibration. Thus a trimolecular collision may be pictured as two bimolecular collisions in rapid succession. A mechanism for such a process may be written as

$$
\begin{array}{lll}
A + B & \rightarrow A\cdots B & k_1 \\
A\cdots B & \rightarrow A + B & k_2 \\
A\cdots B + C & \rightarrow ABC & k_3
\end{array}
\qquad (4.76)
$$

where $A\cdots B$ represents molecules A and B in collision, that is, in a region where $d_{AB} \leqslant d \leqslant d_{A\ldots B}$. Here $d_{A\ldots B}$ represents the distance between centers beyond which the potential $U(d)$ is so small that intermolecular forces are negligible, and d_{AB} represents the smallest distance between centers permitted by the potential. Because the lifetime of species $A\cdots B$ is so short, most such collisions are over before a C molecule can collide with the collision complex. In terms of the mechanism (4.76), the separation of the collision complex $A\cdots B$ is much faster than the bimolecular collision of that complex with C. That is, $k_2[A\cdots B] \gg k_3[A\cdots B][C]$, and the collision complex is essentially in equilibrium with the reactants. This makes it possible to calculate the number of collision complexes per unit volume and hence the binary collision number between $A\cdots B$ and C. Assuming that the intermolecular forces are weak $[U(d) < kT]$ and that $U(d)$

is of the form $\pm K/d^n$, the result is[75]

$$Z_{ABC} = \frac{4}{3} \pi (d^3_{A\cdots B} - d^3_{AB}) \left[1 + \frac{U(d_{A\cdots B}) - U(d_{AB})}{(n-3)kT}\right] \left[\pi \left(\frac{d_{A\cdots B} + d_C}{2}\right)^2\right]$$

$$\left[\frac{8kT}{\pi}\left(\frac{1}{m_{A\cdots B}} + \frac{1}{m_C}\right)\right]^{1/2} \left[\frac{N_A N_B N_C}{V^3}\right] \tag{4.77}$$

Both (4.75) and (4.77) indicate that the collision number is proportional to three concentration factors, and therefore trimolecular reactions are expected to be third order. The rate of disappearance of A molecules per unit volume can be set equal to Z_{ABC} times an efficiency factor q

$$-\frac{1}{V}\frac{dN_A}{dt} = qZ_{ABC}$$

Comparison of this result with the third-order rate expression

$$-\frac{d[A]}{dt} = -\frac{1}{LV}\frac{dN_A}{dt} = k[A][B][C]$$

yields for the rate constant

$$k = \frac{qZ_{ABC}}{L[A][B][C]} \tag{4.78}$$

To avoid detailed calculations of the type carried out for bimolecular reactions, we simply assume that $q = pe^{-E_0/kT}$ as before. Then the rate constant is given by

$$k = \left(\frac{Z_{ABC}}{L[A][B][C]}\right) pe^{-LE_0/RT} \tag{4.79}$$

Although for large values of E_0 the simple exponential would probably be far from correct, it is a fact that most trimolecular reactions are found experimentally to have very small, zero, or even negative Arrhenius activation energies. Consequently the exponential term is close to unity, and no great error results. A possible trimolecular reaction with a high threshold energy is slow both on account of the activation requirement and because trimolecular collisions are rare compared with bimolecular collisions. Therefore some other mechanism is almost certain to provide a faster pathway to the products, and trimolecular reactions with high activation energy should be uncommon except in solution.

PROBLEMS

4.1. Derive a formula for the rate of effusion, dN/dt, of molecules of molar mass M, at temperature T, through a pinhole of area A, if there are N/V molecules per unit volume. [The rate of effusion of gas mixtures has been used by F. E. Harris and L. K. Nash, *Anal. Chem.*, **22**, 1552 (1950), as an analytical method.]

4.2. Derive a formula [analogous to (4.26)] for the line-of-centers rate constant of a bimolecular reaction involving a single gas, that is, for collisions between like molecules. Compare the derived formula with the data of problem **2.5** on the decomposition of nitrogen dioxide to see whether a reasonable value of the molecular diameter d is obtained.

4.3. Verify (4.41a) and (4.41b).

4.4. By differentiating the logarithmic form of (4.24), verify that (4.44) is obtained.

4.5. Derive the relationship between E_a and LE_0, assuming that k is given by (4.25).

4.6. Suppose that a collection of XY molecules is photolyzed by 195-nm ultraviolet radiation, producing X in an excited electronic state with energy 87 kJ mol^{-1} and Y in its ground electronic state. The bond dissociation energy of X–Y is 473 kJ mol^{-1} and the ratio $m_X/m_Y = 6.67$. Calculate the translational energy of Y, assuming that the initial translational energy of XY was negligible.

4.7. Verify that (4.74) follows from (4.73).

REFERENCES

1. W. Kauzmann, *Thermal Properties of Matter*, Vol. I, *Kinetic Theory of Gases*, Benjamin, New York, 1966; E. A. Guggenheim, *Elements of the Kinetic Theory of Gases*, The *International Encyclopedia of Physical Chemistry and Chemical Physics*, Topic 6, Vol. 1, Pergamon, London, 1960; J. Jeans, *Introduction to the Kinetic Theory of Gases*, Cambridge University Press, London, 1959; M. Knudsen, *Kinetic Theory of Gases*, Methuen, London, 1950; L. B. Loeb, *Kinetic Theory of Gases*, McGraw-Hill, New York, 1934; R. C. Tolman, *Statistical Mechanics*, Chemical Catalog Co., New York, 1927.

2. R. H. Fowler and E. A. Guggenheim, *Statistical Thermodynamics*, Cambridge University Press, London, 1960.

3. R. D. Present, *Kinetic Theory of Gases*, McGraw-Hill, New York, 1958.

4. E. H. Kennard, *The Kinetic Theory of Gases*, McGraw-Hill, New York, 1938.

5. See reference 3, Chap. 8.

6. R. D. Levine and J. Jortner, *Molecular Energy Transfer*, Halsted, New York, 1976; R. C. Amme, "Vibrational and Rotational Excitation in Gaseous Collisions," in *The Excited State in Chemical Physics*, J. W. McGowan, Ed., in *Advances in Chemical Physics*, Vol. 28, I. Prigogine and S. A. Rice, Eds., Wiley-Interscience, New York, 1975, pp. 171–265; G. W. Flynn in *Chemical and Biochemical Applications of Lasers*, Vol. 1, C. B. Moore, Ed., Academic, New York, 1974, p. 163; T. L. Cottrell and J. C. McCoubrey, *Molecular Energy Transfer in Gases*, Butterworths, London, 1961; S. H. Bauer, *Chem. Rev.*, **78**, 147 (1978); J. D. Lambert, *Vibrational and Rotational Relaxation in Gases*, Oxford, New York, 1977; A. B. Callear, "An Overview of Molecular Energy Transfer in Gases," in *Gas Kinetics and Energy Transfer*, Vol. 3, P. G. Ashmore and R. J. Donovan, Eds., Specialist Periodical Report, The Chemical Society, London, 1978; R. T. Bailey and F. R. Cruickshank, "Laser Studies of Vibrational, Rotational, and Translational Energy Transfer," *ibid.*

7. E. Weitz and G. Flynn, "Laser Studies of Vibrational and Rotational Relaxation in Small Molecules," *Annu. Rev. Phys. Chem*, **25**, 275 (1974); B. Stevens, *Collisional Activation in Gases*, The *International Encyclopedia of Physical Chemistry and Chemical*

Physics, Topic 19, Vol. 3, Pergamon, London, 1967; C. B. Moore, "Vibration → Vibration Energy Transfer," in *Advances in Chemical Physics,* Vol. 23, I. Prigogine and S. A. Rice, Eds., Wiley-Interscience, New York, 1973, pp. 41–84; J. Ross and E. F. Greene in *Transfert d'énergie dans les gaz,* R. Stoops, Ed., 12th Chemistry Solvay Conference, Inst. Int. Chim. (Douzieme Conseil Chim., 1962), Wiley-Interscience, New York, 1962, p. 363.

8. R. D. Levine and R. B. Bernstein, *Molecular Reaction Dynamics,* Oxford University Press, New York, 1974, Chap. 5.

9. W. H. Flygare, *Acc. Chem. Res.,* **1,** 121 (1968).

10. M. A. Eliason and J. O. Hirschfelder, *J. Chem. Phys.,* **30** (6), 1426 (1959).

11. Reference 2, pp. 495–499.

12. E. Wigner and E. E. Witmer, *Z. Phys.,* **51,** 859 (1928); K. E. Shuler, *J. Chem. Phys.,* **21,** 624 (1953); Y. N. Chiu, *J. Chem. Phys.,* **58,** 722 (1973).

13. E. F. Greene and A. Kuppermann, *J. Chem. Educ.,* **45** (6), 361 (1968); Reference 4, p. 117.

14. P. R. Brooks and E. F. Hayes, Eds., *State-to-State Chemistry,* ACS Symposium Series, No. 56, Washington, DC, 1977.

15. See reference 14, pp. 107–144; G. A. Oldershaw, "Reactions of Photochemically Generated Hot Hydrogen Atoms," in *Gas Kinetics and Energy Transfer,* Vol. 2, P. G. Ashmore and R. J. Donovan, Eds., Specialist Periodical Report, The Chemical Society, London, 1977, pp. 96–122; F. S. Rowland, "Experimental Studies of Hot Atom Reactions," in *Chemical Kinetics, Physical Chemistry, Series One,* Vol. 9, J. C. Polanyi, Ed., *MTP International Review of Science,* Butterworths, London, pp. 109–133; W. A. Chupka, "Ion-Molecule Reactions by Photoionization Techniques," in *Ion–Molecule Reactions,* Vol. 1, J. L. Franklin, Ed., Plenum, New York, 1972, pp. 33–76.

16. A. Kuppermann and J. M. White, *J. Chem. Phys.,* **44,** 4352 (1966).

17. A. Kuppermann, J. Stevenson, and P. O'Keefe, "Dynamics of Reaction of Monoenergetic Atoms in a Thermal Gas," in *Molecular Dynamics of the Chemical Reactions of Gases, Disc. Faraday Soc.,* **44,** 46 (1967).

18. R. G. Gann, W. M. Ollison, and J. Dubrin, *J. Chem. Phys.,* **54,** 2304 (1971); R. G. Gann, W. M. Ollison, and J. Dubrin, *J. Am. Chem. Soc.,* **92,** 450 (1970); R. G. Gann and J. Dubrin, *J. Chem. Phys.,* **50,** 535 (1969).

19. I. W. M. Smith, "Reactive and Inelastic Collisions Involving Molecules in Selected Vibrational States," in *Gas Kinetics and Energy Transfer,* Vol. 2, P. G. Ashmore and R. J. Donovan, Eds., Specialist Periodical Report, The Chemical Society, London, 1977.

20. W. A. Chupka, J. Berkowitz, and M. E. Russell, *Abstracts of Papers, Sixth International Conference on the Physics of Electronic and Atomic Collisions,* MIT Press, 1969, pp. 71–72; R. B. Bernstein, *Isr. J. Chem.,* **9,** 615 (1971).

21. See work cited by R. B. Bernstein, "State-to-State Cross Sections and Rate Constants for Reactions of Neutral Molecules," in reference 14, pp. 1–21.

22. Reference 14, pp. 3–106; K. P. Lawley, Ed., *Molecular Scattering: Physical and Chemical Applications,* in *Advances Chemical Physics,* Vol. 30, I. Prigogine and S. A. Rice, Eds., Wiley-Interscience, New York, 1975; I. W. M. Smith, "The Production of Excited Species in Simple Chemical Reactions," in *The Excited State in Chemical Physics,* J. W. McGowan, Ed., *Advances in Chemical Physics,* Vol. 28, I. Prigogine and S. A. Rice, Eds., Wiley-Interscience, New York, 1975; M. A. D. Fluendy and K. P. Lawley, *Chemical Applications of Molecular Beam Scattering,* Chapman and Hall, London 1973; *Molecular Beam Scattering, Disc. Faraday Soc.,* **55** (1973); M. Henchman, "Rate Constants and Cross Sections," in *Ion–Molecule Reactions,* Vol. 1, J. L. Franklin,

Ed., Plenum, New York, 1972; J. I. Steinfeld and J. L. Kinsey, "The Determination of Chemical Reaction Cross-Sections," in *Progress in Reaction Kinetics*, Vol. 5, G. Porter, Ed., Pergamon, Oxford, 1970; Ch. Schlier, Ed., *Molecular Beams and Reaction Kinetics, Proceedings of the International School of Physics, Enrico Fermi*, Course 44, Academic, New York, 1970; J. Ross, Ed., *Molecular Beams, Advances in Chemical Physics*, Vol. 10, I. Prigogine, Ed., Wiley-Interscience, New York, 1966.

23. D. R. Herschbach, *Pure Appl. Chem.*, **47**, 61 (1976).

24. R. D. Levine and R. B. Bernstein, *Molecular Reaction Dynamics*, Oxford University Press, New York, 1974.

25. J. L. Kinsey, "Molecular Beam Reactions," in *Chemical Kinetics, Physical Chemistry, Series One*, Vol. 9, J. C. Polanyi, Ed., MTP International Review of Science, Butterworths, London, 1972, pp. 173–212.

26. E. H. Taylor and S. Datz, *J. Chem. Phys.*, **23**, 1711 (1955); I. Langmuir and K. H. Kingdon, *Proc. R. Soc. (London)*, **A21**, 380 (1923).

27. R. Weiss, *Rev. Sci. Instr.*, **32**, 397 (1961); Y. T. Lee, J. D. McDonald, P. R. LeBreton, and D. R. Herschbach, *Rev. Sci. Instr.*, **40**, 1402 (1969); J. M. Parson and Y. T. Lee, *J. Chem. Phys.*, **56**, 4658 (1972).

28. J. G. Moehlmann and J. D. McDonald, *J. Chem. Phys.*, **62**, 3052, 3061 (1975).

29. R. N. Zare and P. J. Dagdigian, *Science*, **185**, 739 (1974).

30. J. R. Grover, C. R. Iden, H. V. Lilenfeld, F. M. Kiely, and E. Lebowitz, *Rev. Sci. Instr.*, **47**, 1098 (1976).

31. J. L. Kinsey, *Rev. Sci. Instr.*, **37**, 61 (1966); H. U. Hostettler and R. B. Bernstein, *Rev. Sci. Instr.*, **31**, 872 (1960).

32. T. J. Odiorne, P. R. Brooks, and J. V. V. Kasper, *J. Chem. Phys.*, **55**, 1980 (1971).

33. J. E. Mosch, S. A. Safron, and J. P. Toennies, *Chem. Phys.*, **8**, 304 (1975); D. S. Y. Hsu, G. M. McClelland, and D. R. Herschbach, *J. Chem. Phys.*, **61**, 4927 (1974); P. R. Brooks, "Scattering of K Atoms from Oriented CF_3I Reaction at Both 'Ends'," in *Molecular Beam Scattering, Disc. Faraday Soc.*, **55**, 299 (1973); R. P. Mariella, D. R. Herschbach, and W. Klemperer, *J. Chem. Phys.*, **58**, 3785 (1973).

34. E. Teloy and D. Gerlich, *Chem. Phys.*, **4**, 417 (1974).

35. Reference 24, pp. 191–196; reference 25, pp. 178–180; T. T. Warnock and R. B. Bernstein, *J. Chem. Phys.*, **49**, 1878 (1968); *ibid.*, **51**, 4682 (1969).

36. T. Carrington and J. C. Polanyi, "Chemiluminescent Reactions," in *Chemical Kinetics, Physical Chemistry, Series One*, Vol. 9, J. C. Polanyi, Ed., MTP International Review of Science, Butterworths, London, 1972.

37. K. T. Gillen, A. M. Rulis, and R. B. Bernstein, *J. Chem. Phys.*, **54**, 2831 (1971).

38. M. Polanyi, *Atomic Reactions*, Williams and Norgate, London, 1932; J. L. Magee, *J. Chem. Phys.*, **8**, 687 (1940); D. R. Herschbach, "Reactive Scattering in Molecular Beams," in *Molecular Beams, Advances in Chemical Physics*, J. Ross, Ed., Vol. 10, Wiley-Interscience, New York, 1966, pp. 319–393.

39. J. D. McDonald, P. R. LeBreton, Y. T. Lee, and D. R. Herschbach, *J. Chem. Phys.*, **56**, 769 (1972).

40. G. A. Fisk, J. D. McDonald, and D. R. Herschbach, in *Molecular Dynamics of the Chemical Reactions of Gases, Disc. Faraday Soc.*, **44**, 228 (1967); M. K. Bullit, C. H. Fisher, and J. L. Kinsey, *J. Chem. Phys.*, **60**, 478 (1974).

41. J. T. Cheung, J. D. McDonald, and D. R. Herschbach, *J. Am. Chem. Soc.*, **95**, 7889 (1973).

42. H. G. Bennewitz, W. Paul, and C. H. Schlier, *Z. Phys.*, **141**, 6 (1955); K. H. Kramer and R. B. Bernstein, *J. Chem. Phys.*, **42**, 767 (1965).

43. P. R. Brooks, *Science,* **193,** 11 (1976).

44. P. R. Brooks and E. M. Jones, *J. Chem. Phys.,* **45,** 3449 (1966); R. J. Beuhler, Jr., R. B. Bernstein, and K. H. Kramer, *J. Am. Chem. Soc.,* **88,** 5331 (1966).

45. G. Marcelin and P. R. Brooks, *J. Am. Chem. Soc.,* **95,** 7885 (1973); *ibid.,* **97,** 1710 (1975).

46. D. L. Bunker and E. Goring-Simpson, "Alkali-Methyl Iodide Reactions," in *Molecular Beam Scattering, Disc. Faraday Soc.,* **55,** 93 (1973).

47. J. G. Pruett and R. N. Zare, *J. Chem. Phys.,* **64,** 1774 (1976).

48. C. B. Moore and P. F. Zittel, *Science,* **183,** 541 (1973).

49. A. M. G. Ding, L. J. Kirsch, D. S. Perry, J. C. Polanyi, and J. L. Schreiber, "Effect of Changing Reagent Energy on Reaction Probability and Product Energy-Distribution," in *Molecular Beam Scattering, Disc. Faraday Soc.,* **55,** 252 (1973); D. S. Perry, J. C. Polanyi, and C. W. Wilson, Jr., *Chem. Phys.,* **3,** 317 (1974).

50. D. G. Truhlar, *J. Chem. Educ.,* **55,** 309 (1978).

51. M. Menzinger and R. Wolfgang, *Angew. Chem. Int. Ed.,* **8,** 438, (1969).

52. R. C. Tolman, *J. Am. Chem. Soc.,* **42,** 2506 (1920); see also reference 2.

53. R. L. LeRoy, *J. Phys. Chem.,* **73,** 4338 (1969).

54. B. Shizgal and M. Karplus, *J. Chem. Phys.,* **52,** 4262 (1970); *ibid.,* **54,** 4357 (1971); B. Widom, *J. Chem. Phys.,* **61,** 672 (1974); R. K. Boyd, *Chem. Rev.,* **77,** 93 (1977).

55. H. Johnston and J. Birks, *Acc. Chem. Res.,* **5,** 327 (1972).

56. J. Perrin, *Ann. Phys. (Paris),* [9] **11,** 1–108 (1919).

57. I. Langmuir, *J. Am. Chem. Soc.,* **42,** 2190 (1920); M. Kröger, *Z. Phys. Chem. (Leipz.),* **117,** 387 (1925); J. E. Mayer, *J. Am. Chem. Soc.,* **49,** 3033 (1927); G. N. Lewis and J. E. Mayer, *Proc. Natl. Acad. Sci. USA,* **13,** 623 (1927); F. O. Rice, H. C. Urey, and R. N. Washburne, *J. Am. Chem. Soc.,* **50,** 2402 (1928).

58. F. A. Lindemann, *Trans. Faraday Soc.,* **17,** 598 (1922).

59. J. A. Christiansen, "Reactionskinetiske Studier", Ph.D. Thesis, Copenhagen, 1921.

60. D. C. Tardy and B. S. Rabinovitch, *Chem. Rev.,* **77,** 369 (1977); M. Quack and J. Troe, "Unimolecular Reactions and Energy Transfer of Highly Excited Molecules," in *Gas Kinetics and Energy Transfer,* Vol. 2, P. G. Ashmore and R. J. Donovan, Eds., Specialist Periodical Report, The Chemical Society, London, 1977.

61. R. K. Boyd, *J. Chem. Educ.,* **55,** 84 (1978); K. G. Denbigh and J. C. R. Turner, *Chemical Reactor Theory, An Introduction,* 2nd ed., Cambridge University Press, London, 1971, Chap. 2.

62. C. N. Hinshelwood, *Proc. R. Soc. (London),* **A113,** 230 (1927); C. N. Hinshelwood, *Kinetics of Chemical Change,* Oxford University Press, New York, 1940, pp. 26–29.

63. A. F. Trotman-Dickenson, *Gas Kinetics,* Butterworths, London, 1955, p. 55.

64. S. W. Benson, *J. Chem. Phys.,* **19,** 802 (1951).

65. O. K. Rice and H. C. Ramsperger, *J. Am. Chem. Soc.,* **49,** 1617 (1927); *ibid.,* **50,** 617 (1928).

66. L. S. Kassel, *J. Phys. Chem.,* **32,** 225 (1928); *ibid.,* **32,** 1065 (1928); L. S. Kassel, *Kinetics of Homogeneous Gas Reactions,* Chemical Catalog Co., New York, 1932.

67. N. B. Slater, *Proc. Camb. Philos. Soc.,* **35,** 56 (1939); N. B. Slater, *Theory of Unimolecular Reactions,* Cornell University Press, Ithaca, 1959.

68. R. A. Marcus and O. K. Rice, *J. Phys. Colloid Chem.,* **55,** 894 (1951).

69. R. A. Marcus, *J. Chem. Phys.*, **20**, 359 (1952); G. M. Wieder and R. A. Marcus, *J. Chem. Phys.*, **37**, 1835 (1962); R. A. Marcus, *J. Chem. Phys.*, **43**, 2658 (1965).

70. W. Forst, *Theory of Unimolecular Reactions*, Academic, New York, 1973.

71. P. J. Robinson and K. A. Holbrook, *Unimolecular Reactions*, Wiley-Interscience, New York, 1972; P. J. Robinson, "Unimolecular Reactions," in *Reaction Kinetics*, Vol. 1, P. G. Ashmore, Ed., Specialist Periodical Report, The Chemical Society, London, 1975.

72. W. L. Hase, "Dynamics of Unimolecular Reactions," in *Dynamics of Molecular Collisions*, Part B, W. H. Miller, Ed., Modern Theoretical Chemistry, Vol. 2, Plenum Press, New York, 1976, pp. 121–169.

73. W. L. Hase and D. L. Bunker, Quantum Chemistry Program Exchange Cat. No. QCPE-234, Indiana University, Bloomington.

74. R. C. Tolman, *Statistical Mechanics*, Chemical Catalog Co., New York, 1927.

75. G. G. Hammes, *Principles of Chemical Kinetics*, Academic, New York, 1978, p. 42.

ELEMENTARY PROCESSES: POTENTIAL ENERGY SURFACES AND TRANSITION-STATE THEORY

The theoretical treatment of bimolecular collisions in Chapter 4 is limited in most cases to hard-sphere particles and invariably to species having spherically symmetric potential fields. We show there that reactive collisions can be characterized by a reactive cross section σ_R that depends in general on the relative translational energy of the colliding particles, on the internal energies of the particles, and on the orientation of the particles, but we do not consider explicitly the detailed motion of each atomic nucleus and electron in each of the colliding species. To do so we must apply quantum mechanics to a *supermolecule* that consists of all nuclei and all electrons of all colliding species.

The appropriate wave equation involves a total Hamiltonian made up of kinetic energy operators for all nuclei and electrons as well as potential energy terms for all electron–electron, electron–nucleus, and nucleus–nucleus interactions. Even for very simple collections of nuclei and electrons solution of such a wave equation is extremely difficult. The usual approach is to invoke the Born-Oppenheimer approximation:[1] since the mass of every nucleus is on the order of 2000 or more times the mass of an electron, electronic motion is rapid compared to nuclear motion and electron density can adjust essentially instantaneously to changes in nuclear position. Therefore, unless the nuclei are moving very rapidly, electronic motion can be separated from nuclear motion and the energy can be evaluated for any given arrangement of nuclei with the nuclei being considered at rest. This energy includes all terms except the kinetic energy of the nuclei. It is the potential energy in the dynamical equations of nuclear motion and hence is called potential energy even though it includes kinetic energy of electronic motion. Evaluating this potential energy over the range of possible orientations of nuclei defines a multi-dimensional potential energy hypersurface[2, 3] for a given state of the supermolecule. For many reactions a single electronic eigenfunction can be used to represent the state of the electrons throughout the reactive encounter; the motion is confined to a single potential energy surface and is said to be adiabatic.

Energy minima (dimples or valleys) on a potential energy surface correspond to stable molecular species: reactants, products, or intermediates. For chemical reaction to occur the nuclei must move in such ways that the energy rises and a mountain pass must be surmounted. The nuclear configuration whose coordinates correspond to the col or saddle point at the top of such a pass is referred to as the *activated complex* and is said to be in the *transition state*. Given a potential energy surface together with initial translational and internal energy states of reactants, one can solve the quantum-mechanical or classical equations of motion to determine the path or trajectory of a system of nuclei as it passes from a reactant valley through or near a transition state to a product valley. Solution of the equations of motion for a particular combination of reactant and product states yields a state-to-state cross section for reaction. In principle such state-to-state cross sections can then be integrated or summed over appropriate distributions of initial states to obtain a total cross section and hence a rate constant. In practice, however, so many states are available and solution of the equations is slow enough (even with high-speed computers) that statistically weighted random samples must be chosen from the initial states and used to obtain approximate cross sections and rate constants. By increasing the number of trajectories, calculated errors in these approximate cross sections can be made smaller and smaller.

Lengthy molecular dynamics calculations of the type described in the preceding paragraph can be avoided by making some additional assumptions and concentrating on the minimum-energy path from reactants to activated complex to products. (This minimum-energy path is called the *reaction coordinate*.) The assumptions are: that classical mechanics provides an adequate formalism for calculating trajectories; that reactants and activated complexes maintain equilibrium Boltzmann distributions of energy; and that once a configuration of nuclei on the potential energy surface crosses a critical dividing surface reaction has occurred. The critical dividing surface is chosen to pass through the saddle point and to lie perpendicular to the reaction coordinate. These assumptions suffice[4] for derivation of the transition-state theory (also known as the theory of absolute reaction rates or activated-complex theory) first used for a specific reaction by Pelzer and Wigner in 1932[5] and later developed in general form by Eyring and co-workers.[3, 6, 7] The assumptions of transition-state theory hold quite well in most cases. For reactants in thermal reactions a maximum deviation of 8% from the equilibrium Boltzmann distribution is predicted when $E_a/RT = 5$,[8] and classical mechanics suffices to calculate accurate trajectories except when electrons or light atoms such as H are transferred in a collision. Thus transition-state theory is usually quite successful. Moreover, the transition-state theory leads naturally to a "thermodynamic" approach in which the increase in parameters such as enthalpy, entropy, and Gibbs free energy on going from reactants to transition state may be obtained from rate data. These increases then serve to characterize the activated complex and often help to narrow the selection of possible mechanisms.

POTENTIAL ENERGY SURFACES

The potential energy of a configuration of nuclei of molecules in collision, or of a stable molecule for that matter, may be represented by an f-dimensional surface in an $f+1$ dimensional space, where f is the number of independent variables necessary to specify completely the relative positions of all nuclei. In the case of a diatomic molecule, for example, six variables are required to specify cartesian coordinates of the two nuclei. However, three of these variables may be taken to locate the center of mass and two more indicate the orientation in space of the molecular axis. This leaves one variable – the internuclear distance – to specify the relative positions of the nuclei. Thus $f = 1$ and the potential energy "surface" is the familiar, one-dimensional potential energy curve of a diatomic molecule, shown as the lowest curve in Fig. 5.1. This can readily be generalized to the case of a super-molecule formed by collision of species containing a total of N nuclei, in which case $f = 3N - 6$, three variables being needed to locate the center of mass and (for the general case of a nonlinear supermolecule) three to specify orientation in space.

If there is a change in orientation in space of the supermolecule as well as changes in the relative positions of the nuclei, that is, if the supermolecule rotates during the collision, then the potential energy must be corrected for centrifugal

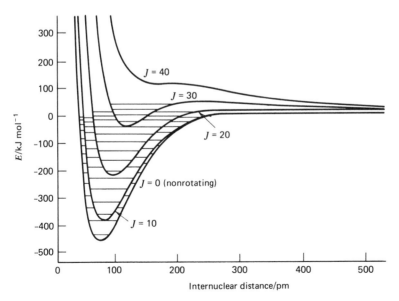

Figure 5.1. Effective potential energy curves for the H_2 molecule in rotational states $J = 0$, 10, 20, 30, and 40. Horizontal lines represent energies of vibrational states. Note how the centrifugal effect of rotation raises the effective potential energy.

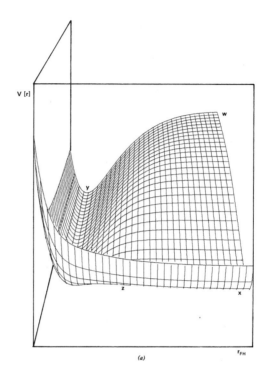

(a)

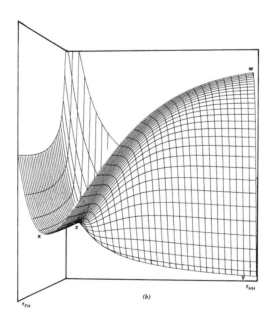

(b)

Figure 5.2.

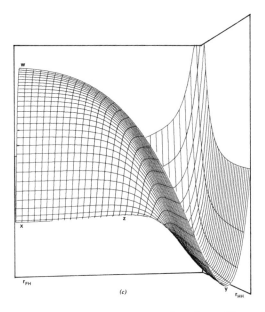

(c)

Figure 5.2. Views of the potential energy surface for collinear F–H–H. Points x and y lie in potential energy minima corresponding to $F + H - H$ and $F - H + H$, respectively; w is a plateau of high potential energy corresponding to $F + H + H$; and z is a saddle point 6.95 kJ mol^{-1} above x. The surface rises rapidly at small values of r_{FH} and r_{HH} and therefore has been cut off at $r_{FH} = 70 \text{ pm}$ and $r_{HH} = 50 \text{ pm}$; the planes corresponding to these values are shown. (a) View of the surface from above. (b) View along the entrance channel. (c) View along the exit channel; the small rise in potential energy at the saddle point is evident.

effects. Again using a diatomic molecule as an example, the effective potential $U(r)$ for rotational state J is:

$$U(r) = V(r) + \frac{J(J+1)h^2}{8\pi^2 \mu r^2}$$

where $V(r)$ is the potential energy as defined above, h is Planck's constant, μ is the reduced mass, and r is the internuclear distance. Effective potential energy curves for various rotational states of H_2 are shown in Fig. 5.1. Centrifugal corrections must be made for accurate calculations of cross sections and rate constants, but they are not necessary for a general understanding of the use of potential energy surfaces. Consequently we confine our attention to $V(r)$.

Consider the potential energy surface for the interaction of three atoms. Here $f = 3N - 6 = 3$, and the three variables might be chosen as the three internuclear distances. A complete representation of the energy would be a three-dimensional surface plotted in a four-dimensional space, but this is difficult to visualize or construct. If one of the variables is fixed or made dependent on the other two, however, a two-dimensional surface can be plotted in three-dimensional space. A series of such plots, one for each value of the fixed bond length or one for each value of a parameter that relates one bond length to the other two, gives an overall picture of the three-dimensional surface.

The most common way of representing the potential energy for a three-atom system is to choose two bond lengths and fix the angle between them. An example of such a *fixed-angle surface* (FAS) for the reaction

$$F + H - H' \rightarrow F - H + H' \tag{5.1}$$

is shown in Fig. 5.2. The surface has been rotated to provide alternative angles of view. The potential energy is plotted in the vertical direction as a function of the F—H internuclear distance r_{FH} and the H—H' distance $r_{HH'}$. The angle θ between F—H and H—H' has been fixed at $180°$. The third interatomic distance $r_{FH'}$ can be expressed in terms of θ and the other two distances as

$$r_{FH'} = (r_{FH}^2 + r_{HH'}^2 - 2r_{FH}r_{HH'} \cos \theta)^{1/2}$$

Although Fig. 5.2 provides a good qualitative view of the potential energy surface for $F + H + H'$, it provides little quantitative information. Values of potential energy as a function of r_{FH} and $r_{HH'}$ are more usefully represented by means of a contour map like Fig. 5.3. In a contour map each curved line represents the locus of all points $(r_{FH}, r_{HH'})$ at which the potential energy has the particular value indicated in the legend. Figure 5.3 is to Fig. 5.2 as a topographic map is to the actual terrain being traversed by a hiker.

Several regions on Figs. 5.2 and 5.3 have been labeled with lowercase letters. Region w represents configurations where F, H, and H' are all well separated and the potential energy varies little with either r_{FH} or $r_{HH'}$. In the topographic analogy region w is a plateau, and you can see from Fig. 5.3 that infinite separation of the three nuclei has been chosen as the zero of energy. Region x corresponds to F far removed from H—H'; hence the potential energy is affected very little by small changes in r_{FH}, and region x is a nearly level valley with steep sides. A vertical cross section through the surface at w to x can be seen from Fig. 5.2b to be very similar to the diatomic potential energy curve for H_2 $(J = 0)$ in Fig. 5.1 since, with F far away, the energy depends primarily on $r_{HH'}$. Region x corresponds to the reactants $F + H - H'$ and is called the entrance channel or entrance valley. In Fig. 5.3 the straight line $r_{HH'} = r_{HH'}^0$ (where $r_{HH'}^0$ is the equilibrium internuclear distance in H_2) is called the entry line. Region y is where H' is far from H—F. Here again there is a flat valley; the cross section at w to y shown in Fig. 5.2c corresponds to the diatomic potential energy curve of HF. Since the bond dissociation energy of HF is greater than for H_2, the exit channel or exit valley at y is deeper than the entrance channel at x. The straight line along the bottom of the exit valley in Fig. 5.3 is called the exit line, and the region near the intersection of the entry line and the exit line is called the corner of the surface.

In region z all three atoms are close together, as in a collision of F with H—H' or of H' with H—F. The entrance and exit valleys connect by way of a saddle point near z, since as F approaches H—H' or H' approaches H—F a van der Waals repulsive force is expected to set in.[9] Although this van der Waals repulsion is small relative to the dissociation energies of H_2 and HF, the increase in potential energy that it causes is evident both in Fig. 5.2c and in Fig. 5.3. (There is also a weak van der Waals attraction that is too small to be seen in the diagrams.) The van der Waals

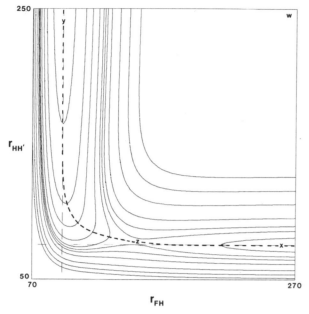

Figure 5.3. Contour diagram representing a fixed-angle surface ($\theta = 180°$) for the reaction $F + H - H' \rightarrow F - H + H'$. Points x and y lie in potential energy minima; w lies on a plateau of high potential energy; and z is a saddle point. Contours have been drawn at -510, -470, -440, -410, -390, -380, -370, -350, -310, -270, and $-230 \, \text{kJ mol}^{-1}$ relative to separated $F + H + H'$. Distances are in picometers. (Data from Polanyi and Schreiber.[17])

repulsion is of the utmost importance because it is responsible for activation energies in general. Without it the path from the entrance valley to the exit valley would be continually downhill. The reaction of F with H_2 and all other exothermic reactions would have zero or very low activation energies and would be immeasurably fast. The repulsion is due to the interaction of filled shells of electrons and to internuclear repulsions as atoms or molecules approach one another. The effect of filled shells is due mainly to the requirement of the Pauli principle that electrons of the same spin cannot have identical space coordinates,[10a] although there has been considerable controversy recently regarding the magnitude of this effect.[10b]

A collision of F with H_2 corresponds to movement along the potential energy surface from region x over the saddle point z to region y. The path of steepest ascent from x to z and of steepest descent from z to y is indicated by a dotted curve in Fig. 5.3. It lies normal to each contour it crosses and is called the *minimum-energy path* or the *reaction coordinate*.[11] The energy as a function of distance along the minimum-energy path is shown in Fig. 5.4. In this case, because (5.1) is an exothermic reaction, the final state is lower in energy than the initial state. The barrier height, that is, the height of the saddle point above the initial point, is closely related to the Arrhenius activation energy and to the threshold energy E_0 defined in Chapter 4. The threshold energy equals the barrier height plus the zero-point vibrational energy of the activated complex minus the total zero-point energy of the reactants.

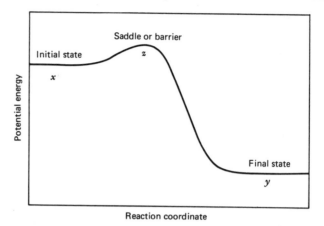

Figure 5.4. Potential energy along the reaction coordinate for the reaction $F + H - H' \rightarrow F - H + H'$.

CALCULATION OF POTENTIAL ENERGY SURFACES

The earliest application of the Born-Oppenheimer approximation to a chemical reaction was by London.[12] On the basis of the Heitler-London valence-bond theory he developed approximation formulas for the energies of systems of three and four univalent atoms in interaction. For three atoms, X, Y, and Z that might be involved in a simple displacement reaction of the type

$$X + Y - Z \rightarrow X - Y + Z$$

the potential energy V is given by the London equation as:

$$V = Q_{XY} + Q_{YZ} + Q_{XZ} \pm \{\tfrac{1}{2}[(J_{XY} - J_{YZ})^2 + (J_{YZ} - J_{XZ})^2 + (J_{XZ} - J_{XY})^2]\}^{1/2} \tag{5.2}$$

Here the coulombic energies Q_{ij} and the exchange energies J_{ij} are integrals that depend on internuclear distances, orbitals on different atoms are assumed to be orthogonal, and the double exchange energy J_{XYZ} has been ignored. Since the complete Heitler-London theory accounts for only two-thirds of the bond energy in the simplest diatomic molecule, H_2, evaluating integrals to obtain the Q_{ij} and J_{ij} in (5.2) is a much less desirable approach to calculating a triatomic binding energy than is using the equation as the basis for a semiempirical procedure. Eyring and Polanyi[13] obtained numerical values for the Q_{ij} and J_{ij} from spectroscopic data for the diatomic molecules XY, YZ, and XZ. Since overlap is still ignored, the energy of a diatomic molecule, say XY, is given by $Q_{XY} \mp J_{XY}$, the plus sign corresponding to a lower energy singlet state (since $J_{XY} < 0$) and the minus sign to a triplet state. The energy of the singlet state $^1E_{XY}$ as a function of internuclear distance r_{XY} can be obtained from a Morse potential,

$$^1E_{XY} = Q_{XY} - J_{XY} = D_{XY}\{\exp[-2\beta_{XY}(r_{XY} - r^0_{XY})] - 2\exp[-\beta_{XY}(r_{XY} - r^0_{XY})]\} \tag{5.3}$$

where D_{XY} is the classical dissociation energy, r_{XY}^0 is the equilibrium internuclear distance, and β_{XY} is a spectroscopic constant that can be calculated from the ground-state vibrational frequency. From the coulomb and exchange integrals calculated by Sugiura[13] for H_2, Eyring and Polanyi noted that $Q_{HH} \simeq 0.10\, V_{HH}$ to $0.15\, V_{HH}$ for all $r_{HH} > 80$ pm, and a choice of $Q_{HH} = 0.14\, V_{HH}$ led to the best fit of the activation energy for the exchange reaction of H with H_2. Accurate values of the potential energy over the full ranges of r_{XY}, r_{YZ}, and r_{XZ} are not to be expected from the London-Eyring-Polanyi (LEP) method, and this approach spuriously predicts a small crater at the saddle point for the H_3 surface. Nevertheless the LEP assumptions suffice to reproduce the qualitative features of potential energy surfaces and to make rough estimates of threshold energies.

Sato[14] extended the LEP method by introducing a parameter S that corresponds to an overlap integral, but is evaluated empirically. In the London-Eyring-Polanyi-Sato (LEPS) method the potential energy is:

$$V_{LEPS} =$$

$$\left(Q_{XY} + Q_{YZ} + Q_{XZ} - \frac{\{\frac{1}{2}\,[(J_{XY} - J_{YZ})^2 + (J_{YZ} - J_{XZ})^2 + (J_{XZ} - J_{XY})^2]\}^{1/2}}{1 + S^2} \right)$$

$$(5.4)$$

and the coulomb and exchange energies are calculated by solving simultaneous equations of the type

$$^1E_{XY} = \frac{Q_{XY} + J_{XY}}{1 + S^2} \tag{5.5a}$$

$$^3E_{XY} = \frac{Q_{XY} - J_{XY}}{1 - S^2} \tag{5.5b}$$

The singlet energies are obtained from Morse curves such as (5.3) and the triplet potential energy curve is taken to be the anti-Morse function

$$^3E_{XY} = \left(\frac{D_{XY}}{2}\right) \{\exp\,[-2\beta_{XY}(r_{XY} - r_{XY}^0)] + 2\,\exp\,[-\beta_{XY}(r_{XY} - r_{XY}^0)]\} \tag{5.6}$$

where the two exponentials are added instead of subtracted. The Sato parameter S^2 can be varied to achieve a particular barrier height or some other desired property of the surface. Kuntz[15] has shown that the effect of varying the Sato parameter is equivalent to setting $S = 0$ and varying the dissociation energy D_{XY} in only the triplet energy expression, that is, obtaining $^1D_{XY}$ from experiment and using $^3D_{XY}$ to adjust the potential energy surface to a known activation energy. The LEPS method does not produce a basin or crater at the saddle point, and it can be extended to provide more adjustable parameters.[16] Recently it has been shown that an extended LEPS surface can be fitted very accurately to the best *ab initio* theoretical surface for the reaction of F with H_2,[17] an important confirmation that adjusting LEPS parameters to reproduce experimental data can be expected to result in a reasonable surface.

Porter and Karplus[18] used a complete valence-bond expression to calculate a potential energy surface for the $H + H_2$ reaction. Overlap integrals were evaluated analytically from atomic $1s$ orbitals optimized for a valence-bond calculation on H_2, and equations (5.5) were used to calculate Q_{HH} and J_{HH} from parameters obtained by fitting (5.3) and (5.6) to the best available potential energy curves[19] for singlet and triplet H_2. The double-exchange integral J_{XYZ} was approximated as a scaled product of the three overlap integrals S_{XY}, S_{YZ}, and S_{XZ}. Except for discontinuities that occur along the line $r_{XY} = r_{YZ}$ as a result of Jahn-Teller instability,[20] the Porter-Karplus surface agrees well with *ab initio* results.[21] The surface also confirms that the lowest-energy path for the reaction involves a linear H—H—H configuration, as originally shown by London.

Although the first *ab initio* calculation of a potential energy surface was carried out on linear, symmetric H_3 in 1936,[22] only during the last decade have *ab initio* methods provided accuracy comparable to that of semiempirical techniques. This is partly because calculations must be repeated for a large number of nuclear configurations, partly because a larger number of atoms and electrons is involved in a collision than in a simple molecule, and partly because calculational techniques that provide adequate results for stable molecules may not be as adequate over the total range of internuclear separations. On the basis of the first two factors alone Murrell[23] predicts an increase of 81 times in the length of calculation for a triatomic versus a diatomic system.

The majority of *ab initio* calculations are based on the Hartree-Fock (HF) approximation, in which each particle is assumed to move in an effective potential field obtained by averaging over the positions of all other particles. This ignores instantaneous correlation of the motion of one electron with the motion of another of opposite spin, resulting in a difference between the HF energy and the exact nonrelativistic energy that is called the correlation energy. For nuclear configurations close to that of a stable molecule the correlation energy remains nearly constant, the HF surface is nearly parallel to the true surface, and vibrational motions, for example, can be predicted reasonably accurately. However, correlation effects are quite important in determining the strengths of weak, long-range interactions such as London forces. Also, whenever the number of paired electrons changes as a function of internuclear distance, excited states are close to the ground state, or a bond formed by a collision differs significantly from one that was broken, major changes in correlation energy are expected. Consequently a HF surface usually does not parallel the exact surface over the complete range of internuclear separations. A simple example is the HF calculation for H_2, which at large internuclear separations gives the energy as the average of $H^+ + :H^-$ and $H \cdot + \cdot H$. This causes the potential energy to increase too steeply along the coordinate leading to H—H bond breaking. The same applies to other bond dissociations or formations.

Configuration interaction (CI)[24] is the most common way of going beyond the HF limit, although unrestricted HF calculations,[25] in which different spatial orbitals may be occupied by the α- and β-spin electrons of a given pair, can also account for some of the correlation energy. The restricted HF method approximates the exact wave function by a single Slater determinant. CI involves taking a linear combination of such Slater determinants, each of which corresponds to a different arrangement of electrons in molecular spin orbitals, that is, to a different electron configuration. The coefficients that minimize the energy of the linear combination are then determined. By including enough configurations, energies very close to the true energy can be obtained. For example, Siegbahn and Liu[26] employed as many as 672 configurations to obtain a potential energy surface for $H + H_2$ that they estimate to be within $3.3\,kJ\,mol^{-1}$ of the exact surface for all $r_{XY} < r_{YZ} \leqslant 180\,pm$. This is the best surface calculated so far for the H_3 system. Obviously a number of configurations as large as 672 requires a great deal of computation. An alternative CI approach is to include only a few configurations – for example, ones that provide for correct dissociation products – and then adjust not only the coefficients in the CI expansion, but also the coefficients in the molecular orbitals themselves to minimize the energy. This multiconfiguration self-consistent field method has been used successfully in a number of cases.[27] Balint-Kurti[28] and Bader and Gangi[29] describe many other potential energy surfaces that have been calculated by *ab initio* and semiempirical methods.

TRAJECTORIES OVER POTENTIAL ENERGY SURFACES

In general a system composed of an atom and a diatomic molecule undergoing reaction does not follow the minimum-energy path over a potential energy surface. For example, Fig. 5.5 shows a possible path or *trajectory* followed by a super-

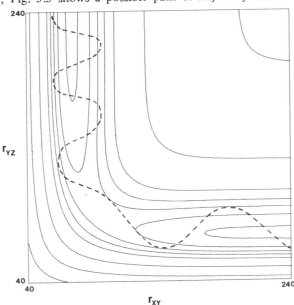

Figure 5.5. Trajectory for a reactive collision on the potential energy surface for H_3 with $H_X\!-\!H_Y\!-\!H_Z$ angle $\theta = 180°$. Oscillations in the trajectory in the entrance and exit channels correspond to vibrations of reactant and product H_2 molecules. Contours have been drawn at -448, -428, -408, -388, -358, -258, -58, and $342\,kJ\,mol^{-1}$ relative to separated $H + H + H$. Distances are in picometers. (Data for surface from Siegbahn and Liu[26] and Truhlar and Horowitz.[30])

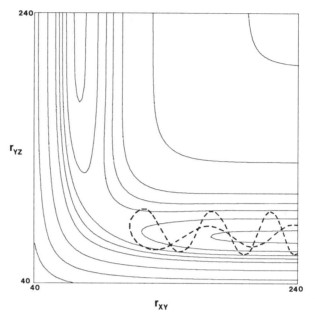

Figure 5.6. Trajectory for a nonreactive, inelastic collision on the potential energy surface for H_3 with H_X—H_Y—H_Z angle $\theta = 180°$. Transfer between translational and vibrational energy occurs upon collision. Energies of the contours are the same as in Fig. 5.5. Distances are in picometers. (Data for contour map from Siegbahn and Liu[26] and Truhlar and Horowitz.[30])

molecule H_3 over the surface of Siegbahn and Liu[26] at a fixed H—H—H angle of $180°$. The oscillation of the trajectory in the entrance channel results from vibration of the reactant H_2 molecule in addition to translational motion of approach of H and H_2. The larger oscillations in the exit channel indicate that the product H_2 molecule is in a more highly excited vibrational state than was the reactant. That the reaction

$$H_X + H_Y\text{—}H_Z \rightarrow H_X\text{—}H_Y + H_Z \qquad (5.7)$$

has occurred along the trajectory of Fig. 5.5 is evident because initially r_{YZ} oscillates about an equilibrium internuclear distance, but at the end of the trajectory r_{YZ} increases steadily while r_{XY} remains small. In a nonreactive trajectory r_{YZ} would not increase above some maximum value. For example, Fig. 5.6 shows a non-reactive but inelastic collision. Here there is a transfer between vibrational and translational energy, but no reaction occurs since r_{XY} is steadily increasing at the end of the trajectory.

It should be noted that both Fig. 5.5 and Fig. 5.6 represent the collinear surface ($\theta = 180°$) for H_3; that is, they can describe only the relative approach of H_X to H_Y—H_Z diagrammed in Fig. 5.7a. Many trajectories do not conform to this restriction. For example, assuming that H_X was not deflected from the straight line path shown in Fig. 5.7b, this collision could be represented by a trajectory on the H_3 surface for $\theta = 135°$ (Fig. 5.8). The approach of H_X to H_Y—H_Z shown in Fig. 5.7c, however, results in a continual change in the H_X—H_Y—H_Z angle θ. Thus

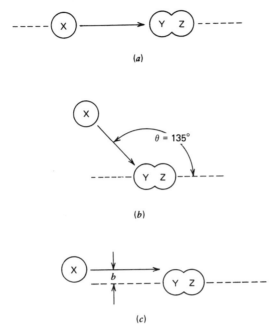

Figure 5.7. Possible approaches of H_X to H_Y-H_Z. (*a*) Collinear approach; (*b*) noncollinear approach, $\theta = 135°$; (*c*) noncollinear approach of H_X parallel to H_Y-H_Z axis, impact parameter $b \neq 0$.

the corresponding trajectory is not confined to a single fixed-angle surface and cannot easily be represented by diagrams such as those in Figs. 5.5, 5.6, and 5.8.

Comparison of Fig. 5.8 with Fig. 5.6 reveals that the barrier height for $\theta = 135°$ is larger than for $\theta = 180°$. Siegbahn and Liu's *ab initio* calculation confirms the prediction first made by London[12] that $\theta = 180°$ corresponds to the minimum barrier height. The greater barrier heights at noncollinear angles introduce a steric effect, corresponding to a reduction of the steric factor p introduced in Chapter 4. Nevertheless, there are many colliding atom–molecule pairs in a typical reaction mixture that have the extra energy needed to achieve reactive trajectories even though their approach is not collinear. In fact calculation of a large number of trajectories on the H_3 surface of Porter and Karplus, which is very similar to the Siegbahn-Liu surface of Figs. 5.5, 5.6, and 5.8, shows that the large number of nonlinear approaches leads to an average approach angle of $160°$ for reactive collisions.[31]

To calculate a trajectory, given a potential energy surface and the initial states of reactant species, requires solution of the equations of motion of the nuclei of the colliding particles. A rigorous quantum-mechanical treatment involves solving a nuclear wave equation in which the potential function for the nuclei is essentially the potential energy described above.[32] This yields a probability that reaction will occur, but not the precise motion of each nucleus during the collision. Furthermore, the quantum-dynamical method is very time-consuming. Fortunately, classical methods provide results that are in reasonable agreement with exact quantum results,[33] apparently because quantum effects are blurred by averaging

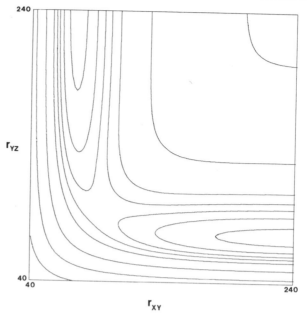

Figure 5.8. Contour map of the potential energy surface for H_3 with $H_X - H_Y - H_Z$ angle $\theta = 135°$. Energies of contours are the same as in Figs. 5.5 and 5.6. Distances are in picometers. Note the increased barrier height — more contours must be crossed by a reactive trajectory than in Fig. 5.5 or 5.6. (Data from Siegbahn and Liu[26] and Truhlar and Horowitz.[30])

over the range of dynamical collision parameters and energies encountered in calculating a large enough number of trajectories to determine a cross section. The principal exception occurs for collision energies at or near the barrier height. Because quantum mechanics is not deterministic there is a finite probability that a transition from reactant state to product state can occur even though the relative energy is less than the barrier height. Such quantum-mechanical tunneling is classically forbidden and cross sections or rate constants derived from classical trajectories are therefore too small in the energy range near the barrier height. Tunneling is especially important for light species such as H, H^+, e^-, and H_2, since as a particle's mass increases classical mechanics becomes a better approximation. Marcus[34] and Miller[35] have developed semiclassical methods that can treat classically forbidden situations like tunneling without giving up the computational simplicity of classical trajectories.

The earliest trajectory calculation[36] was so laborious that more than 20 years passed before the development of high-speed digital computers led to another attempt on the problem. Wall, Hiller, and Mazur[37] performed the first computer simulation of atom–diatomic molecule collisions in 1958, and rapid developments in computer technology have resulted in many more sophisticated calculations since their pioneering work.† Wall, Hiller, and Mazur studied the H + H_2 reaction using a LEP surface with a basin at the saddle point, and in their first paper they con-

---

† For reviews of the literature and details of the calculational technique, see reference 38.

sidered only collinear collisions, features that subsequent work has shown to be untenable. Nevertheless, their paper describes clearly the basic principles of computer simulation of collisions, and we describe their method in some detail here.

The equations of motion for nuclei of atoms of a supermolecule are usually expressed in Hamiltonian form, although Newtonian form also has been used.[39] The total energy for three atoms H_X, H_Y, H_Z at positions x_X, x_Y, x_Z along the x axis and with velocities $\dot{x}_X, \dot{x}_Y, \dot{x}_Z$ is given by the Hamiltonian H:

$$H = E_K + E_P = \left(\frac{m}{2}\right)(\dot{x}_X^2 + \dot{x}_Y^2 + \dot{x}_Z^2) + V(r_{XY}, r_{YZ}, r_{XZ}) \tag{5.8}$$

where m is the mass of a hydrogen nucleus and $V(r_{XY}, r_{YZ}, r_{XZ})$ is the potential energy defined earlier. The motion of the nuclei is governed by the set of six differential equations[38a]

$$\frac{\partial H}{\partial p_i} = \dot{q}_i \qquad (i = 1, 2, 3) \tag{5.9}$$

$$\frac{\partial H}{\partial q_i} = \dot{p}_i \qquad (i = 1, 2, 3) \tag{5.10}$$

where the q_i are generalized coordinates and the p_i are conjugate momenta. (As before $\dot{q}_i$ and $\dot{p}_i$ represent time derivatives.) Wall, Hiller, and Mazur chose the coordinates to represent relative motion of the nuclei:

$$q_1 = x_Z - x_X = r_{XZ}$$

$$q_2 = x_Z - x_Y = r_{YZ}$$

$$q_3 = x_Z \tag{5.11}$$

The conjugate momenta are:[38a]

$$p_1 = \frac{\partial E_K}{\partial \dot{q}_1} = m(\dot{q}_1 - \dot{q}_3) = -m\dot{x}_X$$

$$p_2 = \frac{\partial E_K}{\partial \dot{q}_2} = m(\dot{q}_2 - \dot{q}_3) = -m\dot{x}_Y$$

$$p_3 = \frac{\partial E_K}{\partial \dot{q}_3} = m(3\dot{q}_3 - \dot{q}_1 - \dot{q}_2) = m(\dot{x}_X + \dot{x}_Y + \dot{x}_Z) \tag{5.12}$$

Considering only relative motion by choosing the center of mass to be stationary corresponds to setting $p_3 = 0$, and the Hamiltonian becomes

$$H = \frac{p_1^2 + p_1 p_2 + p_2^2}{m} + V(q_1, q_2) \tag{5.13}$$

where $r_{XY} = q_1 - q_2$. The Hamiltonian is a function of only four variables — two coordinates and two momenta. Values of $\dot{q}_1, \dot{q}_2, \dot{p}_1$, and $\dot{p}_2$ can be obtained by taking appropriate partial derivatives of (5.13) as indicated in (5.9) and (5.10).

Given initial position and momentum values, a numerical integration can be carried out over very small time increments δt using equations such as

$$q_1(t + \delta t) = q_1(t) + \dot{q}_1(t) \cdot \delta t = q_1(t) + \left[\frac{\partial H}{\partial p_1(t)}\right] \cdot \delta t \qquad (5.14)$$

Actually Wall, Hiller, and Mazur used the Runge-Kutta-Gill numerical integration method,[40] which is more accurate than simple equations such as (5.14). They set $\delta t = 2.01 \times 10^{-16}$ s, a time increment short enough relative to the natural period of vibration of H_2 that all motions of the nuclei could be calculated accurately. For the collinear case the calculations were begun with $q_1 = r_{XZ} = 500$ pm, a distance large enough so that interaction between atom and molecule was negligible. The numerical integration continued until r_{XZ} returned to 500 pm or for 1024 iterations $(2.06 \times 10^{-13}$ s), whichever occurred first. The limitation to 1024 iterations was necessary because the spurious basin at the saddle point in the LEP surface resulted in many occurrences of long-lived complexes where all three atoms remained within 500 pm of each other. In their second paper[37b] Wall, Hiller, and Mazur removed the constraint of collinear collisions but continued to use the LEP surface. Of some 700 trajectories in which initial positions and momenta were selected on the basis of pseudorandom numbers generated by the computer (subject to the constraints that $r_{XZ} = 500$ pm and the total energy have a particluar value), only six resulted in reaction. The conclusion that a full analysis of three-dimensional trajectories would be excessively time-consuming seemed justified.

Rapid improvements in computer speed and capabilities soon contradicted that conclusion, however. In 1965 Karplus, Porter, and Sharma[41] carried out a true Monte Carlo investigation of the $H + H_2$ exchange reaction based on the highly accurate potential energy surface of Porter and Karplus.[18] Three-dimensional trajectories were calculated, requiring consideration of three coordinates and three momenta for each nucleus. By fixing the center of mass the number of equations of motion was reduced from 18 to 12. By use of conservation of energy and total angular momentum a further reduction could have been made, but instead constancy of energy and angular momentum were used as partial criteria of the accuracy of the numerical integration. Each trajectory was calculated by choosing coordinates so that the reactant atom and molecule were far enough apart that their interaction was negligible. The molecule was assigned to a particular vibrotational state by selecting vibrational and rotational quantum numbers v and J, and the relative velocity $\dot{r}$ of the atom and molecule was chosen. Then initial values of impact parameter b, orientation of the molecule, momentum of the molecule, and vibrational phase were selected by weighted random sampling from appropriate ranges. For example, since a given impact parameter b represents all approaches of H_X to H_Y-H_Z through an area given by $2\pi b \, db$ (see Fig. 4.5), b should be selected from a distribution uniform in b^2; that is, b would be chosen as the square root of a number s selected at random from the range $0 \leqslant s \leqslant b_{max}^2$, where b_{max} is a value of b such that no reactive collisions occur for $b > b_{max}$. The value b_{max} can be determined by a preliminary calculation of trajectories over a range uniform in b. Once all initial conditions had been selected by this Monte Carlo method, each trajectory was followed until an atom was again far from a molecule.

For each initial choice of $\dot{r}$, v, and J, Karplus, Porter, and Sharma calculated between 200 and 400 trajectories. The reactive cross section $\sigma_R(\dot{r}, v, J)$ was calculated as

$$\sigma_R(\dot{r}, v, J) = (\pi b_{max}^2) \lim_{N \to \infty} \left[\frac{N_R(\dot{r}, v, J)}{N(\dot{r}, v, J)} \right]$$

$$\cong \frac{\pi b_{max}^2 \, N_R(\dot{r}, v, J)}{N(\dot{r}, v, J)} \tag{5.15}$$

where $N_R(\dot{r}, v, J)$ represents the number of trajectories that resulted in reaction and $N(\dot{r}, v, J)$ is the total number of trajectories. The rate constant $k_{v,J}$ for a given vibrotational state can then be calculated from (4.22), or, if Boltzmann distributions are assumed for the velocities of H_X and H_Y-H_Z, from (4.23). The overall temperature-dependent rate constant can then be obtained by a weighted summation over all occupied vibrational and rotational states. If the energy of relative motion is considered (rather than the relative velocity), the final result is the same as (4.29) except that since $\sigma_R(\dot{r}, v, J)$ includes only the initial but not the final vibrotational state, the summation over k, l in (4.29) is not necessary. The final result is:

$$k = L \left[\frac{8}{\pi \mu (RT/L)^3} \right]^{1/2} q_{vj}^{-1} \sum_{v,J} g_J (2J+1) e^{-LE_{v,J}/RT} \int_0^{\infty} \sigma_R(E, v, J) e^{-LE/RT} E \, dE$$

$$\tag{5.16}$$

where g_J is the statistical weight of the Jth rotational state and the other symbols have the same meaning as in (4.29). Karplus, Porter, and Sharma calculated rate constants as a function of temperature from (5.16) over the range $300 \text{ K} \leqslant T \leqslant 1000 \text{ K}$. They found that the values obtained fit the Arrhenius equation with $E_a = 31.1 \text{ kJ mol}^{-1}$ and $A = 4.33 \times 10^{10} \, M^{-1} \, s^{-1}$, in excellent agreement with the experimental estimates of $E_a = 31.4 \text{ kJ mol}^{-1}$ and $A = 5.4 \times 10^{10} \, M^{-1} \, s^{-1}$.

For collinear trajectories the variation of internuclear distances with time can be represented graphically by tracing a path across the two-dimensional potential energy surface, as shown in Figs. 5.5 and 5.6. For the general case of noncollinear collisions such diagrams are not possible, because the angle θ of Fig. 5.7 is constantly changing and the trajectory passes through a series of fixed-angle surfaces. For the general, three-dimensional case it is convenient to plot all three internuclear distances as a function of time, as shown in Fig. 5.9. Such plots indicate vibration of an H_2 molecule as oscillation in the shortest internuclear distance and rotation as crossing of the curves for the two longer internuclear distances.

Karplus, Porter, and Sharma examined a large number of trajectories and from them drew several conclusions about the detailed dynamics of the $H + H_2$ collision. Most collisions are simple or direct: an atom approaches a molecule, collision occurs, and an atom and molecule separate without any long period during which all three internuclear distances are short. Occasionally, as shown in Fig. 5.10, a short-lived collision complex forms, but most collisions require only $\sim 1 \times 10^{-14}$ s. Since the vibration and rotation periods of H_2 are $\sim 0.5 \times 10^{-14}$ and $\sim 20 \times 10^{-14}$ s ($J = 1$), it is obvious that the collision interaction time is too short to

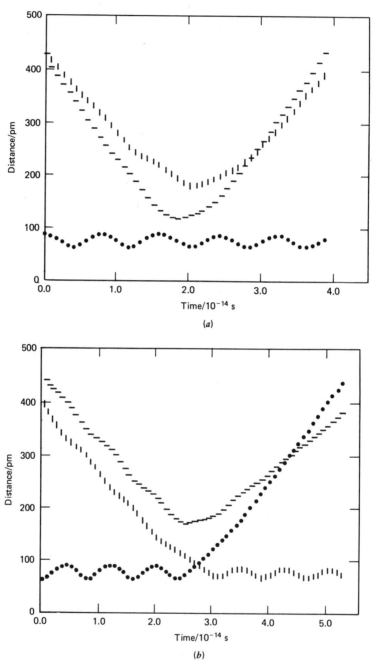

Figure 5.9. Typical trajectories calculated for collision of H with H_2. (*a*) Non-reactive collision; $\dot{r} = 1.96 \times 10^4\ m\,s^{-1}$, $J = 2$, $v = 0$. Crossing of the curves r_{XY} and r_{XZ} is due to rotation of $H_Y - H_Z$. (*b*) Reactive collision; $\dot{r} = 1.32 \times 10^4\ m\,s^{-1}$, $J = 0$, $v = 0$. Note that the collision is inelastic – the product molecule $H_X - H_Y$ has less vibrational energy than the reactant $H_Y - H_Z$ and the product is in an excited rotational state. (Data from Karplus, Porter, and Sharma.[31])

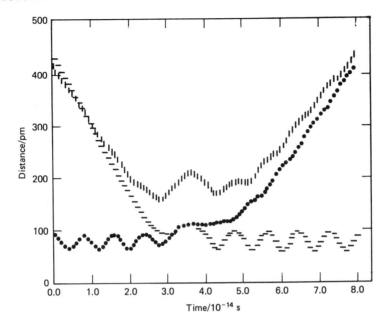

Figure 5.10.   An atypical reactive trajectory calculated for collision of $H + H_2$; $\dot{r} = 1.32 \times 10^4$ m s^{-1}, $J = 2$, $v = 0$. Note that all three nuclei remain close together for $\sim 3 \times 10^{-14}$ s. (Data from Karplus, Sharma, and Porter.[31])

permit equilibration of energy among the various degrees of freedom of the H_3 system. Nonreactive collisions of H with H_2 are essentially elastic. Vibrational energies are the same before and after most collisions, and trajectories such as the one shown in Fig. 5.6 that involves V–T energy transfer are rare. Reactive collisions entail more exchange of vibrational and rotational energy. In Fig. 5.9b, for example, the product molecule has acquired rotational energy and has lost vibrational energy. Even though the barrier height is least for a collision in which the H_X–H_Y–H_Z angle is 180°, the most probable angle for a reactive collision is not 180° because the differential solid angle through which H_X can approach H_Y–H_Z depends on $\sin \theta \, d\theta$ and is a maximum for $\theta = 90°$. As the relative velocity $\dot{r}$ increases this volume factor becomes more important and more reactive trajectories occur for the more probable nonlinear configurations. For a Boltzmann distribution of colliding systems at 1000 K the angle at which the largest contribution to reaction occurs is 160°, a result that is independent of the rotational state of the reactant molecule.

From their trajectory calculations Karplus, Porter, and Sharma found that the reactive cross section σ_R depends on the relative energy of approach of H_X to H_Y–H_Z as shown in Fig. 5.11. The figure also gives the variation of σ_R with E for the assumption that only the relative energy along the line of centers of hard spheres at the instant of impact is effective in surmounting the energy barrier and for the assumption that any relative energy in excess of the threshold value is effective. (Both of these simple cases are discussed in Chapter 4.) For the most part the curve obtained from trajectory calculations lies between the line-of-centers

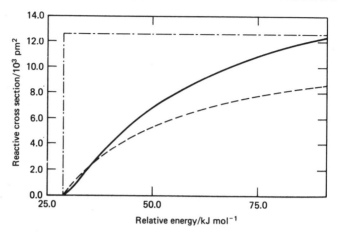

Figure 5.11.   Variation of reactive cross section with relative energy of approach of H_X to $H_Y - H_Z$. (———) $\sigma_R(E, v, J)$ from trajectory calculations; (— — —) σ_R from line-of-centers assumption [k given by (4.26)]; (— · —) σ_R from relative energy assumption [k given by (4.25)]. All curves assume a threshold energy of $28.9\,kJ\,mol^{-1}$ and a maximum impact parameter $b_{max} = 112\,ppm$. (Data from Karplus, Sharma, and Porter.[31])

curve and the total-relative-energy curve, approaching the latter at large relative energies and the former near the threshold energy. However, in the threshold region, which is of greatest importance in determining rate constants of thermal reactions, the cross section predicted by the trajectory calculations is smaller than for either of the hard-sphere models. The smaller initial slope of $\sigma_R(\dot{r}, v, J)$ causes significant differences in the temperature dependence of the rate constant compared to the line-of-centers prediction.

The barrier height of the potential-energy surface used in Karplus, Porter, and Sharma's trajectory calculations is $38.2\,kJ\,mol^{-1}$. Since the zero-point vibrational energy of H_2 is $25.9\,kJ\,mol^{-1}$, one might expect that the energy threshold for reaction in which reactant H_2 is in the $v = 0, J = 0$ state would be $(38.2 - 25.9)$ $kJ\,mol^{-1} = 12.3\,kJ\,mol^{-1}$. This value is much smaller than the minimum energy of $23.8\,kJ\,mol^{-1}$ for which Karplus, Porter, and Sharma found a trajectory that resulted in reaction. The $11.5\,kJ\,mol^{-1}$ difference might be attributed to zero-point energy of the activated complex, but since the calculation of reaction probability was not quantum mechanical but quasiclassical, there was no implicit requirement of such zero-point energy. Instead Karplus, Porter, and Sharma concluded that not all of the translational or vibrational energy can contribute to reaction. To test this conclusion they calculated 1640 collinear trajectories for which the H_2 molecule had zero-point energy and the relative velocity was $6.85 \times 10^3\,m\,s^{-1}$. This corresponds to a total energy that exceeds the barrier height by $3.5\,kJ\,mol^{-1}$. No reactive trajectories occurred.

The unavailability of all of the relative translational energy in a collinear collision can be better understood by considering a skewed and scaled contour map of the H_3 surface as shown in Fig. 5.12. Properly skewing and scaling the axes of a collinear surface results in a situation where the motion of a frictionless point mass over the surface gives the solution of the classical equations of motion for the three

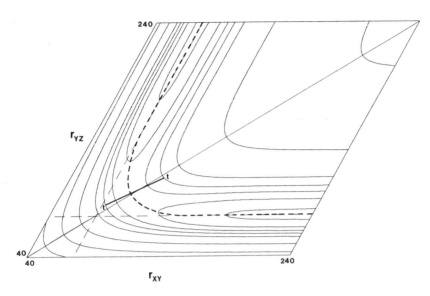

Figure 5.12. Skewed and scaled contour map of the H_3 potential energy surface. Energies of contours are the same as in Figs. 5.5, 5.6, and 5.8. Distances are in picometers. A frictionless mass point moving along the bottom of the entrance channel toward the saddle tends to follow the straight, dashed line rather than curving to follow the minimum-energy path. As explained later in the text, line tt, which is normal to the lower portion of the contour $V = -388\,\text{kJ mol}^{-1}$, can be used to show that transition-state theory is not exact for energies greater than $-388\,\text{kJ mol}^{-1}$.

nuclei.[3, 15, 38b] The angle φ by which the r_{YZ} axis is inclined toward the r_{XY} axis and the scaling factor β by which r_{YZ} must be multiplied must be chosen so as to eliminate the cross term $\dot{r}_{XY}\dot{r}_{YZ}$ in the expression for the kinetic energy of the three particles X, Y, Z:

$$E_K = (\tfrac{1}{2} m_{XYZ})(m_X m_{YZ}\dot{r}^2_{XY} + 2m_X m_Z \dot{r}_{XY}\dot{r}_{YZ} + m_{XY}m_Z\dot{r}^2_{YZ})$$

where $m_{XYZ} = m_X + m_Y + m_Z$ is the total nuclear mass. The cross term can be eliminated by choosing

$$\sin\varphi = \left(\frac{m_X m_Z}{m_{XY}m_{YZ}}\right)^{1/2}$$

and

$$\beta = \left(\frac{m_Z m_{XY}}{m_X m_{YZ}}\right)^{1/2}$$

For the H_3 system $\sin\varphi = 0.5$ and $\beta = 1.0$, resulting in the contour map as shown in Fig. 5.12. The dotted line in the figure is the minimum-energy path over the barrier, and it curves sharply at the corner of the surface. A point mass sliding along the minimum-energy path in the entrance channel with a given velocity corresponds to an atom approaching a molecule that has no vibrational energy. Depending on the precise geometry of the saddle region and the initial velocity, such a point mass will deviate from the minimum-energy path as a result of the path's curvature, passing over a higher barrier than would be predicted on the basis of the minimum-energy path. The path of a point mass becomes more difficult to predict by inspection when the reactant molecule is vibrating, and the point-mass analogy fails entirely for noncollinear trajectories, but similar dynamical effects can still occur. Hence not all of the energy of the colliding atom and molecule can necessarily contribute to reaction.

NONADIABATIC PROCESSES

The potential energy of a collection of nuclei and electrons will actually generate a number of surfaces as a function of changing nuclear positions. These correspond to the ground state (lowest energy surface) and to various excited states of the system. The surfaces bear labels according to their symmetry species and spin properties. When a reacting system jumps from one surface to another, the process is said to be nonadiabatic (or diabatic).

The adiabatic assumption is that electrons move very rapidly compared to nuclear motion. An equivalent statement is that the electronic wave function can change instantaneously for every change of nuclear positions. The result of this behavior is that a system that starts on the ground-state surface should always stay on the lowest surface, despite changes in nuclear configurations, such as reaction to form products. However, this is not always the case when the electronic wave function must change drastically over a short range of nuclear positions.

Figure 5.13 shows two cases of crossing potential energy surfaces; Fig. 5.13a is for states differing in symmetry, spin multiplicity, or both, and Fig. 5.13b is for two states of the same symmetry and spin species. In the latter case, the two surfaces do not cross, but instead are moved apart by an amount of energy, $2\epsilon_{12}$, which is quite large. In general a system on the lower surface can be expected to

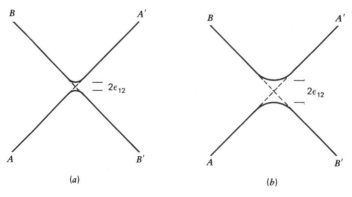

Figure 5.13. (a) Small splitting, ϵ_{12}, for two states differing in symmetry or multiplicity; (b) large splitting for two states of same symmetry and multiplicity. Reprinted with permission from R. G. Pearson, *Symmetry Rules for Chemical Reactions*, Wiley-Interscience, New York, 1976, P. 457.

stay there, unless activated externally. There is a small chance, however, that the system will jump.

The case of Fig. 5.13a is more interesting. Here are two surfaces that one might normally expect to cross, since they do not interact with each other at the lowest level of approximation. However, there is always some level at which they do interact, producing an energy splitting, $2\epsilon_{12}$, that is very small compared to that of Fig. 5.13b. Now the question becomes: Will the system stay on the lower sheet, in other words, behave adiabatically, or will it jump the very small gap between the two sheets? An approximate answer was given by Landau, Zener, and Stueckelberg.[42] The probability of jumping is given by

$$ P = \exp -\left(\frac{4\pi^2 \epsilon_{12}^2}{hv|s_1 - s_2|}\right) \qquad (5.17) $$

where v is the velocity along the reaction coordinate, and s_1 and s_2 are the slopes with which the two surfaces would cross if there were no splitting. The probability of crossing (nonadiabatic behavior) approaches unity as ϵ_{12} becomes very small, and as the velocity becomes very large.

Equation (5.17) is not very accurate, since it is based on an overly simple model; in particular, only one coordinate is considered. There has been a great deal of recent work on expanding the theory, to get rates of intersystem crossing and lifetimes of excited states.[43] The results are very complicated, because it is necessary to consider vibrational motions under conditions where two electronic states are very close. Under these conditions the Born-Oppenheimer approximation breaks down. Vibrational states and electronic states can no longer be separated, but are strongly coupled. Also the theories must evaluate ϵ_{12}, or its various equivalents.

There are two major mechanisms for creating an interaction energy between states of different symmetries and multiplicities. The first is vibronic coupling, often called Herzberg-Teller coupling. This is the mixing into the ground state of excited states of other symmetries as a result of nonsymmetric vibrations of the molecule. It is a mechanism for interaction of states of different symmetry species. The vibrations are now the normal modes that are orthogonal to the reaction coordinate.

The second important mechanism is spin-orbit coupling, which changes both the spin multiplicity and the state symmetry in a synchronous manner. It depends on the inclusion in the Hamiltonian of a magnetic energy due to the spin of an electron and its orbital motion, each of which can produce a magnetic dipole. The effect of this is that every state wave function is always mixed somewhat with states of different multiplicity. This provides a route for changing from one spin state to another.

TRANSITION-STATE THEORY

The transition-state theory is based on the application of statistical mechanics to reactants and activated complexes. Before deriving an expression for the rate of an elementary reaction, it is necessary to review some of the results and terminology of statistical mechanics.[44] According to the Boltzmann law, the probability that one molecule among many at thermal equilibrium is in a particular state having energy E_i and statistical weight g_i is proportional to $g_i e^{-E_i/kT}$. The probability that the molecule will exist in any one or another of its possible states, given by a series of values of the index i, is proportional to

$$\sum_i g_i e^{-E_i/kT} = q \tag{5.18}$$

This sum, designated by q, is the partition function for the molecule.

To show the importance and usefulness of partition functions consider an equilibrium between two isomers A and B:

$$A \rightleftharpoons B$$

There is a set of energy levels for A and another for B. Corresponding to each are partition functions q_A and q_B. Assuming ideal gases or dilute solutions, the equilibrium constant for the reaction is just the ratio of concentrations; but the ratio of concentrations is equal to the ratio of probabilities of the two species:

$$K = \frac{[B]}{[A]} = \frac{q_B}{q_A} \tag{5.19}$$

This is based on the same zero of energy for A as for B. However, it is more convenient to take the lowest energy level of each type of molecule as the zero for that molecule. In this case, if ΔE_0 is the increase in energy from the lowest level of A to the lowest level of B, then q_B must be replaced by $q_B e^{-\Delta E_0/kT}$, since $e^{-\Delta E_0/kT}$ factors out of each term in the partition-function summation. With this notation the equilibrium constant for the isomerization becomes

$$K = \left(\frac{q_B}{q_A}\right) e^{-\Delta E_0/kT} \tag{5.20}$$

The generalization of this equation for the equilibrium constant of any arbitrary reaction is found by expressing the Gibbs free energy change in terms of partition functions and then relating them to the equilibrium constant through the thermodynamic equation $\Delta G^{\ominus} = -RT \ln K_P$,[45] where $\Delta G^{\ominus}$ is the increase in standard free energy for the reaction at temperature T. The result is that for the reaction

$$aA + bB + \cdots = cC + dD + \cdots$$

the concentration equilibrium constant is given by

$$K_c = (LV)^{(a+b+\ldots-c-d-\ldots)} \left[\frac{q_C^c \cdot q_D^d \cdots}{q_A^a \cdot q_B^b \cdots}\right] e^{-L \Delta E_0^{\ominus}/RT} \tag{5.21}$$

where L is the Avogadro constant and V is the volume of the system. The term $L \Delta E_0^{\ominus}$ in the exponential expression is just $\Delta U_0^{\ominus}$, the standard increase in internal energy of the system per mole of reaction at the absolute zero of temperature. If partition functions can be calculated, equilibrium constants can be evaluated theoretically from (5.21).

PARTITION FUNCTIONS FOR TRANSLATION, ROTATION, AND VIBRATION

For the applications of interest here the energy levels of a molecule can be approximated well by a sum of translational, rotational, vibrational, and electronic energies.

$$E = E_t + E_r + E_v + E_e \tag{5.22}$$

Consequently each exponential of (5.18) can be written as a product, such as

$$e^{-E_i/kT} = e^{-E_t/kT} \times e^{-E_r/kT} \times e^{-E_v/kT} \times e^{-E_e/kT} \tag{5.23}$$

and the partition function can be approximated as a product of partition functions for each kind of energy.

$$q = q_t q_r q_v q_e \tag{5.24}$$

The total partition function may have factors due to electron and nuclear spins as well, but these are of little importance in kinetic applications. Also, q_e is usually unity, since most often only the lowest electronic level is occupied at thermal equilibrium, and that is usually a singlet state. Therefore we concentrate our attention on translational, rotational, and vibrational partition functions.

For *translation* discrete energy levels are not usually considered to exist, although on the basis of quantum mechanics they can be calculated for molecules moving in a container of a given volume and of certain shapes. The result is that the energy levels are exceedingly close together. It is more convenient to assume classical motion with continuous energy and to replace the summation in (5.18) by an appropriate integral: the phase integral of classical statistics. For one degree of freedom of translation (one coordinate x) the integral is

$$\frac{1}{h} \int_{-\infty}^{+\infty} \int_0^l e^{-m\dot{x}^2/2kT} m \; d\dot{x} \; dx = \frac{(2\pi mkT)^{1/2}}{h} l \tag{5.25}$$

where h is Planck's constant and l is the distance in the x direction through which the molecule of mass m is permitted to move.

For three degrees of freedom of translation the partition function is a product of three such integrals, and

$$q_t = \frac{(2\pi mkT)^{3/2}}{h^3} V \tag{5.26}$$

where the volume V appears as the product of three distances, say l_1, l_2, l_3, where these are the edges of a rectangular parallelepiped, supposing that to be the shape of the container. For other shapes (5.26) still holds.

For *rotation* the energy levels are discrete but sufficiently close together so that at ordinary temperatures rotation is fully excited. For a *linear molecule* the rotational partition function for two degrees of freedom is

$$q_r = \frac{8\pi^2 IkT}{h^3 \sigma} \tag{5.27}$$

where I is the moment of inertia and σ is the symmetry number, that is, the number of equivalent orientations in space. For a *nonlinear molecule* there are three rotational degrees of freedom and, in general, three different principal moments of inertia, A, B, and C. In this case

$$q_r = \frac{8\pi^2 (8\pi^3 ABC)^{1/2} (kT)^{3/2}}{h^3 \sigma} \tag{5.28}$$

For *vibration* the energy levels are far enough apart so that usually only a very few are occupied to any extent. The partition function q_v must be evaluated strictly as a summation rather than classically as an integral. For the case of a harmonic oscillator with one degree of freedom, as in a *diatomic molecule,* the result is

$$q_v = (1 - e^{-h\nu/kT})^{-1} \tag{5.29}$$

with ν as the fundamental vibrational frequency. This formula is based on the lowest vibrational level as defining the zero of energy, even though it is a half quantum of vibration above the minimum of the potential-energy curve. For s vibrational degrees of freedom the partition function is approximately a product of terms like (5.29):

$$q_v = \prod_{i=1}^{s} (1 - e^{-h\nu_i/kT})^{-1} \tag{5.30}$$

Here the ν_i are the various vibrational frequencies.

Formulas such as (5.23) through (5.30) are necessary if a relatively accurate calculation of chemical equilibrium is desired. However, since there are occasions when only order of magnitude can be obtained practically, it is of value to have rough approximations for the various kinds of partition functions. Such values are conveniently given for a single degree of freedom of each type and designated by f_t, f_r, f_v. The value of f_t is estimated from (5.25) by setting $l = 1$ dm and thus is appropriate for calculating rates per cubic decimeter. The value of f_r is obtained from (5.27) or (5.28) by setting these q_r's equal to f_r^2 or f_r^3 depending on the

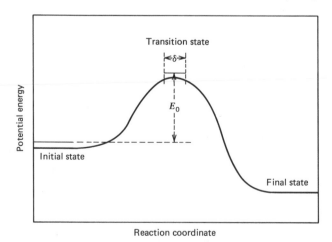

Figure 5.14.   Transition state at the top of the reaction potential-energy barrier.

number of degrees of freedom, and f_v is the same as q_v in (5.29). Substituting typical values of the molecular constants and using a temperature of 300–500 K yields the results of Table 5.1. Note that $f_t \gg f_r > f_v$. This is because the densities of energy levels are related in the same way. The temperature dependence of f_t, f_r, and f_v is also shown because it is important for understanding the temperature dependence of an equilibrium constant and later of a rate constant. The latter involves not only the T in the exponential factor $e^{-LE_0/RT}$, but also temperature

Table 5.1. Approximate values of partition functions per degree of freedom

Type	Designation	Order of magnitude	Temperature dependence
Translation	f_t	$10^9 - 10^{10}$	$T^{1/2}$
Rotation	f_r	$10^1 - 10^2$	$T^{1/2}$
Vibration	f_v	$10^0 - 10^1$	$T^0 - T^1$

dependence of q_t, q_r, and q_v. For vibration f_v is nearly independent of T since vibrational levels are only slightly excited. If, however, the temperature is very high or the vibration frequency ν is very small, then the vibration is fully excited and f_v is proportional to T. (This is T rather than $T^{1/2}$, as in translation and rotation, because there are both kinetic and potential energies associated with the one vibrational degree of freedom.)

DERIVATION OF THE RATE EQUATION

Consider the potential energy of the reacting system as a function of the reaction coordinate, the energy being a minimum with respect to variation in all other coordinates (see Fig. 5.14). The transition state is defined as including those configurations corresponding to the top of the potential-energy barrier within a small but otherwise arbitrary distance δ along the reaction coordinate and including all possible variation in other coordinates, that is, vibrations normal to the reaction coordinate, rotations, and translations of the center of mass of the system. Any system with configuration within the transition state is called an activated complex. The E_0 shown in Fig. 5.14 is the difference in energy between the lowest energy level in the transition state and the lowest energy level in the initial state. These levels are above the potential energy curve because of half quanta of vibration due to motion in other coordinates. E_0 is analogous to ΔE_0 introduced above. Since E_0 is the minimum energy that must be supplied to the reactants to achieve the transition state (and hence to produce products), E_0 corresponds to the threshold energy of collision theory.

   Let the concentration of activated complexes be $c^{\ddagger}$. (The double dagger is used exclusively to designate quantities referring to activated complexes or the transition state.) Activated complexes that are moving along the reaction coordinate from left to right in Fig. 5.14 correspond to the reaction of interest and are considered to have a positive velocity, $\dot{x}$. Those moving from right to left correspond to the reverse reaction and can be ignored. The mean velocity of activated complexes moving to the right can be calculated using the distribution law for a single component of velocity (4.1). The result is

$$\bar{v} = \frac{\int_0^\infty e^{-m^\ddagger \dot{x}^2/2kT} \dot{x}\, d\dot{x}}{\int_{-\infty}^\infty e^{-m^\ddagger \dot{x}^2/2kT}\, d\dot{x}} = \left(\frac{kT}{2\pi m^\ddagger}\right)^{1/2} \tag{5.31}$$

where $m^\ddagger$ is the effective mass of the activated complex for motion along the reaction coordinate and the integral in the numerator is confined to positive velocities $\dot{x}$ to select only those activated complexes that are moving toward products. The average time to move through the transition state to the right is then

$$\frac{\delta}{\bar{v}} = \delta \left(\frac{2\pi m^\ddagger}{kT}\right)^{1/2} \tag{5.32}$$

If we assume that reaction has been achieved once an activated complex has passed through the transition state from left to right, that is, that the system's trajectory is not such that it can somehow travel back across the saddle region to form reactants, then the rate of reaction in terms of number of complexes crossing to the right per unit volume per unit time is

$$\text{rate} = \frac{c^\ddagger}{\delta (2\pi m^\ddagger/kT)^{1/2}}$$

$$= \left(\frac{c^\ddagger}{\delta}\right)\left(\frac{kT}{2\pi m^\ddagger}\right)^{1/2} \tag{5.33}$$

Let us now apply statistical-mechanical theory to evaluate $c^\ddagger$ in terms of experimental reactant concentrations. Suppose that the reaction is bimolecular with the equation

$$A + B \rightarrow M^\ddagger \rightarrow \text{products}$$

where $M^\ddagger$ represents an activated complex. Then the ratio of concentration of activated complexes to concentrations of reactants is

$$\frac{c^\ddagger}{c_A \cdot c_B} = \frac{q_{M^\ddagger}/LV}{(q_A/LV)(q_B/LV)} e^{-E_0/kT} \tag{5.34}$$

Solving for $c^\ddagger$ and substituting into (5.33) yields the rate equation

$$\text{rate} = (LV)\delta^{-1} \left(\frac{kT}{2\pi m^\ddagger}\right)^{1/2} \frac{q_{M^\ddagger}}{q_A q_B} e^{-E_0/kT} c_A c_B \tag{5.35}$$

This shows immediately that the rate of a bimolecular elementary process is second order and it can easily be generalized [in the manner of (5.21)] to show that the order equals the molecularity for any elementary process. Comparing (5.35) with a second-order rate equation and replacing the Boltzmann constant by R/L to avoid confusion with the rate constant yields

$$k = (LV)\delta^{-1} \left(\frac{RT}{2\pi m^\ddagger L}\right)^{1/2} \frac{q_{M^\ddagger}}{q_A q_B} e^{-LE_0/RT} \tag{5.36}$$

and so the rate constant k can be obtained, provided that the partition functions can be evaluated.

For the activated complex the partition function differs from that for a stable molecule in that one degree of freedom — the one along the reaction coordinate — is unusual. This degree of freedom for motion across the barrier and confined within the distance δ (which defines the transition state) is very similar to a translational degree of freedom, in which the potential energy would remain constant. If δ is chosen small enough, the change of potential energy in the transition state may be made as small as desired. Therefore it is a good assumption that this special degree of freedom behaves as a translation and contributes to $q_{M\ddagger}$ a factor

$$(2\pi m^{\ddagger} kT)^{1/2} \left(\frac{\delta}{h}\right)$$

so that

$$q_{M\ddagger} = q_{\ddagger} \left(\frac{2\pi m^{\ddagger} RT}{L}\right)^{1/2} \left(\frac{\delta}{h}\right) \tag{5.37}$$

where $q_{\ddagger}$ includes the contribution from all degrees of freedom except that for the reaction coordinate. Substituting (5.37) into (5.36) yields

$$k = (LV)\delta^{-1} \left(\frac{RT}{2\pi m^{\ddagger} L}\right)^{1/2} \left(\frac{\delta}{h}\right) \left(\frac{2\pi m^{\ddagger} RT}{L}\right)^{1/2} \frac{q_{\ddagger}}{q_A q_B} e^{-LE_0/RT}$$

or

$$k = (LV) \left(\frac{RT}{Lh}\right) \frac{q_{\ddagger}}{q_A q_B} e^{-LE_0/RT} \tag{5.38}$$

where fortunately the unknown quantities $m^{\ddagger}$ and δ have canceled out.

Equation (5.38) can be generalized to an elementary process of any molecularity, say

$$aA + bB + \ldots \rightarrow M^{\ddagger} \rightarrow \text{products}$$

The result is

$$k = (LV)^{(a+b+\ldots-1)} \left(\frac{RT}{Lh}\right) \frac{q_{\ddagger}}{q_A^a q_B^b \ldots} e^{-LE_0/RT} \tag{5.39}$$

The group of factors excluding (RT/Lh) in (5.39) has the form of an equilibrium constant [see (5.21)]. Define

$$K^{\ddagger} = (LV)^{(a+b+\ldots-1)} \frac{q_{\ddagger}}{q_A^a q_B^b \ldots} e^{-LE_0/RT} \tag{5.40}$$

Then

$$k = \frac{RT}{Lh} K^{\ddagger} \tag{5.41}$$

Thus for any elementary process involving ideal gases the rate constant can be thought of as the product of a universal frequency factor RT/Lh that varies only with temperature and a factor $K^{\ddagger}$ that has the form of an equilibrium constant, though it is not strictly an equilibrium constant.

ASSUMPTIONS OF TRANSITION-STATE THEORY

In addition to the Born-Oppenheimer approximation and the adiabatic assumption, which are implicit in the use of a potential energy surface, the preceding derivation contains three major assumptions.[46] The first is that classical mechanics suffices to account for motion along the reaction coordinate, that is, that such motion can be described accurately by a classical trajectory. As in the case of trajectory calculations, this assumption fails for reactions involving species of low molar mass, where a correction for quantum-mechanical tunneling must be applied. Truhlar and Kupperman[47] have published appropriate correction factors for the classical theory, and Miller[48] has developed a quantum-mechanical transition-state theory that closely approximates the results of quantum scattering theory.

The second assumption is that the populations of species in various energy states obey the Boltzmann distribution. This same assumption is usually made in trajectory calculations when a set of exact reactive cross sections is converted to a rate constant by averaging over a distribution of relative speeds and internal energies [see, for example, the discussion preceding (5.16)]. Since activated complexes are the result of collisions among reactant species, the distribution of activated complexes should obey the Boltzmann law so long as the distribution of reactants does so. Provided the rate of reaction is not comparable to the rate at which collisions of reactant molecules redistribute energy among their various degrees of freedom, the rate constant calculated from transition-state theory should be accurate. If there is depletion of highly energetic reactant particles as a consequence of reaction, as in the fall-off and low-pressure regions for unimolecular reactions discussed in Chapter 4, then the predicted rate constant is too large. For thermal reactions a maximum deviation of 8% from the equilibrium Boltzmann distribution of reactant energies is predicted when $E_a/RT = 5$.[8] This is a good deal less than the experimental errors of measurement of frequency factors, and so this second, equilibrium assumption of transition-state theory is a good approximation in most situations.

The third assumption is that any activated complex that crosses the potential-energy barrier will necessarily continue on to form products of the reaction. However, there is the possibility that some trajectories may cross the barrier more than once, with many of these being reflected back to the reactant side as a consequence of the shape of the surface.[49] To account for this possibility the transmission coefficient κ, defined as the fraction of forward crossings that lead to products, was introduced into the transition-state theory. The corrected rate constant becomes

$$k = \kappa \left(\frac{RT}{Lh}\right) K^{\ddagger} \tag{5.42}$$

When a pair of colliding particles has energy only slightly larger than the barrier height, κ should be very close to unity. Once over the barrier such low-energy systems have few or no possible return paths, so that they almost invariably remain on the product side. For high-energy collisions on the other hand, trajectories can cross a wider region of the surface at the top of the barrier and many return paths become possible. This results in a lowered value of the transmission coefficient. Pechukas and McLafferty[50] have described a geometric means of determining a value of the potential energy up to which $\kappa = 1$. Their method is based on a skewed and scaled contour map such as is shown earlier in Fig. 5.12. Referring to this figure we note the line tt, which has been drawn normal to the lower portion of the contour $V = -388 \text{ kJ mol}^{-1}$. Since the minimum value of potential energy on the surface in Fig. 5.12 is -458 kJ mol^{-1} (corresponding to reactants or products), the reactants would require 70 kJ mol^{-1} to reach this contour. A system having initial energy of at least 70 kJ mol^{-1}, then, could conceivably reach and travel along line tt, and, if it did so, would be reflected back to the reactant side of the barrier. Hence κ would be less than unity for systems having 70 kJ mol^{-1} or more initial energy. Absence of normals such as tt that return to the reactant side before intersecting the second portion of a contour implies that a given trajectory can cross the saddle but once and therefore that $\kappa = 1$. Careful examination of the contours in the saddle region for the Siegbahn-Liu surface for H_3 reveals that $\kappa = 1$ up to at least 45 kJ mol^{-1} above the potential energy of the saddle point.

When reactant species obey the Boltzmann law and the temperature is not extremely high, high-energy collisions make only a very small contribution to the reaction rate. Therefore κ can usually be set to unity (that is, ignored) without substantial error.

One exception to this general rule is the case of bimolecular association of two atoms to form a diatomic molecule that is mentioned in Chapter 4. The potential-energy "surface" is just a diatomic potential-energy curve of the sort shown in Fig. 5.1. As the two atoms approach each other the potential energy drops and the kinetic energy increases, causing the system to move past the point of equilibrium until at a small internuclear distance the motion reverses and the atoms fly apart. There is perfect reflection, and κ is essentially zero. Therefore atoms are not likely to combine in such a bimolecular process, but must do so in a trimolecular collision where the third molecule removes energy in the form of translation or vibration. Figure 5.15 shows the difference between the possible bimolecular and trimolecular processes for a system of two atoms A, which can combine to A_2, and an inert atom M, all supposed to be in a linear configuration. The separated atoms are represented by a point such as x in Fig. 5.15. The path $x \to y \to x$ represents a bimolecular collision of A with A that is unsuccessful because of reflection. The path $x \to z \to y$ corresponds to a trimolecular collision that results in combination, with the molecule A_2 left in one of its vibrational levels and the atom and molecule having excess relative kinetic energy.

CALCULATING RATE CONSTANTS

To calculate rate constants from (5.41) or (5.39), detailed expressions for $K^{\ddagger}$ or the q_i must be given in terms of experimental quantities that can be derived from other than rate data. For a stable molecule A, q_A can be calculated using (5.24)–(5.30), the molecular constants such as moments of inertia and vibration frequencies being obtained generally from spectroscopic data. However, although $q_{\ddagger}$ can be written explicitly, the required moments of inertia and vibration frequencies are unknown experimentally. In principle they can be determined theoretically if the potential-energy surface is calculated with sufficient accuracy for the transition state. The moments of inertia depend on the internuclear distances of the activated complex, and the vibration frequencies are functions of the curvature of the potential-energy surface in directions normal to the reaction coordinate. The threshold energy E_0 can be obtained from the barrier height of the potential energy surface by correcting for the zero-point energies of reactants and transition state.

Transition-state-theory predictions of rate constants have been tested in several ways. The most obvious is to perform detailed calculations of the type described in the preceding paragraph and compare the results with experimentally observed rate constants. However, it is difficult to predict correctly the structure of the activated complex, much less its vibrational frequencies, without a reasonably accurate potential-energy surface, and those reactions for which surfaces have been calculated usually involve light species, for which tunneling is important. Consequently there have been few direct tests of transition-state theory, although Miller's quantum-mechanical transition-state theory,[48] in conjunction with accurate potential-energy surfaces, has been quite successful for reactions such as $H + H_2$.

In the absence of an accurate potential-energy surface, it is possible to assume a reasonable structure for the activated complex and estimate vibrational frequencies on the basis of a similar, stable molecule. An early example of such a treatment is the work of Gershinowitz and Eyring[51] on the assumed trimolecular reaction $2NO + O_2 \rightarrow 2NO_2$, where vibrational frequencies were estimated by assuming that

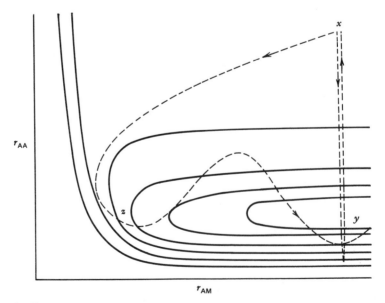

Figure 5.15.   Potential-energy contour diagram to illustrate the action of a third molecule M stabilizing a colliding pair of atoms AA.

the activated complex was very similar to N_2O_4. A similar process was used by Herschbach et al.[52] to estimate order-of-magnitude frequency factors for 12 bimolecular reactions, and many similar calculations have been made.[†]

Several approximate methods have been worked out for calculating frequency factors, particularly for reactions involving abstraction of an atom from a molecule by another atom or a small radical.[53] These depend on simplifications, such as assuming that certain parts of the partition function of the reactants do not change in forming the transition state, and using classical partition functions. The methods are very successful in simple cases. A more detailed procedure has been developed by Benson.[53b] A structure for the transition state is assumed and estimates are made of the entropy changes due to various factors such as ring breaking and ring making. The concept of additivity of bond properties is used. This allows estimates of activation energies to be made also. Various kinds of elementary reactions are considered, but generally the method is most useful for reactions of organic molecules in the gas phase.

Another test of transition-state theory is to calculate rate constants for a series of reactions in which there have been isotopic substitutions, attempting to account for the kinetic isotope effect.[54] Most commonly the isotopic substitution has been D for H, and so the problem of estimating the contribution due to tunneling is present. Within the limits imposed by this problem, comparisons with the results of studies of kinetic isotope effects tend to confirm the validity of transition-state theory.

Because of the general lack of sufficient details about potential energy surfaces for arbitrary reactions, it is of value to make estimates of rate constants, using mere order-of-magnitude values of partition functions as shown in Table 5.1. Consider a bimolecular reaction between two polyatomic molecules A and B of N_A and N_B atoms. Assume that A, B, and the activated complex are nonlinear so that each has three rotational degrees of freedom. The necessary partition functions can be represented as

$$q_A \cong f_t^3 f_r^3 f_v^{3N_A - 6}$$

$$q_B \cong f_t^3 f_r^3 f_v^{3N_B - 6}$$

$$q_{\ddagger} \cong f_t^3 f_r^3 f_v^{3(N_A + N_B) - 7} \qquad (5.43)$$

where for $q_{\ddagger}$ there is one less vibrational degree of freedom than normal because of the reaction coordinate. Substituting (5.43) into (5.39) and canceling f_t's, and so on between different species inasmuch as they represent only order of magnitude,

$$k \cong (LV)\left(\frac{RT}{Lh}\right)\left(\frac{f_v^5}{f_t^3 f_r^3}\right) e^{-LE_0/RT} \qquad (5.44)$$

Substituting $T \cong 300-500\ K$, the values of the f's from Table 5.1, and a total volume $V = 1\ dm^3$ results to within 1 or 2 powers of 10 in

$$k = (10^6\ dm^3\ mol^{-1}\ s^{-1}) e^{-LE_0/RT} \qquad (5.45)$$

[†] See, for example, work cited in reference 54, p. 84.

COMPARISON OF COLLISION AND TRANSITION–STATE THEORIES

The transition-state theory deals statistically with the structures of reactants and activated complexes, whereas the line-of-centers collision theory assumes that the reactants are structureless hard spheres. It is worthwhile to verify that the two theories agree when applied to the same situation. Since only translational energy is considered in the line-of-centers collision theory, applying transition-state theory to the same situation corresponds to considering monatomic molecules as reactants. The activated complex is equivalent to a diatomic molecule with internuclear distance d_{AB} but no vibrational degree of freedom, since instead there is motion along the reaction coordinate. If m_A and m_B are the masses of the two reactant molecules, the moment of inertia works out to be

$$I = \frac{d_{AB}^2 m_A m_B}{m_A + m_B}$$

and the rate constant written in detail is

$$k = (LV)\left(\frac{RT}{Lh}\right) \times$$

$$\frac{\dfrac{[2\pi(m_A + m_B)RT/L]^{3/2} V}{h^3} \cdot \dfrac{8\pi^2 d_{AB}^2 [m_A m_B/(m_A + m_B)] RT/L}{h^2}}{\dfrac{[2\pi m_A RT/L]^{3/2} V}{h^3} \cdot \dfrac{[2\pi m_B RT/L]^{3/2} V}{h^3}} e^{-LE_0/RT}$$

or after cancellation

$$k = L\left[\left(\frac{8RT}{\pi L}\right)\left(\frac{m_A + m_B}{m_A m_B}\right)\right]^{1/2} (\pi d_{AB}^2) e^{-LE_0/RT} \tag{5.46}$$

This is identical with (4.26) (which assumed a steric factor $p = 1$).

For comparison with the approximate expressions of transition-state theory, such as (5.44) and (5.45), it is desirable to have this rigid-sphere model also expressed in the same way.

$$q_A \cong f_t^3 \qquad q_B \cong f_t^3 \qquad q_\ddagger \cong f_t^3 f_r^2$$

This yields

$$k \cong (LV)\left(\frac{RT}{Lh}\right)\left(\frac{f_r^2}{f_t^3}\right) e^{-LE_0/RT}$$

$$\cong (10^{11} \text{ dm}^3 \text{ mol}^{-1} \text{ s}^{-1}) e^{-LE_0/RT} \tag{5.47}$$

which, of course, also agrees with line-of-centers collision theory with $p = 1$. Now, for a reaction between polyatomic or diatomic molecules, transition-state theory gives results (5.44 and 5.45) that are smaller by as much as 10^{-5}. This factor may be likened to the steric factor p, which may then be defined as the rate given by

Table 5.2. Approximate expressions and values for bimolecular rate constants on the basis of transition-state theory

Frequency factor			Steric factor	
Formula	Value/M^{-1} s^{-1}	T exponent	Formula	Value
Atom + atom → diatomic complex				
$\dfrac{VRT}{h}\dfrac{f_r^2}{f_t^3}$	$10^{10}-10^{11}$	$1/2$	1	1
Atom + linear molecule → linear complex				
$\dfrac{VRT}{h}\dfrac{f_v^2}{f_t^3}$	10^8-10^9	$-1/2$ to $1/2$	$\left(\dfrac{f_v}{f_r}\right)^2$	10^{-2}
Atom + linear molecule → nonlinear complex				
$\dfrac{VRT}{h}\dfrac{f_v f_r}{f_t^3}$	10^9-10^{10}	0 to $1/2$	$\left(\dfrac{f_v}{f_r}\right)$	10^{-1}
Atom + nonlinear molecule → nonlinear complex				
$\dfrac{VRT}{h}\dfrac{f_v^2}{f_t^3}$	10^8-10^9	$-1/2$ to $1/2$	$\left(\dfrac{f_v}{f_r}\right)^2$	10^{-2}
Linear molecule + linear molecule → linear complex				
$\dfrac{VRT}{h}\dfrac{f_v^4}{f_t^3 f_r^2}$	10^6-10^7	$-3/2$ to $1/2$	$\left(\dfrac{f_v}{f_r}\right)^4$	10^{-4}
Linear molecule + linear molecule → nonlinear complex				
$\dfrac{VRT}{h}\dfrac{f_v^3}{f_t^3 f_r}$	10^7-10^8	-1 to $1/2$	$\left(\dfrac{f_v}{f_r}\right)^3$	10^{-3}
Linear molecule + nonlinear molecule → nonlinear complex				
$\dfrac{VRT}{h}\dfrac{f_v^4}{f_t^3 f_r^2}$	10^6-10^7	$-3/2$ to $1/2$	$\left(\dfrac{f_v}{f_r}\right)^4$	10^{-4}
Nonlinear molecule + nonlinear molecule → nonlinear complex				
$\dfrac{VRT}{h}\dfrac{f_v^5}{f_t^3 f_r^3}$	10^5-10^6	-2 to $1/2$	$\left(\dfrac{f_v}{f_r}\right)^5$	10^{-5}

(5.44) divided by the rate on the basis of the rigid-sphere model (5.47). For polyatomic molecules

$$p \cong \left(\frac{f_v}{f_r}\right)^5 \tag{5.48}$$

and $p \ll 1$, since generally f_r exceeds f_v by about a factor of 10. Transition-state theory, then, predicts that reactions between polyatomic molecules are slower than expected on the line-of-centers rigid-sphere model, because several rotational degrees of freedom are lost in the transition state. These are replaced by vibrational degrees of freedom, which have a smaller probability associated with them.

Table 5.2 summarizes the results of the approximate transition-state theory, showing formulas and values for the steric factor as defined above and also for the frequency factor, which is everything in the rate constant except the exponential expression containing the activation energy. Also included is the exponent of T in the frequency factor, which is calculated using the information of Table 5.1 with f_v taken in the range T^0 to $T^{1/2}$. This exponent is the m of (2.44) and (2.45).

The general conclusion from Table 5.2 is that, up to a certain point, the more complicated the molecules the smaller the steric factor. The preexponential temperature dependence appears to decrease, perhaps to negative exponents, as complexity increases, but this effect is difficult to test experimentally because of the usually large effect of the exponential.

For *trimolecular reactions* a similar treatment can be given, but only two cases are mentioned here. For reaction of three atoms with a linear complex,

$$k \cong \frac{LV^2RT}{h} \frac{f_t^3 f_r^2 f_v^3}{f_t^9} e^{-LE_0/RT}$$

$$\cong \frac{LV^2RT}{h} \frac{f_r^2 f_v^3}{f_t^6} e^{-LE_0/RT} \tag{5.49}$$

or

$$k \cong (10^8 \ M^{-2} \ s^{-1}) e^{-LE_0/RT}$$

This is in fair agreement with the collision formula, although because of the high exponent on f_t the estimate is very crude. For three diatomic molecules in a nonlinear complex

$$k \cong \frac{LV^2RT}{h} \frac{f_v^8}{f_t^6 f_r^3} e^{-LE_0/RT} \tag{5.50}$$

In this case it is interesting that the temperature dependence would be as $T^{-7/2} e^{-LE_0/RT}$ if vibration were not excited. For a reaction with E_0 small or zero, such a dependence can lead to a rate that decreases as T increases. This constitutes a possible explanation of the peculiar temperature dependence of the nitric oxide–oxygen reaction,[51] but a more likely explanation is that the mechanism is complex (see p. 186).

UNIMOLECULAR REACTIONS

For unimolecular reactions the first-order rate constant based on approximate transition-state theory is

$$k \cong \frac{RT}{Lh} \frac{1}{f_v} e^{-LE_0/RT} \cong (10^{13} \text{ s}^{-1}) e^{-LE_0/RT} \qquad (5.51)$$

This is in good agreement with typical frequency factors that are found in the high-pressure limit. However, as is described in Chapter 4, in the fall-off region and at the low-pressure limit the concentration of molecules having sufficient energy to react is depleted by reaction. Under these circumstances the distribution of energy in activated molecules (and hence in activated complexes) does not obey the Boltzmann law. This violates a fundamental assumption of transition-state theory and thus the simple equation (5.51) does not apply. Nevertheless, it is possible to apply transition-state theory to unimolecular reactions over the full range of pressure. The method has been developed by Marcus[55] and is an extension of the Rice-Ramsperger-Kassel formalism described in Chapter 4. It is known as the Rice-Ramsperger-Kassel-Marcus (RRKM) theory. An excellent summary of the theory is given by Robinson and Holbrook.[56]

To obtain a conceptual overview of the RRKM theory, consider the contour

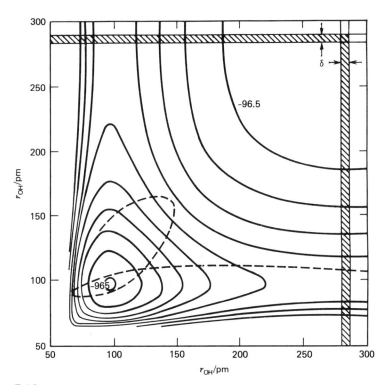

Figure 5.16. Contour map of potential energy of H_2O at a fixed angle of $104.5°$ (the equilibrium bond angle). Contours are at intervals of $-96.5 \text{ kJ mol}^{-1}$ relative to separated $H + H + O$. (Data from Sorbie and Murrell.[57])

map for potential energy of H_2O at a fixed angle of $104.5°$ as shown in Fig. 5.16. Suppose that we wish to calculate the rate of a unimolecular dissociation

$$H_2O \rightarrow OH + H$$

where the H—O—H angle remains at the equilibrium value. Such a reaction corresponds to trajectories (like the one shown) that leave the right-hand edge of Fig. 5.16, or to an equal number of symmetrically related trajectories that leave the top of the diagram (corresponding to fission of the other OH bond). To achieve such a reactive trajectory a molecule must have enough energy to climb out of the potential well at the lower left of the contour map. That is, the molecule must become activated or energized in the sense of the discussion of unimolecular reactions of Chapter 4 by gaining at least an energy E_0, the threshold energy. An energized molecule, represented in general as A^*, is not an activated complex unless it lies within a small region of width δ adjacent to a line on Fig. 5.16 that distinguishes reactants from products. Such a region is shaded in the diagram. (For unimolecular reaction there may or may not exist a saddle over which trajectories must pass. However, the dividing line, or, in general, surface, must occur in, a region where there is essentially no restoring force to prevent further bond extension. It is not necessarily at the exact top of an energy barrier.[58]) Clearly the number of ways of achieving an energized molecule A^* is much larger than the number of ways of achieving an activated complex $A^{\ddagger}$. Also, the relative populations of energized molecules and activated complexes can be expected to depend on the quantity of energy in the system.

To take explicit account of these points the Lindemann mechanism presented in Chapter 4 can be rewritten as follows:

$$A + M \rightarrow A^*_{(E_* \rightarrow E_* + \delta E_*)} + M \qquad \delta k_{1(E_* \rightarrow E_* + \delta E_*)} \qquad (5.52a)$$

$$A^* + M \rightarrow A + M \qquad k_2 \qquad (5.52b)$$

$$A^*_{(E_*)} \rightarrow A^{\ddagger} \qquad k_{3(E_*)} \qquad (5.52c)$$

$$A^{\ddagger} \rightarrow \text{products} \qquad k_{\ddagger} \qquad (5.52d)$$

In general an energized molecule A^* has numerous quantum states within an energy range E_* to $E_* + \delta E_*$, only some of which correspond to energy distributions that can lead to reaction. Even if a state that can dissociate is reached, vibrational modes will not necessarily be appropriately in phase at once. Hence the lifetime of an energized molecule is much greater than a vibrational period — typically in the range 10^{-9} to 10^{-4} s. The rate constant for formation of an energized molecule depends on how highly energized the molecule is. This has been emphasized by defining $\delta k_{1(E_* \rightarrow E_* + \delta E_*)}$ as the contribution to k_1 for energization to the small range E_* to $E_* + \delta E_*$. The rate constant for deactivation k_2 is assumed to be the same for all energy ranges and is obtained from the collision number as $Z_{MA}*/L [M] [A^*]$. (This strong-collision assumption is also made in the other theories of unimolecular reactions in Chapter 4.) On the basis of transition-state theory $k_{3(E_*)}$ is evaluated by summing a set of contributions over possible activated complexes.

In the terminology of the preceding paragraph we can rewrite (4.57), which defines a pseudo-first-order rate constant k for a unimolecular reaction, as

$$\delta k_{(E_* \to E_* + \delta E_*)} = \frac{k_{3(E_*)} \delta k_{1(E_* \to E_* + \delta E_*)}[M]}{k_2[M] + k_{3(E_*)}}$$

$$= \frac{k_{3(E_*)} \delta k_{1(E_* \to E_* + \delta E_*)}/k_2}{1 + k_{3(E_*)}/k_2[M]} \tag{5.53}$$

To obtain k (5.53) must be integrated over all values of E_* that can lead to reaction.

$$k = \int_{E_0}^{\infty} \frac{k_{3(E_*)} dk_{1(E_* \to E_* + dE_*)}/k_2}{1 + k_{3(E_*)}/k_2[M]} \tag{5.54}$$

The quantity $\delta k_{1(E_* \to E_* + \delta E_*)}/k_2$, which is required for evaluation of (5.54), is the ratio of forward and reverse rate constants for energization to the range E_* to $E_* + \delta E_*$. That is, $\delta k_{1(E_* \to E_* + \delta E_*)}/k_2$ is an equilibrium constant for energization, and so it is given by a ratio of partition functions q_A*/q_A. Since both A^* and A have the same zero of energy, no exponential term is needed. The partition function q_A is a normal partition function for all *active* degrees of freedom of the molecule. (An active degree of freedom is one that can contribute energy to the reaction. Translation of the molecule as a whole is inactive. Vibrations and internal rotations are active, except that zero-point vibrational energy does not contribute to reaction. Overall rotation may contribute by way of centrifugal effects, but conservation of angular momentum limits the extent of such contributions.) The function q_A* is the partition function for all energized molecules for which the energy that can contribute to reaction lies within the range E_* to $E_* + \delta E_*$. That is,

$$q_A* = \sum_{i(E_*)}^{i(E_* + \delta E_*)} g_i e^{-E_i/kT} \tag{5.55}$$

where i indexes all states from E_* to $E_* + \delta E_*$. If δE_* is small, then all the E_i may be set to E_* and the exponential terms may be factored out of (5.55).

$$q_A* = \left(\sum_{i(E_*)}^{i(E_* + \delta E_*)} g_i \right) e^{-E_*/kT} \tag{5.56}$$

The summation in (5.56) gives the number of states within an energy range that is δE_* wide and may be replaced by $N_{E_*} \delta E_*$, where N_{E_*} is the density of quantum states (that is, the number of quantum states per unit increment of energy) in the vicinity of energy E_*. The desired quantity $\delta k_{1(E_* \to E_* + \delta E_*)}/k_2$ becomes

$$\frac{\delta k_{1(E_* \to E_* + \delta E_*)}}{k_2} = \frac{q_A*}{q_A} = N_{E_*} \frac{\delta E_* e^{-E_*/kT}}{q_A} \tag{5.57}$$

Implicit in (5.57) is the assumption that quantum states are closely spaced and can be represented by a density function N_{E_*}; this assumption appears to be justified, especially since the short lifetimes of energized molecules result in broadening and overlapping of energy levels as a consequence of the uncertainty principle.[59]

The rate constant $k_{3(E_*)}$ can be obtained by assuming a steady-state concentration of activated complexes, that is, that the rate of formation of activated complexes equals the rate of decomposition of activated complexes to products:

$$k_{3(E_*)}[A^*] = k_{\ddagger}(\tfrac{1}{2}[A^{\ddagger}]) \tag{5.58}$$

where the factor $\tfrac{1}{2}$ enters because of all possible species $A^{\ddagger}$ in the region of Fig. 5.16 that correspond to activated complexes, only half will be moving in the direction of products. Therefore a statistical calculation of $[A^{\ddagger}]$ gives twice the required rate of decomposition. Solving (5.58) for $k_{3(E_*)}$ gives

$$k_{3(E_*)} = \frac{1}{2}\frac{k_{\ddagger}[A^{\ddagger}]}{[A^*]} \tag{5.59}$$

but since $k_{\ddagger}$ depends on the quantity of energy $E_{\ddagger t}$ available for translation along the reaction coordinate, $k_{3(E_*)}$ must be obtained by summing over all possible values of $E_{\ddagger t}$. Let $E_{\ddagger} = E_* - E_0$ be the energy of an activated complex in excess of that needed to achieve the transition state. (It is assumed that the activated complex has been formed from an energized molecule of energy E_*.) The energy $E_{\ddagger}$ can be divided in different ways between vibration and rotation ($E_{\ddagger vr}$) and translation along the reaction coordinate ($E_{\ddagger t}$) such that $E_{\ddagger} = E_{\ddagger vr} + E_{\ddagger t}$. Then the desired rate constant is the sum

$$k_{3(E_*)} = \sum_{E_{\ddagger t}(min)}^{E_{\ddagger}} \frac{1}{2}k_{\ddagger(E_{\ddagger t})}\frac{[A^{\ddagger}_{(E_{\ddagger t}, E_{\ddagger vr})}]}{[A^*_{(E_*)}]} \tag{5.60}$$

[Since $E_{\ddagger vr}$ is quantized it may happen that $E_{\ddagger vr}$ is never exactly equal to $E_{\ddagger}$. Therefore the summation of (5.60) will, in general, begin with some minimum value $E_{\ddagger t}(min)$.]

The rate constant $k_{\ddagger(E_{\ddagger t})}$ for decomposition of activated complexes of translational energy $E_{\ddagger t}$ along the reaction coordinate is evaluated by calculating the time required for a particle of mass $m^{\ddagger}$ to travel a distance δ as is done in (5.32) for normal transition-state theory. In this case, however, an average velocity is not needed, since a $k_{\ddagger(E_{\ddagger t})}$ is calculated for each $E_{\ddagger t}$ and the speed of translation is $(2E_{\ddagger t}/m^{\ddagger})^{1/2}$. The time to cross the barrier is $\delta/(2E_{\ddagger t}/m^{\ddagger})^{1/2}$ and the rate constant is

$$k_{\ddagger(E_{\ddagger t})} = \left(\frac{2E_{\ddagger t}}{m^{\ddagger}\delta^2}\right)^{1/2} \tag{5.61}$$

The ratio of concentrations in (5.60) can be calculated using statistical mechanics as the ratio of partition functions $q_{A\ddagger}/q_A{}^*$, and since each species lies within a narrow energy range this reduces to $\Sigma g_i^{\ddagger}/\Sigma g_i^*$, the identical exponential factor in numerator and denominator canceling out. From the procedure used to obtain (5.57), $\Sigma g_i^* = N_{E_*}\,\delta E_*$, but the situation is more complicated for $\Sigma g_i^{\ddagger}$ since there are relatively few vibrational and perhaps rotational states accessible to the activated complex for a given $E_{\ddagger vr}$. This is because $E_{\ddagger vr}$ is not large so that vibrations are much less highly excited. Consequently only the translational energy $E_{\ddagger t}$ can be treated in terms of a continuous distribution with density of states $N_{E_{\ddagger t}}$. Let $P_{E_{\ddagger vr}}$ be the number of vibration–rotation states whose energy is exactly $E_{\ddagger vr}$. Then the overall number of states in an energy range $\delta E_{\ddagger t}$ is $P_{E_{\ddagger vr}} N_{E_{\ddagger t}} \delta E_{\ddagger t}$, and since $\delta E_* = \delta E_{\ddagger} = \delta E_{\ddagger t}$, (5.60) becomes

$$k_{3(E_*)} = \sum_{E_{\ddagger t (min)}}^{E_{\ddagger}} \frac{1}{2} \left(\frac{2E_{\ddagger t}}{m^{\ddagger} \delta^2} \right)^{1/2} \frac{P_{E_{\ddagger vr}} N_{E_{\ddagger t}}}{N_{E_*}} \tag{5.62}$$

The number of translational levels within the energy range $E_{\ddagger t} \to E_{\ddagger t} + \delta E_{\ddagger t}$ can be obtained from a wave-mechanical treatment of a particle of mass $m^{\ddagger}$ in a box of length δ. The energy levels are

$$E_{\ddagger t} = \frac{n^2 h^2}{8 m^{\ddagger} \delta^2}$$

where n is a quantum number. The density of levels is

$$N_{E_{\ddagger t}} = \frac{dn}{dE_{\ddagger}} = \left(\frac{2 m^{\ddagger} \delta^2}{h^2 E_{\ddagger t}} \right)^{1/2} \tag{5.63}$$

Substituting (5.63) into (5.62) now yields

$$k_{3(E_*)} = \sum_{E_{\ddagger t (min)}}^{E_{\ddagger}} \frac{1}{2} \left(\frac{2E_{\ddagger t}}{m^{\ddagger} \delta^2} \right)^{1/2} \left(\frac{2m^{\ddagger} \delta^2}{h^2 E_{\ddagger t}} \right)^{1/2} \frac{P_{E_{\ddagger vr}}}{N_{E_*}} = \frac{1}{h N_{E_*}} \sum_{E_{\ddagger t (min)}}^{E_{\ddagger}} P_{E_{\ddagger vr}} \tag{5.64}$$

Equation (5.64) is usually corrected by a statistical factor $l_{\ddagger}$ that accounts for the number of equivalent pathways for reaction. For $H_2 O$, $l_{\ddagger} = 2$, corresponding to the fact that trajectories can leave either the top or the right-hand edge of Fig. 5.16. Also, a correction is necessary for the effects of rotations that stay in the same quantum state when a molecule is energized.[55b-55e] This correction is $q_r^{\ddagger}/q_r$, where $q_r^{\ddagger}$ and q_r are partition functions for adiabatic rotations in an activated complex and in a reactant molecule. The final result is

$$k_{3(E_*)} = l_{\ddagger} \frac{q_r^{\ddagger}}{q_r} \frac{\Sigma P_{E_{\ddagger vr}}}{h N_{E_*}} \tag{5.65}$$

To obtain a pseudo-first-order rate constant, substitute (5.65) and (5.57) into (5.54).

$$k = \frac{l_{\ddagger} q_r^{\ddagger}}{h q_r q_A} \int_{E_* = E_0}^{E_* = \infty} \frac{(\Sigma P_{E_{\ddagger vr}})(e^{-LE_*/RT}) \, dE_*}{1 + k_{3(E_*)}/k_2 [M]} \tag{5.66}$$

Provided that the distributions of vibrational and rotational levels are known for reactant and activated complex, (5.66) can be integrated numerically. In the high-pressure limit (5.66) gives the same result as the simple transition-state theory.

"THERMODYNAMIC" TREATMENT OF REACTION RATE

Inasmuch as $K^{\ddagger}$ of (5.40)–(5.42) is similar to an equilibrium constant, it is possible to define quantities analogous to the thermodynamic functions used in connection with ordinary equilibrium constants. The *standard Gibbs free energy of activation* $\Delta G^{\ddagger}$ is defined by the equation

$$\Delta G^{\ddagger} = -RT \ln [K^{\ddagger}(c^{\ominus})^{n-1}] \tag{5.67}$$

where n is the molecularity of the reaction and $c^{\ominus}$ is a standard-state concentration.[60] Generally the standard states of reactants and activated complex are unit concentrations where the concentration unit corresponds to whatever is used in evaluating the rate constant. However, different standard states may be used at different times, and one should be careful to specify which state has been chosen. The notation $\Delta G^{\ddagger}$ does not indicate explicitly that a standard value is involved, but this is implicit in (5.67) and subsequent equations in this section.

Define *standard enthalpy of activation* $\Delta H^{\ddagger}$ by

$$\Delta H^{\ddagger} = RT^2 \frac{d \ln [K_p^{\ddagger}(p^{\ominus})^{n-1}]}{dT} \tag{5.68}$$

where $K_p^{\ddagger}$ is the pressure equilibrium constant analogous to the concentration equilibrium constant $K^{\ddagger}$. Since for ideal gases

$$K_p^{\ddagger}(p^{\ominus})^{n-1} = K^{\ddagger}(c^{\ominus})^{n-1} \left(\frac{c^{\ominus}RT}{p^{\ominus}} \right)^{1-n}$$

(5.68) can be expressed in terms of concentrations as

$$\Delta H^{\ddagger} = RT^2 \frac{d \ln [K^{\ddagger}(c^{\ominus})^{n-1}]}{dT} - (n-1)RT \tag{5.69}$$

Define *standard entropy of activation* $\Delta S^{\ddagger}$ by

$$\Delta S^{\ddagger} = \frac{\Delta H^{\ddagger} - \Delta G^{\ddagger}}{T} \tag{5.70}$$

It follows from these definitions that

$$K^{\ddagger} = (c^{\ominus})^{1-n} e^{-\Delta G^{\ddagger}/RT} = (c^{\ominus})^{1-n} e^{\Delta S^{\ddagger}/R} e^{-\Delta H^{\ddagger}/RT} \tag{5.71}$$

and that the rate constant is

$$k = (c^{\ominus})^{1-n} \left(\frac{RT}{Lh} \right) e^{-\Delta G^{\ddagger}/RT} \tag{5.72}$$

or

$$k = (c^{\ominus})^{1-n} \left(\frac{RT}{Lh} \right) e^{\Delta S^{\ddagger}/R} e^{-\Delta H^{\ddagger}/RT} \tag{5.73}$$

[The units of k are given by $(c^{\ominus})^{1-n} \, s^{-1}$.] The quantities $\Delta G^{\ddagger}$, $\Delta H^{\ddagger}$, and $\Delta S^{\ddagger}$ are the differences in Gibbs free energy, enthalpy, and entropy between 1 mole of activated complexes and 1 mole of each reactant from which the activated complex is formed, all substances being at their standard-state concentrations, usually 1 M. Recall also that the contribution of motion along the reaction coordinate to $G^{\ddagger}$, $H^{\ddagger}$, and $S^{\ddagger}$ for the activated complex is not included in $\Delta G^{\ddagger}$, $\Delta H^{\ddagger}$, and $\Delta S^{\ddagger}$.

RELATIONSHIP OF $\Delta H^{\ddagger}$ AND VARIOUS KINDS OF ACTIVATION ENERGY

The Arrhenius, or empirical, activation energy E_a is defined in (2.42) as

$$\frac{d \ln k}{dT} = \frac{E_a}{RT^2} \tag{2.42}$$

It follows from (5.69) and (5.41) that for a *gas-phase reaction*

$$\Delta H^{\ddagger} = RT^2 \frac{d \ln k}{dT} - RT - (n-1)RT$$

or

$$\Delta H^{\ddagger} = E_a - nRT \tag{5.74}$$

where n is the molecularity (and order) of the elementary reaction. For a reaction in *solution* in the liquid state,

$$\Delta H^{\ddagger} = RT^2 \frac{d \ln K^{\ddagger}}{dT} = RT^2 \frac{d \ln k}{dT} - RT$$

or

$$\Delta H^{\ddagger} = E_a - RT \tag{5.75}$$

The relationship of E_a to the molar threshold energy LE_0 is discussed in Chapter 4. It depends on the variation of reactive cross section as a function of relative collision energy. In terms of transition-state theory LE_0 cannot be related to E_a or $\Delta H^{\ddagger}$ unless the detailed temperature dependence of the partition functions is known. Suppose that

$$k = BT^m e^{-LE_0/RT}$$

where B is independent of temperature and m is as indicated in Table 5.2 for various types of bimolecular reactions. Then as shown in (2.46),

$$E_a = LE_0 + mRT \tag{5.76}$$

and therefore from (5.74) for *gas-phase reactions*

$$\Delta H^{\ddagger} = LE_0 + (m-n)RT \tag{5.77}$$

and from (5.75) for reactions in *solution*

$$\Delta H^{\ddagger} = LE_0 + (m-1)RT \tag{5.78}$$

Because m is usually negative, the three kinds of activation energy are usually related as

$$\Delta H^{\ddagger} < E_a < LE_0 \tag{5.79}$$

INTERPRETATION OF ENTROPY OF ACTIVATION

By using, for example, (5.41) to obtain $K^{\ddagger}$ and then (5.67), (5.69), and (5.70), $\Delta S^{\ddagger}$ can be evaluated from rate data. For *bimolecular reactions* $\Delta S^{\ddagger}$ can be correlated with the steric factor p of the line-of-centers collision theory by comparing the rate-constant expression

$$k = \frac{Z_{AB}}{L\,[A]\,[B]}\, pe^{-LE_0/RT}$$

with (5.73) for $n = 2$. Thus

$$\frac{Z_{AB}}{L\,[A]\,[B]}\, pe^{-LE_0/RT} = \frac{RT}{Lhc^{\ominus}}\, e^{\Delta S^{\ddagger}/R}e^{-\Delta H^{\ddagger}/RT}$$

Substituting $\Delta H^{\ddagger}$ from (5.77) or (5.78) and dividing both sides by $e^{-LE_0/RT}$ yields

$$\frac{Z_{AB}}{L\,[A]\,[B]}\, p = \frac{RT}{Lhc^{\ominus}}\, e^{j}e^{\Delta S^{\ddagger}/R} \tag{5.80}$$

where $j = 2 - m$ or $1 - m$ depending on whether the reaction is in the gas phase or in solution. For $[A] = [B] = 1\,M$ and $c^{\ominus} = 1\,M$,

$$\frac{Z_{AB}}{L\,[A]\,[B]} \cong 10^{11} - 10^{12}\,M^{-1}\,s^{-1}$$

and

$$\frac{RT}{Lhc^{\ominus}} \cong 10^{13}\,M^{-1}\,s^{-1}$$

Since $j \cong 2\text{–}4$ and $e^{j} \cong 10\text{–}50$,

$$p \cong 10^{3}\, e^{\Delta S^{\ddagger}/R} \tag{5.81}$$

This would make $e^{\Delta S^{\ddagger}/R} \cong 10^{-3}$ and $\Delta S^{\ddagger} \cong -57\,J\,K^{-1}\,mol^{-1}$ for a bimolecular reaction with a maximum steric factor $p = 1$. However, the entropy of activation depends on the choice of standard state, that is, on the choice of concentration units in which the rate constant is expressed. For example, if the standard state is chosen to be $1\,mol\,cm^{-3}$, then $p \cong e^{-\Delta S^{\ddagger}/RT}$ and $p = 1$ corresponds to $\Delta S^{\ddagger} = 0$. In any event, $\Delta S^{\ddagger}$ is expected to become more negative for a reaction between two polyatomic molecules than for a reaction between two atoms.

A result more accurate than the approximate relationship (5.81) is obtained by taking the ratio of two equations (5.80) written for each of two similar reactions designated by subscripts 1 and 2:

$$e^{(\Delta S_1^{\ddagger} - \Delta S_2^{\ddagger})/R} = \frac{p_1 Z_1}{p_2 Z_2}$$

or

$$\Delta S_1^{\ddagger} - \Delta S_2^{\ddagger} = R \ln\left(\frac{p_1 Z_1}{p_2 Z_2}\right) \tag{5.82}$$

or, if Z is the same for both reactions,

$$\Delta S_1^{\ddagger} - \Delta S_2^{\ddagger} = R \ln\left(\frac{p_1}{p_2}\right) \tag{5.83}$$

This equation shows that in a series of reactions the more negative $\Delta S^{\ddagger}$ corresponds to the smaller steric factor.

For unimolecular reactions the corresponding treatment gives

$$k = \left(\frac{RT}{Lh}\right) e^{\Delta S^{\ddagger}/R} e^{-LE_0/RT}$$

and in this case $\Delta S^{\ddagger}$ is independent of the choice of standard state. If $(10^{13}\ \mathrm{s}^{-1}) \cdot e^{-LE_0/RT}$ is taken as a "normal" value of a unimolecular rate constant, then, since $(RT/Lh) \cong 10^{13}\ \mathrm{s}^{-1}$, $e^{\Delta S^{\ddagger}/R}$ is a factor that determines whether the reaction goes faster or slower than normal. If $\Delta S^{\ddagger}$ is positive, corresponding to a more probable activated complex, then the reaction is faster than normal. If $\Delta S^{\ddagger}$ is negative, the activated complex is less probable and the rate slower.

ORBITAL SYMMETRY RULES FOR CHEMICAL REACTIONS

Because of its exponential effect, the activation energy usually plays the dominant role in determining rate constants for chemical reactions. For this reason any rules that can be used to predict whether a given reaction mechanism has a large or small energy barrier are of great importance. A great advance was made by Woodward and Hoffmann in devising such rules.[61] They emphasized using the symmetry properties of molecular orbitals of reactants and products.

These symmetry properties are simply the changes in sign that occur for the wave function (MO) at different parts of the molecule. If the molecule has symmetry elements (planes of symmetry, rotational axes, etc.), this information can be given concisely by group theory methods. Similar changes in sign also exist for MOs in molecules that have no symmetry, but they cannot be expressed so conveniently. Woodward and Hoffmann correlated the molecular orbitals of the products with those of the reactants, using identical symmetry as the criterion. If the filled MOs of the reactants correlated with the filled MOs of the products, the reaction was said to be allowed by orbital symmetry (to have a low energy barrier). If the filled MOs of reactants and products did not correlate, the reaction was said to be forbidden by orbital symmetry (to have a large energy barrier).

A somewhat different approach can also be used, which is perhaps easier to visualize.[62] This is based on quantum-mechanical perturbation theory, in which the MOs of the reactants evolve into those of the products as a result of motion along the reaction coordinate.

Chemical reactions consist of the breaking of certain bonds between atoms and the making of new bonds. All MOs correspond to the bonding together of certain atoms, the antibonding of other atoms, and the nonbonding of the remaining atoms. It follows that in a chemical reaction certain molecular orbitals must be vacated of electrons and others must be filled to create the new bonding situation.

The most important of these changes often is a flow of electrons from the highest occupied molecular orbital (HOMO) to the lowest unoccupied molecular orbital (LUMO). The importance of these orbitals has been stressed by Fukui, who in 1952 named them the frontier orbitals.[63] In many cases, however, orbitals other than the HOMO and LUMO are the critical ones.

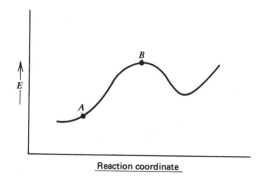

Figure 5.17.   Points A and B are discussed in the text.

Electron movement between two orbitals cannot occur unless the orbitals meet the symmetry requirement. For a bimolecular reaction, the requirement is simply that the two have a net overlap. For a unimolecular reaction, the symmetries of the two must match up with the symmetry of the motion of the nuclei. The nuclear motion can be regarded as equivalent to one or more of the normal modes of vibration of the molecule.

Consider a bimolecular reaction that has reached point A in Fig. 5.17. Two molecules have approached each other with a definite orientation. They have started to interact with each other, but the interaction energy is still small. This means that the MOs of the two separate molecules are still a good starting point for considering the combined system. Those of the same symmetry (positive overlap) will interact more and more strongly as the reaction coordinate is traversed, and at the transition state (point B in Fig. 5.17) quite different MOs will be produced.

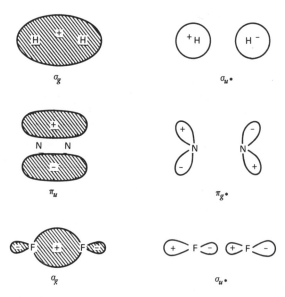

Figure 5.18.   Molecular orbitals of some diatomic molecules. The drawings are schematic, intended to show only the symmetry properties. Occupied MOs are shaded here and elsewhere.

For the reaction to be allowed by symmetry, we must have transfer of electrons from high-energy occupied MOs (φ_i) to low-energy empty MOs (φ_f) that have positive overlap. This minimizes the increase in energy of the system and prevents an excessive energy barrier. Now we can add an additional requirement on φ_i and φ_f using chemical knowledge rather than mathematical or quantum-mechanical arguments. This requirement is that φ_i must represent bonds that are broken and φ_f must represent bonds that are formed during the reaction, for their bonding parts. The reverse statement holds for their antibonding parts. Also we know that some atoms are much more electronegative than other atoms. Therefore electrons move more easily from φ_i to φ_f when they move in the direction of the more electronegative atoms.

To illustrate these principles Fig. 5.18 shows the HOMO and LUMO of several diatomic molecules. Reactions of H_2 are particularly easy to describe. The only MOs of reasonable energy are the bonding σ_g, which is occupied, and the anti-bonding σ_u^*, which is empty. The labels refer to the $\mathbf{D}_{\infty h}$ point group of the single molecule.

One of the simplest of chemical reactions would be isotope exchange between H_2 and D_2 (5.84). Let us assume that (5.84) occurs by a bimolecular mechanism in

$$H_2 + D_2 \rightarrow 2HD \tag{5.84}$$

which H_2 and D_2 collide broadside, giving rise to a four-center transition state:

$$
\begin{array}{c}
\text{H}-\!-\text{H} \\
| \quad\; | \\
| \quad\; | \\
\text{D}-\!-\text{D}
\end{array}
$$

As Fig. 5.19 shows, there is no empty MO of the same symmetry as any of the filled MOs. Hence the reaction is forbidden by orbital symmetry. Figure 5.19 also shows that the requirement of same symmetry is simply that of nonzero overlap.

The earlier statement that the reaction is forbidden simply means that the assumed mechanism has an excessive activation energy. Indeed the energy of the transition state in this case can be calculated quite accurately by *ab initio* quantum-mechanical methods.[64] It lies $515\,\mathrm{kJ\,mol^{-1}}$ *above* the energy of the reactants $H_2 + D_2$. The mechanism is impossible, for all practical purposes. Instead other allowed mechanisms take over for (5.84).

$$D_2 \rightleftharpoons 2D \tag{5.85}$$

$$D + H_2 \rightarrow HD + H \tag{5.86}$$

$$H + D_2 \rightarrow HD + D \tag{5.87}$$

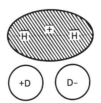

Figure 5.19.   Molecular orbitals in the $H_2 + D_2$ reaction. Symmetry labels are for the $\mathbf{C}_{2v}$ point group.

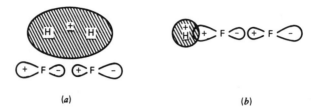

(a)   (b)

Figure 5.20.   Frontier molecular orbitals in the $H_2 + F_2$ reaction (or $H_2 + X_2$, where X is any halogen). (a) Forbidden four-center path. (b) Allowed free-atom path.

The atom—molecule reactions (5.86) and (5.87) are symmetry allowed. Indeed, reactions of free atoms and radicals rarely have serious symmetry restrictions and are often found. Conversely, four-center reactions of diatomic molecules almost always turn out to be forbidden. Figure 5.20 shows the reaction intermediates in the case of (5.88) and (5.89). Note that it clearly must be the empty σ_u^* MO of F_2

$$H_2 + F_2 \rightarrow 2HF \quad \text{(forbidden)} \tag{5.88}$$

$$H + F_2 \rightarrow HF + F \quad \text{(allowed)} \tag{5.89}$$

that accepts electrons in almost any reaction of F_2. This follows from the electronegativity of fluorine. Similar σ_u^* orbitals are the key LUMOs for the other halogen molecules as well.

For years it was considered that the hydrogen—iodine reaction occurred by a bimolecular process in which the two molecules collided broadside. In 1967 Sullivan showed that this was not so.[65] Instead there were mechanisms in which one (high temperature) or two (low temperature) iodine atoms reacted with a hydrogen molecule. These results are now completely understandable in terms of Fig. 5.20.

Figure 5.21 shows that the reaction of nitrogen with oxygen to form nitric oxide is forbidden by orbital symmetry. The important MOs are the filled π orbitals of N_2

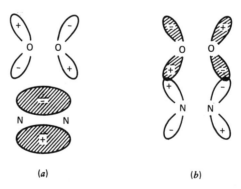

(a)   (b)

Figure 5.21.   Frontier molecular orbitals in the $N_2 + O_2$ reaction. (a) Electron flow from N_2 to O_2 forbidden by symmetry. (b) Opposite electron flow allowed by symmetry, but chemically forbidden.

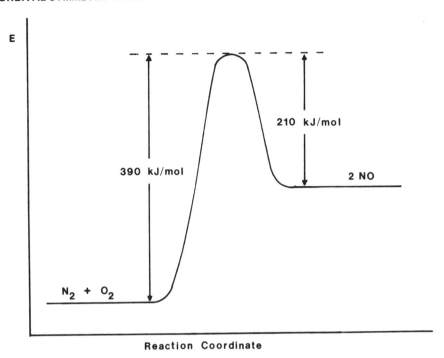

Figure 5.22. A symmetry-imposed barrier to a reaction automatically imposes a barrier for the reverse reaction.

and the π^* MOs of O_2. These antibonding orbitals of oxygen are half-filled and can act as either the HOMO or the LUMO. By symmetry, flow of electrons from the π orbital of N_2 to the π^* orbital of O_2 is forbidden. Electrons could move from O_2 to N_2 (Fig. 5.21b) as far as symmetry is concerned. However, this is chemically unrealistic on electronegativity grounds. Besides, such electron movement would convert the $O=O$ double bond into a triple bond by emptying an antibonding orbital.

Since (5.90) is strongly endothermic, the symmetry barrier is imposed in

$$N_2 + O_2 \rightarrow 2NO \qquad (5.90)$$

$$\Delta H = +180 \text{ kJ mol}^{-1}$$

addition to a thermodynamic one. More important, the reverse of (5.90), which is exothermic, is very slow, the activation energy being 210 kJ mol^{-1}. As Fig. 5.22 shows, a symmetry barrier for a forward reaction automatically creates the same barrier for the reverse reaction. The slowness of the reverse of (5.90) relative to the rate of combination of NO with O_2 to form NO_2 is a major factor leading to formation of photochemical smog in cities.[66]

A number of other examples of forbidden reactions of diatomic molecules have been discussed.[62] The pattern also extends to otherwise saturated groups joined by bonds of any order. Thus (5.91) and (5.92) are also forbidden as concerted

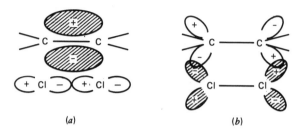

(a)   (b)

Figure 5.23. Frontier molecular orbitals for reactions such as $C_2H_4 + Cl_2$ or $C_2H_4 + H_2$. (a) Electron flow from C_2H_4 to Cl_2 (or H_2) forbidden by symmetry. (b) Opposite electron flow allowed by symmetry, but chemically unrealistic.

processes (Fig. 5.23). Reactions of lower symmetry, such as (5.93), are predicted to

$$C_2H_4 + Cl_2 \rightarrow C_2H_4Cl_2 \tag{5.91}$$

$$C_2H_4 + H_2 \rightarrow C_2H_6 \tag{5.92}$$

be partly forbidden by symmetry. This follows because the overlaps between the

$$C_2H_4 + HCl \rightarrow C_2H_4Cl \tag{5.93}$$

HOMO and LUMO orbitals of these systems are not identically zero, but can be seen to be small (Fig. 5.24). A two-step mechanism (5.94 and 5.95), which is not forbidden, becomes the usual path for addition reactions to multiple bonds.

$$C_2H_4 + H^+ \rightarrow C_2H_5^+ \tag{5.94}$$

$$C_2H_5^+ + Cl^- \rightarrow C_2H_5Cl \tag{5.95}$$

Orbital symmetry rules can be used to show that the reaction of nitric oxide with oxygen probably goes by the complex mechanism[62]

$$NO + O_2 \rightleftharpoons NO_3 \qquad \text{fast} \tag{5.96}$$

$$NO_3 + NO \rightarrow 2NO_2 \qquad \text{slow} \tag{5.97}$$

Reaction (5.96) is essentially at equilibrium and (5.97) is slow compared to the reverse of (5.96). Under these circumstances the observed rate constant becomes $k_{obs} = kK$ where k is the rate constant for (5.97) and K is the equilibrium constant for (5.96). The observed activation energy is given by $E_{obs} = E_a + \Delta H$, which can be negative if ΔH is a substantial negative number.

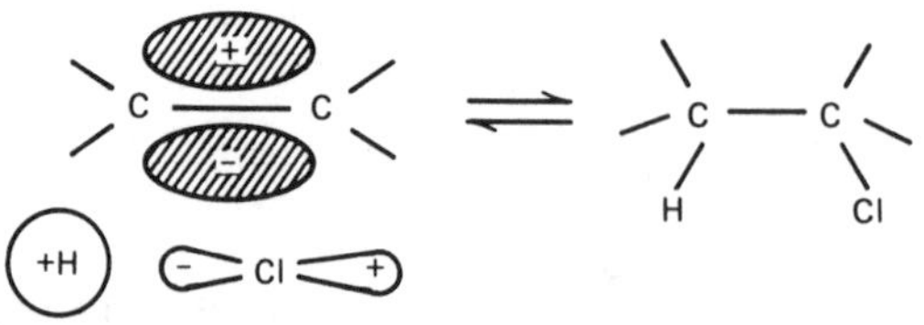

Figure 5.24. One-step addition of HCl to an olefin is partly forbidden. Overlap is small, but not zero.

PROBLEMS

5.1. Determine the correct values of β and φ for skewing and scaling a contour map of the collinear surface for each of the following reactions:

(a) $H + D_2 \rightarrow HD + D$
(b) $H + Br_2 \rightarrow HBr + Br$
(c) $I + H_2 \rightarrow HI + H$
(d) $Br + I_2 \rightarrow IBr + I$

5.2. From the potential energy surface of Siegbahn and Liu[26] the following parameters are obtained for the reaction

$$H + H_2 \rightarrow H_2 + H$$

Classical barrier height $= 41 \text{ kJ mol}^{-1}$; H—H distance in the linear, symmetric transition state $= 93 \text{ pm}$; vibrational frequencies of H_3: $\nu = 6.15 \times 10^{13} \text{ s}^{-1}$ (symmetric stretch), $\nu = 2.73 \times 10^{13} \text{ s}^{-1}$ (degenerate bending); bond length of $H_2 = 74.1 \text{ pm}$; vibrational frequency of H_2: $\nu = 1.32 \times 10^{14} \text{ s}^{-1}$. Use transition-state theory to calculate the rate constant at $T = 400 \text{ K}$. Compare your result with the experimental value obtained from $E_a = 31 \text{ kJ mol}^{-1}$, $A = 5.01 \times 10^{10} M^{-1} \text{ s}^{-1}$. (Hint: the rotational symmetry number for H_3 is 1, not 2.) What phenomenon that can substantially affect the predicted k has been omitted from the transition-state-theory calculation?

5.3. The frequency factors for two unimolecular reactions occurring at $200°C$ are 2.75×10^{15} and $3.98 \times 10^{15} \text{ s}^{-1}$. Calculate the entropy of activation in each case. How do the values obtained depend on the standard state used?

5.4. Carry out the calculation corresponding to problem **5.3** for two bimolecular gas reactions occurring at $300°C$ and having frequency factors of 7.4×10^{10} and $9.6 \times 10^9 M^{-1} \text{ s}^{-1}$. How do the results now depend on the standard state used?

5.5. On the basis of transition-state theory determine the dependence on temperature of the frequency factors for the following types of reactions:

(a) A bimolecular reaction of an atom with a diatomic molecule in which a linear activated complex is formed.

(b) A bimolecular reaction in which two diatomic molecules pass through a nonlinear transition state.

(c) A trimolecular reaction of three diatomic molecules forming a nonlinear complex with no free rotation.

In each case show how the experimental activation energy is related to the threshold energy.

5.6. Assuming translational partition functions to be 10^9 for each degree of freedom, rotational to be 10, and vibrational to be 1, calculate frequency factors in molar and molecular units for the reaction types in problem **5.5**.

5.7. Make the following calculations for the reaction

$$H + HBr \rightarrow H_2 + Br$$

(a) In the linear activated complex $r_{HH} = 150$ pm and the shorter $r_{HBr} = 142$ pm. Calculate the moment of inertia·of the complex and hence the rotational partition function.

(b) The three real vibrational frequencies in the activated complex are at wave numbers of 2340, 460, and 460 cm^{-1}. Calculate the vibrational partition function of the complex.

(c) Calculate the total partition function for the activated complex.

(d) Calculate the total partition functions for the reactants H + HBr, taking $r_{HBr} = 141.4$ pm and the vibrational wave number as 2650 cm^{-1}.

(e) Calculate the rate constant at 300 K, assuming the activation energy to be 5.0 kJ mol^{-1} at that temperature.

5.8. Derive the relationship between steric factor p and entropy of activation $\Delta S^{\ddagger}$ for a bimolecular reaction with the standard state chosen as 1 mol cm^{-3}. Hence verify the statement in the text that under these conditions $p = 1$ for $\Delta S^{\ddagger} = 0$.

REFERENCES

1. M. Born and R. Oppenheimer, *Ann. Phys.*, **84**, 457 (1927); M. Born, *Gott. Nachr., Math. Phys. Kl 1* (1951); M. Born and K. Huang, *Dynamical Theory of Crystal Lattices*, Oxford University Press, New York, 1954; H. Eyring and S. C. Lin, "Potential Energy Surfaces," in *Kinetics of Gas Reactions*, W. Jost, Ed., Physical Chemistry, An Advanced Treatise, Vol. 6A, Academic, New York, 1974.

2. A. Marcelin, *Ann. Phys.*, **3**, 158 (1915).

3. S. Glasstone, K. J. Laidler, and H. Eyring, *The Theory of Rate Processes*, McGraw-Hill, New York, 1941, Chapter 3.

4. B. H. Mahan, *J. Chem. Educ.*, **51** (11), 709 (1974).

5. H. Pelzer and E. Wigner, *Z. Phys. Chem.*, **B15**, 445 (1932).

6. H. Eyring, *J. Chem. Phys.*, **3**, 107 (1935); W. F. K. Wynne-Jones and H. Eyring, *J. Chem. Phys.*, **3**, 492 (1935); H. Eyring, *Chem. Rev.*, **17**, 65 (1935).

7. M. G. Evans and M. Polanyi, *Trans. Faraday Soc.*, **31**, 875 (1935).

8. B. M. Morris and R. D. Present, *J. Chem. Phys.*, **51**, 4862 (1969); B. H. Mahan, *J. Chem. Phys.*, **32**, 362 (1960).

9. A. I. M. Rae, *Chem. Phys. Lett.*, **18**, 574 (1973); R. G. Gordon and Y. S. Kim, *J. Chem. Phys.*, **56**, 3122 (1972); J. E. Lennard-Jones, *Adv. Sci.*, **51**, 136 (1954); R. A. Buckingham, *Proc. R. Soc. (London)*, **A168**, 264 (1938); J. C. Slater and J. G. Kirkwood, *Phys. Rev.*, **37**, 682 (1931); a computer program for calculating van der Waals potential energies of nonreacting, closed-shell atoms and molecules is available: S. Green and R. G. Gordon, QCPE No. 251, Quantum Chemistry Program Exchange, Indiana University, Bloomington, 1974.

10. (a) J. E. Lennard-Jones, *J. Chem. Phys.*, **20**, 1024 (1952); (b) J. L. Bills and R. L. Snow, *J. Am. Chem. Soc.*, **97**, 6340 (1975); M. B. Hall, *J. Am. Chem. Soc.*, **100**, 6333 (1978); H. B. Thompson, M. Wells, and J. E. Weaver, *J. Am. Chem. Soc.*, **100**, 7213 (1978).

11. For more explicit definitions see J. N. Murrell and K. J. Laidler, *Trans. Faraday Soc.*, **64**, 371 (1968); A. Tachibana and K. Fukui, *Theor. Chim. Acta*, **49**, 321 (1978).

12. F. London. *Probleme der modernen Physik (Sommerfeld Festschrift)*, Hirzel, Leipzig, 1928, p. 104; F. London, *Z. Elektrochem.*, **35**, 552 (1929).

13. H. Eyring and M. Polanyi, *Z. Phys. Chem.*, **B12**, 279 (1931); H. Eyring, *Chem. Rev.*, **10**, 103 (1932); Y. Sugiura, *Z. Phys.*, **45**, 484 (1927).

14. S. Sato, *Bull. Chem. Soc., Jap.*, **28**, 450 (1955); S. Sato, *J. Chem. Phys.*, **23**, 592 (1955); S. Sato, *J. Chem. Phys.*, **23**, 2465 (1955).

15. P. J. Kuntz, Report No. WIS-TCI-420, Theoretical Chemistry Insitute, The University of Wisconsin; P. J. Kuntz, "Features of Potential Energy Surfaces and Their Effects on Collisions," in *Dynamics of Molecular Collisions*, Part B, W. H. Miller, Ed., Modern Theoretical Chemistry, Vol. 2, Plenum Press, New York, 1976.

16. P. J. Kuntz, E. M. Nemeth, J. C. Polanyi, S. D. Rosner, and C. E. Young, *J. Chem. Phys.*, **44**, 1168 (1966).

17. J. C. Polanyi and J. L. Schreiber, *Chem. Phys. Lett.*, **29**, 319 (1974).

18. R. N. Porter and M. Karplus, *J. Chem. Phys.*, **40**, 1105 (1964).

19. W. Kolos and C. C. J. Roothaan, *Rev. Mod. Phys.*, **32**, 219 (1960).

20. R. N. Porter, R. M. Stevens, and M. Karplus, *J. Chem. Phys.*, **49**, 5163 (1968).

21. I. Shavitt, R. M. Stevens. F. L. Minn, and M. Karplus, *J. Chem. Phys.*, **48**, 2700 (1968); erratum, *J. Chem. Phys.*, **49**, 4048 (1968).

22. J. O. Hirschfelder, H. Eyring, and N. Rosen, *J. Chem. Phys.*, **4**, 121, 130 (1936).

23. J. N. Murrell, "The Potential Energy Surfaces of Polyatomic Molecules," *Novel Chemical Effects of Electronic Behavior, Struct. Bonding*, **32**, 93 (1977).

24. E. A. Hylleraas, *Z. Phys.*, **48**, 469 (1928).

25. J. A. Pople and R. K. Nesbet, *J. Chem. Phys.*, **22**, 571 (1954).

26. P. Siegbahn and B. Liu, *J. Chem. Phys.*, **68**, 2457 (1978); B. Liu, *J. Chem. Phys.*, **58**, 1925 (1973).

27. P. K. Pearson, W. J. Hunt, C. F. Bender, and H. F. Schaefer III, *J. Chem. Phys.*, **58**, 5358 (1973); H. Basch, *J. Chem. Phys.*, **55**, 1700 (1971); G. Das and A. C. Wahl, *J. Chem. Phys.*, **44**, 87 (1966).

28. G. G. Balint-Kurti, "Potential Energy Surfaces for Chemical Reactions," in *Molecular Scattering: Physical and Chemical Methods*, K. P. Lawley, Ed., Advances in Chemical Physics, Vol. 30, I. Prigogine and S. A. Rice, Eds., Wiley-Interscience, New York, 1975.

29. R. F. W. Bader and R. A. Gangi, "*Ab Initio* Calculation of Potential Energy Surfaces," in *Theoretical Chemistry*, Vol. 2, R. N. Dixon and C. Thomson, Eds., Specialist Periodical Report, The Chemical Society, London, 1975.

30. D. G. Truhlar and C. J. Horowitz, *J. Chem. Phys.*, **68**, 2466 (1978).

31. M. Karplus, R. N. Porter, and R. D. Sharma, *J. Chem. Phys.*, **43**, 3259 (1965).

32. T. F. George and J. Ross, *Annu. Rev. Phys. Chem.*, **24**, 263 (1973).

33. D. Russell and J. C. Light, *J. Chem. Phys.*, **51**, 1720 (1969).

34. R. A. Marcus, *Chem. Phys. Lett.*, **7**, 525 (1970); R. A. Marcus. *J. Chem. Phys.*, **54**, 3965 (1971).

35. W. H. Miller, *J. Chem. Phys.*, **53**, 1949 (1970); W. H. Miller, *Advances in Chemical Physics*, Vol. 25, I. Prigogine and S. A. Rice, Eds., Wiley-Interscience, New York, 1974 p. 69; W. H. Miller, "The Classical *S*-Matrix in Molecular Collisions," in *Molecular Scattering: Physical and Chemical Applications*, K. P. Lawley, Ed., Advances in Chemical Physics, Vol. 30, I. Prigogine and S. A. Rice, Eds., Wiley-Interscience, New York, 1975; J. N. L. Connor, *Chem. Soc. Rev.*, **5**, 125 (1976).

36. J. A. Hirschfelder, H. Eyring, and B. Topley, *J. Chem. Phys.*, **4**, 170 (1936).

37. (a) F. T. Wall, L. A. Hiller, Jr., and J. Mazur, *J. Chem. Phys.*, **29**, 255 (1958); (b) *J. Chem. Phys.*, **35**, 1284 (1961).

38. (a) R. N. Porter and L. M. Raff, "Classical Trajectory Methods in Molecular Collisions," in *Dynamics of Molecular Collisions*, Part B, W. H. Miller, Ed., Mod. Theoret. Chem., Vol. 2, Plenum Press, New York, 1976; (b) J. C. Polanyi and J. L. Schreiber, "The Dynamics of Bimolecular Reactions," in *Kinetics of Gas Reactions*, W. Jost, Ed., Physical Chemistry, An Advanced Treatise, Vol. VIA, Academic, New York, 1974 (c) D. L. Bunker, *Methods Comput. Phys.*, **10**, 287 (1971).

39. K. R. Wilson, "Multiprocessor Molecular Mechanics," in *Computer Networking and Chemistry*, P. Lykos, Ed., ACS Symposium Series, No. 19, American Chemical Society, Washington, DC, 1975; K. R. Wilson, in *Minicomputers and Large-Scale Computation*, P. Lykos, Ed., ACS Symposium Series, No. 57, American Chemical Society, Washington, DC, 1977.

40. S. Gill, *Proc. Cambridge Philos. Soc.*, 47, 96 (1951).

41. M. Karplus, R. N. Porter, and R. D. Sharma, *J. Chem. Phys.*, **43**, 3259 (1965).

42. L. Landau, *Phys. Z. Sowjetunion*, **1**, 88 (1932); **2**, 46 (1932); C. Zener, *Proc. Roy. Soc. (Lond.)*, **A137**, 696 (1932); **A140**, 666 (1933); E. G. C. Stueckelberg, *Helv. Phys. Acta*, 5, 369 (1932).

43. For reviews, see J. Jortner, S. A. Rice, and R. M. Hochstrasser, *Adv. Photochem.*, 7, 149 (1969); E. W. Schlag, S. Schneider, and S. Fischer, *Ann. Rev. Phys. Chem.*, **22**, 465 (1971).

44. T. L. Hill, *Introduction to Statistical Thermodynamics*, Addison-Wesley, Reading, MA, 1960; R. H. Fowler and E. A. Guggenheim, *Statistical Thermodynamics*, Cambridge University Press, London, 1960.

45. See, for example, W. J. Moore, *Physical Chemistry*, 4th ed., Prentice-Hall, Englewood Cliffs, N.J., 1972, p. 297; or I. N. Levine, *Physical Chemistry*, McGraw-Hill, New York, 1978, p. 722.

46. E. Wigner, *Trans., Faraday Soc.*, **34**, 29 (1938).

47. D. G. Truhlar and A. Kupperman, *J. Am. Chem. Soc.*, **93**, 1840 (1971).

48. W. H. Miller, *Acc. Chem. Res.*, **9**, 306 (1976).

49. K. Morokuma and M. Karplus, *J. Chem. Phys.*, **55**, 63 (1971).

50. P. Pechukas and F. J. McLafferty, *J. Chem. Phys.*, **58**, 1622 (1973).

51. H. Gershinowitz and H. Eyring, *J. Am. Chem. Soc.*, **57**, 985 (1935).

52. D. R. Herschbach, H. S. Johnston, K. S. Pitzer, and R. E. Powell, *J. Chem. Phys.*, **25**, 736 (1956).

53. (a) K. S. Pitzer, *J. Am. Chem. Soc.*, **79**, 1804 (1957): O. Sinanoglu and K. S. Pitzer, *J. Chem. Phys.*, **30**, 422 (1959); D. R. Herschbach, H. S. Johnston, and D. Rapp, *J. Chem. Phys.*, **31**, 1652 (1959); (b) S. W. Benson, *Thermochemical Kinetics*, 2nd ed., Wiley, New York, 1976.

54. (a) K. J. Laidler, *Theories of Chemical Reaction Rates*, McGraw-Hill, New York, 1977, pp. 86–99; (b) R. Weston and H. Schwartz, *Chemical Kinetics*, Prentice-Hall, Englewood Cliffs, N.J., 1972, pp. 109–113.

55. (a) R. A. Marcus and O. K. Rice, *J. Phys. Colloid Chem.*, **55**, 894 (1951); (b) R. A. Marcus, *J. Chem. Phys.*, **20**, 359 (1952); (c) G. M. Wieder and R. A. Marcus, *J. Chem. Phys.*, **37**, 1835 (1962); (d) R. A. Marcus, *J. Chem. Phys.*, **43**, 2658 (1965); (e) R. A. Marcus, *J. Chem. Phys.*, **52**, 1018 (1970).

56. P. J. Robinson and K. A. Holbrook, *Unimolecular Reactions*, Wiley-Interscience, New York, 1972, Chapter 4.

57. K. S. Sorbie and J. N. Murrell, *Mol. Phys.*, **29**, 1387 (1975).

58. D. L. Bunker and M. Pattengill, *J. Chem. Phys.*, **48**, 772 (1968).

59. O. K. Rice, *J. Phys. Chem.*, **65**, 1588 (1961).

60. P. J. Robinson, *J. Chem. Educ.*, **55**, 509 (1978).

61. R. B. Woodward and R. Hoffmann, *The Conservation of Orbital Symmetry*, Verlag Chemie, Weinheim, Bergstrasse, 1970.

62. R. G. Pearson, *Symmetry Rules for Chemical Reactions*, Wiley-Interscience, New York, 1976.

63. K. Fukui, T. Yonezawa, and H. Shingu, *J. Chem. Phys.*, **20**, 722 (1952).

64. H. Conroy and G. Malli, *J. Chem. Phys.*, **50**, 5049 (1969).

65. J. H. Sullivan, *J. Chem. Phys.*, **46**, 73 (1967); *ibid.*, **47**, 1566 (1967).

66. B. M. Fung, *J. Chem. Educ.*, **49**, 26 (1972).

SIMPLE GAS-PHASE REACTIONS – INTERPLAY OF THEORY AND EXPERIMENT

One way to test the theories of the preceding chapters is to compare predicted and observed rate constants, frequency factors, and/or activation energies for elementary gas-phase reactions. However, simple reactions in the gas phase are uncommon, most mechanisms being complex. By careful analysis of a complex mechanism it is often possible to get accurate data on the individual steps involved, but it must be remembered that molecularities may not always be as claimed. This is because molecularity depends on an assumed mechanism and is often not directly provable. The order of the reaction is experimentally available and is usually the chief evidence for the molecularity.

A great number of reaction rate data are currently available in the literature. A recent paper[1] lists 18 different compilations of data, and an even more comprehensive listing is available.[2] In some cases rate data are simply collected from the literature and listed, but in other compilations evaluations are made and a best value is selected for each parameter. Factors considered in evaluating rate data fall into two categories:[1,3] assessment of the technique and data processing used by the experimenter;[4] and comparison with other rate measurements, thermodynamic data, or theoretical predictions. In addition to serving as a test of the predictions of theories of reaction rates, carefully evaluated, recommended reaction rate data are extremely useful to engineers, atmospheric chemists, and others who wish to model complex reaction mechanisms such as combustion and photochemical smog. Techniques for computer simulation of such complex systems are discussed in Chapter 8.

Molecular beam experiments and other methods for the study of reactants (and sometimes products) in more or less well defined energy states are described in Chapter 4. Such experiments can provide information about reactive cross sections and reactive collisions that is not available from rate constants for thermal reactions. As we point out in Chapter 4, thermal rate constants are very dependent on the behavior of reactive cross section as a function of energy at energies just above the threshold energy. Consequently comparison of thermal rate data with predicted values provides a good test of theories only in this same energy region. Using molecular beams and other new techniques, reactive cross sections can be measured over a much wider range and compared with the results of trajectory calculations to test the accuracy of potential energy surfaces. Measured energy distributions of products of reactive collisions also provide information about the shape of the potential energy surface. Such experiments often provide a better test of a potential energy surface than would comparison of measured thermal rate constants with theoretical predictions.

This chapter considers the interplay of theory with many of the types of experimental data that are available for gas-phase reactions. Because so many data are available, we concentrate attention on a few reactions that have been very thoroughly studied. No attempt has been made to include or even make reference to all of the rate data that are available.

BIMOLECULAR REACTIONS

A number of second-order and presumably bimolecular reactions occur between radicals and have very small or zero activation energies. For example, the rate constant for recombination of methyl radicals is $2.0 \times 10^{10} \, M^{-1} \, s^{-1}$ at 400 K and $1 \times 10^{10} \, M^{-1} \, s^{-1}$ in the range 1200–1500 K.[5] This and several other radical-recombination reactions are included in Table 6.1. Since each of these reactions is the reverse of a unimolecular decomposition whose first-order rate constant falls off with decreasing pressure, each second-order rate constant must also decrease in the

Table 6.1. Rate constants at 400 K for second-order reactions having zero activation energy

Reaction	$\log (k/M^{-1} \, s^{-1})$			
	Hard-sphere[a]	Lennard-Jones[a]	Observed	Ref.
$CH_3 + CH_3 \rightarrow C_2H_6$	10.6	10.8	10.3	6a
			10.5	6b
			10.4	6c
$C_2H_5 + C_2H_5 \rightarrow C_4H_{10}$	10.5	10.8	10.5	7a
			8.6	7b
			~ 10	7c
$CCl_3 + CCl_3 \rightarrow C_2Cl_6$	10.4	10.7	9.9	8a
			9.7	8b
			9.6	8c
$CF_3 + CF_3 \rightarrow C_2F_6$	10.3	—	10.4	9a
			9.7	9b
$CH_3 + Cl \rightarrow CH_3Cl$	10.4	10.7	11.6	10

[a] For details of calculations, see reference 11.

same pressure range or equilibrium would not be maintained. For the methyl radical recombination there is a factor of 5 decrease when the concentration of radicals decreases from 2×10^{-7} to 2×10^{-9} M.[5] High-pressure values are given in Table 6.1.

Johnston[11] has computed rate constants for the reactions of Table 6.1 using both a hard-sphere, line-of-centers collision theory (4.26), and a modified collision theory that assumes an attractive potential proportional to d_{AB}^{-6}. [This is the attractive portion of the Lennard-Jones 6–12 interaction potential given in (4.31)] The latter calculation is equivalent to transition-state theory for these reactions, the transition state corresponding to configurations in which the centrifugal force of rotation of the two radicals around their center of mass is just balanced by the attractive force (proportional to d_{AB}^{-6}) between the radicals. That is, the approaching radicals must surmount a centrifugal energy barrier (of the type shown in Fig. 5.1) that varies with the rotational state of the supermolecule. Once reactants have crossed the barrier, they are assumed to become products.

Table 6.1 indicates that the two methods of calculation yield nearly the same results, and because the collision diameter (d_{AB} of Fig. 4.3) is nearly the same for all cases, the computed rate constants are nearly identical for all reactions. Both theories reproduce the average of the rate constants in Table 6.1 rather well, but neither can account for the upper and lower ends of the range of experimental values, which differ by a factor of 100.

Another class of reactions to which simple collision theory can be applied fairly successfully comprises those that involve ions. Ion–molecule reactions are important in radiation chemistry, electric discharges, shock waves, flames, short-wavelength photochemistry, and reactions in the upper atmosphere. Their rates can be studied using mass spectrometry, and there has been a great deal of recent research in this area.[12, 13] Positive ions are formed in the ionization chamber and expelled into an accelerating electric field. Here they can react with neutral molecules to produce secondary ions. The experimental quantity observed is the ratio of mass spectral currents for the primary and secondary ions. Also the pressure of the neutral reactant, the length of the path of the ions through the neutral molecules, and the accelerating voltage must be known.

The theory of ion–molecule reactions is based on (4.22) and (4.32). The speed distributions f_A and f_B of (4.22) are not Maxwellian, however, but in one dimension depend on the accelerating voltage and on distance from the electron beam to the exit slit.[14] The cross section σ_R of (4.22) may simply be taken as the line-of-centers value, in which case an equation analogous to (4.26) arises, but since even for nonpolar molecules there are ion-induced dipole forces, it is more realistic to use (4.32). For nonpolar molecules the potential energy is

$$U(d_{AB}) = \frac{-Q^2 \alpha}{(4\pi\epsilon_0)^2 2d_{AB}^4}$$

where α is the molecular polarizability, Q is the charge on the ion, $\epsilon_0 = 8.854 \times 10^{-12}$ C/V is the permittivity of a vacuum, and d_{AB} is the separation of the centers of mass of ion and molecule. Gioumousis and Stevenson[14b] substituted this potential energy into (4.32), integrated (4.22) for the appropriate distribution of speeds, and assumed that every collision resulted in reaction. This theory predicts a second-order rate constant

$$k = L(Q^2\alpha/4\epsilon_0^2\mu)^{1/2}$$ (6.1)

where μ is the reduced mass of the colliding particles.

The rate constant k can also be related to the experimental quantity $R = i_s V/i_p Nl$ where i_s and i_p are the secondary and primary ion currents, N/V is the number density of the molecules, and l is the distance from the electron beam to the exit slit. The relationship is

$$k = L\left(\frac{QEl}{2(4\pi\epsilon_0)m}\right)^{1/2} R$$

where E is the electric field strength and m is the mass of an ion. Thus experimentally determined rate constants can be compared with those predicted by (6.1). Johnston[11] has calculated rate constants for a number of ion–molecule reactions using hard-sphere collision theory and a theory that assumes ion-induced dipole attraction (proportional to d_{AB}^{-4}). Results of the calculations are compared with observed rate constants in Table 6.2. The hard-sphere results are usually 6–10 times smaller than observed rate constants, but when ion-induced dipole attraction is taken into account, much better agreement is achieved. Occasionally, as in the case of reaction of H_2^+ and O_2, the observed rate constant is significantly larger than predicted by either theory. This could be due to attractive forces that increase more rapidly than predicted by d_{AB}^{-4} as the ion and molecule approach.

Moran and Hamill[16] and Su and Bowers[17] have extended the theoretical treatment to include molecules that have permanent dipoles, but their theory still assumes that every collision results in reaction. Light and Lin[18] have eliminated this assumption by applying a statistical theory. This method has given good results for triatomic systems,[19, 20] but is cumbersome to apply to larger ones.

Another very large class of presumably bimolecular reactions comprises those in which an atom or radical interacts with a neutral molecule, usually abstracting an atom and leaving an atom or radical. The selection of rate parameters in Table 6.3 shows that most such reactions have significant activation energies. In many cases, especially those involving reactant molecules of some complexity, the frequency factor A for these reactions is several orders of magnitude smaller than the normal value of $10^{10}–10^{11}$ M^{-1} s^{-1}. This requires a steric factor p that is much less than unity and indicates considerable deviation from hard-sphere line-of-centers collision theory.

Table 6.2. Rate constants for ion–molecule reactions

| Reaction | $\log(k/M^{-1} s^{-1})$ | | | |
	Hard-sphere[a]	a/d_{AB}^4 [a]	Observed	Ref.
$H_2^+ + H_2 \rightarrow H_3^+ + H$	11.5	12.1	12.1	13
$H_2^+ + O_2 \rightarrow HO_2^+ + H$	11.4	12.1	12.7	13
$Kr^+ + H_2 \rightarrow KrH^+ + H$	11.4	11.9	11.5	13
$Ar^+ + H_2 \rightarrow ArH^+ + H$	11.4	12.0	12.0	13
$CH_4^+ + CH_4 \rightarrow CH_5^+ + CH_3$	11.1	11.9	11.9	15

[a] For details of calculations, see reference 11.

Reactions like those in Table 6.3 are most often found as, or suspected of being, steps in complex mechanisms, such as chain reactions. The elementary steps can be observed more or less directly by using some technique, such as photochemical dissociation or electric discharge to produce atoms or radicals at the time the reaction is to be observed. (For a discussion of experimental methods see Wolfrum.[23]) Often rate parameters may be of limited accuracy, since concentrations of free radicals are not always observed directly but are only estimated from other measurements. The usual technique is to estimate the mean lifetime of the free radical by comparison with some other time-dependent factor of the experiment. Mass spectrometry[25] and electron spin resonance[24] do allow direct measurement of radical concentrations and are widely used.

What are readily available for free-radical reactions are ratios of rate constants for competing reactions together with differences in activation energies. Thus the

Table 6.3. Some second-order reactions involving atoms or radicals $X\cdot + RY \rightarrow XY + R\cdot$

Reaction			$E_a{}^a$	$\log (A$	$D_{X-Y}{}^b$	$D_{R-Y}{}^b$
X	R	Y	/kJ mol^{-1}	$/M^{-1}$ s$^{-1})^a$	/kJ mol^{-1}	/kJ mol^{-1}
H	H	H	32	10.64	436	436
H	F	H	146	9.30	436	565
H	Cl	H	15	10.36	436	431
H	Br	H	9	10.79	436	366
H	I	H	3	10.70	436	299
CH$_3$	H	H	51	9.52	435	436
CH$_3$	CH$_3$	H	62	9.00	435	435
CH$_3$	C$_2$H$_5$	H	51	9.3	435	410
CH$_3$	CH$_3$CH$_2$CH$_2$	H	49	9.08	435	410
CH$_3$	CH$_3$CHCH$_3$	H	49	8.85	435	397
CH$_3$	CH$_3$CH$_2$CH$_2$CH$_2$	H	49	9.09	435	410
CH$_3$	CH$_3$CH$_2$CHCH$_3$	H	49	9.12	435	397
CH$_3$	(CH$_3$)$_3$C	H	34	8.47	435	385
CH$_3$	cyclo-C$_5$H$_9$	H	38	9.10	435	397
H	CH$_3$	H	50	11.10	436	435
H	C$_2$H$_5$	H	38	10.89	436	410
H	C$_3$H$_7$	H	35	11.16	436	397
H	i-C$_4$H$_9$	H	31	11.12	436	385
F	H	H	10	10.69	565	436
Cl	H	H	23	10.92	431	436
Br	H	H	82	11.43	366	436
I	H	H	140	11.20	299	436
H	F	F	10	11.08	565	155
H	Cl	Cl	8	11.57	431	243
H	Br	Br	15	11.97	366	194
H	I	I	0	11.6	299	153

[a] For the equation $k = A\,e^{-E_a/RT}$. Data from reference 21.
[b] From reference 22.

parameters in Table 6.3 for methyl-radical reactions are based on competition experiments between the reactions

$$2CH_3 \cdot \xrightarrow{k_1} C_2H_6 \tag{6.2}$$

$$CH_3 \cdot + RH \xrightarrow{k_2} CH_4 + R \cdot \tag{6.3}$$

From the relative amounts of methane and ethane products formed at various temperatures and at various concentrations of the hydrogen-atom donor RH, it is possible to get values of $k_2/k_1^{1/2}$ and $E_2 - \frac{1}{2}E_1$, where E_2 and E_1 are the activation energies for (6.2) and (6.1).[25] Any change in the experimentally determined value of k_1, therefore, affects all the other rate constants that have been determined by competition studies for reactions of the methyl radical.

Such a situation, where a series of related rate constants have been affected by a change in the rate constant for a standard reaction, has arisen recently in the case of ethyl-radical reactions.[7, 26] Note that rate constants reported in Table 6.1 differ by nearly 2 orders of magnitude, which would affect the rate constants for other ethyl-radical reactions by a factor of $10^{0.9}$ The low value of the rate constant for ethyl combination was obtained by an indirect method that depends on the thermochemistry of the ethyl radical.[7b] It now appears that the heats of formation of ethyl, isopropyl, and t-butyl radicals are 10, 10, and 20 kJ mol^{-1} higher than previously accepted values,[20b] a change that brings all values for the rate of recombination of ethyl radicals into agreement.

Bimolecular gas-phase reactions in which both reactants are stable molecules are not nearly as common as those involving atoms or radicals. Some examples are given in Table 6.4 in order of decreasing frequency factor. The entropy of activation, calculated as described in the preceding chapter, decreases roughly with increasing complexity of the reactant molecules as expected from transition-state

Table 6.4. Some second-order gas-phase reactions between stable molecules

Reaction	$E_a{}^a$/kJ mol^{-1}	$\log(A/M^{-1}\,s^{-1})^a$	$\Delta S^{\ddagger b}$/J K^{-1} mol^{-1}
$2HI \rightarrow H_2 + I_2$	184	10.9	-53
$2NOCl \rightarrow 2NO + Cl_2$	100	10.0	-70
$NO_2 + O_3 \rightarrow NO_3 + O_2$	29	9.8	-74
$NO + Cl_2 \rightarrow NOCl + Cl$	85	9.6	-78
$NO + O_3 \rightarrow NO_2 + O_2$	10	8.8	-93
$2C_4H_6 \rightarrow cyclo\text{-}C_8H_{12}$	112	8.1	-106
$C_4H_6 + C_2H_4 \rightarrow$			
$\quad cyclo\text{-}C_6H_{10}$	115	7.5	-118
$O_3 + C_3H_8 \rightarrow C_3H_7O$			
$\quad + HO_2$	51	6.5	-137
$C_4H_6 + CH_2CHCHO \rightarrow$			
$\quad cyclo\text{-}C_8H_9CHO$	82	6.2	-143
$2\,Cyclopentadiene \rightarrow$			
$\quad dicyclopentadiene$	62	4.9	-168

^{a} For the equation $k = A\,e^{-E_a/RT}$
^{b} At 298 K. Standard state is 1 M.

theory. For the Diels-Alder condensations listed near the bottom of Table 6.4, conversion of rotational degrees of freedom into vibrations in the activated complex results in low frequency factors[†] and very negative values for $\Delta S^{\ddagger}$. On the basis of collision theory these reactions require an assignment of a probability factor of 10^{-4} to 10^{-5}.

In Table 6.4 there is a noticeable tendency for the frequency factor to increase as the activation energy increases. Such a parallelism is a very common observation when a group of related reactions is studied.[27] Although different explanations can be involved in various cases one factor is probably always operative in helping to bring about this parallelism. This factor is related to higher density of energy states as the total energy of a complex molecular system increases. Thus, given that the requisite energy exists in the molecule, or group of molecules, there is a greater number of ways of distributing the energy in the system if the energy is large than if it is small. Alternatively we may say that a reaction can occur with a minimum of energy and a precisely defined configuration for the transition state or with an excess of energy and wide range of configurations for the nuclei of the reacting molecules in the transition state.

Benson[28] has suggested a method based on the thermodynamic formulation of transition-state theory for estimating a lower limit to the frequency factor of a bimolecular reaction. For a bimolecular gas-phase reaction it follows from (5.73) and (5.74) that the Arrhenius A factor is

$$A = \left(\frac{RT}{Lhc^{\ominus}}\right) e^2 e^{\Delta S^{\ddagger}/R} \tag{6.4}$$

Hence at a given temperature and for a given standard-state concentration $c^{\ominus}$, the frequency factor can be calculated from $\Delta S^{\ddagger}$. Assuming that the activated complex is tightly bound, the entropy of activation can be estimated by comparing the activated complex with known molecules. If in fact the activated complex is loosely bound, its entropy is greater than the estimated value, but the estimate serves as a lower limit for $\Delta S^{\ddagger}$ and therefore also for A.

As an example of the method consider the reaction

$$CH_3 \cdot + CH_3CH_2CH_3 \rightarrow CH_4 + (CH_3)_2CH \cdot$$

If the activated complex $H_3C\cdots H \cdots CH(CH_3)_2$ is tightly bound, then iso-butane $H_3C-CH(CH_3)_2$ should serve as a good model. The additional H atom in the activated complex is light enough that translational and rotational entropy differences should be negligible, but the external symmetry numbers are 1 for the activated complex and 3 for isobutane, making the entropy of the complex $R \ln 3$ greater than that for isobutane. Electron spin of the complex also contributes $R \ln 2$ to the entropy so that

$$S^{\ddagger}_{\text{complex}} \geqslant S^{\ominus}_{i-C_4H_{10}} + R \ln 6$$

and

$$\Delta S^{\ddagger} = S^{\ddagger}_{\text{complex}} - S^{\ominus}_{CH_3} - S^{\ominus}_{CH_3CH_2CH_3}$$

$$\geqslant S^{\ominus}_{i-C_4H_{10}} - S^{\ominus}_{CH_3} - S^{\ominus}_{CH_3CH_2CH_3} + R \ln 6$$

[†] Experimental errors in determining E_a also tend to make A change in the same direction.

Using values tabulated by Benson[28] for entropies of molecules and radicals this yields $A \geqslant 10^{7.2} M^{-1} s^{-1}$, somewhat lower than the experimental value in Table 6.3. Similar estimates can be made for a wide variety of H-atom transfer reactions, and such a calculation serves as a good check on an experimental frequency factor – any value less than the minimum estimate should immediately engender suspicion.

EMPIRICAL ESTIMATION OF ACTIVATION ENERGIES

A simple equation for estimating activation energies was originally suggested by Evans and Polanyi[29] and further developed by Semenov.[30] The Polanyi-Semenov equation for a series of related reactions is in general

$$E_a = \alpha \Delta E + C \qquad (6.5)$$

For a number of exothermic atom and radical reactions Semenov has suggested $\alpha = 0.25$ and $C = 48 \, kJ \, mol^{-1}$, while for endothermic reactions $\alpha = 0.75$ and $C = 48 \, kJ \, mol^{-1}$. Consider a series of reactions of the type $X \cdot + RH \rightarrow XH + R \cdot$, in which a hydrogen atom is transferred from one atom or radical to another. If X is kept the same and R is varied, the same bond is formed in all cases and (6.5) can be transformed into

$$E_a = \alpha [D_{R-H} - C'] \qquad (6.6)$$

where $C' = D_{X-H} - C/\alpha$. When R is an alkyl radical the R—H bond strength differs for methyl, primary, secondary, and tertiary R groups, and a test of (6.5) is possible for a wide range of atoms and radicals $X \cdot$.[21, 31] As shown in Fig. 6.1, plots of E_a versus D_{R-H} are reasonably linear in many cases. However, the substrates R must be very closely related for linear plots to be obtained.[32] Also, the values of α and C obtained when R = alkyl cover such a broad range as to cast considerable doubt on the use of (6.5) with fixed parameters for several different classes of reactions.

A more ambitious and successful procedure for calculating activation energies is the bond energy—bond order (BEBO) method of Johnston and Parr,[33, 11b] which makes use entirely of information from outside the field of chemical kinetics and contains no adjustable parameters. The original BEBO theory, which is described here, applies only to hydrogen-atom transfer reactions, but a modified form is available for reactions that involve transfer of other atoms.[34] Consider the general reaction

$$A + H{-}B \rightarrow A \cdots H \cdots B \rightarrow A{-}H + B \qquad (6.7)$$

in which A approaches H—B end on, forming an activated complex in which the partially broken $H \cdots B$ bond has bond order n and the partially formed $A \cdots H$ bond has bond order m in the sense of Pauling's concept of bond order.[35] The fundamental assumption of the BEBO method is that the minimum-energy path from reactants to products is that in which the sum of bond orders is unity. This is reasonable because for reactions such as (6.7) the activation energy is on the order of 10% of the relevant bond energies. Hence the energy required to break the H—B bond must be continually supplied by formation of the A—H bond, and there must be strong correlation between changes in one bond and changes in the other during reaction.

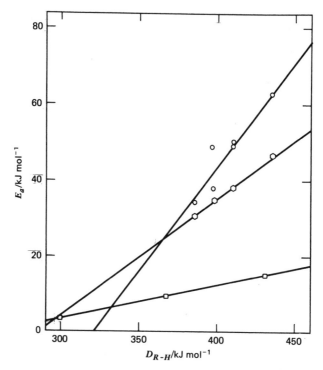

Figure 6.1. Polanyi plot of activation energy versus bond dissociation energy for reactions X• + RH → XH + R•. (○) X = CH₃, R = alkyl; (o) X = H, R = alkyl; (□) X = H, R = Cl, Br, I. Data from Table 6.3.

The total potential energy during reaction can be written as a sum of the energy of H ⋯ B, the energy of A ⋯ H, and a repulsive energy involving A and B. The repulsive term arises because the electron on the H atom being transferred must be able to pair simultaneously with an electron on A and one on B. This can occur only if there is a single electron on A whose spin is parallel to the spin of a single electron on B, and so, if the H atom is ignored, A and B must correspond to a triplet electronic state. The H ⋯ B and A ⋯ H energies can be obtained from bond orders using the relationship

$$E = Dn^p \tag{6.8}$$

where D is the dissociation energy of a single bond, n is the bond order, and p is the empirically determined slope of a log–log plot of bond energy versus bond order. The A ⋯ B repulsive energy is estimated from an anti-Morse function of the type used in the semiempirical LEPS method (5.6). The total potential energy E_{AHB} (relative to the minimum potential energy of reactant HB) can then be represented as

$$E_{AHB} = D_1(1 - n^{p_1}) - D_2(1 - n)^{p_2} + D_3 C(n - n^2)^{0.26\beta} \tag{6.9}$$

where $n = 1 - m$ is the order of the H ⋯ B bond, D_1, D_2, and D_3 are dissociation energies of H–B, A–H, and A–B bonds, p_1 and p_2 are the exponents of (6.8) for H ⋯ B and A ⋯ H, β is the spectroscopic constant of (5.6) for A–B, and $C = 0.5 \exp{[-\beta(r^0_{HB} + r^0_{AH} - r^0_{AB})]}$, where r^0_{HB}, r^0_{AH}, and r^0_{AB} are equilibrium internuclear distances. Equation (6.9) can be evaluated over the range $0 \leqslant n \leqslant 1$, and E_{AHB} is found to pass through a maximum that corresponds to the barrier height of the potential energy surface for the reaction. BEBO calculations for 130 hydrogen-atom-transfer reactions agree within about 10 kJ mol^{-1} with experimentally observed activation energies.[33] This is excellent agreement when it is recalled that the barrier height not only differs from the experimental activation energy by the difference between the sum of zero-point energies of all reactants and the zero-point energy of the activated complex, but also depends on the extent of thermal excitation of reactant and transition state. Table 6.5 compares BEBO calculations and experimental values for activation energies of selected reactions.

Benson[28] has suggested several empirical methods for estimating activation energies of atom-transfer reactions. One involves correlation of the intrinsic activation energy (that is, the activation energy for the reaction in its exothermic direction) of a reaction such as (6.7) with the sum of the electron affinities of the end groups A and B. The best fit for about 30 reactions is given by the equation

$$E_a = \frac{61.9 \text{ kJ mol}^{-1} - 84.0(EA_A + EA_B)}{1 + \Delta H/(167 \text{ kJ mol}^{-1})} \tag{6.10}$$

An even better fit can be obtained by multiplying two factors, one for the bond involving A and one for the bond involving B, as in (6.11)

$$E_a = F_A \cdot F_B \tag{6.11}$$

Benson[28] lists best values of the factors F for a number of end groups, and results using (6.10) and (6.11) are included in Table 6.5.

Table 6.5. Comparison of empirical estimates with experimental activation energies for reactions $A + HB \rightarrow AH + B$

Reaction		$E_a/\text{kJ mol}^{-1}$			
A	B	Expt.[a]	BEBO[b]	EA[c]	Bond factor[d]
H	F	146	138	129	133
H	Cl	15	21	0	7
H	Br	9	8	0	4
H	I	3	4	0	2
H	CH_3	50	46	45	44
H	C_2H_5	38	33	43	36
H	$(CH_3)_2CH$	35	33	—	—
H	$(CH_3)_3C$	31	25	—	—
CH_3	CH_3	62	54	53	51
CH_3	C_2H_5	51	42	51	42
CH_3	$(CH_3)_2CH$	49	38	—	—
CH_3	$(CH_3)_3C$	34	29	—	—

[a] Data from reference 21.
[b] As calculated in reference 33.
[c] Estimated from electron affinities of A and B using (6.10). See reference 28.
[d] Estimated from multiplicative factors for AH and HB bonds using (6.11). See reference 28.

EMPIRICAL POTENTIAL ENERGY SURFACES

A strong argument can be made[36] that the most productive theory or model is one that can be calibrated to experimental results and can be refined and extended as new, perhaps unexpected results become available. Many of the empirical schemes for constructing potential energy surfaces fall into this category, since they contain parameters that can be adjusted to fit the surface to experimental data, such as activation energies. In principle such fitting can also be extended to reproduce observations of the apportionment of energy among translational, vibrational, and rotational modes in products[37] or to the effect of different kinds of reactant energy on observed cross sections for reaction. However, the latter cases require extensive trajectory calculations for each choice of parameters to obtain predicted energy distributions or cross sections, and they also require experimental data that are available in only a small but growing number of cases. A less expensive and more common approach is to use adjustable parameters in an empirical surface to vary the position and thickness of the barrier or the contours of the valleys and, by comparing trajectories on the various surfaces, explore how the overall shape of the surface influences cross sections and hence rate constants. Adjustable parameters can also be used to fit an empirical surface to the results of an *ab initio* calculation. This is important because *ab initio* calculations yield the potential energy only at a limited number of discrete points on the surface, while trajectory calculations require values of potential energy and its derivatives at arbitrarily selected points.

Some empirical surfaces are based entirely on data from outside the field of kinetics and contain no adjustable parameters. An example is provided by the work of Murrell[38] on HCN. The potential energy is a function of the three internuclear distances r_{CH}, r_{CN}, and r_{NH}:

$$V_{HCN}(r_{CH}, r_{CN}, r_{NH}) = V_{CH}(r_{CH}) + V_{CN}(r_{CN})$$

$$+ V_{NH}(r_{NH}) + V_I(r_{CH}, r_{CN}, r_{NH}) \qquad (6.12)$$

The diatomic potential energies V_{CH}, V_{CN}, and V_{NH} are obtained as Rydberg functions

$$V = -D\left(1 + ar + \sum_{j=2}^{n} b_j r^j\right) e^{-ar} \qquad (6.13)$$

with $n = 3$ for V_{CH} and V_{CN} and $n = 1$ for V_{NH}. Parameters D, a, and b_j in (6.13) are those that best reproduce the experimental dissociation energy, equilibrium bond length, harmonic force constant, and, for CH and CN, cubic and quartic force constants. The interaction potential energy V_I is a three-body term that accounts for the difference between the sum $V_{CH} + V_{CN} + V_{NH}$ and the potential energy function deduced from vibrational spectra of HCN and HNC. The surface obtained from (6.12) reproduces the calculated energy[39] of the HCN $\rightarrow$ HNC saddle point within 0.1%, but predicts bond lengths at the saddle point that are 3–9% shorter than those given by the *ab initio* calculation.

Since its interaction term is evaluated from spectral data for triatomic species, Murrell's method applies only to systems having an observable triatomic molecule, unless parameters are to be determined by least-squares fitting to calculated or experimental points on the surface. Kafri and Berry[40] have extended the BEBO method to collinear surfaces for reactions such as (6.7) in which H is transferred from one atom to another. The potential energy along the minimum-energy path is determined as in the BEBO method. For points that do not lie on the minimum-energy path a correction based on Morse curves for $H \cdots B$ and $A \cdots H$ bonds is made. This correction accounts for the effect on potential energy due to bond stretching or compression that occurs when the system is displaced to the point in question from some reference point on the minimum-energy path. For a given point on the surface the reference point is chosen so as to minimize the potential energy. A surface obtained in this way for H_3 is in good agreement with the best available *ab initio* surface.

Another approach to empirical potential energy surfaces is to vary one or more parameters so as to reproduce a barrier height or other experimentally accessible value. One example of this, variation of the Sato overlap integral in the LEPS surface, is mentioned in Chapter 5. Other examples fall into two main categories. One is the switching function,[41] which for a reaction $A + BC \rightarrow AB + C$ consists of Morse potentials for BC and AB, a repulsion term between A and C, and two terms that reduce AB and BC attractions when C and A are near. The second category was originated by Wall and Porter.[42] They generated a fixed-angle surface such as that in Fig. 5.3 by rotating a Morse function (5.3) by 90° around a point in the separated-atoms region, varying the Morse parameters D and β with the angle of

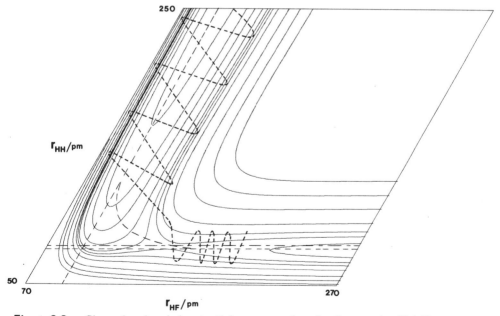

Figure 6.2. Skewed and scaled potential-energy surface for the reaction $F + H_2 \rightarrow$ HF + H. The entry line, the exit line, and the minimum-energy path are dashed, and a line parallel to the entry line has been drawn through the saddle point. The trajectory shown is for a reaction with relative kinetic energy of $5.19 \, \text{kJ mol}^{-1}$ and H_2 in its ground vibrational state. The surface is the same one shown in Figs. 5.2 and 5.3.

rotation to produce a saddle. This method has been extended by Bunker,[43] and recently cubic spline fitting has been used to obtain parameters that reproduce potential energy values calculated *ab initio.*[44]

Trajectory calculations based on prototype surfaces generated using arbitrary functions of the type described above can provide considerable insight into the relationship between surface topography and energy distributions of reactant and product molecules.[45, 46] For an exothermic reaction $A + BC \rightarrow AB + C$ the potential energy surface is classified as attractive if more than half the energy difference between the saddle point and the exit valley is released when the point representing the system moves parallel to the entry line. This corresponds to formation of a relatively stable species ABC followed by weak repulsion of C from AB. Since most of the energy is released while the A–B bond is forming, that is, while A is moving relative to B, much of that energy should appear as vibration of the product molecule AB. This has been confirmed by trajectory calculations.

The opposite situation, a strongly repulsive surface, is shown in Fig. 6.2. Following a line parallel to the entry line from the saddle point to the exit line, only one contour is crossed before the minimum potential energy is reached. However, upon leaving that path, four contours remain to be crossed before the minimum in the exit valley is reached. In Fig. 6.2 there is little drop in energy as A approaches BC. Most energy release occurs as C is repelled from AB. This energy might be expected to appear as translation rather than as vibration, but this is not always the case. Many trajectories such as the one shown in Fig. 6.2 can cut the corner of the surface, fall down the side of the exit valley, and result in vibrationally excited product. This is referred to as mixed energy release.

. For a surface like that of Fig. 6.2, where the saddle point lies in the entrance channel, increased translational energy of reactants is far more effective than increased vibrational energy in promoting reaction. Conversely, when the saddle point lies in the exit channel, vibrational energy is more effective. This is because vibration corresponds to movement parallel to the ordinate of a contour map such as that in Fig. 6.2, and so can result in the system's rounding the corner and surmounting the barrier.

SOME EXTENSIVELY STUDIED REACTIONS

The reaction of hydrogen atoms with hydrogen molecules, or isotopic variations thereof, is probably the most thoroughly studied of all molecular reactions.[47] Certainly it is the most accessible to theoretical treatment, consisting of a three-electron, three-nucleus system for which accurate *ab initio*, as well as semiempirical and empirical, potential energy surfaces can be constructed. It is also the only gas-phase bimolecular reaction for which low-temperature curvature of the Arrhenius plot has been observed and so it serves as a testing ground for theories of tunneling. If tunneling does indeed occur, then the observed rate constant at any temperature is greater than that prediced by classical theory. Moreover, as temperature decreases the relative contribution of tunneling should become larger, since there are still many reactants that cannot quite surmount the barrier but only a few that can. Consequently, the tunneling contribution to the rate constant becomes quite large at low temperatures, and upward curvature in an Arrhenius plot is the result. The goal of tunneling theories (or quantum dynamical calculations) is to reproduce the observed curvature.

Experimental determinations of thermal rate constants for the $H + H_2$ reaction usually involve isotopic labeling ($H + D_2$ or $D + H_2$) or observation of the rate at which *para*-hydrogen reacts to form the equilibrium mixture of 25% *para*-H_2, 75% *ortho*-H_2. (In *o*-H_2 the spins of the two protons are parallel; in *p*-H_2 the spins are antiparallel.) The latter reaction can be carried out by placing pure *p*-H_2 in a quartz reactor and raising the temperature until some H_2 dissociates, often heterogeneously on the walls. The atomic hydrogen so formed reacts with *p*-H_2, and in three cases out of four produces *o*-H_2. Similarly, heating a mixture of $H_2 + D_2$ results in formation of HD, since atomic H or D can react with molecular D_2 or H_2. Such bulk studies have been carried out over the temperature range 720–1023 K, but unfortunately the rate constants obtained are not too reliable. In the *p*-H_2 to *o*-H_2 reaction diffusion of oxygen through the walls of the reactor can produce additional *para*-to-*ortho* conversion; the $H_2 + D_2$ reaction is mechanistically complex and rate constants are calculated from approximate equations that do not correctly give the approach to equilibrium.

More reliable results can be obtained in flow-tube studies. Here atomic H or D is introduced into a stream of H_2 or D_2 and appropriate analysis is made downstream. Atom concentrations can be determined from the heat released as a result of recombination at a Pt wire or with an ESR spectrometer. Molecular concentrations can be determined from thermal conductivity of the gas or by gas chromatography. Flow-tube studies cover a temperature range from 167 to 745 K. Results of both types of measurement for the reaction $D + H_2 \rightarrow DH + H$ are summarized in the Arrhenius plot of Fig. 6.3.

Other experimental studies have involved nonthermal energy distributions in reactants. We describe in Chapter 4 the work of Kuppermann and White,[49] who used hot D atoms produced by photolysis of DI to establish a threshold energy of $31.8 \pm 1.9 \, kJ \, mol^{-1}$ for the reaction

$$D + H_2 \rightarrow HD + H \qquad (6.14)$$

Additional results obtained in a similar way have been used to obtain the reactive cross section as a function of energy in the range $(33.6 \leqslant LE \leqslant 106) \, kJ \, mol^{-1}$.[50] The cross section at $67 \, kJ \, mol^{-1}$ is $1.28 \, \text{Å}^2$ and at $106 \, kJ \, mol^{-1}$ it is $2.24 \, \text{Å}^2$. Crossed-molecular-beam experiments at relative energies corresponding to an effective $T = 1400 \, K$[51] indicate that HD is scattered backward in the center-of-mass coordinate system, that a linear collision complex is favored, and that the translational energy of products equals that of reactants. Based on a threshold energy $LE_0 = 25 \, kJ \, mol^{-1}$ and the assumption that cross section is proportional to $(1 - E_0/E)$, the best fit to the molecular beam results is

$$\sigma_R = 0.35 \, \text{Å}^2 \left(1 - \frac{25 \, kJ \, mol^{-1}}{LE} \right)$$

However, this predicts significantly smaller cross sections than were observed in the hot-atom studies, perhaps because of the assumption that the data should fit a line-of-centers functional form.

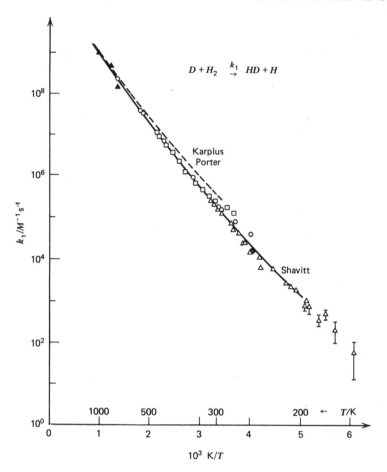

Figure 6.3. Arrhenius plot for the reaction $D + H_2 \rightarrow HD + H$. Solid points are results of bulk experiments, and open points are from flow experiments.[48] Lines show results of trajectory calculations (Karplus, Porter and Sharma, see Chap. 5 ref. 31) and transition-state theory (Shavitt,[53]).

The thermal rate constant for the $H + H_2$ reaction is quite sensitive to the shape of the potential energy surface near the saddle point,[52] and so transition-state theory has been applied extensively to the calculation of thermal rate constants. For temperatures above 400 K transition-state theory with transmission coefficient $\kappa = 1$ and based on accurate potential-energy surfaces reproduces the temperature dependence of the rate constant (see Fig. 6.3) and the relative rates of isotopic variations of the reaction quite well.[53] However, because of the low masses of H, D, and T, κ must be corrected for tunneling contributions to low-temperature rate constants, and such corrections are unreliable. Comparison of exact quantum-mechanical transition probabilities with transition-state theory calculations for collinear and three-dimensional H_3 reveal discrepancies, especially at low temperatures.[54] At 200 K and below the divergence is of the order of a factor of 10–100. Most tunneling corrections assume that tunneling occurs along the minimum-energy path, but choosing a tunneling path that cuts the corner of the surface[54b] results in agreement within a factor of 2 with exact quantum calculations. Such a path is the shortest possible and it connects points of maximum vibrational amplitude on a classical trajectory.

Trajectory calculations for $H + H_2$ on the Porter-Karplus potential-energy surface are described in Chapter 5. For reaction (6.14) at $LE = 106 \, kJ \, mol^{-1}$ these quasiclassical trajectories yielded $\sigma_R = 1.62 \, Å^2$ as compared with the experimental value of $2.24 \, Å^2$ mentioned above. The temperature dependence of the rate constants predicted by these trajectory calculations is shown in Fig. 6.3 to be rather close to experimental values with slightly less positive curvature at low temperatures than the experimental data. More recently quasiclassical trajectory calculations on the Porter-Karplus and other surfaces have revealed further aspects of the reaction.[55] The cross section for the $H + H_2$ reaction peaks at $1.82 \, Å^2$ at $LE = 289 \, kJ \, mol^{-1}$ and declines to $0.11 \, Å^2$ at $LE = 1819 \, kJ \, mol^{-1}$. At the latter energy the cross section for dissociation to 3H is $0.67 \, Å^2$. At low energy the incoming H atom reacts with the atom in H_2 that is closest, and the product molecule rebounds backward. At higher energy the dominant reaction is displacement, in which the incoming H knocks away the closest atom in the molecule and then bonds to the other one. At even higher energy, collisions with low impact parameter b produce dissociation, while those with higher b result in the incoming atom stripping off the closest atom in the molecule and carrying that atom along in a forward direction. Thus the maximum angular scattering shifts forward as E increases, a result in reasonable agreement with molecular-beam experiments.

Exact quantum calculations of reaction probabilities in which the H_3 supermolecule is permitted any three-dimensional configuration are extremely time-consuming, but they provide the ultimate test of a potential-energy surface, since other trajectory calculations and transition-state theory all involve simplifying assumptions. A very recent report of such calculations based on the highly accurate Siegbahn-Liu *ab initio* potential-energy surface gives the reaction probability as a function of energy shown in Fig. 6.4.[56] This can be compared with the quasiclassical trajectory cross-section results in Fig. 5.11. Of particular interest is that quantum calculations reveal several maxima and minima in reaction probability as a function of energy. Quantum calculations on the Porter Karplus surface show these same maxima and minima at nearly the same energies as on the Siegbahn-Liu surface, but the threshold energy is approximately $4.8 \, kJ \, mol^{-1}$ higher on the more accurate surface. This is apparently due to the greater thickness, as well as the slightly greater height, of the Siegbahn-Liu barrier. Greater barrier thickness reduces the effectiveness of tunneling.

Figure 6.5 summarizes a number of parameters obtained from experimental and theoretical studies of (6.14). The activation energy of $31.8 \, kJ \, mol^{-1}$ is obtained from the slope of Fig. 6.3. Zero-point energies were obtained from Johnston's tabulation[57] and the barrier height of $40.5 \, kJ \, mol^{-1}$ is from the Siegbahn-Liu *ab initio* potential energy surface. Several values have been reported for the threshold energy. Based on the hot-atom studies mentioned earlier[49] Kuppermann and White reported $LE_0 = 31.8 \, kJ \, mol^{-1}$, but the lowest-energy trajectory that resulted in a reactive collision on the Karplus-Porter surface had relative translational energy of only $23.8 \, kJ \, mol^{-1}$. The former value appears to be nicely consistent with the

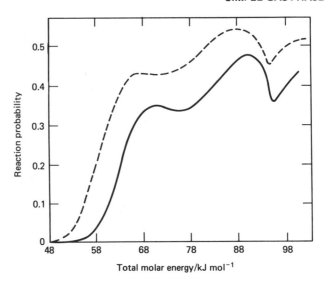

Figure 6.4.   Reaction probability versus total collision energy for $H + H_2$ reaction as obtained from quantum-dynamical calculations by Walker, Stechel, and Light.[56] Solid curve based on Siegbahn and Liu's potential-energy surface; dashed curve based on the Porter-Karplus surface. [Reprinted with permission from R. B. Walker, E. B. Stechel, and J. C. Light, *J. Chem. Phys.*, **69**, 2922 (1978).

remainder of Fig. 6.5, and it should be recalled that in the quasiclassical trajectory method of Porter, Karplus, and Sharma there was no requirement that the activated complex have any zero-point energy at all (see discussion in Chapter 5).

As a prototype of an exothermic bimolecular reaction we now consider

$$F + H_2 \rightarrow FH + H \tag{6.15}$$

a reaction that has probably received more experimental study of its microscopic details than any other. Product energy and angular distributions have been investigated by chemical laser, infrared chemiluminescence, and molecular beam techniques, and total rate constants have been determined by mass spectrometry and hot-atom experiments.[58] A high degree of vibrational excitation of the product molecule is characteristic of (6.15) and its isotopic analogues, with roughly two-thirds of the available energy of (6.15) appearing as vibrational excitation. Rotational excitation of the product is quite small, and the product angular distribution shows backward scattering primarily, but with a higher proportion of forward scattering as relative translational energy of reactants increases. Enhanced translational energy of reactants is channeled almost entirely into translation and rotation of products. Studies involving specific rotational states of reactants reveal a very small decrease in rate with increasing rotation, but the fraction of available energy in product vibration is smaller for $j = 1$ than for either $j = 0$ or $j = 2$.

Backward scattering is characteristic of a repulsive potential energy surface. Together with the low rotational excitation of the product, backward scattering also suggests an approximately collinear encounter. High product vibrational excitation results from mixed energy release of the kind diagrammed in Fig. 6.2.

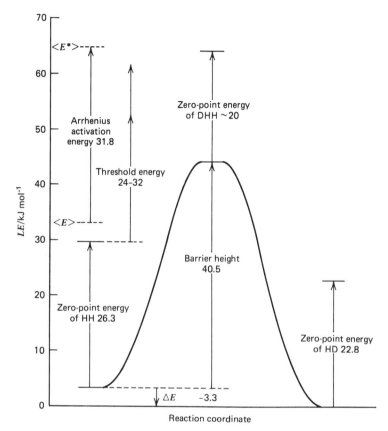

Figure 6.5. Parameters for the reaction $D + H_2 \rightarrow DH + H$. Sources of estimates given in text.

These conclusions have been supported by numerous trajectory studies on a number of surfaces. One classical three-dimensional study on a surface whose parameters were adjusted to provide the best fit to observed product energy distributions is especially revealing.[32] Except for the lower vibrational excitation from $j = 1$ compared to $j = 0$ or $j = 2$, the theoretical treatment is in good accord with experiment. The trajectory study also predicts that 81% of enhanced reagent vibration will appear as product vibration and that reagent translational energy is more effective than vibrational energy in producing reactive collisions. Product rotational energy derives mainly from release of repulsion in bent configurations, and the axis of rotation of the product molecule is $180°$ from the axis of orbital motion of the product atom. The constancy of reactive cross section as a function of reactant rotational state may imply a small potential energy well in the entry channel, before the saddle point is reached. Such a well has been found when spin–orbit interaction of F is included in the calculation of the surface.[59] Trajectory calculations have also been successful in reproducing the relative rates of reaction of F with either H or D in the $F + HD$ reaction.[60] The interplay of theory and experiment has been especially fruitful in the case of the reaction of F with H_2.

There are few gas-phase bimolecular reactions where both reactants are simple molecules. As we point out in Chap. 5, for diatomic reactants such processes are usually forbidden on the basis of orbital topology, and so more complex mechanisms are required. In many cases one of the reactant molecules dissociates to atoms. These then react by way of bimolecular or trimolecular processes that are allowed by orbital topology. In a few cases there is evidence that trimolecular combination of three diatomics is preferable to a forbidden bimolecular reaction.

An example of the former situation is the $H_2 + I_2$ reaction. In his initial studies of this reaction Bodenstein[61] proposed two possible mechanisms:

Simple bimolecular

$$H_2 + I_2 \xrightarrow{\ k_1\ } 2HI \tag{6.16}$$

Dissociation

$$I_2 \underset{}{\overset{K}{\rightleftharpoons}} 2I \tag{6.17a}$$

$$2I + H_2 \xrightarrow{\ k_2\ } 2HI \tag{6.17b}$$

Both mechanisms yield a second-order rate law, with $k = k_1$ for (6.16) and $k = Kk_2$ for (6.17). For a long time (6.16) was assumed to be the mechanism, and transition-state theory was applied to a planar trapezoidal activated complex, producing a frequency factor that agreed within an order of magnitude with experiment.[62] In 1967 Sullivan[63] measured the reaction rate at temperatures low enough so that the rate should have been immeasurably small except that the concentration of I atoms was maintained by continual photodissociation of I_2. Using an Arrhenius plot to extrapolate the rate constant k_{ph} of the photochemical reaction to the temperature range of the thermal reaction, Sullivan found that the thermal rate constant $k_{th} = Kk_{ph}$. If both mechanisms are assumed to operate, $k_{th} = k_1 + Kk_2$. but since $k_{th} = Kk_{ph}$ is found experimentally, $k_1 \ll Kk_2$ and (6.16) makes a negligible contribution, as required by orbital topology. Hammes and Widom[64] argue that the one-step mechanism is not eliminated entirely from consideration because the absorption of light may increase the concentration of energized (and hence reactive) iodine molecules in direct proportion to the increase in $[I]^2$. Such a view requires that the reactive, excited iodine molecules be in rapid equilibrium with iodine atoms, and there is reason to doubt that this in fact occurs.[65]

An example of the second type of alternative to a forbidden bimolecular reaction of diatomic molecules is low temperature hydrogen–deuterium isotopic exchange

$$H_2 + D_2 \rightarrow 2HD \tag{6.18}$$

At temperatures below 1500 K dissociation of H_2 and D_2 is orders of magnitude too slow to account for the rate of (6.18) and hydrogen-atom concentrations are negligible.[66] Consequently a trimolecular process has been proposed to account for the exchange.[67] Such a proposal is supported by single-collision molecular-beam studies of the analogous exchange reaction of Br_2 and Cl_2,[68] in which bimolecular collisions gave no reaction while collision of Br_2 with the van der Waals dimer

$Cl_2 \cdots Cl_2$ produced exchange at thermal energies. On the basis of SCF calculations, including configuration interaction,[66] a hexagonal H_6 transition state lies 289 kJ mol^{-1} above $3H_2$ and 163 kJ mol^{-1} below $2H_2 + 2H$; the transition state H_4 corresponding to a bimolecular process lies well above $2H_2 + 2H$. Applying transition-state theory based on calculated properties of hexagonal H_6 yields $E_a = 275 \text{ kJ mol}^{-1}$ and $\log (A/M^{-1} s^{-1}) = 5.38$ at 1000 K. The observed activation energy is 170 kJ mol^{-1},[69] but interpretation of the experimental results is complicated by the fact that vibrational excitation of reactant molecules significantly affects the rate. Thus the experimental E_a may apply to vibrational "ladder climbing" rather than isotope exchange. Nevertheless, trimolecular exchange appears to be too slow to account entirely for observed rates.

Diels-Alder reactions provide relatively simple examples of bimolecular processes in which both reactants are molecules. Several mechanisms, all of which are allowed by orbital topology, have been proposed. A concerted Diels-Alder reaction is one that occurs in a single step. For a concerted mechanism a further subdivision can be made on the basis of whether both new sigma bonds are equally formed in the transition state. If they are, the mechanism is synchronous; if not, it is two stage. If one bond forms completely before the other begins to form at all, a biradical or bipolar intermediate occurs and the reaction has a two-step mechanism.

As an example of the application of experimental measurements and theoretical calculations to eludication of mechanism, consider the Diels-Alder addition of ethylene and butadiene. The three mechanisms described above can be diagrammed as:

Concerted, synchronous

(6.19)

Concerted, two-stage

(6.20)

Two-step

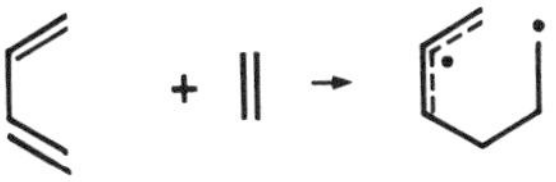

(6.21a)

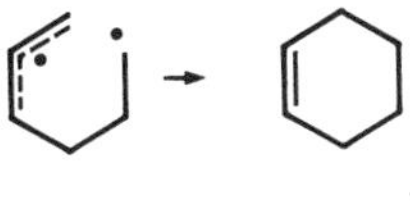

(6.21b)

A generally applicable approach to the mechanistic problem is to attempt to place all stable systems, intermediates, and transition states on a potential energy surface, and then include or exclude certain mechanisms on energetic grounds. Frey and Pottinger have done this empirically for the ethylene–butadiene reaction,[70] with the results shown in Fig. 6.6. Since TS4 lies below either of the biradical species and well below all other activated complexes, it is reasonable to eliminate (6.21) and conclude that the reaction of ethylene with butadiene is concerted. Such a conclusion need not necessarily apply to all Diels–Alder reactions, however, and in some cases there is evidence that it does not.[71]

Several methods were used to obtain the energies summarized in Fig. 6.6. Heats of formation for ethylene, butadiene, and cyclohexene are available in the literature. Frey and Pottinger estimated heats of formation for vinylcyclobutane and the *cis*- and *trans*-biradicals using group additivity.[72] This consists of summing contributions over all polyvalent atoms in a molecule, each contribution depending on the nature of the ligands bonded to an atom as well as on the atom itself. Frey and

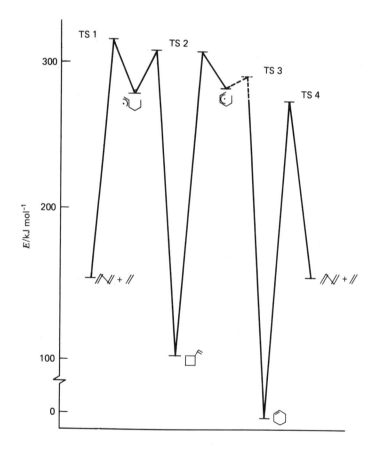

Figure 6.6. Energy diagram for ethylene + butadiene, cyclohexene, and vinyl-cyclobutane systems. (Reprinted with permission from H. M. Frey and R. Pottinger, *J. Chem. Soc., Faraday Trans., I.,* **74**, 1827 (1978).)

Pottinger obtained experimental activation energies for decomposition of vinyl-cyclobutane to ethylene and butadiene and for isomerization to cyclohexene. These values were used to place TS1 and TS2 on Fig. 6.6. TS3 was not located exactly, and TS4 was obtained from previously reported activation energies for decomposition of cyclohexene to ethylene and butadiene. This last assignment agrees within $3\,kJ\,mol^{-1}$ with the value calculated from the activation energy of the ethylene—butadiene reaction.

The empirical results cited in the preceding paragraph do not answer the question of synchronous versus two-stage mechanism, and this subject remains controversial despite several attempts to resolve it theoretically. Semiempirical molecular orbital theory supports a two-stage process,[73] while self-consistent field MO calculations with limited configuration interaction produce a surface on which a concerted, synchronous reaction is preferred.[74] Semiempirical MO calculations contain parameters whose values have been chosen to reproduce as accurately as possible the properties of stable molecules, and for this purpose semiempirical methods are often more accurate than more rigorous SCF calculations and require far less computational effort. However, extrapolation of semiempirical methods to properties of transition states may require different parameterization.[74b] On the other hand, optimizing energy over all the degrees of freedom of any but the smallest activated complex is such an enormous task for rigorous quantum calculations as to be almost out of the question. Consequently subtle distinctions, as between mechanisms (6.18) and (6.19), have yet to be made unambiguously by either method.

UNIMOLECULAR REACTIONS

The literature contains many studies of first-order gas-phase reactions and several recent reviews are available.[75] Some rate parameters appear in Table 6.6. First-order gas-phase reactions often are complicated, sometimes being partly heterogeneous and sometimes involving free-radical chains (see Chapter 10), and there is no guarantee that all reactions in Table 6.6 have single-step unimolecular mechanisms. For a multistep process the observed rate constant may be a product of constants rather than the coefficient for an elementary unimolecular process, and so some care must be taken to be sure that a reaction is simple or to interpret the rate data correctly for a complex mechanism.

Since concentrations of radicals in thermal reaction systems are expected to be near or below the detection limit for techniques like mass spectrometry and electron spin resonance, the distinction between molecular mechanisms and free-radical chains has often been based on the action of inhibitors such as O_2, NO, and propene. However, NO can actually initiate chain reactions in some cases,[76] and so it is dangerous to assume that chemical change occurring in the presence of excess inhibitor must involve a purely molecular mechanism. On the other hand, the presence of free radicals does not preclude experimental determination of the rate constant for a unimolecular step. For example, Shilov and Sabirova[77] have postulated for pyrolysis of chloromethane the non-chain radical mechanism:

Table 6.6. Some first-order gas-phase reactions[a]

Reaction	$\log(A/s^{-1})$	$E_a/\text{kJ mol}^{-1}$	$\Delta S^{\ddagger}/\text{J K}^{-1}\text{ mol}^{-1}$
Isomerizations			
Cyclopropane → propene	15.5	274	43.5
Methylcyclopropane → but-1-ene	14.4	253	22.4
Chlorocyclopropane →			
3-chloropropene	14.8	235	30.1
Cyclopropene → propyne	12.1	147	− 21.5
Vinylcyclopropane → cyclopentene	13.5	208	5.2
Cyclobutene → buta-1,3-diene	13.3	137	1.4
	15.9	241	51.2
Hexafluoro Dewar benzene →			
hexafluorobenzene	13.2	356	− 0.5
trans-CHDCHD → *cis*-CHDCHD	13.0	272	− 4.3
cis-But-2-ene → *trans*-but-2-ene	13.8	263	11.0
cis-CHClCHCl → *trans*-CHClCHCl	12.8	234	− 8.2
trans-CHClCHCl → *cis*-CHClCHCl	12.7	231	− 10.1
Hexa-1,3,5-triene →			
cyclohexa-1,3-diene	11.9	125	− 25.4
Hexa-1,3-diene → hexa-2,4-diene	10.8	136	− 46.5
cis-Hepta-1,5-diene →			
3-methylhexa-1,5-diene	10.7	145	− 48.4
$CH_3NC \rightarrow CH_3CN$	13.6	161	7.1
$C_2H_5NC \rightarrow C_2H_5CN$	13.8	160	11.0
Vinyl allyl ether → pent-4-enal	11.7	128	− 29.2
Decompositions			
Cyclobutane → ethylene	15.6	262	45.4
Methylcyclobutane → ethylene			
+ propene	15.4	256	41.6
→ ethylene + CH_2 CO	14.6	218	26.3
Cyclohexene → buta-1,3-diene			
+ ethylene	15.3	280	39.7
Cyclohexa-1,3-diene → benzene			
+ H_2	12.0	178	− 23.5
Ethylene oxide →			
CH_3CHO, CH_2O, CH_2CO	14.1	238	16.7
→ N_2 + $MeCHCH_2$	14.0	139	14.8
Ethyl fluoride → HF + ethylene	13.4	251	3.3
Ethyl chloride → HCl + ethylene	14.0	244	14.8
Ethyl bromide → HBr + ethylene	13.5	226	5.2

Table 6.6. Some first-order gas-phase reactions (continued)

Reaction	$\log(A/s^{-1})$	$E_a/\text{kJ mol}^{-1}$	$\Delta S^{\ddagger}/\text{J K}^{-1}\,\text{mol}^{-1}$
Ethyl iodide $\rightarrow$ HI + ethylene	14.1	221	16.7
Ethyl acetate $\rightarrow$ acetic acid + ethylene	12.5	200	-13.9
Methylene diacetate $\rightarrow$ HCHO + $(CH_3CO)_2O$	9.2	138	-77.1
Methylene dibutyrate $\rightarrow$ HCHO + $(PrCO)_2O$	9.2	138	-77.1
Isopropyl ether $\rightarrow$ propene + i-PrOH	14.6	266	26.2
Vinyl ethyl ether $\rightarrow$ C_2H_4 + MeCHO	11.4	193	-35.0
$CHF_2CH_2SiF_3 \rightarrow CHFCH_2$ + SiF_4	12.3	137	-17.8
$CHF_2CH_2SiF_2Me \rightarrow CHFCH_2$ + $MeSiF_3$	11.3	136	-36.9
$(CH_3CO)_2O \rightarrow CH_3COOH + CH_2CO$	12.1	144	-21.6

[a] Entropy of activation calculated at 298 K.

$$CH_3Cl \rightarrow CH_3 + Cl \qquad (6.22a)$$

$$CH_3 + CH_3Cl \rightarrow CH_4 + CH_2Cl \qquad (6.22b)$$

$$Cl + CH_3Cl \rightarrow HCl + CH_2Cl \qquad (6.22c)$$

$$CH_2Cl + CH_3Cl \rightarrow CH_2Cl_2 + CH_3 \qquad (6.22d)$$

$$2CH_2Cl \rightarrow CH_2ClCH_2Cl \qquad (6.22e)$$

$$CH_2ClCH_2Cl \rightarrow HCl + CH_2CHCl \qquad (6.22f)$$

$$CH_2CHCl \rightarrow HCl + C_2H_2 \qquad (6.22g)$$

At 800°C (6.22e) and (6.22f) are more important than (6.22d), and so the rate of the slow step (6.22a) can be measured as one-third the rate of production of HCl. [Three molecules of HCl are produced rapdily by steps (6.22c), (6.22f), and (6.22g) for each methyl radical and Cl atom produced by step (6.22a).]

The usual test for heterogeneous reaction is to pack the reaction vessel with glass wool, beads, or tubing and to determine the effect of surface/volume ratio on the reaction rate. However, absence of an effect does not guarantee homogeneous reaction, since the rate of a radical chain that is both initiated and terminated at the surface can be unaffected. A better test of homogeneity is often to change the nature of the surface. Also, conditioning the surface of a reaction vessel with a carbonaceous film can lead to surfaces containing high concentrations of free electrons.[78] Such surfaces are active catalysts, promoting *cis-trans* isomerizations and possibly initiating free-radical chains. Some early measurements of rates of *cis-trans* isomerizations gave unusually low frequency factors that were attributed to small transmission coefficients due to change in electron multiplicity in the transition state, but these now appear to be complicated by surface effects or free-radical reactions.[79] Any conditioned surface should be demonstrated to be inert, not just assumed to be so.

Bauer[80] has suggested a criterion for determining whether an elimination reaction where the reactant is a ring compound occurs in a single concerted step. If a small molecule such as N_2, CO, CO_2, and C_2H_2 is eliminated by means of a cyclic transition state, the bond lengths and angles in the transition state are unlikely to correspond to those in the product molecule. Consequently, for a concerted process, the small molecule is produced in a structurally distorted and hence vibrationally excited state. If, on the other hand, the elimination has a stepwise mechanism in which one bond breaks to form an intermediate with a lifetime of several vibrational periods, rapid energy transfer in the intermediate permits structural relaxation before the small molecule is emitted, and a vibrationally excited product is not expected. Infrared chemiluminescence or laser techniques can be used to detect vibrational excitation of products and hence to distinguish mechanism.

Many of the frequency factors in Table 6.6 are in the neighborhood of 10^{13} s^{-1}, in agreement with transition-state theory if the entropy of activation is zero. Several special cases may be noted. If a cyclic intermediate is formed from a noncyclic reactant, the entropy of activation is negative and the frequency factor is reduced, because internal rotations in the reactant become vibrations in the activated complex with a loss of entropy. Some examples are decompositions of acetic anhydride, methylene diesters, and fluoroalkylsilanes, as well as isomerization of allyl vinyl ether, cyclization of hexa-1,3,5-triene, and the Cope rearrangement of hepta-1,5-diene. Szwarc and Murawski[81] proposed the cyclic transition state

for production of ketene and acetic acid from acetic anhydride, and Shuler and Murphy[82] proposed that isomerization of allyl vinyl ether proceeds as

A four-center transition state has been proposed for fluoroalkylsilane decomposition:[83]

$$CHF_2CH_2SiF_3 \rightarrow \left[\begin{array}{c} F\text{-----}SiF_3 \\ | \quad\quad | \\ | \quad\quad | \\ CHF\text{==}CH_2 \end{array} \right]^{\ddagger} \rightarrow SiF_4 + CHFCH_2$$

Conversely, decomposition of a ring compound should be accompanied by an increase in entropy and a high frequency factor,[84] as in the case of isomerization of cyclopropane and decomposition of cyclobutanes.

If a molecule dissociates into two fragments each of which contains several atoms, and if these two groups rotate freely in the activated complex, then a high frequency factor is found. Evidence for such free rotation comes from the data of Table 6.1. The high rates of recombination of some of these radicals and the success of hard-sphere line-of-centers collision theory in predicting those rates can only mean that, for example, the orientation of two combining methyl groups is of no importance. There must then be free rotation of the methyl groups in the transition state for ethane decomposition, since the transition state must be the same as for combination.[85] RRKM calculations give best agreement with experiment for ethane decomposition when based on a model with free rotation in the transition state,[86] but recent work[87] questions whether RRKM is capable of reproducing the temperature dependence of the rate constants over the experimental range of 250–2500 K. In any case the reverse reactions of Table 6.1, either measured directly or calculated from the forward rate and the equilibrium constant, do have high frequency factors ($10^{14} - 10^{17}$ s^{-1}).

POLAR REACTIONS IN GAS PHASE

A novel interpretation of a series of gas-phase reactions has been given by Maccoll for the first-order decomposition of organic halides.[88] An elimination of hydrogen halide occurs and olefin is formed.

$$R_1\text{---}\underset{\underset{H}{|}}{\overset{\overset{R_2}{|}}{C}}\text{---}\underset{\underset{X}{|}}{\overset{\overset{R_3}{|}}{C}}\text{---}R_4 \longrightarrow \underset{R_1}{\overset{R_2}{>}}C\text{==}C\underset{R_4}{\overset{R_3}{<}} + HX \quad (6.23)$$

Many of these reactions for substituted halides appear to be simple unimolecular processes not involving free radicals or chain reactions. Table 6.7 gives rate data for several bromides and the corresponding acetates in which elimination of acetic acid occurs.

Maccoll calls attention to the large increases in rate caused by α-methyl substitution (R_3, R_4) in the bromides. These increases are very similar to the effect on rates of solvolysis and elimination reactions of those halides in solutions. For example, activation energies for solvolysis of ethyl, isopropyl, and *tert*-butyl

Table 6.7. Relative rates of elimination of HX from $CHR_1 R_2 CR_3 R_4 X$ at 600 K[a]

R_1	R_2	R_3	R_4	X = Br	X = OAc
H	H	H	H	1	1
CH_3	H	H	H	3.0	—
CH_3	CH_3	H	H	6.6	—
H	H	CH_3	H	222	12
CH_3	H	CH_3	H	430	8.5
H	H	CH_3	CH_3	39,000	1000
CH_3	CH_3	CH_3	CH_3	240,000	—
H	H	C_6H_5	H	14,000	24

[a] Data from reference 75a.

bromides in 80% ethanol–water are 126, 112, and 97 kJ mol^{-1}. For the gas-phase elimination reaction the corresponding energies are 222, 200, and 173 kJ mol^{-1}. A number of other correspondences exist in that substituents such a phenyl, vinyl, and oxygen increase both the rates of elimination in the gas phase and solvolysis in solution. The effect of β-methyl (R_1, R_2) substitution is small compared to α-methyl substitution.

The first-order solution reactions are commonly believed to involve ionic intermediates. Because of the striking resemblances between the gas-phase and solution behavior, Maccoll proposed that an ion-pair structure for the transition state in the gas phase is most probable:[88c]

$$\begin{array}{ccc} \overset{>}{C}-\overset{<}{C} & \quad\text{or}\quad & \overset{>}{C}=\overset{<}{C} \\ \ \ \overset{|}{H}\ \ \overset{+}{Br^-} & & \ \ \overset{+}{\underset{}{H}}\ \ Br^- \end{array}$$

More recently a semi-ion-pair transition state has also been suggested.[89] As is discussed in the next chapter, free ions are normally not formed in the gas phase because of the excessive energy required. However, in this case the ions are not separated and so the energy required is much less.

O'Neal and Benson[28, 90] have assigned bending, stretching, and torsional frequencies to the activated complexes for a large number of alkyl halide pyrolyses. From these frequencies they calculated entropies of activation and hence frequency factors, obtaining agreement with experimental values of log A that was generally within ± 0.3. From their frequency factors and observed rate constant values O'Neal and Benson calculated a consistent set of activation energies and found that α-methylation produces an average lowering of 24 kJ mol^{-1} per methyl group, while β-methylation lowers the activation energy by only about 7 kJ mol^{-1}. This agrees with the trends observed by Maccoll.

For acetates pyrolysis is much less similar to the solvolysis reactions. Hence the transition state is considered to be much less polar. The cyclic mechanism and transition state proposed by Hurd and Blunck[91] seem entirely reasonable, particularly since they explain the sterochemical fact of *cis* elimination in suitably substituted acetates.

EFFECT OF PRESSURE ON UNIMOLECULAR REACTIONS

It is shown in Chapter 4 that an expected feature of a true unimolecular reaction would be a decrease in the first-order rate constant with decreasing pressure and an eventual change to second-order kinetics at very low pressure. All the theories of unimolecular reaction predict that the pressure at which falloff occurs should depend on the number of vibrational modes involved, with more complex molecules exhibiting falloff at lower pressures. For unimolecular reactions having activation energies from 170 to $250 \, kJ \, mol^{-1}$ and for temperatures in the range from 500 to 700 K, the falloff region is easily accessible only for reactant molecules containing fewer than a dozen atoms. Nevertheless a large number of studies have been made in the low-pressure region. These involve mainly cyclopropane and its derivatives, cyclobutane and substituted cyclobutanes, halogenated ethanes and propanes, methyl and ethyl isocyanides, and a few other small molecules.[92]

Isomerization of cyclopropane to propene provides an excellent example of the low-pressure behavior of a unimolecular reaction.

$$\tag{6.24}$$

The usual tests for free radicals are negative and the reaction does not appear to be heterogeneous. Figure 6.7 summarizes experimental results[93] and theoretical predictions[94] for isomerization of cyclopropane. The logarithm of k/k_∞, the ratio of the first-order rate constant at any pressure to the limiting high-pressure constant, is plotted against the logarithm of pressure. The measured rate constant falls off by a factor of 10 in going from 100 to 0.001 kPa. At very low pressures falloff stops and the reaction again becomes first order.[93c] The rate constant under very low pressure conditions depends on the surface/volume ratio because molecules are predominantly energized on the walls rather than by collisions with other gas molecules.

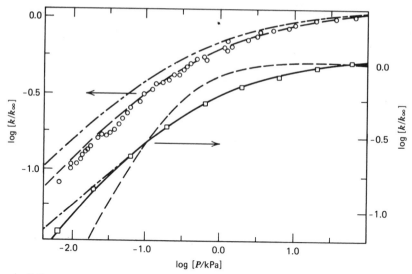

Figure 6.7. Experimental and theoretical fall-off curves for isomerization of cyclopropane. Upper curves (left ordinate): (O) experimental points; (—·—·—) RRKM prediction (Wieder and Marcus[94]); (— — —) RRKM prediction (Lin and Laidler[94]). Lower curves (right ordinate): (———) best curve through experimental points: (□) Slater theory prediction; (— — —) Lindemann-Hinshelwood theory prediction; (—·—·—) RRK theory.

Pritchard et al.[93b] compared their measured rate constants with values calculated using the Lindemann-Hinshelwood theory, RRK theory, and Slater's theory. The simple Lindemann-Hinshelwood theory does not fit the experimental results at all well. The RRK theory provides quite a good fit with the parameter s of (4.69) arbitrarily chosen to be 13 and a collision diameter of 3.9 Å, and a somewhat better fit probably could have been obtained. Slater[94a] carried out calculations based on (4.73) with $s = 13$ and was able to reproduce the shape of the fall-off curve quite accurately. However, Slater's calculated values had to be shifted by -0.3 along the log P axis in order to coincide with the experimental rate constants, and Slater's calculated frequency factor $A_\infty = 4.0 \times 10^{14}$ s^{-1} was somewhat lower than the experimental[95] $A_\infty = 3.2 \times 10^{15}$ s^{-1}. Slater's treatment assumes that reaction occurs when a hydrogen atom on one carbon atom approaches within a critical distance of another carbon atom. Since each of six equivalent hydrogen atoms can move toward either of two carbon atoms, reaction can occur in 12 ways, and a statistical factor of 12 is included in the theoretical A_∞ value quoted above.

Wieder and Marcus[94b] first applied RRKM theory to cyclopropane isomerization. They considered two activated complexes:

the second of which has an enantiomeric form. For complex **1** the calculated curve was displaced to lower pressures, and agreement with experiment required the assumption that only one-fourth of the collisions of an activated molecule result in deactivation. Calculations based on complex **2** gave similar results and also predicted incorrect temperature dependence for the fall-off curve. Lin and Laidler[94c] considered a mechanism involving a biradical intermediate (6.25) and used the

$$(6.25)$$

high-pressure Arrhenius parameters of Falconer et al.[95] instead of those of Chambers and Kistiakowski.[93a] This produced much better agreement with experiment, apparently because of the small change in Arrhenius parameters, but not because of the change in activated complex. In general the fall-off curve is not very sensitive to the choice of reaction mechanism. This is because predictions in the fall-off region depend on inherent assumptions of RRKM theory as well as on the structure of the activated complex. Beyond the fall-off region the low-pressure second-order rate constant is entirely independent of the activated complex, depending only on the properties of the energized molecule.

Slater's theory would not work at all well if the biradical mechanism (6.25) obtains and there is considerable evidence in support of such a mechanism. *trans*-Cyclopropane-d_2 gives 25% rearrangement to *cis*-cyclopropane-d_2 (6.26) during the time required for 8% conversion to propene,[96] and no cyclopropane-d_2 is found in

$$(6.26)$$

which both deuterium atoms are on the same carbon. This is consistent with rupture of a C—C bond, internal rotation, and reclosing of the trimethylene radical, which corresponds to the equilibrium step in (6.25). Arrhenius parameters for geometrical isomerization (6.26) are nearly the same as for structural isomerization (6.24), and the rate constants for both reactions behave similarly in the fall-off region, lending credence to the biradical mechanism. Studies of optically active

1,2-disubstituted cyclopropanes[97] and of cis-2,3-dideuteriovinylcyclopropane[98] also support the biradical mechanism. Attempts to identify the intermediate have not been successful,[99] but there is reason to believe that its lifetime is $< 3 \times 10^{-11}$ s at 450°C,[100] making identification by normal methods unlikely. Molecular orbital calculations of energy as a function of reaction coordinate[101] divide along the lines mentioned earlier for the Diels-Alder reaction, with semiempirical calculation showing a minimum corresponding to the biradical and SCF calculations indicating a concerted process. However, a *triplet* biradical is calculated to provide a lower-energy path for reaction,[102] provided that the transition to the triplet energy surface is facile.

ENERGY RANDOMIZATION AND ENERGY TRANSFER

The major feature that distinguishes Slater's theory of unimolecular reactions from RRK or RRKM theory is the assumption of the latter that energy is redistributed rapidly and randomly among the various degrees of freedom of an energized molecule before that molecule either reacts or is deenergized. Comparison of calculated and observed rate parameters for thermal reactions may be influenced by assumptions other than energy randomization and hence is not a sensitive tool for ascertaining the validity of this assumption. Methods such as chemical activation, hot-atom reactions, photoactivation, and crossed molecular beams can produce non-Boltzmann distributions of energy in energized species, and they often provide direct tests of the hypothesis of intramolecular energy randomization.[103]

Chemical activation, in which a chemical reaction produces a highly excited species that subsequently undergoes unimolecular reaction, has been the most commonly used of the techniques mentioned above. Butler and Kistiakowsky[104] carried out the first chemical activation study, isomerization of methylcyclopropane formed by reaction of methylene biradicals with either propene or cyclopropane:

$$CH_2: \ + \ CH_3CH=CH_2 \qquad\qquad\qquad M \qquad\qquad (6.27)$$

$$CH_2: \ + \ \triangle \ \longrightarrow \ \triangle^* \ \longrightarrow \ \text{butenes}$$

Both of the formation reactions are highly exothermic and so the methylcyclopropane is formed with more than enough energy to isomerize to various butenes. Isomerization therefore occurs unless the excited methylcyclopropane molecule transfers its excess energy by colliding with a bath molecule M. Butler and Kistiakowsky found that the proportion of each butene isomer in the product mixture is independent of the way in which vibrationally excited methylcyclobutane is formed. Since the distribution of vibrational energy at the instant of formation of a methylcyclobutane molecule is expected to be quite dependent on the formation reaction, Butler and Kistiakowsky's result implies that energy must be completely redistributed before isomerization occurs. Numerous other experiments provide similar support for the RRKM assumption of rapid randomization of excitation energy.

However, there are cases where the time required for energy redistribution can be measured. Rynbrandt and Rabinovitch[105] reacted CD_2 with hexafluorovinylcyclopropane to form hexafluorobicyclopropyl (**3**) in which the newly formed ring

$$
\begin{array}{c}
CF_2 \\
| \quad \quad CF_2 \\
CH_2 \quad CF - CF \quad | \quad * \\
\quad \quad \quad \quad \quad CD_2 \\
\mathbf{3}
\end{array}
\quad
\longrightarrow
\quad
\begin{array}{c}
CF_2 \\
| \quad CF - CF = CD_2 \; + \; CF_2 \\
CH_2 \quad \quad \mathbf{4} \\
\\
\quad \quad \quad \quad \quad CF_2 \\
CH_2 = CF - CF \; | \quad + \; CF_2 \\
\quad \quad \quad \quad \quad CD_2 \\
\quad \quad \quad \mathbf{5}
\end{array}
$$

was labeled with deuterium. The products **4**, in which the newly formed ring has decomposed, and **5**, in which excitation energy has been redistributed to break the ring at the other side of the molecule, were detected by mass spectrometry. At pressures below 1 atm products **4** and **5** are formed in equal amounts, in accord with the assumption of energy randomization. However, increasing the pressure causes the ratio [**4**]/[**5**] to increase, showing that the newly formed ring opens preferentially. When **3** is produced by reacting CH_2 with dideuterohexafluorovinylcyclopropane, increasing pressure causes [**4**]/[**5**] to decrease, also showing that the newly formed ring opens preferentially. Increasing pressure decreases the time available for **3** to react, because at higher pressure the average lifetime of **3** before a collision deenergizes it is shorter. From the product ratio as a function of pressure Rynbrandt and Rabinovitch calculated that randomization of energy in hexafluorobicyclopropyl is 99% complete in 4×10^{-12} s. It should be noted that the two ends of the hexafluorobicyclopropyl molecule are linked by only a single carbon–carbon bond. This may act as a bottleneck, preventing easy transfer of energy from one ring to another and thus making nonrandomization detectable at relatively low pressures.

Randomization of energy in a molecule is due essentially to vibrational anharmonicity, a phenomenon amply demonstrated by vibrational spectra. Since vibrations are anharmonic, there is coupling among normal modes, and resolution into independent normal vibrations is not strictly possible. If the formal concept of normal modes is retained, anharmonic coupling leads to variation of amplitude of each normal vibration over time, and since amplitude is proportional to $(\text{energy})^{1/2}$ this can formally be interpreted as energy exchange among the normal modes.[106] At least some of the vibrations of a molecule that has been sufficiently energized to decompose must be in high quantum states, where anharmonicity (and hence vibrational coupling) is considerable, and so rapid redistribution of energy is expected. Thiele and Wilson[106b] numerically integrated the equations of motion for a linear triatomic system containing unequal masses governed by Morse potentials and found rapid energy flow between the two vibrational modes. Bunker[107] carried out trajectory calculations for model triatomic systems and found that internal energy was randomized in $< 10^{-11}$ s except in cases such as that of H_2O, where relative masses are greatly different. Trajectory studies of a real system, isomerization of CH_3NC to CH_3CN, lead to the opposite view, however; intramolecular vibrational relaxation was found to be slow.[108] Experimental studies of methylisocyanide isomerization are generally in accord with RRKM theory, and there is some reason to question the accuracy of the potential energy surface used for the trajectory calculations. A recent study of infrared multiphoton dissociation of CH_3NC adduces considerable evidence that the reaction proceeds by way of isomerization to CH_3CN followed by dissociation of the latter. However, CN produced from CH_3NC was found to be rotationally and vibrationally hotter than CN produced by direct dissociation of CH_3CN. This could be due to slow vibrational relaxation, indicating non-RRKM behavior.[109]

Reddy and Berry[115] have applied a unique intracavity continuous-wave dye laser technique to photoactivate allyl isocyanide. They were able selectively to energize each of the three types of C—H bond in the molecule by one-photon activation, and to determine true unimolecular rate constants (corresponding to k_3 of the Lindemann mechanism given in Chapter 4). For example, irradiation of allyl isocyanide at exactly 715.0 nm deposits 5 quanta of C—H stretch excitation at the =CH— site. This is sufficient energy to promote isomerization. Observed rate constants differ from statistical expectations by up to a factor of 3, implying that redistribution of energy is not complete by the time isomerization takes place.

Another important aspect of unimolecular reaction rate theory that has received considerable attention is the strong-collision assumption. RRKM theory treats energization and deenergization as single steps rather than ladder-climbing processes in which molecules acquire or lose energy in a series of small jumps from one quantum level to another. If ladder climbing obtains it is possible that a limited rate of transfer between levels below the critical energy may affect the concentration of energized molecules and hence the rate. Ladder climbing requires more detailed mathematical treatment[110] than if the strong-collision assumption is made. For thermal reactions in the normal temperature range most of the energized molecules lie no more than $40 \, kJ \, mol^{-1}$ above the critical energy, and so it is not necessary that a single collision remove an excessively large quantity of energy. When molecules are highly excited by chemical activation or photoactivation, on the other hand, several collisions may be required for deactivation.

A crude allowance for limited energy transfer can be made by introducing a collision-efficiency parameter β into the expression for calculating k_2, the rate constant for deenergization. Then the factor $k_2[M]$ in (5.66) becomes $\beta Z_{MA^*}/L[A]$, and the result is to shift the fall-off curve ($\log k/k_\infty$ versus $\log P$) to higher pressure by $\log \beta$. The reciprocal collision efficiency $1/\beta$ is the average number of collisions required to deenergize an energized molecule. Actually β should depend on the excess energy in an energized molecule, a greater number of collisions being required the greater the excess, but the crude model of a constant β produces a reasonable approximation to the fall-off curves.

Studies of isomerization of methyl isocyanide to acetonitrile provide the most extensive data on energy transfer in thermal unimolecular systems. Rabinovitch and coworkers[111] studied this isomerization in the presence of over 100 different diluent gases at pressures corresponding to the second-order region of the fall-off curve. From the various shifts along the $\log P$ axis they found β values ranging from 0.24 to 1.0. A plot of β versus the number of atoms per molecule of diluent gas is shown in Fig. 6.8. Collision efficiency increases with size of the colliding molecule, reaching a constant value for molecules containing more than 10 atoms.

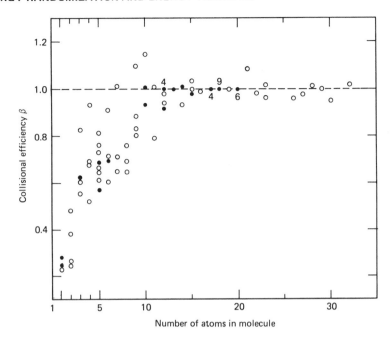

Figure 6.8. Collisional efficiencies β for energization of methyl isocyanide at 554 K. Filled points represent coincident data for two (or more, as indicated) bath gases. (Data from Rabinovitch et al.[111]).

Other studies indicate that these larger molecules transfer $20 \, \text{kJ mol}^{-1}$ or more energy per collision, and since this is comparable to the expected excess energy over the critical energy, it is reasonable to assume strong collisions ($\beta = 1$) for these large molecules. Even for light, structureless species such as He no more than four collisions are required for deenergization.

Further analysis of the collision-efficiency data reveals a good correlation of β with polarizability of the diluent species as well as systematic deviations that depend on molecular dipole moments. This indicates that long-range intermolecular attractions are important, and Rabinovitch et al. find a general correlation of β with boiling points. Within the general correlation three classes of molecules can be distinguished: monatomics, diatomic and small linear molecules, and nonlinear polyatomics. The same three classes are evident in studies of the effectiveness with which various molecules mediate bimolecular recombination of iodine atoms.[112] All monatomics have $\beta \cong 0.25$; for linear species β ranges from 0.25 to 0.6, increasing with increasing boiling point; and for nonlinear polyatomics β increases from 0.6 to 1.0, leveling off at the latter value for boiling points above 300 K. Rabinovitch et al. attribute these three categories of behavior to the fact that the number of transitional vibrational modes, that is, the number of modes of the collision complex that correlate with rotational or relative translational modes of the separated collision partners, is three, five, and six for monatomics, linear molecules, and nonlinear polyatomics. Lin and Rabinovitch[113] have developed this idea in detail.

Table 6.8. Examples of third-order reactions

Reaction	$\log(k/M^{-2}\,s^{-1})$	T/K	$\log(A/M^{-2}\,s^{-1})$	$E_a/kJ\,mol^{-1}$
$2H + H_2 \rightarrow 2H_2$	$10.0 \pm .1$	300	—	—
$2O + O_2 \rightarrow 2O_2$	$9.0 \pm .1$	300	—	—
$2N + N_2 \rightarrow 2N_2$	$9.2 \pm .3$	300	—	—
$N + O + N_2 \rightarrow NO + N_2$	9.5	350	—	—
$2Cl + Cl_2 \rightarrow 2Cl_2$	10.0 ± 1.5	300	—	—
$2Br + Ar \rightarrow Br_2 + Ar$	9.4	300	—	—
$2I + He \rightarrow I_2 + He$	—	293	9.5	-6.3
$2I + Ne \rightarrow I_2 + Ne$	—	333	8.3	-6.3
$2I + Ar \rightarrow I_2 + Ar$	—	293	9.9	-8.8
$2I + I_2 \rightarrow 2I_2$	—	—	8.2	-21.8
$2I + H_2 \rightarrow I_2 + H_2$	—	293	10.0	—
$2I + CO_2 \rightarrow I_2 + CO_2$	—	293	10.4	-8.4
$2I + C_6H_6 \rightarrow I_2 + C_6H_6$	—	293	11.2	-10.1
$2I + C_2H_5I \rightarrow I_2 + C_2H_5I$	—	293	11.7	—
$H + O_2 + Ar \rightarrow HO_2\cdot + Ar$	—	—	8.67	-6.7
$O + O_2 + Ar \rightarrow O_3 + Ar$	—	—	6.5	-9.6
$O + O_2 + N_2 \rightarrow O_3 + N_2$	—	—	7.0	-7.1
$H + NO + H_2 \rightarrow HNO + H_2$	—	—	10.0	-2.5
$O + NO + N_2 \rightarrow NO_2 + N_2$	—	—	9.2	-7.9
$O + NO + O_2 \rightarrow NO_2 + O_2$	—	—	8.9	-7.5
$CH_3\cdot + NO + CH_3COCH_3 \rightarrow CH_3NO + CH_3COCH_3$	11.5	473	—	—
$2NO + O_2 \rightarrow 2NO_2$	—	—	3.0	-4.6
$2NO + Cl_2 \rightarrow 2NOCl$	—	—	3.7	15.5
$2NO + Br_2 \rightarrow 2NOBr$	3.5	—	—	—
$2I + H_2 \rightarrow 2HI$	—	—	7.8	22.2

TRIMOLECULAR REACTIONS

Third-order gas-phase reactions are rare. The majority of them involve recombination of atoms, small radicals, or small molecules in which a chemically inert third body carries off excess energy. A second category includes the reaction $2I + H_2$ discussed earlier in this chapter and several reactions of nitric oxide. Table 6.8 gives experimental data on some third-order gas-phase reactions.

Rates of the recombination reactions can be predicted successfully by using (4.75), letting the steric factor $p = 1$, choosing a collision distance δ of about 1 Å, and calculating the rate from

$$-\frac{1}{V}\frac{dN_a}{dt} = pZ_{ABC}e^{-LE_0/RT} \qquad (6.28)$$

This gives reasonable confirmation of the trimolecular nature of these reactions. The alternative "sticky" collision treatment discussed in Chapter 4 may also be applied to recombination reactions. Two possibilities arise. If the sticky collision is between the combining species, with the third body M then carrying off excess energy, the mechanism is the reverse of the Lindemann mechanism for unimolecular reaction in its second-order region. This is called the energy-transfer mechanism for recombination. If the sticky collision complex involves the third body M, the mechanism is called the atom–molecule complex mechanism. In general, as indicated by the data for I-atom recombination,[112, 114] the more complex the structure of the third body M the more efficient M is in removing excess energy from a pair of colliding species. By the same token such complex molecules M are more efficient in transferring energy to another species, as in activation of a molecule for unimolecular reaction. For the atom, molecule, and radical recombination reactions activation energies are negative. As temperature increases the quantity of excess kinetic energy increases, and so the efficiency of deactivating collisions is less. Transition-state theory predicts a negative temperature dependence for trimolecular reaction with E_0 small or zero (as would be expected for recombination), since according to (5.50) the frequency factor depends on $T^{-7/2}$ if vibration is not excited in the transition state.

The reaction $2I + H_2$ and the reactions in Table 6.8 that involve NO are the only third-order reactions in which all three reactants are altered chemically. When (6.28) is applied to the I-atom reaction the steric factor $p \cong 10^{-2}$, and for reactions involving NO $p \cong 10^{-7}$. Here the sticky collision mechanism has considerable support; a preequilibrium to form excited, reactive I_2^* is discussed earlier.

PROBLEMS

6.1. Obtain or estimate collision diameters for the reactants in the methyl-radical reactions of Table 6.3. Hence calculate rate constants for each of these reactions at 400 K using hard-sphere line-of-centers collision theory. Compare your results with the experimental data given in the table.

6.2. For the competitive reactions (6.1) and (6.2) show that the ratio $k_2/k_1^{1/2}$ can be determined from experimentally accessible concentrations, while k_2/k_1 cannot.

Also show how $E_2 - \frac{1}{2}E_1$ can be calculated from experimentally measured quantities.

6.3. Use the Polanyi-Semenov equation to predict activation energies for the reactions in Table 6.3 based on the values of α and C given in the text.

6.4. Given the entropy values

$$S^{\ominus}(C_2 H_6) = 230 \, J \, K^{-1} \, mol^{-1}$$

$$S^{\ominus}(H) = 115 \, J \, K^{-1} \, mol^{-1}$$

at $300 \, K$ and for a standard state of 1 atm, estimate the frequency factor for the reaction

$$H + C_2 H_6 \rightarrow H_2 + C_2 H_5$$

Compare your result with the value reported in Table 6.3. (Hint: Be sure to correct $\Delta S^{\ddagger}$ so that it applies to a standard state of $1 \, M$ using the equation $\Delta S_{p\ominus} = \Delta S_{c\ominus} + (\Delta n) - R \ln (c^{\ominus} RT / p^{\ominus})$.)

6.5. Hydrogen gas at room temperature and above is an equilibrium mixture of 25% *para*-hydrogen and 75% *ortho*-hydrogen. If almost pure *para*-hydrogen is prepared at very low temperatures, the conversion to the equilibrium mixture is a slow process in the absence of a catalyst. The rate of conversion follows a first-order law so that in any one run

$$\ln \frac{(\text{initial \% } para - 25)}{(\% \, para - 25)} = kt$$

The first-order k varies with total pressure of hydrogen at $923 \, K$ as follows [A. Farkas, *Z. Phys. Chem.*, B10, 419 (1930)]:

P/mm Hg	50	100	200	400
$k \times 10^3 /s^{-1}$	1.06	1.53	2.17	3.10

Calculate the order with respect to total hydrogen. What mechanism does this suggest? Using thermodynamic data and the above rate data, calculate the rate constant at $923 \, K$ for the reaction

$$H + p-H_2 \rightarrow o-H_2 + H$$

6.6. The reaction of nitric oxide and hydrogen

$$2NO + 2H_2 \rightarrow N_2 + 2H_2 O$$

$P_{H_2}^0 = P_{NO^0}$	Total pressure/mmHg	$t_{1/2}/s$
	354	81
	288	140
	243	176
	202	224

$P_{H_2}^0 = $ constant	P_{NO^0}/mmHg	Initial rate/mmHg s^{-1}
	359	1.50
	300	1.03
	152	0.25

has been studied by C. N. Hinshelwood and T. E. Green [*J. Chem. Soc.*, **129**, 730 (1926)], who determined the half-life of the reaction and the initial rates as a function of the pressure of the reactants. From the data given calculate the total order of the reaction and the order with respect to nitric oxide.

6.7. Use the BEBO method and data in Tables 4-4 and 11-1 of H. S. Johnston, *Gas Phase Reaction Rate Theory*, Ronald Press, New York, 1966, to estimate activation energies for

(a) $HCl + CH_3 \rightarrow Cl + CH_4$

(b) $H_2 + CH_3 \rightarrow OH + CH_4$

(c) $CH_4 + CD_3 \rightarrow CH_3 + CHD_3$

REFERENCES

1. Robert F. Hampson, Jr., and David Garvin, *J. Phys. Chem.*, **81**, 2317 (1977).

2. NBS List of Publications Number 73. *Chemical Kinetics Tables, Data Evaluations, and Bibliographies. A Guide to the Literature*, revised December 1977, Chemical Kinetics Information Center, National Bureau of Standards, Washington, D.C..

3. D. L. Baulch and D. C. Montague, *J. Phys. Chem.*, **83**, 42 (1979); Norman Cohen and Karl Westberg, *J. Phys. Chem.*, **83**, 46 (1979).

4. CODATA Task Group on Data for Chemical Kinetics, "The Presentation of Chemical Kinetics Data in the Primary Literature," *CODATA Bull.*, **13** (1974); R. J. Cvetanović, D. L. Singleton, and G. Paraskevopoulos, *J. Phys. Chem.*, **83**, 50 (1979) and references therein.

5. K. Glaenzer, M. Quack, and J. Troe, *Chem. Phys. Lett.*, **39** [2], 304 (1976).

6. (a) R. E. March and J. C. Polanyi, *Proc. R. Soc. (London)*, **A273**, 360 (1963); (b) R. Gomer and G. B. Kistiakowski, *J. Chem. Phys.*, **19**, 85 (1951); (c) D. A. Parkes, D. M. Paul, and C. P. Quinn, *J. Chem. Soc., Faraday Trans. I*, **72**, 1935 (1976).

7. (a) A. Shepp and K. O. Kutschke, *J. Chem. Phys.*, **26**, 1020 (1957); (b) R. Hiatt and S. W. Benson, *J. Am. Chem. Soc.*, **94**, 6886 (1972); (c) D. M. Golden, K. Y. Choo, M. J. Perona, and L. W. Piszkiewicz, *Int. J. Chem. Kinet.*, **8**, 381 (1976).

8. (a) I. A. Matheson, H. W. Sidebottom, and J. M. Tedder, *Int. J. Chem. Kinet.*, **6**, 493 (1974); (b) M. L. White and R. R. Kuntz, *Int. J. Chem. Kinet.*, **5**, 187 (1973); (c) G. R. DeMarè and G. H. Huybrechts, *Trans. Faraday Soc.*, **64**, 1311 (1968).

9. (a) P. B. Ayscough, *J. Chem. Phys.*, **24**, 944 (1956); (b) R. Hiatt and S. W. Benson, *Int. J. Chem. Kinet.*, **4**, 479 (1972).

10. R. Eckling, P. Goldfinger, G. Huybrechts, G. Martens, L. Meyers, and S. Smoes, *Chem. Ber.*, **93**, 3014 (1960).

11. (a) Harold S. Johnston and Paul Goldfinger, *J. Chem. Phys.*, **37**, 700 (1962); (b) Harold S. Johnston, *Gas-Phase Reaction Rate Theory*, Ronald Press, New York, 1966, Chapter 8.

12. I. Koyano, "Ion-Molecule Reactions," in *Comprehensive Chemical Kinetics*, Vol. 18, C. H. Bamford and C. F. H. Tipper, Eds., Elsevier, Amsterdam, 1976; Eldon E. Ferguson, "Ion–Molecule Reactions," *Ann. Rev. Phys. Chem.*, **26**, 17 (1975); W. S. Koski, "Scattering of Positive Ions by Molecules," in *Molecular Scattering, Physical and Chemical Applications*, K. P. Lawley, Ed., Advances in Chemical Physics, Vol. 30, I. Prigogine and S. A. Rice, Eds., Wiley-Interscience, New York, 1975; J. L. Franklin, Ed., *Ion-Molecule Reactions*, Vols. 1 and 2, Plenum Press, New York, 1972; M. T. Bowers, Ed., *Gas Phase Ion Chemistry*, Vols. 1 and 2, Academic, New York, 1979.

13. E. W. McDaniel, V. Cermak, A. Dalgarno, E. E. Ferguson, and L. Friedman, *Ion–Molecule Reactions*, Wiley, New York, 1970.

14. (a) P. Langevin, *Ann. Chim. Phys.*, **5**, 245 (1905); (b) G. Gioumousis and D. P. Stevenson, *J. Chem. Phys.*, **29**, 294 (1958).

15. A. Giardini-Guidoni and L. Friedman, *J. Chem. Phys.*, **45**, 937 (1966).

16. T. F. Moran and W. H. Hamill, *J. Chem. Phys.*, **39**, 1413 (1963).

17. T. Su and M. T. Bowers, *J. Chem. Phys.*, **58**, 3027 (1973).

18. J. C. Light and J. Lin, *J. Chem. Phys.*, **43**, 3209 (1965).

19. J. C. Light, *J. Chem. Phys.*, **40**, 3221 (1964).

20. F. A. Wolf, *J. Chem. Phys.*, **44**, 1619 (1966).

21. J. A. Kerr, "Metathetical Reactions of Atoms and Radicals," in *Comprehensive Chemical Kinetics*, Vol. 18, C. H. Bamford and C. F. H. Tipper, Eds., Elsevier, Amsterdam, 1976.

22. J. A. Kerr and A. F. Trotman-Dickenson, *Handbook of Chemistry and Physics*, 54th ed., Chemical Rubber Co., Cleveland, 1974.

23. Juergen Wolfrum, "Atom Reactions," in *Kinetics of Gas Reactions*, W. Jost, Ed., Physical Chemistry, An Advanced Treatise, Vol. 6B, H. Eyring, D. Henderson, and W. Jost, Eds., Academic, New York, 1975.

24. A. A. Westenberg, in *Progress in Reaction Kinetics*, Vol. 7, G. Porter, Ed., Pergamon, New York, 1975.

25. (a) A. F. Trotman-Dickenson and E. W. R. Steacie, *J. Am. Chem. Soc.*, **72**, 2310 (1950); (b) L. M. Dorfman and R. Gomer, *Chem. Rev.*, **46**, 499 (1950).

26. (a) D. A. Parkes and C. P. Quinn, *J. Chem. Soc., Faraday Trans. I*, **72**, 1952 (1976); (b) Wing Tsang, *Int. J. Chem. Kinet.*, **10**, 821 (1978).

27. R. A. Fairclough and C. N. Hinshelwood, *J. Chem. Soc.*, **1937**, 538, 1573; J. E. Leffler, *J. Org. Chem.*, **20**, 1202 (1955); D. A. Blackadder and C. N. Hinshelwood, *J. Chem. Soc.*, **1958**, 2728.

28. S. W. Benson, *Thermochemical Kinetics*, Wiley-Interscience, New York, 1976.

29. M. G. Evans and M. Polanyi, *Trans. Faraday Soc.*, **34**, 11 (1938).

30. N. Semenov, *Some Problems of Chemical Kinetics and Reactivity*, Vol. 1, Pergamon Press, New York, 1958, pp. 27 and 58.

31. J. A. Kerr, *Chem. Rev.*, **66**, 465 (1966).

32. S. W. Benson and W. B. DeMore, *Ann. Rev. Phys. Chem.*, **16**, 397 (1965).

33. H. S. Johnston and C. Parr, *J. Am. Chem. Soc.*, **85**, 2544 (1963).

34. S. W. Mayer, L. Shieler, and H. S. Johnston, *11th Symposium (International) on Combustion*, The Combustion Institute, Pittsburgh, PA, 1967, p. 837.

35. L. Pauling, *J. Am. Chem. Soc.*, **69**, 542 (1947).

36. D. L. Bunker, *Acc. Chem. Res.*, **7**, 195 (1974).

37. J. C. Polanyi and J. L. Schreiber, in *Potential Energy Surfaces, Disc. Faraday Soc.*, **62**, 267 (1977); J. L. Schreiber, Ph. D. Thesis, University of Toronto, Toronto, Canada, 1973.

38. J. N. Murrell, "Potential Energy Surfaces for Studying the Reactions and Molecular Dynamics of Small Polyatomic Molecules," in *Gas Kinetics and Energy Transfer*, Vol. 3, P. G. Ashmore and R. J. Donovan, Eds., Specialist Periodical Report, The Chemical Society, London, 1978.

39. P. K. Pearson, H. F. Schaeffer III, and U. Wahlgren, *J. Chem. Phys.*, **62**, 350 (1975).

40. O. Kafri and M. J. Berry, in *Potential Energy Surfaces, Disc. Faraday Soc.*, **62**, 127 (1977).

41. N. C. Blaise and D. L. Bunker, *J. Chem. Phys.*, **37**, 2713 (1962); L. M. Raff and M. Karplus, *J. Chem. Phys.*, **44**, 1212 (1966).

42. F. T. Wall and R. N. Porter, *J. Chem. Phys.*, **36**, 3256 (1962).

43. D. L. Bunker and N. C. Blaise, *J. Chem. Phys.*, **41**, 2377 (1964); D. L. Bunker and C. A. Parr, *J. Chem. Phys.*, **52**, 5700 (1970).

44. J. N. L. O'Connor, W. Jakubetz, and J. Manz, *Mol. Phys.*, **29**, 347 (1975); J. M. Bowman and A. Kuppermann, *Chem. Phys. Lett.*, **34**, 523 (1975); S. K. Gray, J. S. Wright, and X. Chapuisat, *Chem. Phys. Lett.*, **48**, 155 (1977); R. W. Hamming, *Numerical Methods for Scientists and Engineers,* 2nd ed., McGraw-Hill, New York, 1973, pp. 349–356.

45. M. G. Evans and M. Polanyi, *Trans. Faraday Soc.*, **35**, 178 (1939); P. J. Kuntz, E. M. Nemeth, J. C. Polanyi, S. D. Rosner, and C. E. Young, *J. Chem. Phys.*, **44**, 1168 (1966).

46. P. J. Kuntz, "Features of Potential Energy Surfaces and Their Effect on Collisions," *Dynamics of Molecular Collisions,* Part B, in W. H. Miller, Ed., Modern Theoretical Chemistry, Vol. 2, Plenum Press, New York, 1976.

47. D. G. Truhlar and R. E. Wyatt, *Ann. Rev. Phys. Chem.*, **27**, 1 (1976); J. Wolfrum, "Atom Reactions," in *Kinetics of Gas Reactions,* W. Jost, Ed., Physical Chemistry, An Advanced Treatise, Vol. 6B, H. Eyring, D. Henderson, and W. Jost, Eds., Academic, New York, 1975.

48. A. Farkas and L. Farkas, *Proc. R. Soc. (London),* **A152**, 124 (1935); B. A. Ridley, W. R. Schulz, and D. J. LeRoy, *J. Chem. Phys.*, **44**, 3344 (1966); A. A. Westenberg and N. deHaas, *J. Chem. Phys.*, **47**, 1393 (1967); D. N. Mitchell and D. J. LeRoy, *J. Chem. Phys.*, **58**, 3449 (1973).

49. A. Kuppermann and J. M. White, *J. Chem. Phys.*, **44**, 4352 (1966).

50. A. Kuppermann, *Isr. J. Chem.*, **7**, 303 (1969).

51. R. Gengenbach, Ch. Hahn, and J. P. Toennies, *J. Chem. Phys.*, **62**, 3620 (1975) and references therein.

52. H. Eyring and M. Polanyi, *Z. Phys. Chem.*, (Leipzig) **B12**, 279 (1931).

53. I. Shavitt, *J. Chem. Phys.*, **49**, 4048 (1968); K. A. Quickert and D. J. LeRoy, *J. Chem. Phys.*, **53**, 1325 (1970); **54**, 5444 (1971); G. W. Koeppl, *J. Chem. Phys.*, **59**, 3425 (1973).

54. (a) D. G. Truhlar and A. Kuppermann, *J. Am. Chem. Soc.*, **93**, 1840 (1971); (b) R. A. Marcus and M. E. Coltrin, *J. Chem. Phys.*, **67**, 2609 (1977) and references therein.

55. J. T. Adams and R. N. Porter, *J. Chem. Phys.*, **59**, 4105 (1973), **60**, 3354 (1974); A. C. Yates and W. A. Lester, *Chem. Phys. Lett.*, **24**, 305 (1974); M. Karplus in *Molecular Beams and Reaction Kinetics,* Ch. Schlier, Ed., Academic, New York, 1970, p. 407; D. J. Malcome-Lawes, *J. Chem. Soc., Faraday Trans. II,* **71**, 1183 (1975).

56. R. B. Walker, E. B. Stechel, and J. C. Light, *J. Chem. Phys.*, **69**, 2922 (1978).

57. Reference 11*b*, page 189.

58. M. R. Levy, *Prog. React. Kinet.*, **10**, 1 (1979) and references therein; see also ref. 32 and references therein.

59. R. L. Jaffe, K. Morokuma, and T. F. George, *J. Chem. Phys.*, **63**, 3417 (1975).

60. R. L. Wilkins, *J. Phys. Chem.*, **77**, 3081 (1973).

61. M. Bodenstein, *Z. Phys. Chem. (Leipz.),* **13**, 56 (1894); **22**, 1 (1897); **29**, 295 (1898).

62. S. Glasstone, K. J. Laidler, and H. Eyring, *The Theory of Rate Processes,* McGraw-Hill, New York, 1941, p. 234.

63. J. H. Sullivan, *J. Chem. Phys.*, **46**, 73 (1967).

64. G. G. Hammes and B. Widom, *J. Am. Chem. Soc.*, **96**, 7621 (1974): see also J. B. Anderson, *J. Chem. Phys.*, **61**, 3390 (1974).

65. R. M. Noyes, *J. Am. Chem. Soc.*, **96**, 7623 (1974); S. B. Jaffe and J. B. Anderson, *J. Chem. Phys.*, **51**, 1057 (1969).

66. D. A. Dixon, R. M. Stevens, and D. R. Herschbach, in *Potential Energy Surfaces, Disc. Faraday Soc.*, **62**, 110 (1977) and references therein; see also S. H. Bauer, *Annu. Rev. Phys. Chem.*, **30**, 271 (1979).

67. J. S. Wright, *Chem. Phys. Lett.*, **6**, 476 (1970); *Can. J. Chem.*, **53**, 549 (1975).

68. D. A. Dixon, D. L. King, and D. R. Herschbach, in *Molecular Beam Scattering, Disc. Faraday Soc.*, **55**, 375 (1973); D. L. King, D. A. Dixon, and D. R. Herschbach, *J. Am. Chem. Soc.*, **96**, 3328 (1974); D. A. Dixon and D. R. Herschbach, *J. Am. Chem. Soc.*, **97**, 6268 (1975).

69. S. H. Bauer, D. M. Lederman, E. L. Resler, and E. R. Fisher, *Int. J. Chem. Kinet.*, **5**, 93 (1973), and references therein.

70. H. M. Frey and R. Pottinger, *J. Chem. Soc., Faraday Trans. I*, **74**, 1827 (1978).

71. Reference 28, pp. 136–140; but see also J. J. Gajewski, *J. Am. Chem. Soc.*, **101**, 4393 (1979).

72. Reference 28, pp. 26–28.

73. M. J. S. Dewar, S. Olivella, and H. S. Rzepa, *J. Am. Chem. Soc.*, **100**, 5650 (1978) and references therein.

74. (a) R. E. Townshend, G. Ramunni, G. Segal, W. J. Hehre, and L. Salem, *J. Am. Chem. Soc.*, **98**, 2190 (1976); (b) M. V. Basilevski, A. G. Shamov, and V. A. Tikhomirov, *J. Am. Chem. Soc.*, **99**, 1369 (1977).

75. (a) P. J. Robinson and K. A. Holbrook, *Unimolecular Reactions,* Wiley-Interscience, New York, 1972; (b) P. J. Robinson, "Unimolecular Reactions," in *Reaction Kinetics,* Vol. 1, P. G. Ashmore, Ed., Specialist Periodical Report, The Chemical Society, London, 1975; (c) D. C. Tardy and B. S. Rabinovitch, *Chem. Rev.*, **77**, 369 (1977).

76. B. W. Wojciechowski and K. J. Laidler, *Trans. Faraday Soc.*, **59**, 369 (1963).

77. A. E. Shilov and R. D. Sabirova, *Zh. Fiz. Chem.*, **33**, 1365 (1959).

78. K. A. Holbrook, *Proc. Chem. Soc.*, **1964**, 418; M. R. Bridge and J. L. Holmes, *J. Chem. Soc. (B)*, **1966**, 713; J. L. Holmes and L. S. M. Ruo, *J. Chem. Soc. (A)*, **1968**, 1231.

79. R. B. Cundall, *Prog. React. Kinet.*, **2**, 165 (1964); M. C. Lin and K. J. Laidler, *Can. J. Chem.*, **46**, 973 (1968).

80. S. H. Bauer, *J. Am. Chem. Soc.*, **91**, 3688 (1969).

81. M. Szwarc and J. Murawski, *Trans. Faraday Soc.*, **47**, 269 (1951).

82. F. W. Schuler and G. W. Murphy, *J. Am. Chem. Soc.*, **72**, 3155 (1950).

83. R. N. Haszeldine, P. J. Robinson, and J. A. Walsh, *J. Chem. Soc. (B)*, **1970**, 578.

84. O. K. Rice and H. Gershinowitz, *J. Chem. Phys.*, **3**, 479 (1935); B. G. Gowenlock, *Q. Rev.*, **14**, 133 (1960).

85. R. A. Marcus, *J. Chem. Phys.*, **20**, 359 (1952).

86. M. C. Lin and K. J. Laidler, *Trans. Faraday Soc.*, **64**, 79 (1968); E. V. Waage and B. S. Rabinovitch, *Int. J. Chem. Kinet.*, **3**, 105 (1971).

87. D. B. Olson and W. C. Gardiner, Jr., *J. Phys. Chem.*, **83**, 922 (1979).

88. (a) A. Maccoll, *Chem. Rev.*, **69**, 33 (1969); (b) A. Maccoll, *Adv. Phys. Org. Chem.*, **3**, 91 (1965); (c) A. Maccoll in *Theoretical Organic Chemistry* (Kekule Symposium of the Chemical Society, 1958), Butterworths, London, 1959, p. 230 ff.

89. S. W. Benson and A. N. Bose, *J. Chem. Phys.*, **39**, 3463 (1963).

90. H. E. O'Neal and S. W. Benson, *J. Phys. Chem.*, **71**, 2903 (1967).

91. C. D. Hurd and F. H. Blunck, *J. Am. Chem. Soc.*, **60**, 2419 (1938).

92. See reference 75*a*, pp. 240–242 for a summary of experimental studies.

93. (a) T. S. Chambers and G. B. Kistiakowsky, *J. Am. Chem. Soc.*, **56**, 399 (1934); (b) H. O. Pritchard, R. G. Sowden, and A. F. Trotman-Dickenson, *Proc. R. Soc., (London)*, **A217**, 563 (1953); (c) A. D. Kennedy and H. O. Pritchard, *J. Phys. Chem.*, **67**, 161 (1963).

94. (a) N. B. Slater, *Proc. R. Soc. (London)*, **A218**, 224 (1953); (b) G. M. Wieder and R. A. Marcus, *J. Chem. Phys.*, **37**, 1835 (1962); (c) M. C. Lin and K. J. Laidler, *Trans. Faraday Soc.*, **64**, 927 (1968).

95. W. E. Falconer, T. F. Hunter, and A. F. Trotman-Dickenson, *J. Chem. Soc.*, **1961**, 609.

96. B. S. Rabinovitch, E. W. Schlag, and K. B. Wiberg, *J. Chem. Phys.*, **28**, 504 (1958).

97. R. G. Bergman and W. L. Carter, *J. Am. Chem. Soc.*, **91**, 7411 (1969).

98. M. R. Willcott and V. H. Cargle, *J. Am. Chem. Soc.*, **91**, 4310 (1969).

99. M. C. Flowers and H. M. Frey, *J. Chem. Soc.*, **1960**, 2758.

100. S. W. Benson and P. S. Nangia, *J. Chem. Phys.*, **38**, 18 (1963).

101. R. Hoffmann, *J. Am. Chem. Soc.*, **90**, 1475 (1968); J. A. Horsley, Y. Jean, C. Moser, L. Salem, R. M. Stevens, and J. S. Wright, *J. Am. Chem. Soc.*, **94**, 279 (1972); P. J. Hay, W. J. Hunt, and W. A. Goddard, *J. Am. Chem. Soc.*, **94**, 638 (1972).

102. L. Salem and C. Rowland, *Angew. Chem., Int. Ed.*, **11**, 92 (1972).

103. See reference 75a, Chap. 8, reference 75b, pp. 96–115, and reference 75c, pp. 391–405 for detailed discussions of these methods.

104. J. N. Butler and G. B. Kistiakowsky, *J. Am. Chem. Soc.*, **82**, 759 (1960).

105. J. D. Rynbrandt and B. S. Rabinovitch, *J. Phys. Chem.*, **74**, 4175 (1970); **75**, 2164 (1971).

106. (a) R. H. Tredgold, *Proc. Phys. Soc. (London)*, **A68**, 920 (1955); (b) E. Thiele and D. J. Wilson, *J. Chem. Phys.*, **35**, 1256 (1961); (c) O. K. Rice, *J. Phys. Chem.*, **65**, 1588 (1961).

107. D. L. Bunker, *J. Chem. Phys.*, **37**, 393 (1962); **40**, 1946 (1964); D. L. Bunker and M. Pattengill, *J. Chem. Phys.*, **48**, 772 (1968).

108. H. H. Harris and D. L. Bunker, *Chem. Phys. Lett.*, **11**, 433 (1971); D. L. Bunker, *J. Chem. Phys.*, **57**, 332 (1972); D. L. Bunker and W. L. Hase, *J. Chem. Phys.*, **59**, 4621 (1973).

109. K. W. Hicks, M. L. Lesiecki, S. M. Riseman, and W. A. Guillory, *J. Phys. Chem.*, **83**, 1936 (1979).

110. See reference 75a, Chap. 10.

111. S. C. Chan, B. S. Rabinovitch, J. T. Bryant, L. D. Spicer, T. Fujimoto, Y. N. Lin, and P. S. Pavlou, *J. Phys. Chem.*, **74**, 3160 (1970); S. C. Chan, J. T. Bryant, L. D. Spicer, and B. S. Rabinovitch, *J. Phys. Chem.*, **74**, 2058 (1970); L. D. Spicer and B. S. Rabinovitch, *J. Phys. Chem.*, **74**, 2445 (1970); F. M. Wang, T. Fujimoto, and B. S. Rabinovitch, *J. Phys. Chem.*, **76**, 1935 (1972); see also reference 75c.

112. K. E. Russell and J. Simons, *Proc. R. Soc. (London).*, **A217**, 271 (1953).

113. Y. N. Lin and B. S. Rabinovitch, *J. Phys. Chem.*, **74**, 3151 (1970); see also B. Stevens, *Mol. Phys.*, **3**, 589 (1960).

114. D. L. Bunker and N. Davidson, *J. Am. Chem. Soc.*, **80**, 5085, 5090 (1958).

115. K. V. Reddy and M. J. Berry, *Chem. Phys. Lett.*, **66**, 223 (1979).

SEVEN

REACTIONS IN SOLUTION

Although reactions in the gas phase are simpler to deal with theoretically, the fact remains that most chemistry encountered in practice occurs in solutions of one kind or another. The first question that arises is whether there is any fundamental difference between the kinetics of reaction in solution and in the gaseous state. Briefly, when a reaction occurs by the same mechanism in solution and in the gas, the kinetics are often not changed appreciably. However, because of the increased interactions in condensed media, the mechanism is usually changed completely and the kinetics are greatly altered. There is a wide variety of reactions that do not occur in the gas phase at all, under normal conditions, but that occur more or less readily in various solvents. The favored mechanisms in solution are ionic ones, involving formation and interaction of charged particles. Such reactions occur best in polar solvents. Reactions of ions occur in the gas phase only under special, high energy conditions.

In fact it is difficult to find examples of reactions that have been studied both in the gas phase and in solution. That is, a given set of reactants will behave quite differently in the two media. The favored reactions in the gas phase involve free radicals. Even if the same free radicals are formed in solution, it is likely that their reactions will be different. In such cases it is not always clear that the same reactions are being observed, even in part.

As an example, Table 7.1 shows some results on the decomposition of di-t-butyl peroxide.[1] In the gas phase between 110 and 280°C, the stoichiometry is given quite accurately by,

$$(CH_3)_3COOC(CH_3)_3 = 2(CH_3)_2CO + C_2H_6 \qquad (7.1)$$

if a large amount of glass surface is available. The reaction is cleanly first order and is homogeneous, since the rate is unchanged in packed and unpacked reaction vessels. The reaction rates and the activation energies are almost unchanged in several liquid solvents, as shown in the table.

However, the products are quite different, considerable amounts of t-butyl

Table 7.1. Decomposition of di-t-butyl peroxide

Solvent	A/s^{-1}	E_a/kJ mol^{-1}	Amount t-butyl alcohol	Amount acetone
Vapor	3.2×10^{16}	163	0	2.0
i-Propylbenzene	0.6×10^{16}	157	1.61	0.39
t-Butylbenzene	1.1×10^{16}	159	0.75	1.25
Tri-n-butyl amine	0.35×10^{16}	155	~1.9	~0.1

Source. Raley, Rust, and Vaughan.[1]

alcohol being formed in solution, at the expense of acetone. The results can be explained if the rate-determining step in all cases is the cleavage of the peroxide bond.

$$(CH_3)_3 COOC(CH_3)_3 \rightarrow 2(CH_3)_3 CO\cdot \tag{7.2}$$

The dissociation energy for this bond may be estimated as 159 kJ mol^{-1}, which agrees with the activation energy in all media.

In the vapor state, the ensuing reactions are

$$(CH_3)_3 CO\cdot \rightarrow (CH_3)_2 CO + CH_3\cdot \tag{7.3}$$

$$2CH_3\cdot \rightarrow C_2 H_6 \text{ surface} \tag{7.4}$$

with (7.4) requiring a surface to carry away the substantial heat of recombination, which might otherwise cause immediate redissociation. In a hydrogen-bearing solvent, hydrogen atom abstraction is a ready process, forming t-butyl alcohol.

$$(CH_3)_3 CO\cdot + RH \rightarrow (CH_3)_3 COH + R\cdot \tag{7.5}$$

The radicals R$\cdot$, and any methyl radicals, can then undergo various reactions.

Clearly the cleavage of the peroxide bond involves little change in polarity and is unaffected by the presence of a solvent.

An interesting study has been made of the thermolysis of di-t-butyl peroxide in very high pressures of cyclopropane. (2.6–3.6 mol liter^{-1}), over a range of temperatures.[2] The rate constants in the range of 125–160°C fit smoothly onto those in the range 100–124°C, even though the former are above the critical point of cyclopropane and the latter are below the critical point. A plot of log k versus $1/T$ is a straight line with no discontinuity at the critical temperature. There is a discontinuity in product ratio at the critical point, much more t-butyl alcohol being formed below 125°C. This shows that a liquid phase reaction is occurring.

Similarly the rate of decomposition of acetyl peroxide is little affected by going from the gas phase to various nonpolar, or even polar, solvents. The rate of racemization of α-pinene is very nearly the same in the vapor and in various non-polar solvents.[3] In this case the rate-determining step is the breaking of a strained carbon–carbon bond. A number of bimolecular reactions also have similar rates and activation energies in the gas phase and in inert solvents. Most of these are of the Diels-Alder type, such as the self-condensation of cyclopentadiene. Such reactions involve little change in polarity.

Table 7.2. Rate constants at $25°C$ for hydrogen-atom abstraction reactions

Substrate	$k_{H_2O}/M^{-1} s^{-1}$	$k_{gas}/M^{-1} s^{-1}$
CH_3COCH_3	1.7×10^6	3.3×10^4
CH_3OH	1.6×10^6	8.2×10^2
C_2H_5OH	2.5×10^7	4.1×10^4
$n\text{-}C_3H_7OH$	2.7×10^7	3.6×10^5
$i\text{-}C_3H_7OH$	7.8×10^7	5.2×10^5
C_2H_6	2.5×10^6	1.8×10^4
C_3H_8	2.2×10^7	1.1×10^5
$n\text{-}C_4H_{10}$	3.9×10^7	1.6×10^5
$n\text{-}C_6H_{14}$	1.5×10^8	3.5×10^4
$C_2H_4^a$	3×10^9	1×10^5

Source. Aqueous data from Neta, Holdren, and Schuler.[4] Gas-phase data compiled by Kondratiev.[5]

[a] This reaction is one of H-atom addition to the double bond.

If the solvent has quite different properties from the reactants, larger effects can be expected. Table 7.2 shows the rate constants at $25°C$ for H atom reacting with a number of organic substrates, both in the gas phase and in water. The reaction is one of atom abstraction, for example,

$$H\cdot + RH \rightarrow H_2 + R\cdot \tag{7.6}$$

except for ethylene, where the H atom adds. When RH is an alcohol, the hydrogen atom abstracted is bound to carbon. The rate data in aqueous media were obtained by the technique of pulse radiolysis, in which ionizing radiation produces hydrated electrons.[6] These react with protons to produce free H atoms. Most of the data in Table 7.2 were obtained from competition experiments, but the absolute rate constant for at least one reaction must be known.

It can be seen that these free-radical reactions are much more rapid in water than in the gas phase. These results can be understood in terms of the structure of water in the presence of hydrocarbon solutes.[7] Water relies chiefly on hydrogen bonding, or dipole–dipole attractions, for its intermolecular interactions. Dispersion forces are small, because of the low polarizability of H_2O. Hydrocarbons, conversely, rely almost entirely on dispersion forces. The two kinds of molecules are of maximum incompatibility. As a result hydrocarbon molecules exist in aqueous solution in cavities that are bounded by icelike hydrogen-bonded water molecules. The hydrogen atom may have a similar environment.

If the two cavities containing the nonpolar reactants come together, the reactants are trapped in a single ice-bound cavity and normally react before separating again. The activation energies in the gas phase are in the range of $20–40$ kJ mol^{-1}. The corresponding figures in water are not known, but they would be very much affected by the heats of forming hydrogen bonds in the solvent. Frequency factors in solution would be similarly affected by the entropies of hydrogen-bond formation. In both cases it would be a dynamic situation and not a static one, since the movement of the cavities by making and breaking of hydrogen bonds would be important.

THE COLLISION THEORY IN SOLUTION

Let us consider the application of our theories of kinetics, developed in preceding chapters, to reactions in solution. We start with the collision theory, but first we need to consider in more detail the structure of liquids. Except for water, the molar volume of a liquid is roughly 10% greater than that of the corresponding solid. However, x-ray diffraction studies show that short-range order persists in the liquid state and that nearest neighbor distances are almost unchanged. Also we know that the potential energy of a liquid is higher than that of its solid by about 10%. That is, the heat of fusion is roughly 10% of the heat of sublimation. This leads to a picture of the liquid state in which each molecule has an environment very much like that of the solid, but with some nearest neighbors missing. In their place are vacancies, or holes.[8] Roughly 1 neighbor in 10 is missing. Figure 7.1 shows a schematic representation of liquid state structure.

The molecules are not moving freely, as in the gas, but instead move in the potential field of their neighbors. This motion approximates the oscillations of a three-dimensional harmonic oscillator. For example, the molar heat capacity of monatomic liquids is about $3R$, instead of the $\frac{3}{2}R$ found for gases. The contribution to C_p from the translational component of other simple liquids is also $3R$. Rotation may either be free, as in the gas, or strongly hindered, becoming vibrations. The far infrared spectra of liquids show broad absorptions in the $20-200$ cm^{-1} region corresponding to these lattice modes.

Since the mean vibrational energy is $3RT$ for the hindered translation, the mean kinetic energy is $\frac{3}{2}RT$, the same as for the gas phase. Hence the root-mean-square velocities are the same in the two phases. We can attempt to calculate the collision number by calculating the volume swept out per second and multiplying this by the number of other molecules in unit volume. However, a correction must be made for the volume occupied by the molecules themselves, or conversely we must take only the free volume, V_f, available for molecular motion. This correction affects the concentration of the other molecules and raises the number of collisions by a factor

$$\frac{Z_{AB}(\text{solution})}{Z_{AB}(\text{gas})} = \frac{V}{V_f} \tag{7.7}$$

This ratio could be $10:1$, or even larger since the free volume is not clearly defined.

However, the preceding is not a very satisfactory model for collisions in the liquid state. The molecules do not move freely in the free volume, but are subject to a varying potential energy. The holes in the solvation shell allow for molecular motion leading to a net displacement from the original position. The motion consists of a series of discontinuous displacements, each of which is followed by a residence in the solvent cage in which the motion is one of rapid collisions with the molecules that make up the cage. Thus if two molecules become neighbors by random displacements, they tend to collide a large number of times before they separate again. Such an event is called an "encounter."[9]

For noninteracting particles these encounters are determined by the random,

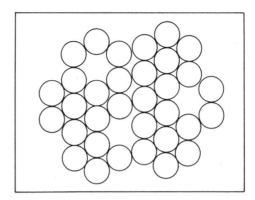

Figure 7.1. Schematic structure of the liquid state showing short range order, long range disorder, and holes.

Brownian movements of the particles through the viscous medium, and hence are diffusion controlled. The basic equations for diffusion are Fick's two laws:

$$\phi = -D \, \nabla \, \rho \qquad (7.8)$$

$$\frac{\partial \rho}{\partial t} = \nabla \cdot \phi = -D \nabla^2 \rho \qquad (7.9)$$

Here ϕ, the flux, is the number of particles moving across a boundary of unit area per second, D is the diffusion coefficient, and $\rho = N/V$ is the number density of particles.

Consider the case of spherical symmetry with d the distance from the origin and boundary conditions $\rho = \rho_0$ for all values of d at $t = 0$, and $\rho = \rho_0$ at $d = \infty$ and $\rho = 0$ at $d = d_{AB}$ when $t > 0$. (Here d_{AB} is the hard-sphere collision distance defined in Chapter 4.) Equations (7.8) and (7.9) can be solved to yield

$$\rho = \rho_0 \left\{ 1 - \frac{d_{AB}}{d} \, \text{erfc} \left[\frac{d - d_{AB}}{2(Dt)^{1/2}} \right] \right\} \qquad (7.10)$$

$$\phi = \frac{D\rho_0}{d_{AB}} \left[1 + \frac{d_{AB}}{(\pi Dt)^{1/2}} \right] \qquad (7.11)$$

In (7.10) erfc (x) stands for the complement of the error function of x, which is tabulated for various values of x. In (7.11) the second term usually becomes small in comparison to the first term in a very short time. This corresponds to $(\partial \rho / \partial t)$ being equal to zero, a stationary state. At any instant of time ϕ is the flux of B molecules into a single A molecule and ρ_0 is the instantaneous bulk number density of B molecules. The diffusion coefficient $D = D_A + D_B$ because of the simultaneous movement of both particles.

The total rate is clearly $4\pi d_{AB}^2 \phi \rho_A$ and accordingly the second-order rate constant for encounters becomes, using only the first term of (7.11),

$$k_e = 4\pi d_{AB} DL \qquad (7.12)$$

a result first obtained by Smoluchowski.[10] Putting in reasonable values of $d_{AB} = 5 \times 10^{-8}$ cm and $D = 10^{-5}$ cm^2 s^{-1} gives a rate constant of 4×10^9 M^{-1} s^{-1}. It may be noted that an approximate equation for the diffusion coefficient of a spherical molecule of radius σ is

$$D = \frac{RT}{6\pi\eta\sigma L} \tag{7.13}$$

where η is the viscosity of the medium. Hence a viscous medium slows down the rate of encounters. Furthermore, such processes have an activation energy because of the temperature dependence of the viscosity.

If a bimolecular reaction occurred at every collision, its rate constant would be given by (7.12). Such reactions are said to be diffusion, or encounter, controlled. Their rate constants define the maximum rate attainable in a liquid medium. However, most reactions have substantial energy barriers, or strict steric requirements, and many collisions are needed before a reaction will ensue. For such reactions we need to know the duration of an encounter before the escape of one partner from the solvent cage terminates it, and the frequency of vibrational collision during an encounter.

An estimate of the duration can be obtained by solving (7.8) for a single species with the boundary condition $\rho_0 \neq 0$ at $d = 0$ and $\rho_0 = 0$ for all other d.

$$\rho = \frac{S}{(4\pi D t)^{3/2}} e^{-d^2/4Dt} \tag{7.14}$$

where S is a normalization constant. Setting it equal to one converts (7.14) to the probability of a given molecule migrating a distance d in time t, after multiplying by the volume element in spherical coordinates. It can then be used to solve for the mean square displacement at any time.

$$\overline{d^2} = 6Dt \tag{7.15}$$

Let $t = \tau_{AB}$, the duration of an encounter between A and B molecules. Equation (7.15) then gives

$$\tau_{AB} = \frac{\overline{\lambda^2}}{6D} \tag{7.16}$$

where $D = D_A + D_B$, as before, and $\overline{\lambda^2}$ is the mean square displacement needed to break off an encounter. As an order of magnitude, λ may be set equal to $\sqrt{2}d_{AB}$, a distance great enough to ensure that the molecules no longer are in the same solvent cage. Alternatively, λ may be equal to the diameter of a solvent molecule.

In terms of the steric factor, p, and the energy requirement to surmount the barrier, the probability of a reaction occurring during an encounter is given by

$$P_r = \frac{vpe^{-LE/RT}}{vpe^{-LE/RT} + 1/\tau_{AB}} \tag{7.17}$$

where v is the frequency of collisions between A and B during an encounter. The rate constant for the reaction then becomes

$$k = 4\pi d_{AB} D L P_r \qquad (7.18)$$

If $1/\tau_{AB}$ is much smaller than $vpe^{-LE/RT}$, then the rate constant is given by (7.12), and the reaction is diffusion controlled. If $1/\tau_{AB}$ is much larger than $vpe^{-LE/RT}$, then the rate constant is given by

$$k = \frac{2\pi d_{AB} \overline{\lambda^2} vpe^{-LE/RT} L}{3} \qquad (7.19)$$

This equation is the liquid-phase equivalent of the gas-phase result

$$k = L \left(\frac{8\pi RT}{L\mu}\right)^{1/2} d_{AB}^2 pe^{-LE/RT} \qquad (7.20)$$

Equations equivalent to (7.18) and (7.19) can be derived in another way.[11] The molecules A and B are assumed to form an associated pair A:B, which can either react further or dissociate again to A and B.

$$A + B \underset{k_2}{\overset{k_1}{\rightleftarrows}} A:B \xrightarrow{k_3} \text{product} \qquad (7.21)$$

Applying the steady-state assumption to the concentration of A:B, leads to the rate equation

$$\text{rate} = \frac{k_1 k_3}{k_2 + k_3} [A][B] \qquad (7.22)$$

The rate constant is $k_1 k_3/(k_2 + k_3)$, which is identical with (7.18), k_1 being the Smoluchowski rate constant k_e. In the limit of $k_2 \gg k_3$, the rate constant becomes

$$k = \frac{k_1 k_3}{k_2} = K k_3 \qquad (7.23)$$

where K is the equilibrium constant for the formation of A:B. If A and B have no special interaction, this constant is simply $4\pi d_{AB}^3 L/3$, the volume corresponding to an encounter complex divided by the total (unit) volume. If $\overline{\lambda^2} = 2d_{AB}^2$ and if k_3 is again written as $vpe^{-LE/RT}$, then (7.23) is identical with (7.19).

The frequency v may be estimated from the total vibrational energy $6RT$, for both molecules by setting it equal to $Lh\bar{v}$, where $\bar{v}$ is the mean vibrational frequency. Only one collision in m is between A and B molecules, if m is the total number of nearest neighbors. If m is taken as 12 and if $d_{AB} = 5 \times 10^{-8}$ cm, $\overline{\lambda^2} = 2d_{AB}^2$ and $T = 298$ K, the rate constant is found to be

$$k = \frac{2\pi RT d_{AB}^3 pe^{-LE/RT}}{3h} \cong 2.7 \times 10^{12} \, pe^{-LE/RT} \, M^{-1} \, s^{-1} \qquad (7.24)$$

This may be compared to $k = 2.6 \times 10^{11} \, pe^{-LE/RT} \, M^{-1} \, s^{-1}$, calculated from (7.20).

The preceding discussion predicts that Arrhenius frequency factors will be somewhat larger in solution than in the vapor state for bimolecular reactions. There is not much evidence to support this prediction. The most probable, as well as the average, value of the frequency factor for second-order reactions in solution is $10^{11} \, M^{-1} \, s^{-1}$, the same as in the gas phase.[12]

The validity of (7.12) for reactions that occur on every collision has been amply verified. As an example, the rate of recombination of methyl radicals in water may be cited.[13]

$$2CH_3 \cdot (aq) \rightarrow C_2H_6(aq) \tag{7.25}$$

The rate constant, expressed in terms of disappearance of radicals, is 3.2×10^9 $M^{-1} \, s^{-1}$ at 25°C. While the exact collision radius and diffusion constant are not known, this number is certainly of the right magnitude. Furthermore, the activation energy of 16 kJ mol^{-1} is very nearly the value of 19 kJ mol^{-1} for the variation of the viscosity of water with temperature, using (7.13). The rate constant may be compared with the gas-phase value of $2.0 \times 10^{10} \, M^{-1} \, s^{-1}$, with zero activation energy (p. 193). This figure is the high-pressure limit.

The rate of recombination of iodine atoms in carbon tetrachloride,

$$2I \cdot \xrightarrow{CCl_4} I_2 \tag{7.26}$$

has a second-order rate constant of $8.2 \times 10^9 \, M^{-1} \, s^{-1}$ at 25°C.[14] In this case the diffusion constant is known to be $4.0 \times 10^{-5} \, cm^2 \, s^{-1}$ for the atom and d_{AB} may be estimated as 4.30×10^{-8} cm from the van der Waals' radius of iodine atoms. The value of k_e is calculated to be $13 \times 10^9 \, M^{-1} \, s^{-1}$ from these numbers. The crossover from diffusion control to activation control has been discussed by Noyes.[15] By varying the viscosity of the medium, it has been possible to demonstrate this crossover experimentally.[16] That is, a sufficiently viscous medium must eventually lead to diffusion control of rates.

We consider next unimolecular reactions in liquid solutions. It is obvious that we are in the high pressure limit of the theories of such processes. Collisions with the solvent guarantee an equilibrium between ordinary molecules and activated molecules. The solvent acts as a heat bath coupled to the reactants. Conservation of energy and of momentum, which can be factors in gas-phase reactions, are of little concern in solution.

The initial reaction of a molecule, A—B, often produces two fragments A and B, which are trapped in the same cage of solvent molecules. The fragments may recombine to form A—B, may escape from the cage and then react further, or may react in the cage to form new fragments A$'$ and B$'$. The system is given by the equations

$$A–B \underset{k_2}{\overset{k_1}{\rightleftharpoons}} (A+B) \xrightarrow{k_3} A+B \tag{7.27}$$
$$\downarrow k_4$$
$$(A'+B')$$

The symbol $(A + B)$ represents the caged initial product called a geminate pair.[15] It can be seen that k_2 is the same as $\nu p e^{-LE/RT}$, and k_3 is $1/\tau_{AB}$ as used earlier. In the typical cases discussed above, $\nu \cong 3 \times 10^{12}$ s^{-1} and $1/\tau_{AB} \cong 1 \times 10^{10}$ s^{-1}. This suggests that, if $p = 1$ and $E = 0$, only one time in 300 will a geminate pair separate to form independent A and B molecules. Compared to the gas phase there could be a 300-fold reduction in the frequency factor of the unimolecular rate constant.

This conclusion can be arrived at in another way by considering the equilibrium constants in solution and in the gas.

$$A{-}B \underset{k_r}{\overset{k_f}{\rightleftharpoons}} A + B \qquad K = \frac{k_f}{k_r} \tag{7.28}$$

Here k_f and k_r are the observable first-order and second-order rate constants.

Consider an idealized case where k_r is controlled by diffusion, and ΔH is the same in solution and in the gas. The entropy change for (7.28) is positive in the gas phase because of the extra translational entropy of the products. But $\Delta S^{\ominus}$ in solution is much smaller, because of the restricted motion of molecules in the liquid state. Consideration of the differences in $\Delta S^{\ominus}$ lead to the conclusion (see below) that $K_{gas} > K_{solution}$ by a factor of $10^2 - 10^3$. Since k_r is already smaller in solution by a factor of 10 or so, it follows that k_f in solution must be smaller than in the gas by a factor of 10^3.

However, these considerations are too simple. The fragments A and B are usually not structureless, nonreactive spheres. Except for atoms, they have a structure and a conformation that can change in the available time scale. Such changes can prevent ready recombination to A$-$B. Furthermore, A and B must be reactive fragments, since they have binding power for each other. Depending on whether bond cleavage is homolytic or heterolytic,

$$A{:}B \rightarrow (A{\cdot} + B{\cdot}) \quad \text{homolytic} \tag{7.29}$$

$$A{:}B \rightarrow (A + {:}B) \quad \text{heterolytic} \tag{7.30}$$

free radicals or Lewis acids and bases are formed. These can interact with even inert solvents such as paraffin hydrocarbons by more than van der Waals forces. Binding to a solvent molecule forms a new species that is stabilized against recombination to A$-$B.

The escape rate constant $1/\tau_{AB}$ can be expressed in terms of the viscosity of the medium by using (7.13). Assuming that $r_A = r_B = d_{AB}/2$

$$\frac{1}{\tau_{AB}} = \frac{RT}{4\pi L r^3 \eta} \tag{7.31}$$

This is just the equation for the rotational relaxation time of a spherical molecule in a viscous medium as calculated by Debye.[17] Clearly rotation can compete with escape from the solvent cage. Other processes such as inversion at a chiral center or rotation about a single bond can occur even more rapidly, as can further decompositions of A and B. Ample evidence exists that all these processes can occur in the solvent cage.[18]

As an example of what may happen when two free radicals are formed within a solvent cage, some results with acetyl peroxide in the solvent isooctane are illuminating.[19] The carbonyl group of the starting peroxide was labeled with oxygen-18. After partial decomposition at $80°C$, the unreacted peroxide had the label extensively scrambled

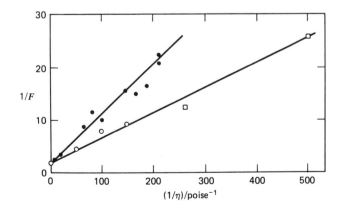

$$(7.32)$$

Approximately 38% of all radical pairs formed recombine to give scrambled starting material. In addition the results show that 12.4% of $CH_3CO_2CH_3$ and 3% of C_2H_6 are formed within the cage by the initiating reaction:

$$CH_3CO_2 \cdot \rightarrow CH_3 \cdot + CO_2 \qquad (7.33)$$

Only 47% of the products are formed outside of the cage.

The fraction of radicals F that react with each other within the cage should be given by

$$F = \frac{k_r}{k_r + 1/\tau_{AB}} = \frac{k_r \eta}{k_r \eta + c} \qquad (7.34)$$

where k_r is the rate constant for the radical recombination and $c = RT/4\pi Lr_a^3$. A plot of $1/F$ versus $1/\eta$ should be a straight line with an intercept of unity and a slope of $RT/4\pi Lr_a^3 k_r$, if k_r is constant for solvents of different η. Figure 7.2 shows such a plot for the recombination of $t\text{-}C_4H_9O\cdot$ radicals from the reaction of the t-butyl perester of oxalic acid.[20]

Figure 7.2. Plot of $1/F$ versus $1/\eta$ for the recombination of t-butoxyl radicals in several solvents at $45°C$. Closed circles are aromatic solvents; open circles and squares are aliphatic solvents. [Reproduced by permission from E. Niki and Y. Kamiya, *J. Am. Chem. Soc.*, **96**, 2129 (1974), copyright 1974, American Chemical Society.]

$$(CH_3)_3COO-\overset{\overset{\displaystyle O}{\|}}{C}-\overset{\overset{\displaystyle O}{\|}}{C}-OOC(CH_3)_3 \rightarrow 2(CH_3)_3CO\cdot + 2CO_2$$

$$2(CH_3)_3CO\cdot \rightarrow (CH_3)_3COOC(CH_3)_3 \tag{7.35}$$

The slopes are greater for aromatic solvents than for aliphatic solvents. This is an indication that k_r is less in aromatic solvents. Since aromatics stabilize free radicals better than aliphatic solvents,[21] such a result is not unexpected. The intercept is closer to 2 than to 1, however. This is a consequence of the radicals being formed at some distance from each other.[15] Some radicals escape before reacting even in an extremely viscous solvent.

There are also unimolecular reactions in which no reactive intermediates are formed, and therefore escape from the solvent cage is not a factor. An example is the rearrangement of a bicycloheptene into 1,3-cycloheptadiene.[22]

$$\tag{7.36}$$

This is an example of a reaction forbidden by orbital symmetry[23] and may involve a diradical formed by breaking the bridging carbon–carbon bond. However, the radical centers cannot escape from each other in this case. The rate, frequency factors (10^{14} s^{-1}), and activation energies (186 kJ mol^{-1}) are the same in the gas phase and in dimethylphthalate, a viscous solvent. It may be mentioned finally that the average experimental value of the frequency factor for unimolecular reactions in solution is 10^{13} s^{-1}, just as in the gas phase.

THE TRANSITION-STATE THEORY IN SOLUTION

The theory of absolute reaction rates using partition functions is not particularly useful in solution. The evaluation of partition functions in the liquid state is too difficult, compared to evaluation in the gaseous state. However, the pseudothermodynamic treatment of the theory has proved very useful since we can take advantage of known effects of solvents on thermodynamic properties. For example, consider the equilibrium constants for the reaction

$$A + B \ldots \rightleftharpoons C + D \ldots \tag{7.37}$$

expressed in mol liter^{-1} both in the vapor phase and in liquid solution. If the vapor state can be considered ideal, the ratio of equilibrium constants can be expressed as

$$\frac{K \text{ (solution)}}{K \text{ (gas)}} = \frac{\pi_A^0 \pi_B^0 \ldots}{\pi_C^0 \pi_D^0 \ldots} \left(\frac{RT}{V_0}\right)^{\Delta n} \frac{\gamma_A \gamma_B \ldots}{\gamma_C \gamma_D \ldots} \tag{7.38}$$

where $\pi_A^0 = P_A/x_A$ is the appropriate Henry's law constant for substance A, Henry's law being obeyed as the solution becomes sufficiently dilute. The activity coefficient γ_A is a measure of the deviation from Henry's law and approaches unity at infinite dilution. V_0 is the volume per mole of solution, and Δn is the change in the amount of substance for (7.37). The ratio may also be written as

$$\frac{K \text{ (solution)}}{K \text{ (gas)}} = \frac{e^{-\Delta H'_A/RT} e^{-\Delta H'_B/RT} e^{\Delta S'_A/R} e^{\Delta S'_B/R} \cdots}{e^{-\Delta H'_C/RT} e^{-\Delta H'_D/RT} e^{\Delta S'_C/R} e^{\Delta S'_D/R} \cdots} \tag{7.39}$$

where $\Delta H'_A$ and $\Delta S'_A$ are the heat and entropy of vaporization of A not from pure liquid A but from the solution itself.

The immediate deduction from (7.38) is that reactions in solution are favored when the reactants are more volatile than the products. In the gas phase the reverse is true. Equation (7.39) emphasizes the point that normally the most important factor is the change in the heat of reaction in solution compared to the vapor. Since the latter has zero intermolecular interactions by definition, if ideal, it is the difference in interaction of molecules of A, B, C, D, and so on with the solvent that determines the effect. The most important feature of these molecules for a given solvent is their polarity, the limit occurring when some of the molecules are ions.

If we now have the kinetic situation

$$A + B \ldots \rightleftharpoons M^{\ddagger} \rightarrow \ldots \tag{7.40}$$

where A and B are reactants and $M^{\ddagger}$ is an activated complex, then the specific rate constant for this reaction in the gas phase is $k_g = (RT/Lh)K_g^{\ddagger}$, where $K_g^{\ddagger}$ is the "equilibrium constant" between the complex and the reactants, assuming the vapors to be ideal. In solution the analogous equation must be corrected to take into account deviations from ideal behavior. The thermodynamic "equilibrium constant" should be defined as a ratio of activities.

$$K_s^{\ddagger} = \frac{a^{\ddagger}}{a_A a_B \cdots} = \frac{c^{\ddagger}}{c_A c_B \cdots} \frac{\gamma^{\ddagger}}{\gamma_A \gamma_B \cdots} \tag{7.41}$$

Consequently if the *rate* of a reaction is proportional to the *concentration* of the activated complex, the rate constant is dependent on the ratio of activity coefficients.

$$\text{rate} = \left(\frac{RT}{Lh}\right) c^{\ddagger} = \left(\frac{RT}{Lh}\right) K_s^{\ddagger} c_A c_B \cdots \frac{\gamma_A \gamma_B \cdots}{\gamma^{\ddagger}} \tag{7.42}$$

$$k_s = \left(\frac{RT}{Lh}\right) K_s^{\ddagger} \frac{\gamma_A \gamma_B \cdots}{\gamma^{\ddagger}} \tag{7.43}$$

The activity coefficients can be referred to any convenient standard state, the usual one being that of infinite dilution for the solutes. The rate constant in the standard state is then equal to $(RT/Lh)K_s^{\ddagger}$. A comparison of the rates in the gas phase and in solution based on (7.43) leads to a repetition of (7.38).

For a bimolecular reaction the ratio of rate constants becomes

$$\frac{k_s}{k_g} = \frac{\pi_A^0 \pi_B^0}{\pi_{\ddagger}^0} \left(\frac{V_0}{RT}\right) \frac{\gamma_A \gamma_B}{\gamma_{\ddagger}} \tag{7.44}$$

Also the heats and entropies of activation are changed.

$$\Delta H_s^{\ddagger} = \Delta H_g^{\ddagger} - \sum \Delta H' \text{ (vaporization)}$$

$$(7.45)$$

$$\Delta S_s^{\ddagger} = \Delta S_g^{\ddagger} + \sum \Delta S' \text{ (vaporization)}$$

If the activated complex had a large heat of vaporization, which would be true if it were a very polar aggregate, the reaction in solution would be favored over that in the gas phase. On the other hand, if the reactants were ions of opposite charge so that the activated complex was less polar, reaction in the vapor would be favored. For nonpolar reactants and transition state, the entropy favors reaction in solution by a factor of about 100.

Normally the negative of the heats of vaporization, called the heats of solution, are used. These quantities are always negative since some form of intermolecular attraction must exist in the liquid state. Figure 7.3 shows how the effective energy barrier in solution can either be raised or lowered in the liquid phase compared to the gas. Reactions with no potential barrier would have rates controlled by rates of diffusion. Such reactions are not covered by transition-state theory.

For a unimolecular reaction, (7.38) leads to

$$\frac{k_s}{k_g} = \frac{\pi_A^0 \gamma_A}{\pi_{\ddagger}^0 \gamma_{\ddagger}}$$

$$(7.46)$$

If the chemical change does not lead to a large change in polarity, it is expected that the activated complex will resemble the reactant and the ratio k_s/k_g will be close to unity. The transition-state theory should be valid in solution, but the comparison must be made to the high-pressure limit of the gas-phase reaction. There is no indication in (7.46) that the solvent cage plays any role. Escape from the solvent cage being diffusion controlled is not part of the theory as normally used. Diffusion itself, as an activated process, could be treated by transition-state theory. The activated complex would be a molecule in the act of making a jump from one position to another.

THE INFLUENCE OF THE SOLVENT

The preceding arguments show that in certain cases there is not much difference between reactions in the gaseous state and those in solution. For such reactions the nature of the solvent seems to make little difference. For the great majority of reactions, however, which occur in solution but not in the gas phase, the specific properties of the solvent are important in determining not only the rate but also the equilibrium. A discussion is now presented of the effect of the properties of the solvent, particularly those leading to deviations from ideal behavior.

Scatchard[24] and particularly Hildebrand[25] have developed expressions for the activity coefficients of nonelectrolytes as solutes in various liquid solvents. On the assumption that the heat of mixing is responsible for all deviations from ideal

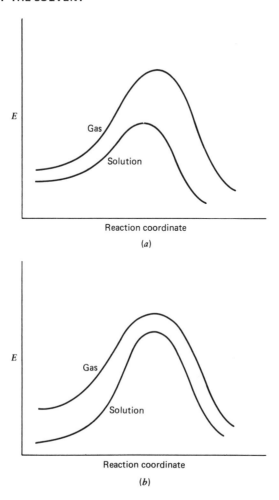

Figure 7.3. Effect of solvation on potential energies of reactants and activated complexes: (*a*) reaction speeds up in solution relative to gas phase; (*b*) reaction slows down in solution relative to gas phase.

behavior and that the interaction energy of a solute molecule and a solvent molecule is the geometric mean of solute–solute and solvent–solvent interactions, it is possible to derive the following equation:

$$RT \ln f_2 = V_2 \phi_1^2 \left[\left(\frac{\Delta E_2}{V_2} \right)^{1/2} - \left(\frac{\Delta E_1}{V_1} \right)^{1/2} \right]^2 \qquad (7.47)$$

where V_2 is the molar volume of solute (as a liquid); ΔE_2 is the molar energy of vaporization of the solute (as a pure liquid), V_1 and ΔE_1 are the same quantities for the solvent; ϕ_1 is the volume fraction of the solvent, equal to unity for a dilute solution; and f_2 is the activity coefficient of the solute referred to a standard state of pure liquid solute (not infinite dilution). For dilute solutions (7.47) may be written as

$$RT \ln f_2 = V_2(\delta_2 - \delta_1)^2 \tag{7.48}$$

where $\delta_2^2 = \Delta E_2/V_2$ is a parameter that may be called the internal pressure or the cohesive energy density. Equation (7.48) has been useful in predicting solubilities of nonelectrolytes as a function of the differences in internal pressures (Hildebrand and Scott).

For maximum solubility (complete miscibility) the solute and solvent should have identical values of δ. In the theory of regular solutions, as it is called, only positive deviations from ideal behavior are possible. Negative deviations would also lead to total solubility. For solids the heat of fusion must be taken into account to calculate solubility.

The theory may now be applied to a kinetic problem by writing (7.43) in the form

$$k = k_0 \frac{f_A f_B}{f_\ddagger} \tag{7.49}$$

for a bimolecular reaction where k_0 is the rate constant of a given reaction in an ideal solution. Hence k_0 is independent of the solvent whose properties are brought in only as they affect the ratio of activity coefficients. From (7.48) and (7.49) we now have

$$\ln k = \ln k_0 + \frac{V_A}{RT}(\delta_1 - \delta_A)^2 + \frac{V_B}{RT}(\delta_1 - \delta_B)^2 - \frac{V_\ddagger}{RT}(\delta_1 - \delta_\ddagger)^2 \tag{7.50}$$

If the activated complex has a greater solubility in the solvent than the reactants, the rate is large compared to the rate in an ideal solution. If the reactants resemble the solvent more than the activated complex does, the reverse is true. Briefly, reactions producing products more polar than the reactants are favored in polar solvents; reactions giving products less polar than the reactants are favored in nonpolar solvents. The concept of internal pressure should only be applied to reactions of neutral molecules. Ions form solutions that are far from regular.

Table 7.3 shows the range of values of δ for a number of common solvents. For the gas phase $\delta = 0$. For liquid helium and hydrogen, δ is 0.5 and 2.5, respectively.[25] The largest values of δ are for liquid metals, where δ ranges from 16 for cesium to 145 for tungsten.[25] Excluding the metals, it can be seen that the internal pressure is a good measure of solvent polarity, or the ability to dissolve polar solutes. Values of the dielectric constant, ϵ_r, are also included in Table 7.3, since this is another measure of solvent polarity.

Equation (7.50) is awkward to use since the δ values of the reactants are usually unknown. However, if we assume that $V_\ddagger = V_A + V_B$, which seems reasonable, a linear equation is obtained.[26]

$$RT \ln\left(\frac{k}{k_0}\right) = 2\delta_1(V_\ddagger\delta_\ddagger - V_A\delta_A - V_B\delta_B) + (V_A\delta_A^2 + V_B\delta_B^2 - V_\ddagger\delta_\ddagger^2) \tag{7.51}$$

Table 7.3. Some properties of common solvents[a]

Solvent	δ[b]	ϵ_r[c]	$\log k$[d]
Water	24.0	80.0	—
Ethylene glycol	14.5	38.7	—
Methanol	14.5	33.6	−3.67
Ethanol	12.9	25.1	−3.80
Nitromethane	12.6	37.5	−1.73
1-Propanol	12.0	20.6	−3.91
Acetonitrile	11.8	36.8	−2.11
2-Propanol	11.5	18.3	—
1-Butanol	10.7	17.1	−4.11
Bromoform	10.6	4.4	−2.33
Pyridine	10.6	12.3	—
t-Butyl alcohol	10.5	10.9	—
Nitrobenzene	10.4	35.8	−2.10
Acetophenone	10.4	17.6	−2.16
Carbon disulfide	10.0	2.6	−4.38
1,2-Dichloroethane	9.91	10.6	−2.20
Methyl iodide	9.9	7.0	—
Methylene chloride	9.88	9.1	−2.33
Dioxane	9.73	2.2	−3.21
Acetone	9.66	21.1	−2.60
Chlorobenzene	9.50	5.7	−2.93
Tetrahydrofuran	9.32	7.6	−3.31
Tetrachloroethylene	9.3	2.3	−5.00
Chloroform	9.24	4.8	−2.67
Benzene	9.15	2.3	−3.52
Ethyl acetate	9.04	6.2	−3.44
Toluene	8.91	2.4	−3.80
Mesitylene	8.8	2.3	−4.40
Carbon tetrachloride	8.58	2.2	−4.63
Cyclohexane	8.18	2.0	−5.93
Ethyl ether	7.74	4.4	−4.70
Hexane	7.24	1.9	−6.8
Isopropyl ether	7.14	3.9	—
2,2,4-Trimethylpentane	6.85	1.9	—
Gas phase	0	1.0	—
Dimethyl sulfoxide	12.80	46.5	—
Dimethylformamide	11.82	36.7	−2.00

[a] For a more extensive listing of solvent properties see T. R. Griffiths and D. C. Pugh, *Coord. Chem. Rev.*, **29**, 129 (1979).

[b] At 25°C. Reference 25.

[c] At 20°C. Reference 28b.

[d] At 20°C. Reaction (7.53). Reference 30.

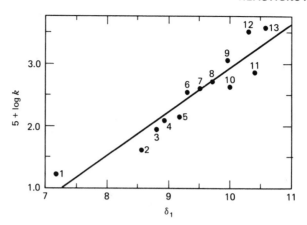

Figure 7.4. Reaction of pyridine with methyl iodide. Log k versus the square root of the internal pressure of the solvent: (*1*) isopropyl ether, (*2*) carbon tetrachloride, (*3*) mesitylene, (*4*) toluene, (*5*) benzene, (*6*) chloroform, (*7*) chlorobenzene, (*8*) bromobenzene, (*9*) iodobenzene, (*10*) dioxane, (*11*) anisole, (*12*) benzonitrile, (*13*) nitrobenzene. [Reproduced by permission from A. P. Stefani, *J. Am. Chem. Soc.,* **90,** 1694 (1968), copyright 1968, American Chemical Society.]

Figure 7.4 shows a plot of log k against δ_1 for the reaction of pyridine with methyl iodide in a number of solvents.

$$C_5H_5N + CH_3I \rightarrow C_5H_5NCH_3^+ + I^- \qquad (7.52)$$

The transition state should resemble a molecule with a large dipole moment, because of the incipient departure of the iodide ion. The figure shows that polar solvents favor (7.52). Also there is a rough linearity between log k and δ_1, as predicted by (7.51). The positive slope means that $\delta_\ddagger$ is larger than the mean of δ_A and δ_B. A negative slope is found for reactions where the reverse may be true.[27]

Numerous other parameters have been suggested as a means of classifying solvents according to their solvating ability.[28] Several of these are based on solvatochromic shifts in electronic spectra.[29] The solvent will cause a shift in the absorption maximum of a band corresponding to a ground state and an excited state that have different polarities. The magnitude of the shift on an energy scale characterizes the solvent. This approach has the advantage of being applicable to mixed solvents, where δ would be hard to obtain.

Other scales are based on a series of log k values for a standard reaction run in a series of solvents. As Fig. 7.4 suggests, there is an approximately linear relationship among all these different scales of solvent polarity.[26]

Table 7.3 also includes the log k values for the reaction

$$(n-C_3H_7)_3N + CH_3I \rightarrow (n-C_3H_7)_3NCH_3^+ + I^- \qquad (7.53)$$

for many of the solvents listed.[30] As can be seen there is only a rough relationship between the order of log k and the δ values. Other scales of solvent polarity would not do much better. The dielectric constant is also a poor guide.

SOLVATION OF IONS

The failure of generalized scales of solvent polarity is chiefly due to the failure to consider specific interactions between the reactant molecules and the solvent. These are short-range chemical interactions. The most important by far are those that may be classified as donor–acceptor interactions, or generalized acid–base interactions.[31]

$$A + :B \rightleftharpoons A:B \qquad (7.54)$$

A reactant molecule or ion often functions primarily as a Lewis acid or a base. Good solvents can function as both. For example, water is a base by virtue of the lone pairs on oxygen, and an acid by way of the positive hydrogen atoms, which can form hydrogen bonds to bases. Acetone is a base at the oxygen atom, and a Lewis acid by way of the positive carbon atom of the carbonyl group.

These positions are often called nucleophilic (nucleus-loving) and electrophilic (electron-loving) centers, respectively. Thus a nucleophilic reagent is a Lewis (or Brønsted) base, and an electrophilic reagent is a Lewis acid. It has been agreed to reserve the terms nucleophilic and electrophilic reactivity to kinetic phenomena. For equilibrium results, the terms acidity and basicity should be used. Whether in a transition state or in stable molecules, the strength of interaction between A and :B is a complex function. It cannot be expressed in terms of a single parameter for A and another one for :B.

As an example of the consequences of generalized acid–base behavior, it can be seen in Table 7.3 that the alcohols give abnormally low rate constants for (7.53) compared to other solvents of similar δ and ϵ_r values. The alcohols are polar, protic solvents. They form strong hydrogen bonds to the nitrogen atom of $(n-C_3H_7)_3N$, which is then deactivated towards reaction with CH_3I. The linearity shown in Fig. 7.4 for the very similar reaction (7.52) exists only because the solvents are all nonprotic, or aprotic.

Ionic solutes make special demands on the solvent because of the magnitude of the interaction. For small ions, the acid–base character plays a dominant role, cations acting as Lewis acids and anions as bases. The hydration energy of a cation such as Fe^{2+} can be broken up into two parts,[32] the first of which is the energy of interaction with a primary coordination layer, a molecular process.

$$Fe^{2+}(g) + 6H_2O(g) \rightarrow Fe(H_2O)_6^{2+}(g) \qquad \Delta H = -1438 \text{ kJ mol}^{-1} \qquad (7.55)$$

The second is the energy of interaction, often called the Born energy, of the complex ion thus formed.

$$Fe(H_2O)_6^{2+}(g) \rightarrow Fe(H_2O)_6^{2+}(aq) \qquad \Delta H = -777 \text{ kJ mol}^{-1} \qquad (7.56)$$

Here the solvent is treated as a continuum.

The latter energy can often be estimated using classical electrostatics, the Born equations, for an ion of charge Ze and radius r.

$$\Delta G_{el} = -L \frac{Z^2 e^2}{8\pi\epsilon_0 r} \left(1 - \frac{1}{\epsilon_r}\right) \qquad (7.57)$$

$$\Delta S_{el} = -L \frac{Z^2 e^2}{8\pi\epsilon_0 \epsilon_r r} \left(\frac{\partial \ln \epsilon_r}{\partial T} \right)_P \tag{7.58}$$

$$\Delta H_{el} = \Delta G_{el} + T \Delta S_{el} \tag{7.59}$$

The value of -777 kJ mol^{-1} in (7.56) was calculated from (7.57)–(7.59) by using $r = (0.83 + 2.76) = 3.59$ Å as the complex ion radius. The value 2.76 Å is the diameter of the water molecule (from the closest O—O distance in ice), and 0.83 Å is the crystal radius of Fe^{2+}. For water $(\partial \ln \epsilon_r/\partial T)_P = -0.0046$ K^{-1}, the dielectric constant decreasing with increasing temperature.

If an ion were sufficiently large, and if the charge were distributed with spherical symmetry, only the Born energy would need to be considered. However, most ions are not large enough, or spherical. If the molecule can be represented by a series of point charges, for example, the nuclei with suitable effective charges, Q_i, Q_j, due to electronic shielding, a generalization of Born's equation may be used.[33]

$$\Delta G_{el} = -L \frac{1}{4\pi\epsilon_0} \left(1 - \frac{1}{\epsilon_r} \right) \left(\sum_{i>j} \frac{Q_i Q_j}{r_{ij}} + \sum_i \frac{Q_i^2}{2r} \right) \tag{7.60}$$

Here r_{ij} is the distance between the charges. The self energy terms must also be included, and for these terms the distance is taken as the radius of the ion. Equation (7.60) could also be applied to the solvation energy of a neutral molecule. The sum of the effective charges must add up to zero in this case.

Of course the only exact calculation of the solvation energy of a molecule or ion would be an accurate quantum-mechanical calculation. At present this is much too difficult to do, though some progress is being made with the interaction of small solute molecules with a limited number of solvent molecules.[34] Attempts have also been made to give a quantum-mechanical version of the classical equation (7.60).[35] In these cases the reciprocal distances appropriate for point charges are replaced by Coulomb integrals appropriate for average values, $1/r_{ij}$, for charge clouds.

Generally speaking, in the solution process it is necessary to break bonds between solvent molecules to create a cavity to hold the solute molecule or ion. A correction should be made for the energy of cavity formation.[36] As a rough approximation, it may be calculated by multiplying the surface area of the solute molecule by the surface tension of the solvent, γ_l. If the solute is spherical,

$$\Delta G_{cav} = 4\pi r^2 \gamma_l \tag{7.61}$$

If $r = 5 \times 10^{-8}$ cm and $\gamma_l = 20$ dynes cm^{-1}, $\Delta G_{cav} = 40$ kJ mol^{-1}, which is a very considerable quantity comparable to heats of vaporization of simple molecules. The surface tension of a curved surface, however, is only about 30% of that for a plane surface, so that (7.61) overestimates the surface free energy by a factor of about 3.[37] The equation is useful in reminding us that it is harder to dissolve a solute in a solvent with a high internal pressure. The latter is closely related to surface tension.[25]

The best way to study the effect of solvents on the properties of ions is to study reactions of ions in the gas phase. The reason for this paradoxical statement is that any thermodynamic data obtained in the gas phase may be compared with the large amount of data already available in solutions. The information wanted is equilibrium constants, heats of reaction, and entropies of reaction for reactions involving ions in the gas phase. Fortunately, in recent years several experimental techniques have become available for such studies. They include high-pressure mass spectrometry,[38] ion cyclotron resonance,[39] and the flowing afterglow method.[40]

One of the most important applications of these new techniques has been the determination of pK_a and $\Delta H^{\ominus}$ values for gas-phase Brønsted acidities of many molecules

$$HX(g) \rightleftharpoons H^+(g) + X^-(g) \tag{7.62}$$

A sampling of the data available is given in Table 7.4. Also included are values of pK_a in water and in dimethyl sulfoxide, a useful aprotic solvent. Because of its small autoprotolysis constant, much weaker acids can be measured in DMSO than in water.[41]

The most noticeable feature of Table 7.4 is that ionizations such as (7.62) are energetically prohibited in the gas phase, whereas in polar solvents they occur to a degree that is readily measurable. A closer analysis of one case (7.63) is useful.

$$CH_3COOH(g) \rightleftharpoons CH_3COO^-(g) + H^+(g) \tag{7.63}$$

Table 7.4. Gas-phase and solution acidities at $25°C$

Acid	pK_a, gas	pK_a, H_2O	pK_a, DMSO	$\Delta H^{\ominus}/\text{kcal mol}^{-1}$, gas
H_2O	280	15.7	>34	391
CH_3OH	272	15.1	29.1	381
C_2H_5OH	270	15.9	29.5	378
$i\text{-}C_3H_7OH$	268	17.1	30.1	376
$t\text{-}C_4H_9OH$	267	19.2	31.3	375
C_6H_5OH	251	10.0	18.4	346
CH_3COOH	247	4.8	11.9	344
HF	267	3.5	–	371
HCN	252	9.3	12.9	353
H_2S	253	6.9	–	352
$C_6H_5NH_2$	263	>17	30.7	367
C_5H_6	255	15.6	28.8	353
CH_3CN	266	>17	31.3	372
CH_3COCH_3	264	>17	26.5	371
$CH_3SO_2CH_3$	257	>17	31.1	367
CH_3NO_2	257	10.2	17.2	359
$C_6H_5C{\equiv}CH$	265	>17	28.8	370

Source. J. E. Bartmess, J. A. Scott, and R. T. McIver, Jr., *J. Am. Chem. Soc.*, **101**, 6046, 6056 (1979).

In the gas $\Delta H^{\ominus} = + 1439 \text{ kJ mol}^{-1}$ and $\Delta S^{\ominus} = + 96 \text{ J mol}^{-1} \text{ K}^{-1}$. The entropy change is favorable and comes almost entirely from the translational entropy of the proton. However, it is overwhelmed by the large energy requirement, due mainly to the high heat of formation of the proton.

The same reaction in aqueous solution,

$$CH_3COOH(aq) \rightleftharpoons H^+(aq) + CH_3COO^-(aq) \tag{7.64}$$

has $\Delta H^{\ominus} = - 0.40 \text{ kJ mol}^{-1}$ and $\Delta S^{\ominus} = - 92 \text{ J mol}^{-1} \text{ K}^{-1}$, so that it is essentially thermally neutral. The tremendous change in $\Delta H^{\ominus}$ resides in the difference between the heats of hydration of the ionic products and that of molecular acetic acid, which is only $- 53 \text{ kJ mol}^{-1}$. Accordingly

$$\Delta H^{\ominus}_{hyd}(H^+) + \Delta H^{\ominus}_{hyd}(ac^-) = - 1493 \text{ kJ mol}^{-1} \tag{7.65}$$

It is not possible to measure the heat of hydration of a single ion experimentally, but various plausible assumptions do allow for the construction of tables of individual ionic heats of hydration.[42] The value of $\Delta H^{\ominus}_{hyd}(H^+)$ is $- 1090 \text{ kJ mol}^{-1}$ from these assumptions. This leaves $- 403 \text{ kJ mol}^{-1}$ for the heat of hydration of the acetate ion.

The second important feature of Table 7.4 is that the order of acid strengths in the gas phase is not at all the same as in aqueous solution. For example, carbon acids such as acetone, acetonitrile, and dimethylsulfone (and even toluene) are much stronger acids than water in the vapor state, whereas in aqueous solution they have acidities too weak to measure. This means that the different anions, X^-, have quite different heats of hydration. In particular, OH^- is very well stabilized by water, as is H^+. Other small ions, where the charge is concentrated on an electronegative atom, or hard bases, are also well solvated by water. Large ions, especially those where the charge is well dispersed by resonance, or where the donor atom is of low electronegativity, that is, soft bases, are much more poorly solvated.

The superior power of water for solvating hard bases is due to its ability to form hydrogen bonds. The pK_a values in DMSO, a solvent that cannot form hydrogen bonds, demonstrate this very clearly. Water is too weak an acid to measure in DMSO, whereas the carbon acids, such as acetone and phenylacetylene, are reasonably acidic. While individual heats of solvation for ions in dimethylsulfoxide are not known, relative values are available.[43] These show that the hydroxide ion, because of its small size, is more strongly solvated in DMSO than are large carbanions. However, the difference in solvation energies is much less than in water.

In the gas phase the Brønsted acidity is strongly influenced by the size of the ion X^-. The larger X^- is, the stronger is the acid HX, though the nature of the atom to which the proton is bonded and delocalization possibilities also play a role. The size effect follows directly from classical electrostatics, the Born charging energy in a vacuum being $Z^2 e^2 / 8\pi\epsilon_0 r$. As an example, it can be seen in Table 7.4 that the relative acidities of the alcohols, CH_3OH, C_2H_5OH, $i\text{-}C_3H_7OH$, and $t\text{-}C_4H_9OH$, are inverted in the gas phase compared to water. The effect of increasing size stabilizes the ion in the gas phase, but hinders solvation in solution.[44]

The same effects of size show up for positive ions. The series NH_3, CH_3NH_2, $(CH_3)_2NH$, and $(CH_3)_3N$ increases in basicity steadily in the gas state.[45]

$$R_3N(g) + H^+(g) \rightleftharpoons R_3NH^+(g) \tag{7.66}$$

In solution the order is irregular, because alkyl substitution reduces the solvation energy of the ammonium ion. While there is certainly justification for calling gas-phase data intrinsic data because complications due to solvation are absent, the fact that an ion is self-stabilized by spreading its charge over all the available atoms is a kind of internal solvation that favors large ions.

Ionization in solution (7.64) is accompanied by a large decrease in entropy, compared to the increase found in the gas phase. This is a common phenomenon and not just for the specific case of acetic acid. The decrease in entropy is connected with the orientation of solvent molecules around the solute ions and a consequent loss of mobility of these molecules. A useful analogy has been made between the orientation of solvent molecules and the "freezing" of solvent.[46] Since the entropy of freezing of water is -23 J mol^{-1} K^{-1}, it appears that four water molecules are frozen in forming the ions in (7.64). There is also a decrease in heat capacity, $\Delta C_p^{\ominus}$, for reactions forming ions in water. Since ice has a lower heat capacity than water, this is also consistent with the freezing model of solvation.

Table 7.5 shows that $\Delta S^{\ominus}$ is more negative in solvents less polar than water. Dioxane is a nonpolar molecule and its mixtures with water have a lower dielectric constant. The equilibrium constant gets smaller as ϵ_r decreases. Surprisingly, $\Delta H^{\ominus}$ changes in the unexpected direction and the unfavorable change in free energy is due entirely to the entropy effect. Similar results are found for other ionic equilibria. This may be explained in part by classical electrostatics since (7.58) and (7.59) predict such an effect.[47] The model of frozen solvent molecules also predicts the entropy results. Nonpolar molecules are relatively free in the liquid state and the entropy of freezing such molecules is typically twice as large as for water (unless the solid state retains free rotation). Hence such molecules bound tightly to ions experience a larger entropy decrease than do polar, associated molecules. In effect, the latter are already partly "frozen" in the bulk liquid.

Table 7.5. Ionization of acetic acid in water–dioxane mixtures at $25°C$

X^a	K_a	$\Delta H^{\ominus}$/kJ mol^{-1}	$\Delta S^{\ominus}$/J mol^{-1} K^{-1}	ϵ_r
0	1.75×10^{-5}	-0.40	-92.4	78
20	5.11×10^{-6}	-0.20	-102.0	61
45	4.93×10^{-7}	-1.84	-126.7	38.5
70	5.78×10^{-9}	-2.55	-167.7	17.7
8´	3.10×10^{-11}	-5.60	-212.5	9.5

Source. Data from H. S. Harned and B. B. Owen, *Physical Chemistry of Electrolytic Solutions,* 3rd ed., Reinhold, New York, 1958.

^{a} X = weight percent dioxane.

KINETICS OF IONIZATION

Table 7.6 gives the experimental values of the activation energies and entropies of activation for a number of reactions involving the formation of ions from neutral molecules in a variety of solvents. It appears that two generalizations can be made: (1) the activation energy depends on the type of reaction and does not change rapidly from solvent to solvent, and (2) the entropy of activation is almost always negative and changes with the solvent, becoming more negative as the polarity of the solvent decreases. It follows from this that the rates of reactions producing ions in solution increase with the polarity of the solvent and that the increase is governed largely by the change in the entropy of activation.

It seems reasonable again to relate the entropy decrease in going from reactants to activated complex to the freezing of solvent molecules around the incipient

Table 7.6. Activation energies and entropies of activation for ionizations

Reaction	Solvent	E_a/kJ mol^{-1}	$\Delta S^{\ddagger}$/J K^{-1} mol^{-1}
$C_6H_5NH_2 + C_6H_5\overset{O}{\underset{\|}{C}}CH_2Br \rightarrow$	Benzene	33.8	-234
	Chloroform	45.2	-192
	Acetone	46.5	-163
$C_6H_5-\overset{O}{\underset{\|}{C}}-CH_2-NH_2C_6H_5^+ + Br^{-a}$	Nitrobenzene	56.5	-138
	Methanol	52.0	-138
	Ethanol	58.2	-117
$p\text{-}NO_2\!\!-\!\!\bigcirc\!\!-\!\!CH_2Br + H_2O \rightarrow$			
	50% dioxane	79.6	-126
	70% dioxane	72.0	-154
$p\text{-}NO_2\!\!-\!\!\bigcirc\!\!-\!\!CH_2OH + H^+ + Br^{-b}$	90% dioxane	64.5	-185
$HA + H_2O \rightarrow H_3O^+ + A^{-c}$	Water		
Acetoacetic ester		59.5	-109
α-Methylacetylacetone		75.4	-67.0
Nitromethane		94.6	-84.5
Nitroethane		96.0	-77.0
Ethyl nitroacetate		67.0	-71.1
$(CH_3)_3CCl \rightarrow (CH_3)_3C^+ + Cl^{-d}$	Water	99.5	-51.0
	Methanol	106	-12.9
	Ethanol	111	-13.3
	Acetic acid	110	-10.4
	90% dioxane	93.0	-77.5
	90% acetone	94.0	-70.3
	50% ethanol	90.4	-5.8
	90% ethanol	95.1	-36.3
$cis\text{-}Pt(NH_3)_2Cl_2 + H_2O \rightarrow$ $cis\text{-}Pt(NH_3)_2ClH_2O^+ + Cl^{-e}$	Water	83.6	-59.0
$trans\text{-}Pt(NH_3)_2Cl_2 + H_2O \rightarrow$ $trans\text{-}Pt(NH_3)_2ClH_2O^+ + Cl^{-e}$	Water	83.6	-46.0

[a] H. E. Cox, *J. Chem. Soc.*, **119**, 142 (1921).

[b] J. W. Hackett and H. C. Thomas, *J. Am. Chem. Soc.*, **72**, 4962 (1950). Rate constant second order.

[c] R. G. Pearson and R. L. Dillon, *J. Am. Chem. Soc.*, **75**, 2439 (1953). Rate constant first order.

[d] S. Winstein and A. H. Fainberg, *J. Am. Chem. Soc.*, **79**, 5937 (1957). Rate constant first order.

[e] L. F. Grantham, T. S. Elleman, and D. S. Martin, Jr., *J. Am. Chem. Soc.*, **77**, 2965 (1955).

ions. The activated complex is almost an ion pair at its distance of closest approach or at least an exceedingly polar complex approaching an ion pair. Each end of the polar complex has already accumulated a layer of solvent molecules, whose presence is necessary to allow the process of separating the ions completely to continue. As the separation occurs the layer of solvent molecules is completed so that usually a further decrease in entropy occurs in going from the transition state to the products. This last statement has an important bearing on the rate of reverse reaction, the formation of neutral molecules from ions, which is discussed shortly. From the same arguments given in regard to equilibria, the less polar solvents have a greater loss in freedom in becoming frozen to the ions than do the more polar solvents, hence the lower rates of reaction in the less polar solvents. It should be mentioned at this point that there is considerable difficulty in studying rates of ionization reactions in very nonpolar solutions such as those in hexane or benzene. Very small quantities of polar materials, or salts, can cause large changes in reaction rate.[48] Also, the reactions may be reversible and may not occur appreciably unless the product precipitates or a material is added to combine with product irreversibly. The rate may then depend on the impurities present, the added substances, or even the surface of the crystallized salt and does not depend on the solvating properties of the medium. There is also evidence that a polar reactant in a nonpolar solvent will tend to act as a solvating agent as well.[49] For this reason reactions run in nonpolar solvents must be measured at low concentrations of reactants if the effect of the solvent is being studied.

The explanation of the slowness of reactions producing ions was one of the early stumbling blocks to the application of the collision theory of reaction rates to solutions. The large negative entropies of activation correspond to very low probability factors in the equation

$$k = pZe^{-E_a/RT}$$

Values of p down to 10^{-9} are found experimentally for such reactions. An adequate qualitative explanation of these values can be given in terms of the solvated, polar activated complex. The collision theory is based on a p value of unity if reaction occurs every time two reactant molecules with the requisite energy collide. If in addition it is postulated that the collision occurs simultaneously with the presence of several suitably oriented solvent molecules or that the collision is actually an n-body collision instead of a two-body collision, then very low probability factors become reasonable. The variation of the p factor with solvent may be explained, again qualitatively, by the reasonable assumption that a polar molecule will more frequently be coordinated with the reactant molecule in the approximate position necessary for participation in the reaction than will a nonpolar molecule.

A theory for the influence of the dielectric constant of the medium on the free energy of a polar molecule has been given by Kirkwood.[50] If only electrostatic forces are considered, the difference in free energy of a dipole in a medium with dielectric constant ϵ_r and with a dielectric constant of unity is given by

$$\Delta G_{el} = -L \, \frac{\mu^2(\epsilon_r - 1)}{r^3(2\epsilon_r + 1)4\pi\epsilon_0} \tag{7.67}$$

where μ is the dipole moment and r is the radius of the molecule. Applying this to the transition-state theory for the reaction $A + B \rightleftharpoons M\ddagger$ where A, B, and $M\ddagger$ are polar species, and remembering that

$$k = \left(\frac{RT}{Lh}\right) e^{-\Delta G\ddagger/RT}$$

we obtain

$$\ln k = \ln k_0 - \frac{L(\epsilon_r - 1)}{4\pi\epsilon_0 RT(2\epsilon_r + 1)} \left[\frac{\mu_A^2}{r_A^3} + \frac{\mu_B^2}{r_B^3} - \frac{\mu_\ddagger^2}{r_\ddagger^3}\right] \qquad (7.68)$$

where k is the rate constant in the medium of dielectric constant ϵ_r, and k_0 is the rate constant in a condensed medium of dielectric constant unity, where the nonelectrostatic forces are the same for the activated complex as for the reactants. Equation (7.68) predicts that, if the activated complex is more polar than the reactants (as would be true if the products were ions), the rate of the reaction increases with the dielectric constant of the medium. For some such reactions in mixtures of two solvents of variable composition so that the dielectric constant can be changed, a straight line can be obtained by plotting $\log k$ against $(\epsilon_r - 1)/(2\epsilon_r + 1)$.

If a reaction is studied in several different pure solvents, however, (7.68) is not very reliable. Figure 7.5 shows a plot of $\log k$ versus $(\epsilon_r - 1)/(2\epsilon_r + 1)$ for the ionization of p-methoxyneophyl-p-toluenesulfonate.[51]

$$CH_3 OC_6 H_4 C(CH_3)_2 CH_2 OSO_3 C_6 H_4 CH_3 \rightarrow R^+ + OTs^- \qquad (7.69)$$

$$(ROTs)$$

The final product of the solvolysis reaction depends on the solvent, but the initial, and rate-determining, step is believed to be the ionization to a carbonium ion stabilized by a specific interaction with the aromatic ring. Only a rough trend is seen in Fig. 7.5, but families of related solvents, such as the carboxylic acids, tend to form nearly straight lines.

Equation (7.68) is derived for a point dipole at the center of a sphere. A more realistic equation would be (7.60), which also gives the classical electrostatic contribution to the solvation energy of any molecule. In this case the application to transition-state theory gives

$$\ln k = \ln k_0 - \frac{L(\epsilon_r - 1)}{4\pi\epsilon_0 RT\epsilon_r} \cdot \text{constant} \qquad (7.70)$$

The constant represents the changing charge distribution in going from reactants to the activated complex. Equation (7.70) is actually very similar to (7.68) and would still not include specific solvation effects.

In Table 7.6 it may be seen that the ionization of t-butyl chloride in water is accompanied by an increase in entropy. In this reaction, as in (7.69), a carbonium ion is formed; that is, we have an $S_N 1$ mechanism (the widely used label $S_N 1$ refers to a nucleophilic substitution reaction that is unimolecular). This differs

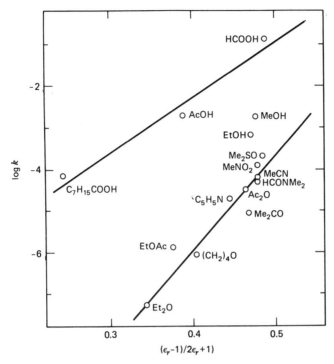

Figure 7.5. Plot of log k for p-methoxyneophyl tosylate versus function of dielectric constant. [Reproduced with permission from S. G. Smith, A. H. Fainberg, and S. Winstein, *J. Am. Chem. Soc.*, **83**, 618 (1961), copyright 1961, American Chemical Society.]

from the solvolysis of methyl chloride in which a water (or solvent) molecule is believed to play a specific role as a nucleophilic reagent.

$$H_2O + CH_3Cl \rightleftharpoons H_2O^+ \text{--} CH_3 \text{--} Cl^- \rightarrow CH_3OH_2^+ + Cl^- \qquad (7.71)$$

Such a mechanism is called an S_N2 process (the term S_N2 means nucleophilic substitution, bimolecular).

The activated complex now contains one strongly bound water molecule in addition to solvent needed to stabilize the developing dipole. For (7.71) in water $\Delta S^\ddagger$ is $-18 \text{ J K}^{-1} \text{ mol}^{-1}$.[52] The difference between this value and the $+51$ $\text{J K}^{-1} \text{ mol}^{-1}$ for t-butyl chloride reflects both the entropy loss of the bound water molecule and the difference due to the polarity and size of the transition state.

Whether or not the solvent plays a specific role as a reactant, its large excess in concentration usually guarantees that the kinetics of solvolytic reactions show simple first-order behavior. This raises the question of whether the observed first-order rate constant should be divided by the concentration of solvent to get a second-order constant. Such a constant would be useful in comparing the nucleophilic reactivity, or basicity, of the solvent to that of other molecules and ions added as solutes. Several difficulties arise, one being that it is not always possible to tell if the solvent is indeed playing a specific role as reactant.

A second difficulty has been pointed out by Langford and Tong.[53] In solution a bimolecular reaction can be broken down into two parts: first an encounter complex is formed by diffusion as is discussed earlier and then this complex reacts to give the products. As shown in (7.21)–(7.23), this gives the observed second-order rate constant as equal to Kk_3. The second-order nature results from the equilibrium constant K, since k_3 is first order. But in solvolysis reactions there is no need to form the encounter complex, it is always present. The rate constant is simply given by k_3. If we wished to compare other reactant molecules with the solvent, the correct procedure would be to evaluate K for the other reactants, and then compare the various values of k_3. Unfortunately, except for certain cases, it is not possible to determine the value of K.

Similar arguments apply to the case of proton transfer reactions

$$H_2O + HA \xrightarrow{k} H_3O^+ + A^- \tag{7.72}$$

where one solvent molecule plays a specific role, admitting the reality of the hydronium ion H_3O^+. The data quoted in Table 7.6 treat k as a first-order rate constant. If it were divided by the solvent concentration, $\Delta S^{\ddagger}$ would be 33.4 units more negative ($-R \ln 55.5$).

Methods have been developed to determine the role of the solvent for certain cases. For example, the solvolysis of an organic halide, or tosylate, may be either S_N1 or S_N2, as is mentioned above. An assumption is made that solvents can be characterized by two parameters, one for its general ionizing power, and one for its nucleophilicity. Model compounds such as methyl chloride and t-butyl chloride can then be used to assign such parameters to various solvents and solvent mixtures. An unknown halide would then be solvolyzed in these solvents to see if it behaved in an S_N1 or S_N2 fashion. Such methods are based on the existence of linear free energy relationships and are discussed in more detail in Chapter 9.

REACTIONS BETWEEN IONS

Some reactions between ions are certainly very fast, as when a precipitate forms on mixing two reagents. Such reactions are often controlled by rates of diffusion, though in other cases rates of nucleation are rate limiting. After nuclei of colloidal size form, their growth and their aggregation rates are usually diffusion controlled. Even slow reactions between ions would have as their first step the diffusion-controlled formation of an encounter complex. The usual diffusion laws, (7.8) and (7.9), must be modified to allow for the forces that exist between the ions. These electrostatic forces lead to relative motion of the particles, either toward or away from each other. The flux can be written as $\phi = \rho \bar{r}$, where $\bar{r}$ is the relative velocity. There is a contribution to this relative velocity due to the force. When acceleration is negligible, the viscous force equals the force of attraction (or repulsion). This leads to

$$\phi = \frac{\rho DL}{RT} \nabla U - D \nabla n \tag{7.73}$$

The solution of (7.9) when (7.73) is inserted is extremely difficult in general. However, one important case can be solved[54] for the stationary state, that is when $(\partial \rho / \partial t) = 0$. This is the case of two ions where the potential energy is expressed by Coulomb's equation.

$$U = \frac{Q_A Q_B}{4\pi\epsilon_0 \epsilon_r d} \tag{7.74}$$

where ϵ_r is used for the dielectric constant of the medium. The solution for the bimolecular rate constant becomes

$$k = \frac{Q_A Q_B L^2 D}{\epsilon_0 \epsilon_r RT [\exp (Q_A Q_B L / 4\pi\epsilon_0 \epsilon_r d_{AB} RT) - 1]} \tag{7.75}$$

The limiting values of (7.75) are of interest. When the electrostatic energy $Q_A Q_B L / 4\pi\epsilon_0 \epsilon_r d_{AB}$ is small compared to RT, the thermal energy, (7.75) reduces simply to (7.12). When the electrostatic energy is large compared to the thermal energy and negative (oppositely charged ions), (7.75) becomes

$$k = \frac{Q_A Q_B L^2 D}{\epsilon_0 \epsilon_r RT} = \frac{\Lambda}{\epsilon_0 \epsilon_r} \tag{7.76}$$

The molar conductance Λ of the electrolyte that the ions A and B may be considered to form has been substituted in the right-hand side of (7.76). This is the equation first derived by Langevin[55] and often used to estimate the maximum rate of reaction of oppositely charged ions. For H^+ and OH^- the molar conductance at infinite dilution $\Lambda_0 = (349.82 + 198.0)\,\Omega^{-1}$ cm^2 mol^{-1} = 547.82 Ω^{-1} cm^2 mol^{-1}. Substituting into (7.76) we have

$$k = \frac{547.82\ \Omega^{-1}\ \text{cm}^2\ \text{mol}^{-1}}{(8.854 \times 10^{-12}\ \text{CV}^{-1}\ \text{m}^{-1})(78.3)} = 7.90 \times 10^{11}\ \frac{(\text{VA}^{-1})^{-1}\ \text{cm}^2\ \text{mol}^{-1}}{\text{AsV}^{-1}\ \text{m}^{-1}}$$

$$= 7.90 \times 10^{11}\ \frac{\text{cm}^2\ \text{m}}{\text{mol s}} \times \frac{\text{dm}^2}{100\ \text{cm}^2} \times \frac{10\ \text{dm}}{\text{m}} = 7.90 \times 10^{10}\ M^{-1}\ \text{s}^{-1}$$

This is somewhat less than the observed value.

If the electrostatic energy is large and positive (similarly charged ions), then (7.75) predicts a small value for k. For example, if A and B are univalent, $d_{AB} = 3.5$ Å, and the solvent is water, the value of the rate constant is calculated to be about one-third of that for a pair of neutral molecules with the same value of D. For highly charged ions or solvents of low dielectric constant, the slowing down due to electrostatic repulsion can become very much larger.

Table 7.7 presents data for a number of reactions of oppositely charged ions whose rates are given reasonably well by (7.76), and which are presumably diffusion controlled. Most of these very rapid reactions are studied by relaxation methods that are discussed in Chapter 8. The reaction between the benzyl cation and bromide ion was studied by pulse radiolysis. The cation was formed from benzyl chloride by a pulse of high-energy electrons lasting only a few hundred nanoseconds. The reaction with excess bromide ion was then monitored directly by following the UV absorption spectrum for a few microseconds.

Table 7.7. Rate constants in water at $25°C$ for diffusion-controlled reactions between ions

Reaction	$k/M^{-1} \, s^{-1}$
$H^+ + OH^- \rightarrow H_2O^a$	1.4×10^{11}
$H^+ + F^- \rightarrow HF^a$	1.0×10^{11}
$H^+ + HS^- \rightarrow H_2S^a$	7.5×10^{10}
$H^+ + CH_3COO^- \rightarrow CH_3COOH^a$	4.5×10^{10}
$NH_4^+ + OH^- \rightarrow NH_3 \cdot H_2O^a$	4.0×10^{10}
$Mg^{2+} + SO_4^{2-} \rightarrow Mg^{2+}, SO_4^{2-\,b}$	2.8×10^{10}
$Mn^{2+} + SO_4^{2-} \rightarrow Mn^{2+}, SO_4^{2-\,b}$	4.2×10^{10}
$C_6H_5CH_2^+ + Br^- \rightarrow C_6H_5CH_2Br^c$	5.2×10^{10}

[a] M. Eigen and L. DeMaeyer, *Z. Elektrochem.*, **59**, 986 (1955); M. Eigen, K. Kustin, and G. Maas, *Z. Phys. Chem.*, **30**, 130 (1961).
[b] G. Atkinson and S. Petrucci, *J. Phys. Chem.*, **70**, 3122 (1966).
[c] R. L. Jones and L. M. Dorfman, *J. Am. Chem. Soc.*, **96**, 5715 (1974); in 1,2-dichloroethane.

The reactions of Mn^{2+} and Mg^{2+} with sulfate ion were followed by ultrasonic absorption techniques. These detect several stages of reaction of which only the first is reported in the table. The several stages of reaction were first postulated by Eigen, and later demonstrated by Eigen and Tamm.[56]

$$M^{2+}(aq) + X^{2-}(aq) \underset{k_{21}}{\overset{k_{12}}{\rightleftharpoons}} \underset{1}{[M^{2+}WWX^{2-}]} \underset{k_{32}}{\overset{k_{23}}{\rightleftharpoons}} \underset{2}{[M^{2+}WX^{2-}]} \underset{k_{43}}{\overset{k_{34}}{\rightleftharpoons}} MX \quad (7.77)$$

The symbol W stands for water and intermediates **1** and **2** refer to ion pairs in which each ion is still solvated by its first coordination shell (**1**) and in which X^{2-} has become part of the second coordination shell of the metal ion (**2**). There is still controversy as to whether **1** and **2** can be distinguished experimentally. The final product, MX, is a complex in which X^{2-} is part of the first coordination shell. The species **1** and **2** are often called solvent-separated ion pairs, and MX is considered an intimate ion pair, in the case of their organic analogues.

The rate constants of Table 7.7 refer to k_{12} of (7.77), and it is for this step that (7.76) should be valid. Very similar numbers have been found for a number of divalent cations combining with divalent anions.[56] This is expected since the molar conductances should be much the same. By contrast the rate constant k_{34} should be strongly dependent on the nature of M^{2+} and X^{2-}. It has been found to range from $10^2 \, s^{-1}$ for Be^{2+} to greater than $10^9 \, s^{-1}$ for Ba^{2+} and Hg^{2+}.[57]

For large spherical ions, such as $Mg(H_2O)_6^{2+}$ and $SO_4(H_2O)_4^{2-}$ we would expect classical electrostatics to be most reliable. This can be further tested by calculating the equilibrium constant for the formation of **1** and/or **2**. The original theory for the formation of ion pairs is due to Bjerrum.[58] A simpler form has been developed by Fuoss and Eigen.[59] For the general case of spherical molecules

$$A + B \rightleftharpoons A, B$$

$$K_{AB} = \left(\frac{4\pi L d_{AB}^3}{3}\right) e^{-LU/RT} \tag{7.78}$$

The formula A, B implies that the original bonds in A and B have not been broken. U is the potential energy of interaction, and for ions it is given by (7.74).

Equation (7.78) has been found to be nearly correct for a large number of ion-pair equilibria in both water and nonaqueous solvents. For example, the ion pair $Co(NH_3)_6^{3+}$, Cl^- has an experimental K_{AB} of 74 M^{-1} in water at 25°C. This requires a value of about 4.0 Å for d_{AB}.[60] The crystallographic distance is estimated as 4.3 Å.

The very rapid proton transfers shown in Table 7.7 suggest that once an ion pair has been formed between the hydronium ion and the anion of a weak acid, the proton moves rapidly to its final binding site. Even if one or two additional water molecules intervened, proton jumps could occur by way of the Grotthus chain mechanism

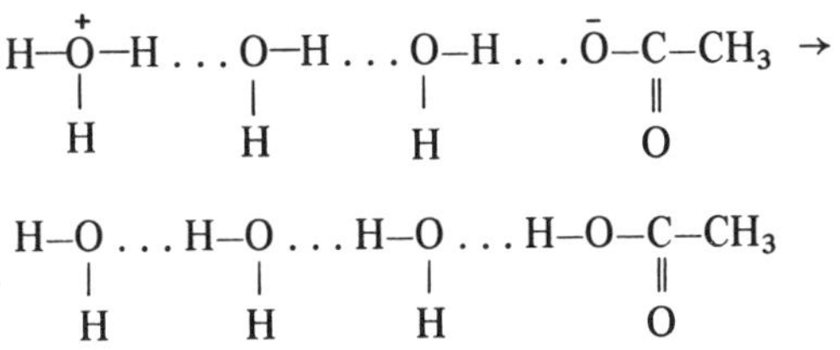

(7.79a)

The hydroxide ion could behave in a similar way:

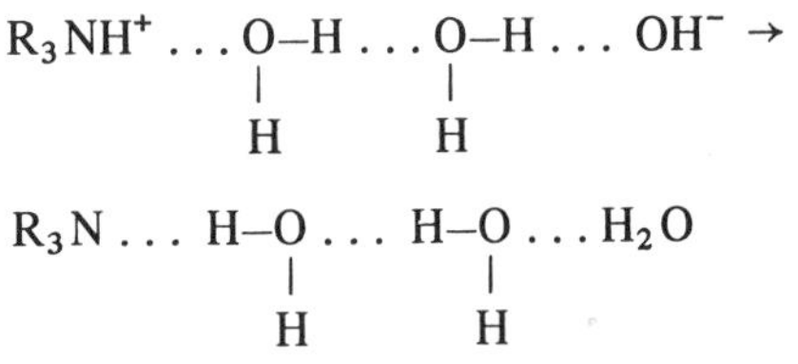

(7.79b)

In (7.79) dotted lines indicate hydrogen bonds. Charge transfer requires primarily electron rearrangement with only minor displacement of protons and therefore takes place very rapidly.

Not all proton transfers are diffusion controlled, even if thermodynamically favored. For example, the rate constant for the reaction

$$H_3O^+ + CH_2NO_2^- \rightarrow H_2O + CH_3NO_2 \tag{7.80}$$

is only $6.8 \times 10^2\ M^{-1}\ s^{-1}$ (see Table 7.6). This is the result of the strong rearrangement in charge density between the nitromethane molecule and its anion. As a result, protonation occurs first at the oxygen atom, probably at a rate near diffusion controlled:

$$H_3O^+ + CH_2NO_2^- \rightleftharpoons H_2O + CH_2{=}NO_2H \tag{7.81}$$

Table 7.8. Reactions between ions in water at 25°C

Reactants	$k/M^{-1}\,s^{-1}$	$\Delta H^{\ddagger}/kJ\,mol^{-1}$	$\Delta S^{\ddagger}/JK^{-1}\,mol^{-1}$
$Co(NH_3)_5Cl^{2+} + OH^-$ [a]	1.6	109	$+132$
$Cr(H_2O)_6^{3+} + NCS^-$ [b]	1.1×10^{-5}	106	$+17$
$NH_4^+ + CNO^-$ [c]	5.7×10^{-5}	96.1	$+13$
$(C_2H_5)_3NCH_2CO_2C_2H_5^+$			
$\quad + OH^-$ [d]	32	50.6	-47
$BrCH_2CO_2^- + S_2O_3^{2-}$ [e]	4.3×10^{-3}	63.2	-62
$S_2O_3^{2-} + SO_3^{2-}$ [f]	5.5×10^{-5}	58.1	-115
$Cr(NH_3)_5Cl^{2+} + Hg^{2+}$ [g]	8.0×10^{-2}	59.8	-56
$Pt(NH_3)_3Cl^+ + Br^-$ [h]	7.1×10^{-4}	71.1	-67

[a] G. Bushnell and G. C. Lalor, *J. Inorg. Nucl. Chem.*, **30**, 219 (1968). Zero ionic strength.
[b] C. Postmus and E. L. King, *J. Phys. Chem.*, **59**, 1216 (1955). Zero ionic strength.
[c] W. J. Svirbely and J. C. Warner, *J. Am. Chem. Soc.*, **57**, 1883 (1935). Zero ionic strength.
[d] R. P. Bell and F. J. Lindars, *J. Chem. Soc.*, **1954**, 4601. Zero ionic strength.
[e] P. A. H. Wyatt and C. W. Davies, *Faraday Soc., Trans.*, **45**, 774 (1949). Zero ionic strength.
[f] D. P. Ames and J. W. Willard, *J. Am. Chem. Soc.*, **73**, 164 (1951). Zero ionic strength.
[g] J. H. Espenson and S. R. Hubbard, *Inorg. Chem.*, **5**, 686 (1966). Ionic strength 2 *M*.
[h] U. Belluco, R. Ettore, F. Basolo, R. G. Pearson, and A. Turco, *Inorg. Chem.*, **5**, 591 (1966). Ionic strength 0.1 *M*.

The nitronic acid thus formed rearranges slowly to the more stable nitromethane, by way of redissociating into the ions.

Not all reactions between ions are extremely rapid, even if oppositely charged ions are the reactants. Just as great a variation in rate constants is found as for neutral reactants. Table 7.8 gives some rate data for reactions between ions of different charges in water. It can be seen that reactions between ions of the same charge are characterized by very negative entropies of activation, and reactions between ions of opposite charge usually have positive entropies of activation. Bear in mind that a bimolecular reaction with a normal frequency factor of 10^{11} $M^{-1}\,s^{-1}$ would have a negative entropy of activation of about 40 J K^{-1} mol^{-1}.

In terms of the collision theory, reactions between ions have probability factors, p, as small as 10^{-8} for ions of the same charge, and as large as 10^8 for ions of opposite charge. It was pointed out by Christiansen[61] that the collision theory should be modified to allow for the forces between the ions. There would be fewer collisions between ions of the same charge, and more collisions between ions of opposite charge. The same conclusions can be reached more readily by the transition-state theory.[62] In forming the activated complex, the two ions are brought from infinity to a distance $r_{\ddagger}$ in the activated complex. The free energy of activation has an electrostatic contribution.

$$\Delta G_{el}^{\ddagger} = \frac{LZ_A Z_B e^2}{4\pi\epsilon_0 \epsilon_r r_{\ddagger}}$$

(7.82)

There are also contributions to the heat of activation and to the entropy of activation. For water at $25°C$ and with $r_{\ddagger} = 2$ Å,

$$\Delta S_{el}^{\ddagger} = -40 Z_A Z_B \text{ J K}^{-1} \text{ mol}^{-1}$$

(7.83)

For oppositely charged ions this term is positive and accounts for large positive entropies of activation.

Equation (7.83) does not calculate all of the entropy of activation, but only a contribution to it. Still it accounts in a general way for the results shown in Table 7.8. The interpretation of $\Delta S^{\ddagger}$ in terms of solvent freezing about ions is also applicable. Depending on whether the activated complex is of greater charge than any of the reactants or of lesser charge, $\Delta S^{\ddagger}$ increases or decreases. Note that (7.58) predicts a dependence on the square of the ionic charge.

The classical electrostatic theory may also be used to predict the dependence of the rate constant on the dielectric constant of the solvent.[63]

$$\ln k = \ln k_0 - \frac{LZ_A Z_B e^2}{4\pi\epsilon_0 \epsilon_r R T r_{\ddagger}}$$

(7.84)

Here k_0 is the rate constant in a medium of infinite dielectric constant. The equation predicts a linear plot of $\log k$ against $1/\epsilon_r$ with a positive slope if the ions are of opposite charge, and a negative slope if the reactants have the same sign. Indeed such linear plots are often found for mixtures of fairly similar liquids, such as water and alcohol, or even water and dioxane. However, the linearity usually fails as the dielectric constant becomes small. A common procedure has been to use the slope of such a plot to calculate $r_{\ddagger}$, the only unknown in (7.84). Unfortunately, the values found in this way are often 1 or 2 Å, which makes the whole procedure of doubtful utility. Classical electrostatics would not be valid at such small distances between charges. Since the numbers are otherwise of reasonable magnitude, it is clear that a major effect is indeed the classical one.

Some of the reactions in Table 7.8 have rather complex mechanisms, such as prior ion-pair formation. The most extreme example is the ammonium cyanate reaction to form urea.

$$NH_4^+ + CNO^- \rightarrow (NH_2)_2 CO$$

(7.85)

The rate law is second order, suggesting a simple encounter of the two ions.

$$\text{rate} = k [NH_4^+] [CNO^-]$$

(7.86)

However, a mobile equilibrium exists between the ions and molecular products,

$$NH_4^+ + CNO^- \rightleftharpoons NH_3 + HCNO$$

(7.87)

so that the rate law may equally be interpreted as

$$\text{rate} = k'[NH_3] [HCNO] = \left(\frac{k' K_w}{K_a K_b}\right) [NH_4^+] [CNO^-]$$

(7.88)

where K_w/K_aK_b is the equilibrium constant for (7.87). A great deal of effort has been devoted to trying to decide between (7.86) and (7.88),[64] including the use of electrostatic theory to prove that the reactants were ions. For example, a plot of $\log k$ against $1/\epsilon_r$ is linear with a positive slope.

A little reflection will show that such tests cannot distinguish between ionic reactants and neutral reactants that are in rapid equilibrium with them. The results of kinetic experiments give only the composition of the activated complex and cannot give details as to its structure. Kinetically it cannot matter if the activated complex is formed from NH_4^+ and CNO^- or from NH_3 and HCNO. Because of (7.87), the effect of changing the dielectric constant is exactly the same in both cases. From a mechanistic point of view it seems almost certain that reaction is between the molecules. A simple nucleophilic attack by ammonia on the carbonyl group can be invoked.

$$
\begin{array}{ccccc}
\text{H-N=C=O} & \to & \text{H-N-C-O}^- & \to & \text{H-N-C=O} \\
\text{H-\ddot{N}-H} & & \text{H-N}^+\text{-H} & & \text{H N-H} \\
\text{H} & & \text{H} & & \text{H}
\end{array}
\qquad (7.89)
$$

REACTIONS BETWEEN IONS AND NEUTRAL MOLECULES

Inasmuch as reactions in which ions are formed or destroyed have abnormal frequency factors due to strong solvation, we may expect that reactions in which charge is conserved may have more normal values of pZ. This expectation is borne out by the sample data presented in Table 7.9. In most of these reactions charge is conserved, for example,

$$CH_3Br + I^- \to CH_3I + Br^- \qquad (7.90)$$

In a few cases charge is created, for example,

$$Co(NH_3)_5Cl^{2+} + H_2O \to Co(NH_3)_5H_2O^{3+} + Cl^- \qquad (7.91)$$

Reactions of this type have lower frequency factors as a result of stronger solvation of the developing charges.

In a few other cases low frequency factors are found, even though no increase in the number of ions occurs. Examples are the base hydrolysis of ethyl acetate and other esters, and the reaction

$$\textit{trans}\text{-Pt(piperidine)}_2Cl_2 + SCN^- \to$$

$$\textit{trans}\text{-Pt(piperidine)}_2ClSCN + Cl^- \qquad (7.92)$$

In these cases the large value of $-\Delta S^{\ddagger}$ must be attributed to the reaction mechanism. There is complete bonding of the nucleophilic reagent to the substrate before the leaving group departs. A four-coordinate tetrahedral intermediate is formed in ester hydrolysis and a five-coordinate intermediate is obtained in the platinum complex reaction. The latter effect also shows up in the reaction of

Table 7.9. Reactions between ions and neutral molecules

Reaction[a]	Solvent	E_a/kJ mol^{-1}	$\log (A/M^{-1}\,s^{-1})$
$CH_3Br + I^-$	H_2O	76.3	10.22
$CH_3Br + I^-$	CH_3OH	76.3	10.35
$CH_3Br + I^-$	$C_2H_4(OH)_2$	74.8	10.66
$CH_3Br + I^-$	$(CH_3)_2CO$	63.3	11.02
$CH_3I + F^-$	H_2O	105.3	11.32
$CH_3I + I^-$	H_2O	75.7	9.95
$CH_3I + OH^-$	H_2O	92.9	12.09
$CH_3I + CN^-$	H_2O	85.5	11.78
$CH_3I + SCN^-$	H_2O	83.4	11.17
$CH_3I + S_2O_3^{2-}$	H_2O	78.9	12.30
$CH_3I + Ag^+$	H_2O	81.1	11.64
$CH_3I + CH_3O^-$	CH_3OH	91.7	12.30
$C_2H_5I + C_6H_5O^-$	C_2H_5OH	92.0	11.60
$i\text{-}C_3H_7I + C_6H_5O^-$	C_2H_5OH	92.4	11.46
$(CH_3)_2SO_4 + CH_3O^-$	CH_3OH	87.0	12.40
$CH_3CO_2C_2H_5 + OH^-$	H_2O	75.2	7.38
$(CH_3)_2COHCH_2COCH_3$ $+ OH^-$ [b]	H_2O	47.5	11.12
$CO_2 + OH^-$ [c]	H_2O	56.0	13.81
$trans\text{-}Pt(piperidine)_2Cl_2$ $+ SCN^-$ [d]	CH_3OH	41.8	6.68
$Co(NH_3)_5Cl^{2+} + H_2O$ [e]	H_2O	95.7	9.28[f]
$Cr(H_2O)_5NO_3^{2+} + H_2O$ [e]	H_2O	89.4	9.81[f]

[a] Unless otherwise indicated data referenced in E. A. Moelwyn-Hughes, *Chemical Statics and Kinetics in Solution*, Academic, New York, 1971, Chaps. 8 and 9.
[b] G. M. Murphy, *J. Am. Chem. Soc.*, **53**, 977 (1931).
[c] B. R. W. Pinsent, L. Pearson, and F. J. W. Roughton, *Trans. Faraday Soc.*, **52**, 1512 (1956).
[d] U. Belluco, L. Cattalini, R. Ettore, and M. Martelli, *Gazz. Chim. Ital.*, **94**, 356 (1964).
[e] J. O. Edwards, F. Monacelli, and G. Ortazzi, *Inorg. Chim. Acta*, **11**, 47 (1974); this is a review with many other data.
[f] Converted to second-order units by dividing by 55.5 M.

$Pt(NH_3)_3Cl^+$ with Br^- in Table 7.8. The loss of translational entropy outweighs the favorable electrostatic factor.

To see the electrostatic effect one can compare the reaction of ethyl acetate with that of $(C_2H_5)_3NCH_2CO_2C_2H_5^+$ (Table 7.8). The cationic ester reacts 200 times as fast at 25°C. Its entropy of activation is only -47 J K^{-1} mol^{-1}, compared to that of -98 J K^{-1} mol^{-1} for ethyl acetate. The activation energy is slightly greater for the positively charged ester. This is predicted by the electrostatic theory:[65]

$$\Delta H_{el}^{\ddagger} = L \frac{Z_A Z_B e^2}{4\pi\epsilon_0 \epsilon_r r_{\ddagger}} \left[1 + T\left(\frac{\partial \ln \epsilon_r}{\partial T}\right)_P\right]$$ (7.93)

The entire term in brackets is a small negative number for water solvent.

Since most molecules are polar, there can be electrostatic effects between ions and neutral molecules. The interaction can lead to a larger value of the equilibrium constant for the formation of an encounter complex than the case where U is zero. The energy is given (for large values of r) by

$$U = - \frac{|Ze| \mu \cos \theta}{4\pi\epsilon_0 \epsilon_r r^2} \qquad (7.94)$$

where Ze is the charge on the ion, μ is the dipole moment of the molecule, r is the distance from the center of the ion to the center of the dipole, and θ is the angle of approach of the ion to the line of the dipole (θ is zero when the ion approaches the oppositely charged end of the dipole head-on and $180°$ when it approaches the end of the dipole with the same charge). This energy is small compared to the energy of two ions. Furthermore, we can expect that the total solvation of an ion and a molecule will be not much more than the solvation of the activated complex formed by their union. This activated complex will have a charge equal, of course, to the charge of the ion.

We may estimate the effect of the solvent on the rate of reaction between an ion and a neutral molecule from (7.94). Since the attraction (assuming correct orientation for the dipole) will be somewhat greater, the rate of the reaction will be larger in a medium of lower dielectric constant. Another method that has been used for discussing the effect of the solvent involves the Born equation for the charging of an ion in a continuous dielectric. Because of the difference in the radius r for the reactant ion and the activated complex, there is a difference in free energies that adds to the free energy of activation:

$$\Delta G_{el}^{\ddagger} = L \frac{Z^2 e^2}{8\pi\epsilon_0 \epsilon_r} \left(\frac{1}{r_{\ddagger}} - \frac{1}{r} \right) \qquad (7.95)$$

Accordingly, the rate constant may be written

$$\ln k = \ln k_0 + \frac{LZ^2 e^2}{8\pi\epsilon_0 \epsilon_r RT} \left(\frac{1}{r} - \frac{1}{r_{\ddagger}} \right) \qquad (7.96)$$

where k_0 is again the rate constant in a medium of infinite dielectric constant. Since $r_{\ddagger}$ is larger than r, the rate again should be somewhat greater in a medium of lower dielectric constant.[62] This seems to be true in most cases, but (7.96) cannot be taken too seriously as a quantitative explanation.

Specific solvent effects and the actual charge distribution in nonspherical ions may produce as large, or larger, effects than those of the dielectric constant. Table 7.10 shows some relevant data on these factors. The reaction is the unimolecular decomposition of substituted benzisoxazole-3-carboxylate anion (7.97).[66]

Table 7.10. Rates of decarboxylation at $30°C$ of 6-nitrobenzisoxazole-3-carboxylate ion in various solvents

Solvent	k/s^{-1}	Solvent	k/s^{-1}
H_2O	7.3×10^{-6}	CH_3CN	2.9
CH_3OH	2.5×10^{-4}	$(CH_3)_2SO$	10
$HCONH_2$	7.4×10^{-4}	$(CH_3)_2CO$	24
C_2H_5OH	1.0×10^{-3}	$HCON(CH_3)_2$	37
C_6H_6	4.8×10^{-3}	$CH_3CON(CH_3)_2$	160
CH_3NO_2	5.8×10^{-1}	NMP^a	250
C_6H_5CN	2.5	$HMPA^b$	700

Source. Data from reference 66.
[a] *N*-Methyl-2-pyrollidone.
[b] Hexamethylphosphoramide.

$$(7.97)$$

Decarboxylation produces a smaller anion so that more polar solvents might be expected to produce the highest rates. In fact, the rate is slowest in water and in methanol, speeds up in benzene, and speeds up even more in polar, aprotic solvents. The reactant ion is one with the negative charge highly localized on the hard oxygen atoms of the carboxylate group. Hydrogen bonding solvents stabilize this structure, leading to a low rate of reaction. The immediate product is a soft anion with the charge well delocalized. This structure is stabilized most effectively by polar, but nonhydrogen-bonding solvents. The reaction in benzene is strongly influenced by ion pairing with the cation, which acts to stabilize the localized charge.

A great deal is known emprically about the nucleophilic reactivity of various anions as a function of solvent.[67] Fortunately, since methods are available for studying ion–molecule reactions in the gas phase, rate data for similar reactions in the complete absence of solvent are available.[68] These reactions, if exothermic, are very much faster than in solution. For example, the reaction

$$OH^- + CH_3Cl \rightarrow CH_3OH + Cl^- \qquad (7.98)$$

has a rate constant of $9.6 \times 10^{11}\ M^{-1}\ s^{-1}$ in the gas phase at room temperature. The corresponding value in water is only $6.7 \times 10^{-6}\ M^{-1}\ s^{-1}$. The relative

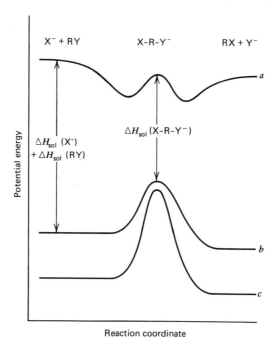

Figure 7.6.   Representative reaction coordinate diagrams for a nucleophilic displacement reaction in the (*a*) gas phase and in (*b*) dipolar aprotic and (*c*) protic solvents. [Reproduced with permission from W. N. Olmstead and J. I. Brauman, *J. Am. Chem. Soc.*, **99**, 4219 (1977), copyright 1977, American Chemical Society.]

reactivities toward CH_3Cl in the gas phase follow the order

$$OH^- \sim NH_2^- > F^- \sim H^- \sim CH_3O^- > CH_3S^- > Cl^- > CN^- > Br^-$$

In water or methanol there would be a strong, but not complete, inversion of this ordering.[69]

The gas-phase reactions occur by an ion–molecule collision that is not random but controlled by rather strong ion–dipole and ion-induced dipole forces (see p. 194). A collision complex that has a bonding energy of some $40-80\,kJ\,mol^{-1}$ is formed. Since deactivating collisions are rare, the lowering of the potential energy must be balanced by a rise in the translational, rotational, and vibrational energy. This internal energy can then be used to surmount any modest potential energy barrier to reaction. Figure 7.6 shows a schematic of the energy-reaction coordinate profile in the gas phase compared to solution for a typical S_N2 reaction. Both protic and aprotic solvents are shown, with the assumption that both the entering and leaving groups are charge-localized ions.

Figure 7.6 suggests that the activation energies for many reactions in solution are due entirely to solvation effects. This is borne out in a few cases by accurate quantum-mechanical calculations of the gas-phase potential energy profiles. For example, the reaction of CO_2 with OH^- to form HCO_3^- has no barrier in the gas phase according to such calculations.[70] In water the barrier is $56\,kJ\,mol^{-1}$, as shown in Table 7.9.

We might then suppose that in many reactions involving ions the activation energy may be due entirely to the energy required to partly desolvate the reactant ion, to free it for reaction with the substrate. This view was first espoused by Moelwyn-Hughes, who pointed out that activation energies for many simple reactions of anions averaged about one-quarter of their solvation energies.[71] Ritchie has also found that the rate constants of many nucleophiles with a series of similar electrophilic substrates can be given by the equation,[72]

$$\log k = \log k_0 + N_+ \qquad (7.99)$$

The constant k_0 depends only on the substrate and N_+ depends only on the nucleophile, but is also a function of the solvent. Such a simple result is expected if $\log k_0$ is a measure of desolvation of the substrate, and N_+ depends only on the desolvation of the nucleophile.

However, (7.99) is valid only for a limited number of substrates that are of rather similar nature, usually organic cations. While desolvation of the reactants must play some role in determining activation energies in solution, many other factors also are involved. Even in the gas phase there are intrinsic energy barriers for most chemical reactions. These are determined for the most part by the detailed interaction of the reactants as they are brought together to form the activated complex. Also, in solution the amount of solvent reorganization needed is not constant, but depends on the particular reaction.

The properties of mixed solvents present a special problem, even if average properties such as dielectric constant can be measured. For most solutes, and especially ions, the environment around each molecule can be quite different from the average environment. For example, silver nitrate in an equimolar mixture of water and acetonitrile has the nitrate ion surrounded preferentially by water and the silver ion preferentially by acetonitrile.[73] This information can be obtained from classical transference number studies.

Sometimes more modern methods can be used not only to probe the immediate environment of a solute molecule, but also to explain otherwise mysterious rate data. A case in point is the solvolysis of a chromium (III) complex.[74]

$$Cr(NCS)_6^{3-} + H_2O \rightarrow Cr(NCS)_5 H_2O^{2-} + SCN^- \qquad (7.100)$$

Water can be replaced both as a solvent and as a reactant by alcohols, pyridine, and dimethylformamide. However, acetonitrile and nitromethane are too poorly coordinating to take part as reactants. Figure 7.7 shows the rate constant for (7.100) as a function of composition in mixtures of water and acetonitrile.

It can be seen, by observing the circles, that a small amount of acetonitrile depresses the rate very markedly. In this case nuclear magnetic resonance relaxation times can be used to measure the composition of the solvent in immediate contact with the paramagnetic complex ion. The results give a parameter, n/n_0, which gives the number of water molecules in close encounter with the ion in the mixed solvent (n), in comparison to the number (n_0) in pure water. The rate constant is also plotted in Fig. 7.7 as a function of n/n_0. The straight line represents what would be found if the water entered the coordination sphere of chromium only on a statistical basis as one NCS$^-$ ligand departed. This kind of mechanism is called an I_d, or dissociative interchange, mechanism.[75] The water molecule is

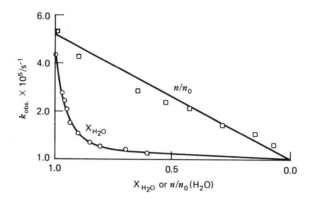

Figure 7.7. Hydrolysis rates for $Cr(NCS)_6^{3-}$ in water–CH_3CN mixtures at $25°C$ plotted (○) as a function of mole fraction water and (□) as a function of the probability of encounter, n/n_0. [Reproduced by permission from C. H. Langford and J. P. K. Tong, *Acc. Chem. Res.*, **10**, 259 (1977), copyright 1977, American Chemical Society.]

necessary for the reaction to occur, but it plays only a minor role in causing the thiocyanate ion to leave. That is, the activation energy is governed mainly by bond breaking requirements.

INFLUENCE OF IONIC STRENGTH

The preceding discussions are based on the assumption that the properties of the solution are those of infinite dilution, that is, that the presence of solute molecules or ions does not affect the properties of the reactants or activated complex. In practice real solutions have deviation from ideality at moderate concentration. If ions are present, the deviation becomes apparent even at low concentrations. The effect on the rate is given by (7.43), which may be written as

$$k = k_0 \frac{\gamma_A \gamma_B \cdots}{\gamma_\ddagger} \tag{7.101}$$

where k_0 is the rate constant at infinite dilution in the given solvent. This choice selects the standard state as a hypothetical state of unit concentration with a partial molal heat content, volume and heat capacity for each component being equal to that at infinite dilution. The value of k_0 is experimentally available by measuring k at a series of finite concentrations and extrapolating the results to infinite dilution.

The most important application of (7.101) occurs when one or more of the reactants are ions. According to the Debye-Hückel theory, the relation between the activity coefficient of an ion and the ionic strength is given for dilute solutions (less than $0.01\,M$) by

$$-\ln \gamma_i = \frac{Z_i^2 \, \alpha\sqrt{\mu}}{1 + \beta a_i \sqrt{\mu}} \tag{7.102}$$

where μ is the ionic strength, a_i is the distance of closest approach of another ion to the ith ion, and α and β are constants for a given solvent and temperature. The numerical values are $\alpha = 0.509 \times 2.303$ and $\beta = 0.329 \times 10^8$ in water at $25°C$. Equation (7.101) can be written, using (7.102), as

$$\ln k = \ln k_0 - \frac{Z_A^2 \alpha\sqrt{\mu}}{1 + \beta a_A \sqrt{\mu}} - \frac{Z_B^2 \alpha\sqrt{\mu}}{1 + \beta a_B \sqrt{\mu}} + \frac{(Z_A + Z_B)^2 \alpha\sqrt{\mu}}{1 + \beta a_\ddagger \sqrt{\mu}} \tag{7.103}$$

or, if we adopt a mean value a for the distance of closest approach,

$$\ln k = \ln k_0 + \frac{2Z_A Z_B \alpha\sqrt{\mu}}{1 + \beta a\sqrt{\mu}} \cong \ln k_0 + 2Z_A Z_B \alpha\sqrt{\mu} \tag{7.104}$$

The approximation in (7.104) is valid only for dilute solution where μ is small. Equations (7.101) and (7.104) were first derived by Brønsted[76a] and by Bjerrum.[76b] They follow directly from the transition-state theory, although that theory was not used as such in the original derivation. The assumption must be made that the activated complex has its equilibrium ionic atmosphere in spite of its short lifetime ($\sim 10^{-13}$ s), which is much smaller than the times of relaxation of ionic atmospheres. However, since reacting ions approach each other relatively slowly, their individual ionic atmospheres have time to adjust themselves to very nearly the equilibrium distribution for the activated complex.

An equation identical to (7.104) can be derived on the basis of the collision theory.[77] This is done simply by considering the perturbing effect of the ionic atmosphere on the calculated values of the electrostatic free energy needed to bring together the two charged reactants. Thus collisions between oppositely charged ions are reduced and collisions between similarly charged ions are enchanced by the presence of the atmospheres, which tend to reduce the electrostatic attraction in the first case and the repulsion in the second.

Equation (7.104) predicts a linear relationship if log k is plotted against the square root of the ionic strength, with a slope proportional to the product $Z_A Z_B$. This is found to be true qualitatively at least for a number of reactions between ions. The quantitative agreement is excellent in a number of samples, the observed slope agreeing with the Debye-Hückel theoretical slope. However, the relationship has been much abused, experimental data being plotted against the square root of ionic strength for concentrated solutions. Since the limiting law is only valid for solutions below 0.01 molal for 1–1 electrolytes, and for even lower concentrations for higher charged ions, such procedure has no theoretical justification.

Furthermore, even at quite low concentrations there is often evidence for complex formation between an ion reactant and an added ion of opposite sign, which renders (7.104) invalid. Particularly for multiply charged ions and for ions of the transition metals that form strong complexes, it is found that addition products are formed. Such complexes may be reasonably stable or exist only in small concentration. In either case they usually have an influence on the reaction rate, in some cases so great as to be classified as catalytic. Even if complexes are not formed, at higher concentrations specific effects of added ions are certain to be found.

There have been improvements in the theory of electrolyte solutions since the days of Debye and Hückel.[78] Extended equations for the activity coefficients of ions are available and could be applied to (7.101). However, these equations are still based on the principle that the ionic strength determines the activity coefficient. As Olson and Simonson first emphasized,[79] the activity, in a kinetic sense at least, is mainly determined by the ions of opposite charge in solution. Following the ideas of Davies,[80] if there is a reaction between ions A^+ and B^- in the presence of an "inert" electrolyte, MX, we must consider at least three independent reactions:

$$A^+ + B^- \xrightarrow{k_1} \text{products}$$

$$AX + B^- \xrightarrow{k_2} \text{products} + X^- \qquad (7.105)$$

$$MB + A^+ \xrightarrow{k_3} \text{products} + M^+$$

This means that there are also two equilibrium constants to consider, for the formation of AX and MB.

In aqueous solution this complexity can be removed by working in sufficiently dilute solution. In nonaqueous solutions of low dielectric constant, it is nearly impossible to avoid the problem of ion pairing. Some help can be gained by using very large ions, free from exposed groups that are chemically active. For example, in acetone solution LiCl is extensively associated, and considerably less reactive than free chloride ion. A quaternary ammonium salt, such as $(C_4H_9)_4NCl$, would be much more dissociated.[81]

The use of the cation complexing agents called crown ethers and cryptands can effectively tie up counterions that would otherwise reduce the rate.[82] In tetrahydrofuran ($\epsilon_r = 7.6$) the salt $Na_2Fe(CO)_4$ reacts with alkyl halides as the tight ion pair $NaFe(CO)_4^-$. This is 20,000 times less reactive than the solvent separated ion pair $Na^+SFe(CO)_4^{2-}$, which forms in N-methyl pyrrolidone. "Free" $Fe(CO)_4^{2-}$ can be formed in the latter solvent by adding 2 mol of a cryptand. It is about three times as reactive as $Na^+SFe(CO)_4^{2-}$.[85]

If one of the reactants is a neutral molecule so that $Z_B = 0$, then (7.104) predicts no effect of the ionic strength. This appears to be true for very dilute solutions. However, at higher ionic concentration the rate constant may change because of changes in the activity coefficients of the ions not given by the Debye-Hückel theory and because the activity coefficients of neutral molecules are affected by higher ionic strengths. For example, Hückel[83a] proposed a term in the first power of the ionic strength to reproduce the activity coefficient of an ion at higher concentrations:

$$-\ln \gamma_i = \frac{Z_i^2 \alpha \sqrt{\mu}}{1 + \beta a_i \sqrt{\mu}} - b_i \mu \qquad (7.106)$$

Also the activity coefficient of a neutral molecule may be expressed in terms of[83b]

$$\ln \gamma_0 = b_0 \mu \qquad (7.107)$$

We may then write

$$\ln k = \ln k_0 + (b_0 + b_A - b_{\ddagger})\mu \qquad (7.108)$$

so that the logarithm of the rate constant becomes a linear function of the first power of the ionic strength. This seems to be approximately true for a number of reactions involving two neutral molecules or a neutral molecule and an ion. Unfortunately, the b_0, b_A, and $b_{\ddagger}$ coefficients are largely empirical, and it is difficult to predict even the sign of the effect on the rate constants.

Figure 7.8 gives some representative results for a reaction between a neutral molecule and an ion. The reaction is the acid-catalyzed hydrolysis of γ-butyrolactone to γ-hydroxybutyric acid. The rate is proportional to the concentration of hydrogen ion and of lactone. The dotted lines show $\log k/k_0$ plotted against the ionic strength for sodium chloride and sodium perchlorate solutions. The value of k_0 is determined by extrapolation to zero ionic strength. The salt effects are seen to be in opposite directions. For comparison the logarithm of the activity coefficient of the lactone is plotted against the ionic strength for

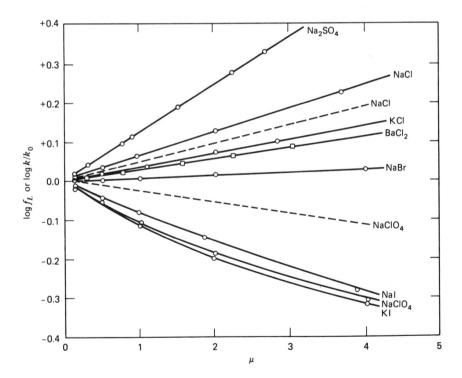

Figure 7.8. Plot of ——— logarithm of the activity coefficient and ——— of the relative rate of hydrolysis of γ-butyrolactone in various salt solutions. [Reproduced by permission from Long, McDevit, and Dunkle, *J. Phys. Colloid Chem.*, **55**, 819 (1951).]

a number of salts. These were determined independently from distribution experiments. It is seen that the opposite salt effects of sodium chloride and perchlorate on the rates are related to the opposite effects of these salts on the activity coefficient of the lactone.

If two neutral molecules are reacting to form ions, there should be an ionic strength effect on the rate of development of the incipient charges. A theory can be given that predicts a linear dependence of $\ln k$ on the ionic strength.[84] Such an effect is found in dilute, polar solvents. In nonpolar solvents much more pronounced effects are found. It has been shown that the addition of 0.1 M LiClO$_4$ to ethyl ether increases the rate of ionization of an organic sulfonate by a factor of 10^5. The empirical equation,

$$k = k_0 + b\,[\text{salt}] \tag{7.109}$$

where k_0 is the rate constant in the absence of added salt, has been found to work well in solvents of low polarity, such as glacial acetic acid.[85]

HIGH-PRESSURE EFFECTS ON RATES

In an earlier section the internal pressure, $\Delta E_{\text{vap}}/V$, of a liquid is mentioned. Typical values of this solvent property are in the range of 2000–8000 atm. Interestingly, it is also possible to generate such high pressures by mechanical means and to see what influence they have on rates of reaction. From thermo-dynamics and the transition-state theory, we can write[86]

$$\left(\frac{\partial \ln k}{\partial P}\right)_T = -\frac{\Delta V^{\ddagger}}{RT} \tag{7.110}$$

where $\Delta V^{\ddagger}$ is the difference in molar volumes between the activated complex and the reactants. If $\Delta V^{\ddagger}$ is ± 15 ml mol^{-1}, then a pressure of 1000 atm increases the rate constant by a factor of 1.8 at 25°C. A positive value of the volume of activation corresponds to a reaction that slows down at high pressure.

The experimental results are usually represented by the empirical equation

$$\ln k = \ln k_0 + bP + cP^2 \tag{7.111}$$

where k_0 is the rate constant at 1 atm pressure. Normally one calculates $\Delta V^{\ddagger} = -bRT$ as the volume of activation at zero pressure. Table 7.11 shows a few results on $\Delta V^{\ddagger}$ from the literature.[87]

Numbers such as those in the table have been widely used to obtain mechanistic information. The assumption is that various features such as bond breaking, bond making, and creation of charges, will make contributions to the observed $\Delta V^{\ddagger}$. Estimates of the most important of these factors are included in the table. It can be seen that the organic examples have values of $\Delta V^{\ddagger}$ that are consistent with bond making or bond breaking in the transition state, or the development of charges. The latter causes a decrease in volume by virtue of the electrostriction of the solvent about the charges.

Table 7.11. Activation volumes for selected reactions

Reaction	Solvent	$\Delta V^\ddagger/\text{cm}^3\ \text{mol}^{-1}$
$(CH_3)_3COOC(CH_3)_3 \rightarrow 2(CH_3)_3CO\cdot$	Carbon tetrachloride	$+13.3$[a]
cyclobutane–$CHC_2H_5 \rightarrow C_2H_4 + CH_2{=}CHC_2H_5$	Gas phase	$+28.2$
(ene reaction structure with $CHCO_2C_4H_9$ / CH_2)		
(bicyclic structure, $CHCO_2C_4H_9$)	n-Butyl chloride	-29.6
$(C_2H_5)_3N + C_2H_5I \rightarrow (C_2H_5)_4N^+I^-$	Benzene	-50.2
$CH_3Br + H_2O \rightarrow CH_3OH_2^+ + Br^-$	Water	-14.5
$Co(NH_3)_5H_2^*O^{3+} + H_2O \rightarrow Co(NH_3)_5H_2O^{3+} + H_2^*O$	Water	$+1.2$
$Cr(NH_3)_5H_2^*O^{3+} + H_2O \rightarrow Cr(NH_3)_5H_2O^{3+} + H_2^*O$	Water	-5.8
$Co(NH_3)_5Cl^{2+} + H_2O \rightarrow Co(NH_3)_5H_2O^{3+} + Cl^-$	Water	-10.6
$Cr(NH_3)_5Cl^{2+} + H_2O \rightarrow Cr(NH_3)_5H_2O^{3+} + Cl^-$	Water	-10.8

Estimates of contributions to $\Delta V^\ddagger$

Bond cleavage	$+10$ ml	Ionization	-20 ml
Bond formation	-10 ml	Charge neutralization	$+20$ ml
	Diffusion control	$+20$ ml	

[a] C. Walling and G. Metzger, *J. Am. Chem. Soc.*, **81**, 5365 (1959). Other data from reference 87.

The inorganic examples are not so readily interpreted. The water exchange of $Co(NH_3)_5 H_2O^{3+}$ is believed to occur by an I_d mechanism. One might have expected a greater increase in volume than 1.2 ml. The $\Delta V^{\ddagger}$ of -5.8 ml for $Cr(NH_3)_5 H_2O^{3+}$ water exchange is consistent with the belief that substitution reactions of chromium (III) require more nucleophilic assistance from the entering ligand than do those of cobalt (III).[88] If this is so, it is not clear why $\Delta V^{\ddagger}$ is nearly the same for the aquation of both $Cr(NH_3)_5 Cl^{2+}$ and $Co(NH_3)_5 Cl^{2+}$.

It has been pointed out that when the coordination number of a metal ion is reduced, the bond distances to the remaining ligands decrease,[89] causing a reduction in volume. The decrease depends on the electronic structure of the metal ion, being larger for chromium (III) than for cobalt (III). Generally speaking, the mechanistic conclusions that can be drawn from experimental values of $\Delta V^{\ddagger}$ are similar to those that would emerge from a consideration of $\Delta S^{\ddagger}$. In both cases considerable caution must be used in forming conclusions because the measured quantities can depend on many factors.

PROBLEMS

7.1. The rate of dimerization of cyclopentadiene and the reverse rate of dissociation of the dimer have been studied in the gas phase and in the pure liquid [B. S. Khambata and A. Wasserman, *Nature*, **137**, 496 (1936); A Wasserman, *Trans. Faraday Soc.*, **34**, 128 (1938)]. The values of the frequency factors and activation energies are as follows (concentrations in mol liter^{-1}, time in seconds):

	Dimerization second order		Dissociation first order	
Gas phase	$10^{6.1}$	16.7 kcal mol^{-1}	$10^{13.1}$	35.0 kcal mol^{-1}
Pure liquid	$10^{5.7}$	16.0 kcal mol^{-1}	$10^{13.0}$	34.5 kcal mol^{-1}

Calculate the equilibrium constant in each case and $\Delta H^{\ominus}$ and $\Delta S^{\ominus}$. What properties of the activated complex can be deduced, assuming the mechanism to be the same in both cases?

7.2. The hydrolysis of benzoyl chloride has been studied in mixtures of water and acetone [B. L. Archer and R. F. Hudson, *J. Chem. Soc.*, **1950** 3259]:

$$C_6 H_5 COCl + H_2 O \rightarrow C_6 H_5 COOH + H^+ + Cl^-$$

The reaction was found to be first order in all cases and the rate constants and activation energies were determined. From the data calculate the entropy of activation at 25°C for each of the mixtures investigated and give a reasonable interpretation of the variations. Discuss the possibility of studying this reaction in the gas phase.

Volume % water	E_a/kcal mol^{-1}	$k \times 10^4$/s^{-1} (25°C)
40.0	16.9	25.25
33.3	16.3	14.11
25.0	14.9	6.94
15.0	13.6	3.00
5.0	11.0	0.55

7.3. Predict the influence of (a) increasing dielectric constant and (b) increasing ionic strength on the rates of the following reactions:

$$CH_3Br + H_2O \rightarrow CH_3OH + H^+ + Br^-$$

$$Co(NH_3)_5Br^{2+} + Hg^{2+} + H_2O \rightarrow Co(NH_3)_5H_2O^{3+} + HgBr^+$$

$$Co(NH_3)_5Br^{2+} + NO_2^- \rightarrow Co(NH_3)_5NO_2^{2+} + Br^-$$

$$OH^- + CH_3Br \rightarrow CH_3OH + Br^-$$

$$OH^- + BrCH_2CH_2CH_2CH_2CH_2CO_2^- \rightarrow HOCH_2CH_2CH_2CH_2CH_2CO_2^- + Br^-$$

7.4. The reaction between persulfate ion and iodide ion is as follows:

$$S_2O_8^{2-} + 2I^- \rightarrow 2SO_4^{2-} + I_2$$

Kinetically the reaction is second order, first order each in iodide ion and persulfate ion. The variation of the second-order rate constant with the total concentration of reactants is considerable as shown by the data below. Plot the data and compare with the theoretical predictions of the Brønsted-Bjerrum-Christiansen equation. Initial concentration of $K_2S_2O_8$, 0.00015 M; temperature 25°C.

c_{KI}/M	Log $(k/M^{-1}$ min$^{-1})$
0.0016	$\bar{1}$.013
0.0020	$\bar{1}$.021
0.0032	$\bar{1}$.049
0.0040	$\bar{1}$.065
0.0060	$\bar{1}$.072
0.0080	$\bar{1}$.100
0.0100	$\bar{1}$.124
0.0120	$\bar{1}$.143
0.0180	$\bar{1}$.199
1.0240	$\bar{1}$.228

Source: C. V. King and M. B. Jacobs, *J. Am. Chem. Soc.*, **53**, 1704 (1931).

7.5. The rate constant for the electron transfer reaction

$$IrBr_6^{3-} + IrCl_6^{2-} \xrightarrow{k} IrBr_6^{2-} + IrCl_6^{3-}$$

is given by $k = 2.2 \times 10^{11} e^{-22,100 J/RT \, mol} M^{-1} \, s^{-1}$. Calculate the rate constant at $25°C$ and compare to (7.76), assuming $D = 1 \times 10^{-5} \, cm^2 \, s^{-1}$.

7.6. The reaction in water

$$NH_3 + C_2H_5NO_2 \underset{k_2}{\overset{k_1}{\rightleftharpoons}} NH_4^+ + CH_3CHNO_2^-$$

has an equilibrium constant of 29.8 at $0°C$. The value of $\Delta H^\ominus$ is $-41.8 \, kJ \, mol^{-1}$ and $\Delta S^\ominus$ is $-135 \, J \, mol^{-1} \, K^{-1}$. The rate constant k_1 is $9.50 \times 10^{-4} \, M^{-1} \, s^{-1}$ at $0°C$ and $1.62 \times 10^{-3} \, M^{-1} \, s^{-1}$ at $5.2°C$. Though the temperature difference is too small for accuracy, calculate k_2 at $0°C$ and also $\Delta H_2^\ddagger$ and $\Delta S_2^\ddagger$.

REFERENCES

1. J. H. Raley, F. F. Rust, and W. E. Vaughan, *J. Am. Chem. Soc.*, **70**, 88, 1336, 2767 (1948).

2. C. K. Yip and H. O. Pritchard, *Can. J. Chem.*, **49**, 2290 (1971).

3. R. L. Burwell, *J. Am. Chem. Soc.*, **73**, 4461 (1951).

4. P. Neta, G. R. Holdren, and R. H. Schuler, *J. Phys. Chem.*, **75**, 449, 1654 (1971).

5. V. N. Kondratiev, *Rate Constants for Gas Phase Reactions*, National Bureau of Standards Publication, Washington, DC, 1972.

6. For rate data of hydrated electrons see NSRDS-NBS 43 and NSRDS-NBS 43 Supplement compiled by M. Anbar, M. Bambenek, and A. B. Ross, 1973 and 1975. Available from Superintendent of Documents, U.S. Government Printing Office, Washington, DC.

7. H. S. Frank and M. W. Evans, *J. Chem. Phys.*, **13**, 507 (1945); H. Scheraga, *Acc. Chem. Res.*, **12**, 7 (1979); D. N. Glew, *J. Phys. Chem.*, **66**, 605 (1962).

8. For other theories of the liquid state see D. Henderson, W. Jost, H. Eyring, Eds., *Physical Chemistry: An Advanced Treatise*, Vol. 8A Academic, New York, 1976. For a further discussion of the hole theory see H. Eyring and M. S. Jhon, *Significant Liquid Structures*, Wiley-Interscience, New York, 1969.

9. E. Rabinowitch, *Trans. Faraday Soc.*, **33**, 1225 (1937), was the first to recognize the importance of the cage effect.

10. M. V. Smoluchowski, *Phys. Z.*, **17**, 557, 583 (1916).

11. A. M. North, *The Collision Theory of Chemical Reactions in Liquids*, Methuen, London, 1964.

12. E. A. Moelwyn-Hughes, *The Kinetics of Reaction in Solutions*, 2nd ed., The Clarendon Press, Oxford 1947.

13. B. Hickel, *J. Phys. Chem.*, **79**, 1054 (1975).

14. R. M. Noyes, *J. Am. Chem. Soc.*, **86**, 4529 (1964).

15. R. M. Noyes, *Prog. React. Kinet.*, **1**, 129 (1961).

16. E. F. Caldin and B. B. Hasinoff, *J. Chem. Soc., Faraday Trans. I*, **71**, 515 (1975).

17. P. Debye, *Polar Molecules*, Dover, New York, 1929.

18. E. Grunwald, K. C. Chang, and J. E. Leffler, *Ann. Rev. Phys. Chem.*, **27**, 369 (1976).

19. J. W. Taylor and J. C. Martin, *J. Am. Chem. Soc.*, **88**, 3650 (1966); **89**, 6904 (1967).

20. (*a*) E. Niki and Y. Kamiya, *J. Am. Chem. Soc.*, **96**, 2129 (1974); (*b*) H. Kiefer and T. G. Traylor, *ibid.*, **89**, 6667 (1967).

21. G. A. Russell, *J. Am. Chem. Soc.*, **79**, 2977 (1957).

22. G. R. Branton, H. M. Frey, D. C. Montague, and I. D. R. Stevens, *Trans. Faraday Soc.*, **62**, 659 (1966).

23. R. B. Woodward and R. Hoffmann, *J. Am. Chem. Soc.*, **87**, 395 (1965).

24. G. Scatchard, *Chem. Rev.*, **8**, 321 (1931).

25. J. H. Hildebrand and R. L. Scott, *Solubility of Non-Electrolytes,* 3rd ed., Dover, New York, 1950.

26. H. F. Herbrandson and F. R. Neufeld, *J. Org. Chem.*, **31**, 1140 (1966).

27. A. P. Stefani, *J. Am. Chem. Soc.*, **90**, 1694 (1968).

28. For reviews see (*a*) C. Reichardt, *Angew. Chem. Int. Ed. (Engl.),* **4**, 29 (1965); (*b*) I. A. Koppel and V. A. Palm, in *Advances in Linear Free Energy Relationships,* N. B. Chapman and J. Shorter, Eds., Plenum Press, London, 1972.

29. This approach was pioneered by E. M. Kosower, *J. Am. Chem. Soc.*, **80**, 3253, 3261, 3267 (1958).

30. C. Lassau and J. C. Jungers, *Bull. Soc. Chim. Fr.*, 2678 (1968). Many other solvents are also included.

31. R. F. Hudson and B. Saville, *J. Chem. Soc.*, **1955**, 4114, 4121; R. S. Drago and K. F. Purcell, *Prog. Inorg. Chem.*, **6**, 217 (1965); R. G. Pearson, Ed., *Hard and Soft Acids and Bases,* Dowden, Hutchinson, and Ross, Stroudsburg, PA, 1973.

32. F. Basolo and R. G. Pearson, *Mechanisms of Inorganic Reactions,* 2nd ed., Wiley, New York, 1968, Chap. 2.

33. G. J. Hoijtink, E. deBoer, P. H. Van der Meij, and W. P. Wiejland, *Rec. Trav. Chim.*, **75**, 487 (1956); I. Fischer-Hjalmars, A. Henriksson-Enflo, and C. Herrmann, *Chem. Phys.*, **24**, 167 (1977).

34. For example, see P. Schuster. W. Jakubetz, and W. Marius, *Top. Curr. Chem.*, **60** (1975); A. Pullman, B. Pullman, and H. Berthod, *Theor. Chim. Acta,* **47**, 175 (1978).

35. I. Jano, *C. R.,* **261**, 103 (1965); R. Constanciel and O. Tapia, *Theor. Chim. Acta,* **48**, 75 (1978).

36. H. H. Uhlig, *J. Phys. Chem.*, **41**, 1215 (1937); D. D. Eley, *Trans. Faraday Soc.*, **35**, 1281, 1421 (1939).

37. D. S. Choi, M. S. Jhon, and H. Eyring, *J. Chem. Phys.*, **53**, 2608 (1970).

38. P. Kebarle and E. W. Godbole, *J. Chem. Phys.*, **39**, 1131 (1963); M. T. Bowers, D. H. Aue, H. M. Webb, and R. T. McIver, *J. Am. Chem. Soc.*, **93**, 4314 (1971).

39. J. L. Beauchamp, *Ann. Rev. Phys. Chem.*, **22**, 527 (1971); J. D. Baldeschwieler and S. S. Woodgate, *Acc. Chem. Res.*, **4**, 114 (1971).

40. P. Kebarle, *Ann. Rev. Phys. Chem.*, **28**, 445 (1977); V. M. Bierbaum, M. F. Golde, and F. Kaufman, *J. Chem. Phys.*, **65**, 2715 (1976); D. K. Bohme, R. M. Hemsworth, H. W. Rundle, and H. I. Schiff, *ibid.*, **58**, 3504 (1973).

41. (a) A great deal of useful pK_a data in DMSO has been obtained by Bordwell and his collaborators; F. G. Bordwell and F. J. Cornforth, *J. Org. Chem.*, **43**, 1763 (1978); (b) for a review see *Solute−Solvent Interactions,* Vol. 2, J. F. Coetzee and C. D. Ritchie, Eds., M. Dekker, New York, 1976, Chap. 12.

42. For a discussion of the assumptions see C. V. Krishnan and H. L. Friedman in reference 41b. For tables of heats of hydration of ions see reference 32, p. 81; D. R. Rosseinsky, *Chem. Rev.*, **65**, 467 (1965); E. M. Arnett, L. E. Small, D. Oancea, and D. Johnston, *J. Am. Chem. Soc.*, **98**, 7346 (1976); D. H. Aue and M. T. Bowers, *Gas Phase Ion Chemistry*, Vol. 2, Academic, New York, 1979, Chap. 9.

43. E. M. Arnett, D. E. Johnston, and L. E. Small, *J. Am. Chem. Soc.*, **97**, 5598 (1975).

44. J. I. Brauman and L. K. Blair, *J. Am. Chem. Soc.*, **90**, 5636 (1968).

45. M. S. B. Munson, *J. Am. Chem. Soc.*, **87**, 2332 (1965); D. H. Aue, H. M. Webb, and M. T. Bowers, *J. Am. Chem. Soc.*, **98**, 311, 318 (1976).

46. J. L. Magee, T. Ri, and H. Eyring, *J. Chem. Phys.*, **9**, 419 (1941).

47. R. G. Pearson, *J. Chem. Phys.*, **20**, 1478 (1952).

48. R. G. Pearson and D. C. Vogelsong, *J. Am. Chem. Soc.*, **80**, 1038, 1048 (1958); C. G. Swain and E. E. Peques, *ibid.*, 812.

49. S. D. Ross and R. C. Peterson, *J. Am. Chem. Soc*, **80**, 2447 (1958); R. F. Hudson and I. Stelzer, *J. Chem. Soc. (B)*, **1966**, 775; J. E. Pardue and G. R. Dobson, *Inorg. Chim. Acta*, **20**, 207 (1976).

50. J. G. Kirkwood, *J. Chem. Phys.*, **2**, 351 (1934).

51. S. G. Smith, A. H. Fainberg, and S. Winstein, *J. Am. Chem. Soc.*, **83**, 618 (1961).

52. R. L. Heppolette and R. E. Robertson, *Proc. R. Soc. (London)*, **A252**, 273 (1959).

53. C. H. Langford and J. P. K. Tong, *Acc. Chem. Res.*, **10**, 258 (1977).

54. P. Debye, *Trans. Electrochem. Soc.*, **82**, 265 (1942); J. Q. Umberger and V. K. LaMer, *J. Am. Chem. Soc.*, **67**, 1099 (1945).

55. P. Langevin, *Ann. Chim. Phys.*, **5**, 245 (1905); L. Onsager, *J. Chem. Phys.*, **2**, 599 (1934).

56. M. Eigen and K. Tamm, *Z. Elektrochem.*, **66**, 107 (1962).

57. M. Eigen, *Ber. Bunsenges. Phys. Chem.*, **67**, 753 (1963); G. Geier, *ibid.*, **69**, 617 (1965).

58. N. Bjerrum, *K. Dan. Vidensk. Selsk. Mat. Fys. Medd.*, **9**, 7 (1926).

59. R. M. Fuoss, *J. Am. Chem. Soc.*, **80**, 5059 (1958); M. Eigen, *Z. Phys. Chem. (Neue Folge)*, **1**, 176 (1954).

60. See reference 32, Chap. 1.

61. J. A. Christiansen, *Z. Phys. Chem.*, **113**, 35 (1924).

62. W. F. K. Wynne-Jones and H. Eyring, *J. Chem. Phys.*, **3**, 492 (1935); K. J. Laidler and H. Eyring, *Ann. N. Y. Acad. Sci.*, **39**, 303 (1940).

63. G. Scatchard, *Chem. Rev.*, **10**, 229 (1932).

64. The reaction has been reviewed in A. A. Frost and R. G. Pearson, *Kinetics and Mechanism*, 2nd ed., Wiley, New York, 1961, Chap. 12.

65. See R. P. Bell and B. A. W. Coller, *Trans. Faraday Soc.*, **61**, 1445 (1965), for a further discussion of electrostatic effects in ester hydrolysis.

66. D. S. Kemp and K. Paul, *J. Am. Chem. Soc.*, **92**, 2553 (1970).

67. A. J. Parker, *Chem. Rev.*, **69**, 1, (1969).

68. (a) D. K. Bohme and L. B. Young, *J. Am. Chem. Soc.*, **92**, 7354 (1970); (b) W. N. Olmstead and J. I. Brauman, *ibid.*, **99**, 4219 (1977).

69. R. G. Pearson, H. Sobel, and J. Songstad, *J. Am. Chem. Soc.*, **90**, 319 (1968).

70. B. Jönsson, G. Karlström, and H. Wennerström, *J. Am. Chem. Soc.*, **100**, 1658 (1978).

71. E. A. Moelwyn-Hughes, *Proc. R. Soc. (London)*, **A196**, 540 (1949); D. N. Glew and E. A. Moelwyn-Hughes, *ibid.*, **A211**, 254 (1952).

72. C. D. Ritchie and M. Sawada, *J. Am. Chem. Soc.*, **99**, 3754 (1977).

73. See H. Schneider, reference 41b, Chap. 11.

74. S. Behrendt, C. H. Langford, and L. S. Frankel, *J. Am. Chem. Soc.*, **91**, 2236 (1969).

75. C. H. Langford and H. B. Gray, *Ligand Substitution Processes*, Benjamin, New York, 1965, Chap. 1.

76. (a) J. N. Brønsted, *Z. Phys. Chem.*, **102**, 169 (1922); (b) N. Bjerrum, *ibid.*, **108**, 82 (1924).

77. J. A. Christiansen, *Z. Phys. Chem.*, **113**, 35 (1924).

78. K. S. Pitzer, *Acc. Chem. Res.*, **10**, 371 (1977).

79. A. R. Olson and T. R. Simonson, *J. Chem. Phys.*, **17**, 348, 1167, 1322 (1949).

80. C. W. Davies, *Prog. React. Kinet.*, **1**, 161 (1961); see also B. Perlmutter-Hayman, *ibid.*, **6**, 240 (1971).

81. A useful book is C. W. Davies, *Ion Association*, Butterworths, London, 1962.

82. J. P. Collman, R. G. Finke, J. N. Cawse, and J. I. Brauman, *J. Am. Chem. Soc.*, **99**, 2515 (1977).

83. (a) E. Hückel, *Phys. Z.*, **26**, 93 (1925); (b) P. Debye and J. McAuley, *Phys. Z.*, **26**, 22 (1925).

84. L. C. Bateman, M. G. Church, E. D. Hughes, C. K. Ingold, and N. A. Taber, *J. Chem. Soc.*, **1940**, 979.

85. S. Winstein, S. Smith, and D. Darwish, *J. Am. Chem. Soc.*, **81**, 5511 (1959); A. H. Fainberg and S. Winstein, *ibid.*, **78**, 2763 (1956).

86. M. G. Evans and M. Polanyi, *Trans. Faraday Soc.*, **31**, 875 (1935).

87. An extensive compilation of results is given by T. Asano and W. J. LeNoble, *Chem. Rev.*, **78**, 407 (1978).

88. T. W. Swaddle, *Coord. Chem. Rev.*, **14**, 217 (1974).

89. C. H. Langford, *Inorg. Chem.*, **18**, 3288 (1979).

COMPLEX REACTIONS

Most reactions occurring in nature or in the laboratory do not take place at a single collision between reactant molecules, but have mechanisms that involve several such *elementary processes* or *reaction steps*. Such reactions are called *complex reactions*. If a mechanism involves a reactant that can undergo two or more reactions independently and concurrently, the reactions are called *parallel* or *side* reactions. If, on the other hand, there is a set of reactions such that the product of one reaction is the reactant for the next, the reactions are called *series* or *consecutive* reactions. Complex reaction mechanisms may involve various combinations of series and parallel reactions. Certain of these combinations are common enough to be given special names, for example, competitive reactions, reversible reactions, coupled reactions, chain reactions, and some types of catalysis such as general acid catalysis and Michaelis-Menten enzyme kinetics.

In this chapter we treat many of the important general types of complex reactions, giving examples of actual reactions as observed in the laboratory. The very important category of homogeneous catalytic reactions is the subject of the next chapter, and chain reactions are presented in Chapter 10. Heterogeneous catalysis can be discussed from the present viewpoint by making certain suppositions about possible intermediate compounds between reactants and the catalyst. However, because of its great complexity we do not treat heterogeneous catalysis.

As is described in Chapter 2, the most common means for obtaining rate constants from experimental data involves the integrated form of the rate law. For complex reactions integration is often more difficult than for the simple cases of Chapters 2 and 3. Integrated expressions are available for many possible mechanisms in addition to those treated here,[1,2] but in some cases — often those of great practical interest — closed-form integration of a rate expression is impossible. For example, to better understand combustion in a spark-ignited automobile engine Trumpy et al.[3] combined known thermodynamic and kinetic data with new experimental observations on combustion of ethane to conclude that more than 30 elementary reactions were required to explain observations at temperatures below 1300 K. Using computerized numerical integration of 29 coupled differential rate equations, Trumpy et al. predicted within ± 10% the observed time of engine knock following spark ignition and reproduced the main features of the pressure — temperature — time history of the combustion. Without the aid of a high-speed computer and advanced numerical methods it would be well-nigh impossible to begin to elucidate such complex mechanisms. Therefore we devote a portion of this chapter to numerical integration.

PARALLEL FIRST-ORDER REACTIONS

Let the mechanism be

$$A \xrightarrow{k_1} U \qquad (8.1a)$$

$$A \xrightarrow{k_2} V \qquad (8.1b)$$

$$A \xrightarrow{k_3} W \qquad (8.1c)$$

Then, using A, U, V, and W to represent the corresponding concentrations, A_0, and so on for initial concentrations, and $k = k_1 + k_2 + k_3$,

$$-\frac{dA}{dt} = k_1 A + k_2 A + k_3 A$$

$$= (k_1 + k_2 + k_3)A = kA \qquad (8.2)$$

and

$$\ln\left(\frac{A_0}{A}\right) = kt$$

or

$$A = A_0 e^{-kt} \qquad (8.3)$$

The reaction is simple first order as far as A is concerned. Also

$$\frac{dU}{dt} = k_1 A = k_1 A_0 e^{-kt}$$

and

$$U = \frac{-k_1 A_0}{k} e^{-kt} + \text{constant}$$

or

$$U = U_0 + \frac{k_1 A_0}{k}(1 - e^{-kt}) \qquad (8.4a)$$

$$V = V_0 + \frac{k_2 A_0}{k}(1 - e^{-kt}) \qquad (8.4b)$$

$$W = W_0 + \frac{k_3 A_0}{k}(1 - e^{-kt}) \qquad (8.4c)$$

If $U_0 = V_0 = W_0 = 0$, the equations simplify and

$$\frac{V}{U} = \frac{k_2}{k_1} \qquad \frac{W}{U} = \frac{k_3}{k_1} \tag{8.5}$$

The products are in a constant ratio to each other, independent of time and initial concentration of the reactant A, as shown in Fig. 8.1. Even if the rate of a set of reactions of this type is immeasurably fast, relative rate constants can be determined by measuring relative concentrations of products and using (8.5).

TWO PARALLEL FIRST-ORDER REACTIONS, PRODUCING A COMMON PRODUCT

$$A \xrightarrow{k_1} C + \ldots$$

$$B \xrightarrow{k_2} C + \ldots$$

Such reactions, of course, take place independently as far as the reactants are concerned. But, if the concentration of the common product C or the sum of the concentrations of A and B is measured experimentally, the task of finding k_1 and k_2 becomes more complex. This situation commonly arises in radioactivity when the radiation from mixed radioactive substances is observed. Observations of Brown and Fletcher[4] on hydrolysis of mixed tertiary aliphatic chlorides provide a chemical example.

The rate equations $-dA/dt = k_1 A$ and $-dB/dt = k_2 B$ integrate to

$$A = A_0 e^{-k_1 t} \quad \text{and} \quad B = B_0 e^{-k_2 t}$$

But, since $C_\infty = A_0 + B_0$

$$C = A_0 - A + B_0 - B$$

$$= C_\infty - A_0 e^{-k_1 t} - B_0 e^{-k_2 t}$$

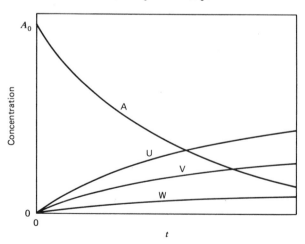

Figure 8.1. Concentration–time curves for parallel first-order reactions. A is the reactant; U, V, W are products.

so that

$$\log{(C_\infty - C)} = \log{(A_0 e^{-k_1 t} + B_0 e^{-k_2 t})}$$

If there were just one reactant A or B, or if $k_1 = k_2$, a plot of $\log{(C_\infty - C)}$ versus t would be linear. If, on the other hand, both A and B were present, $\log{(A + B)}$ or $\log{(C_\infty - C)}$ were plotted versus t, and $k_1 \neq k_2$ there would be a curve. The results of Brown and Fletcher on the hydrolysis of what was supposed to be diethyl-t-butylcarbinyl chloride are plotted in Fig. 8.2 and show such curvature. After some time the curve becomes linear because the more reactive component, say A, has disappeared and the expression for $\log{(C_\infty - C)}$ becomes

$$\log{B} = \log{(C_\infty - C)} = \log{B_0} - \frac{k_2 t}{2.303}$$

Therefore, from the slope and intercept of the line, B_0 and k_2 may be determined and B can be found at any time. Then A may be calculated by difference,

$$A = C_\infty - C - B$$

and plotting $\log{A}$ versus t yields A_0 and k_1. Figure 8.3 shows plots for A and B separately. These compounds are interpreted as being isomers that arise in the synthesis of the tertiary chloride. The relative values of A_0 and B_0 indicate that A is 35% and B is 65% of the original mixture.

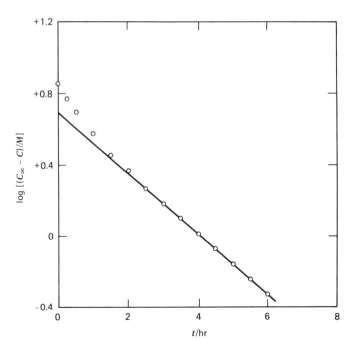

Figure 8.2. Deviation from a simple first-order plot for the hydrolysis of di-ethylbutyl-carbinyl chloride (Data from Brown and Fletcher.[4])

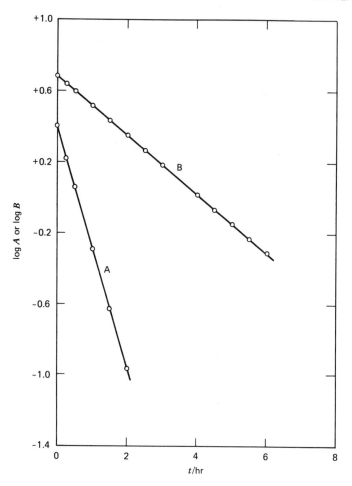

Figure 8.3. Analysis of reaction in Fig. 8.2 into two parallel hydrolyses of isomers. (Data from Brown and Fletcher.[4])

PARALLEL HIGHER-ORDER REACTIONS, ALL OF THE SAME ORDER

Suppose that the mechanism is:

$$aA + bB \xrightarrow{k_1} U + \ldots$$

$$aA + bB \xrightarrow{k_2} V + \ldots$$

$$aA + bB \xrightarrow{k_3} W + \ldots$$

Then

$$\frac{-dA}{dt} = (k_1 + k_2 + k_3)A^a B^b \qquad (8.6)$$

As far as reactants A and B are concerned, the complex reaction behaves kinetically as each of its components, but with an effective rate constant $k = k_1 + k_2 + k_3$.

Although the integrated expressions for the concentrations as a function of time may be complicated, the products appear in constant ratio despite the order of reaction; since

$$\frac{dU}{dt} = k_1 A^a B^b \quad \text{and} \quad \frac{dV}{dt} = k_2 A^a B^b \qquad (8.7)$$

then

$$\frac{dV}{dU} = \frac{k_2}{k_1} \quad \text{and} \quad \frac{V}{U} = \frac{k_2}{k_1} \qquad (8.8)$$

provided $U_0 = V_0 = 0$. It is therefore possible to determine relative rate constants, for the formation of isomers, for example, without knowing the order of the reaction. From the temperature dependence of these relative rate constants, differences of heats of activation and entropies of activation can be calculated. This is of interest in connection with the study of the effect of substituents on reaction rate.

PARALLEL FIRST- AND SECOND-ORDER REACTIONS

Hydrolysis of an organic halide may take place as an $S_N 1$ or first-order reaction or by an $S_N 2$ or second-order reaction. In certain cases both mechanisms may occur side by side.[5] Assuming the first-order mechanism

$$A \xrightarrow{k_1} D + E$$
$$E + B \xrightarrow{\text{fast}} C$$

and the second-order mechanism

$$A + B \xrightarrow{k_2} C + D$$

where A is the organic halide and B is hydroxide ion, the rate of reaction in terms of the extent of reaction variable x defined in (2.5) is

$$\frac{dx}{dt} = k_1 (a - x) + k_2 (a - x)(b - x)$$

where a and b are initial concentrations of A and B. The integrated expression is

$$\frac{1}{(k_1/k_2) + b - a} \ln \left[\frac{a}{(k_1/k_2) + b} \cdot \frac{(k_1/k_2) + b - x}{a - x} \right] = k_2 t$$

but this is not very useful for determining k_1 and k_2 or testing the fit to experimental data. Young and Andrews[6] applied the differential equation in the form

$$\frac{dx/dt}{a - x} = k_1 + k_2 (b - x)$$

to data on hydrolysis of primary butenyl chloride by plotting $(dx/dt)/(a - x)$ versus $(b - x)$.

Several cases are known where systems that at first appeared to involve parallel first- and second-order reactions were found to be second order only.[7] For example, Stein has reexamined racemization of 1-phenylbromoethane by lithium bromide in acetone, a reaction that was originally investigated by Hughes et al.[8] in a study that provided strong support for the S_N2 mechanism. Over a limited range of concentration of LiBr the kinetics appear to be parallel first and second order, but when a wide range of concentrations is studied the rate is found not to be linear in LiBr concentration. This effect is apparently due to incomplete dissociation of LiBr in acetone. When constant ionic strength is maintained by keeping the sum of concentrations of LiBr and $LiNO_3$ equal to $0.1 M$, a plot of rate of racemization versus concentration of LiBr is linear and passes through the origin.

SERIES FIRST-ORDER REACTIONS

Consider the case

$$A \xrightarrow{k_1} B$$
$$B \xrightarrow{k_2} C$$

This mechanism is exemplified by certain hydrolyses, by radioactive series, and by the reaction of potassium permanganate, manganous sulfate, and oxalic acid, as in the original work of Harcourt and Esson. Esson[9] first integrated the differential equations, which are:

$$\frac{dA}{dt} = -k_1 A \tag{8.9a}$$

$$\frac{dB}{dt} = k_1 A - k_2 B \tag{8.9b}$$

$$\frac{dC}{dt} = k_2 B \tag{8.9c}$$

The equation for A readily integrates to

$$A = A_0 e^{-k_1 t} \tag{8.10}$$

This substituted into (8.9b) yields

$$\frac{dB}{dt} = k_1 A_0 e^{-k_1 t} - k_2 B \tag{8.11}$$

This linear first-order equation may be integrated by the usual integrating factor methods to give (if $B_0 = 0$):

$$B = \frac{A_0 k_1}{k_2 - k_1} (e^{-k_1 t} - e^{-k_2 t}) \tag{8.12}$$

C may be found most conveniently from the stoichiometric relationship[†] obtained by integrating the sum of (8.9a), (8.9b) and (8.9c):

$$\frac{dA}{dt} + \frac{dB}{dt} + \frac{dC}{dt} = 0 \tag{8.13}$$

Therefore

$$A + B + C = A_0 \tag{8.14}$$

if $B_0 = C_0 = 0$. Then

$$C = A_0 - A - B$$

$$= A_0 \left[1 + \frac{1}{k_1 - k_2} (k_2 e^{-k_2 t} - k_1 e^{-k_1 t})\right] \tag{8.15}$$

These integrations may also be performed by the general determinantal method discussed in a later section. Figure 8.4 shows concentrations of A, B, and C as a function of time for the typical case where $k_1 = 0.1$ s^{-1}, $k_2 = 0.05$ s^{-1}.

Equations (8.10), (8.12), and (8.15) can be put in simpler form by introducing dimensionless parameters. Let $\alpha = A/A_0$, $\beta = B/A_0$, $\gamma = C/A_0$. These are concentrations relative to the initial value A_0 and can vary from 0 to 1. Also let $\tau = k_1 t$ and $\kappa = k_2/k_1$. Then the equations become

$$\alpha = e^{-\tau} \tag{8.16}$$

$$\beta = \frac{1}{\kappa - 1} (e^{-\tau} - e^{-\kappa\tau}) \tag{8.17}$$

$$\gamma = 1 + \frac{1}{1 - \kappa} (\kappa e^{-\tau} - e^{-\kappa\tau}) \tag{8.18}$$

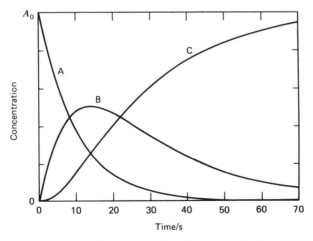

Figure 8.4. Concentration–time curve for substances A, B, and C in series first-order reactions.

[†] See reference 103 for a discussion of the relationship between mechanism and stoichiometry.

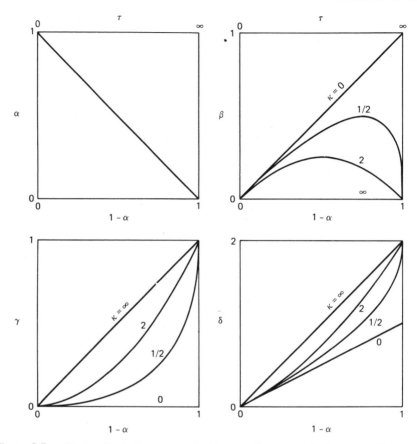

Figure 8.5. Plots of relative concentrations α, β, γ, and δ for series first-order reactions as a function of $1 - \alpha$ for various values of κ, the relative rate constant.

To visualize the concentration–time relationships most effectively it is useful to consider plots of α, β, and γ as functions of τ for various values of κ as shown in Fig. 8.5. Instead of using a linear scale for τ, $1 - e^{-\tau}$ or $1 - \alpha$ is plotted. This shows the whole range of time from 0 to ∞ as $1 - \alpha$ goes from 0 to 1, and causes certain relationships to be linear. Note that the concentration of the intermediate B, as measured by β, goes through a maximum whose position depends on κ, the relative value of the rate constants. By using (8.17) and setting $d\beta/d\tau = 0$ it is found that

$$\tau_{\max} = \frac{1}{\kappa - 1} \ln \kappa$$

The maximum shifts to smaller τ as κ gets larger and larger. Also the value of β at the maximum is

$$\beta_{\max} = \kappa^{\kappa/(1-\kappa)} \tag{8.19}$$

which shows that the maximum becomes less pronounced as κ gets larger.

The rate of formation of C is slow at first. This is an example of an *induction period*. The duration of the induction period, arbitrarily taken as the time to reach the inflection point on the C versus t curve of Fig. 8.4, equals the time for B to reach its maximum value since, from (8.9), when $d^2C/dt^2 = 0$, $dB/dt = 0$.

Jensen and Lenz[10] describe a situation where observation of an induction period for formation of C led to the unexpected detection of an intermediate B. The reaction is hydrolysis of acetals of benzaldehydes

$$\underset{\overset{|}{ArCHOR_1}}{\overset{OR_2}{|}} + H_2O \xrightarrow{k_1} \underset{\overset{|}{ArCHOR_1}}{\overset{OH}{|}} + R_2OH \qquad (8.20a)$$

$$\underset{\overset{|}{ArCHOR_1}}{\overset{OH}{|}} \xrightarrow{k_2} ArCHO + R_1OH \qquad (8.20b)$$

Reaction (8.20a) is first order in H^+ as well as in the acetal; (8.20b) is catalyzed by either H^+ or OH^- and is first order in hemiacetal. Under conditions of constant pH in aqueous solution, (8.20a) and (8.20b) are series pseudo-first order and can be followed by spectrophotometric observation of the benzaldehyde formed in (8.20b). Jensen and Lenz observed that the usual first-order plot deviated from linearity during the early part of the reaction (see Fig. 8.6), indicating an induction period. That the mechanism is given by (8.20) is supported by the fact that if the reaction is quenched after about 10% completion by rapidly injecting base, there is a rapid increase in absorbance. This is presumably due to rapid, base-catalyzed conversion of hemiacetal to benzaldehyde according to (8.20b). Combining the absorbances before and after quenching with the absorbance upon completion of the reaction, Jensen and Lenz obtained [hemiacetal]/[acetal] at t_q, the time of quenching, for various t_q. From (8.9b), when $dB/dt = 0$, that is, when B reaches its maximum value where B is [hemiacetal],

$$\frac{[\text{hemiacetal}]}{[\text{acetal}]} = \frac{k_1}{k_2}$$

Since k_1 can be obtained from the slope of line A of Fig. 8.6, this relationship serves to determine k_2. The same value of k_2 can be obtained by extrapolating line A of Fig. 8.6 to zero time, obtaining the difference between observed $A_\infty - A$ values and line A, and plotting that difference versus time. This results in line B of Fig. 8.6. Jensen and Lenz conclude on the basis of this type of analysis that there are many acetal hydrolyses for which k_2 is small enough relative to k_1 so that there is a significant buildup of hemiacetal.

Included in Fig. 8.5 is a plot of the function δ, which is defined as $\delta = \beta + 2\gamma$.

$$\delta = 2 - \frac{1 - 2\kappa}{1 - \kappa} e^{-\tau} - \frac{1}{1 - \kappa} e^{-\kappa\tau} \qquad (8.21)$$

As the reaction proceeds δ varies from 0 to 2. It is more closely related to what is usually measured in a reaction than is β or γ alone. In the hydrolysis of a dihalide δ measures the amount of halide ion produced; in the saponification of a diester with a large excess of base, δ measures the amount of base consumed relative to the original diester. In general, whenever each step of the reaction

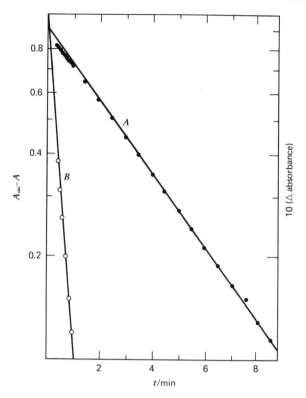

Figure 8.6. Plot of $\log(A_\infty - A)$ versus time for p-methylbenzaldehyde diethyl acetal in acetate buffer (ionic strength $= 0.5\,M$ KCl). Line A is experimental data. Line B is $10 \cdot (\Delta$ absorbance) versus time, where Δ absorbance is the difference between line A (extrapolated from the second and third half-lives) and experimental data recorded early in the reaction [Reprinted with permission from J. L. Jensen and P. A. Lenz,[10] *J. Am. Chem. Soc.,* **100,** 1291 (1978).]

produces the same change in some reactant or by-product, δ is a useful measure of the extent of reaction.

The practical problem of determination of rate constants from experimental data has been handled in some detail by Swain.[11a] However, there are two alternatives that are more convenient. The first is an extension of the Powell plot[12] discussed in Chapter 2. Plotting percent reaction, as given by 50δ, versus $\log \tau$ gives the family of curves shown in Fig. 8.7, each identified by a particular value of κ. Since $\tau = k_1 t$ and $\log \tau = \log k_1 + \log t$, a plot of experimental values of 50δ versus $\log t$ has the same shape as one of the curves in Fig. 8.7, but is shifted horizontally by $-\log k_1$. Besides determining k_1 from the shift of $\log t$ with respect to $\log \tau$ and k_2 from k_1 and κ, this technique provides convincing evidence that the reaction is series first order. It is unlikely that a similar plot of some other complex mechanism would show just the same form as one of the family of curves calculated from (8.21).

A time-ratio method originally derived to determine rate constants of competing

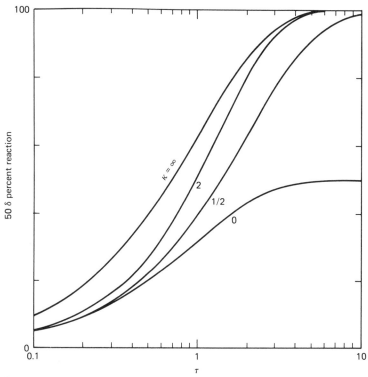

Figure 8.7. Percent reaction versus time parameter τ on logarithmic scale for series first-order reactions.

second-order reactions[13] is of general applicability. Applied to a series first-order reaction it gives more accurate results than the Powell plot. Let τ_1 and τ_2 correspond to δ_1 and δ_2, which are determined by certain percentages of reaction. The ratio τ_2/τ_1 equals t_2/t_1, the ratio of actual times to reach those percentages. But (8.21) shows that τ_2/τ_1, and therefore also the time ratio t_2/t_1, depends only on κ for fixed δ_2, δ_1. Therefore a graph or table of values of κ versus t_2/t_1 can be used to evaluate κ, following which a graph or table of κ versus τ for a particular δ will yield k_1. This result should be verified by using additional pairs of times for other combinations of percentage reaction. Table 8.1 gives values of τ and time ratios for various κ's at 15, 35, and 70% reaction, that is, for $\delta = 0.30$, 0.70, and 1.40. Applied to the data of Kaufler[11b] on hydrolysis of 2, 7-dicyanonaphthalene the time ratio method gives the same result obtained by Swain.[11a]

Alcock, Benton, and Moore[14] describe a computer program for least-squares analysis of spectrophotometric data from series first-order reactions. Usually the extinction coefficient of the intermediate B is unknown, and so three parameters, k_1, k_2, and ϵ_B must be fitted to the experimental data. Alcock, Benton, and Moore found that in general two sets of k_1, k_2, and ϵ_B gave equally good fits to the observations, although in some cases negative or extremely large parameters serve to indicate the physically reasonable solution. These two sets correspond to either rapid formation of a weakly absorbing intermediate or slow formation of

Table 8.1. Series first-order reactions. Time–percentage reaction relations for various relative rate constants

κ	τ_{15}	τ_{35}	τ_{70}	$\log t_{35}/t_{15}$	$\log t_{70}/t_{15}$	$\log t_{70}/t_{35}$
100	0.168	0.436	1.21	0.415	0.858	0.443
50	0.172	0.441	1.21	0.407	0.847	0.440
20	0.188	0.457	1.23	0.385	0.815	0.430
10	0.209	0.484	1.26	0.366	0.781	0.415
5	0.236	0.536	1.32	0.356	0.748	0.392
2	0.277	0.664	1.54	0.367	0.745	0.378
1.5	0.289	0.686	1.65	0.376	0.757	0.381
1.1	0.300	0.734	1.80	0.388	0.779	0.391
0.9	0.308	0.766	1.93	0.395	0.796	0.401
0.7	0.315	0.806	2.11	0.409	0.826	0.417
0.5	0.324	0.863	2.41	0.425	0.871	0.446
0.2	0.342	0.999	3.81	0.465	1.047	0.582
0.1	0.349	1.078	6.19	0.490	1.249	0.759
0.05	0.353	1.132	11.10	0.506	1.497	0.991
0.02	0.355	1.173	26.55	0.519	1.874	1.355
0.01	0.356	1.188	52.09	0.524	2.166	1.642

Source. Modified from Swain.[11]

a strongly absorbing intermediate. An important corollary is that a rapid increase in absorbance followed by a slower decrease does not necessarily imply that the first reaction step (governed by k_1) is fast and the second step is slow. If k_1 and k_2 are nearly equal, the least-squares refinement may oscillate between the two solutions and never converge.

GENERAL FIRST-ORDER SERIES AND PARALLEL REACTIONS

Consider a set of m substances with concentrations $A_1, A_2, \ldots, A_m$, and suppose that each substance may react by a first-order process to form each of the other substances. Let the rate constants be k_{ij}, where the first subscript refers to the substance reacting and the second to the substance formed. Assume that one molecule is produced for each one reacting. For $m = 4$ the situation would be as shown in Fig. 8.8.

The m rate equations are then of the form

$$\frac{dA_i}{dt} = -k_{i1}A_i - k_{i2}A_i \ldots + k_{1i}A_1 + k_{2i}A_2 + \ldots$$

or

$$\frac{dA_i}{dt} = (-k_{i1} - k_{i2} \ldots - k_{im})A_i + k_{1i}A_1 + k_{2i}A_2 + \ldots k_{mi}A_m \quad (8.22)$$

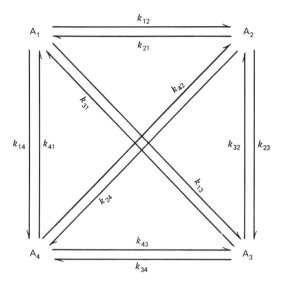

Figure 8.8.   General first-order series/parallel reaction scheme for four species.

This set of m differential equations in m dependent variables A_i may be integrated in closed form.[15] The notation of Matsen and Franklin is used here.

Equations (8.22) may be put into the form

$$\frac{dA_i}{dt} + \sum_{j=1}^{m} K_{ij}A_j = 0 \tag{8.23}$$

where $K_{ij} = -k_{ji}$ except when $i = j$, and

$$K_{ii} = \sum_{p=1}^{m} k_{ip}$$

except when $p = i$.

Now assume a *particular* solution of the form

$$A_i = B_i e^{-\lambda t} \tag{8.24}$$

where the B_i are constants and λ is a parameter to be determined. Substitute (8.24) into (8.23) and cancel out the exponential to obtain the set of simultaneous homogeneous linear equations in the B_i:

$$-\lambda B_i + \sum_{j=1}^{m} K_{ij}B_j = 0 \tag{8.25}$$

or

$$\sum_{j=1}^{m} (K_{ij} - \delta_{ij}\lambda)B_j = 0 \qquad i = 1, 2, \ldots, m \tag{8.26}$$

where $\delta_{ij} = 1$ if $i = j$ or 0 if $i \neq j$. The condition for a nontivial solution for the B_i is that the determinant of the coefficients is zero.

$$|K_{ij} - \delta_{ij}\lambda| = 0 \tag{8.27}$$

This determinantal equation, or secular equation, is an m th degree algebraic equation in the parameter λ. It has as solutions m values of λ. Let these be λ_r, where $r = 1, 2, \ldots, m$. Corresponding to each λ_r there is a set of relative values B_j that are solutions of (8.26). Let these B_j's be designated B_{jr}.

These particular solutions are not usually the solutions desired since they do not satisfy the initial conditions of concentration. It is necessary to have a *general* solution, which may be written as a linear combination of particular solutions. Let

$$A_i = \sum_{r=1}^{m} B_{ir} Q_r^0 e^{-\lambda_r t} \tag{8.28}$$

where Q_r^0 are coefficients in the linear combination and may be determined from the initial conditions. Any linear combination of particular solutions is also a solution of the differential equations (8.23), since the latter are linear equations.

As an example, consider two consecutive first-order reactions

$$A_1 \underset{k_{21}}{\overset{k_{12}}{\rightleftarrows}} A_2 \underset{k_{32}}{\overset{k_{23}}{\rightleftarrows}} A_3 \tag{8.29}$$

This situation has been discussed by several authors.[16,17] The secular equation is

$$\begin{vmatrix} k_{12} - \lambda & -k_{21} & 0 \\ -k_{12} & k_{21} + k_{23} - \lambda & -k_{32} \\ 0 & -k_{23} & k_{32} - \lambda \end{vmatrix} = 0 \tag{8.30a}$$

or when expanded

$$-\lambda^3 + \lambda^2 (k_{12} + k_{21} + k_{23} + k_{32}) - \lambda (k_{12} k_{23} + k_{21} k_{32} + k_{12} k_{32}) = 0 \tag{8.30b}$$

The three solutions for λ_r are

$$\lambda_1 = 0 \tag{8.31a}$$

$$\lambda_2 = \frac{1}{2}(p + q) \tag{8.31b}$$

$$\lambda_3 = \frac{1}{2}(p - q) \tag{8.31c}$$

where $p = (k_{12} + k_{21} + k_{23} + k_{32})$ and $q = [p^2 - 4(k_{12} k_{23} + k_{21} k_{32} + k_{12} k_{32})]^{1/2}$. Substituting these λ_r into (8.26) gives three equations, any two of which are independent and which can be solved for B_{2r} and B_{3r}, making the arbitrary assumption that the various B_{1r} are unity. Alternatively, the B_{2r} or B_{3r} could be set equal to unity. Letting

$$B_{1r} = 1 \tag{8.32a}$$

$$B_{2r} = \frac{k_{12} - \lambda_r}{k_{21}} \tag{8.32b}$$

$$B_{3r} = \frac{k_{23}(k_{12} - \lambda_r)}{k_{21}(k_{32} - \lambda_r)} \tag{8.32c}$$

the general solution is then

$$A_1 = \sum_r Q_r^0 e^{-\lambda_r t} \tag{8.33a}$$

$$A_2 = \sum_r Q_r^0 \frac{(k_{12} - \lambda_r)}{k_{21}} e^{-\lambda_r t} \tag{8.33b}$$

$$A_3 = \sum_r Q_r^0 \frac{k_{23}(k_{12} - \lambda_r)}{k_{21}(k_{32} - \lambda_r)} e^{-\lambda_r t} \tag{8.33c}$$

Suppose that, at time $t = 0$, $A_1 = A_1^0$ and $A_2 = A_3 = 0$, that is, only the first reactant is present initially. Then (8.33a)–(8.33c) become

$$A_1^0 = \sum_r Q_r^0 \tag{8.34a}$$

$$0 = \sum_r Q_r^0 \frac{k_{12} - \lambda_r}{k_{21}} \tag{8.34b}$$

$$0 = \sum_r Q_r^0 \frac{k_{23}(k_{12} - \lambda_r)}{k_{21}(k_{32} - \lambda_r)} \tag{8.34c}$$

These when solved for Q_r^0 result in

$$Q_1^0 = A_1^0 \frac{k_{21} k_{32}}{\lambda_2 \lambda_3} \tag{8.35a}$$

$$Q_2^0 = A_1^0 \frac{k_{12}(\lambda_2 - k_{23} - k_{32})}{\lambda_2(\lambda_2 - \lambda_3)} \tag{8.35b}$$

$$Q_3^0 = A_1^0 \frac{k_{12}(k_{23} + k_{32} - \lambda_3)}{\lambda_3(\lambda_2 - \lambda_3)} \tag{8.35c}$$

And finally, by substituting into (8.33) and simplifying

$$A_1 = A_1^0 \left[\frac{k_{21} k_{32}}{\lambda_2 \lambda_3} + \frac{k_{12}(\lambda_2 - k_{23} - k_{32})}{\lambda_2(\lambda_2 - \lambda_3)} e^{-\lambda_2 t} \right.$$
$$\left. + \frac{k_{12}(k_{23} + k_{32} - \lambda_3)}{\lambda_3(\lambda_2 - \lambda_3)} e^{-\lambda_3 t} \right] \tag{8.36a}$$

$$A_2 = A_1^0 \left[\frac{k_{12} k_{32}}{\lambda_2 \lambda_3} + \frac{k_{12}(k_{32} - \lambda_2)}{\lambda_2(\lambda_2 - \lambda_3)} e^{-\lambda_2 t} \right.$$
$$\left. + \frac{k_{12}(\lambda_3 - k_{32})}{\lambda_3(\lambda_2 - \lambda_3)} e^{-\lambda_3 t} \right] \tag{8.36b}$$

$$A_3 = A_1^0 \left[\frac{k_{12} k_{23}}{\lambda_2 \lambda_3} + \frac{k_{12} k_{23}}{\lambda_2(\lambda_2 - \lambda_3)} e^{-\lambda_2 t} \right.$$
$$\left. - \frac{k_{12} k_{23}}{\lambda_3(\lambda_2 - \lambda_3)} e^{-\lambda_3 t} \right] \tag{8.36c}$$

These tedious calculations can be shortened by using matrix methods.[17c] Lowry and John[16] point out that the concentration of the intermediate A_2 may or may not go through a maximum, depending on the relative values of the rate constants. If A_2 does go through a maximum, the ratio $\kappa = k_{23}/k_{12}$ may be obtained from (8.19). If the equilibrium constants $K_1 = k_{12}/k_{21}$ and $K_2 = k_{23}/k_{32}$ are also known, (8.36a)–(8.36c) can be expressed in terms of a single rate constant, and that constant can be obtained by a graphical method.[17a]

The problem of determining rate constants for the general case of first-order series and parallel reactions where none of the initial concentrations is zero is extremely difficult. If concentrations of all components can be measured at suitable times, the problem can be solved.[18] Wei and Prater[19] have described a method based on the structure of the reacting system by which rate constants may be evaluated from concentrations measured at unspecified times. McLaughlin and Rozett[20] use Laplace-Carson transforms to eliminate differentials from the rate equations for first-order series and parallel reactions involving up to four components, obtaining a set of simultaneous, nonhomogeneous linear equations that can be solved by computer.

COMPETITIVE, CONSECUTIVE SECOND-ORDER REACTIONS

Consider the complex reaction

$$A + B \xrightarrow{k_1} C + E$$

$$A + C \xrightarrow{k_2} D + E$$

with

$$\frac{dA}{dt} = -k_1 AB - k_2 AC \tag{8.37}$$

$$\frac{dB}{dt} = -k_1 AB \tag{8.38}$$

$$\frac{dC}{dt} = k_1 AB - k_2 AC \tag{8.39}$$

$$\frac{dD}{dt} = k_2 AC \tag{8.40}$$

By material balance there results

$$A - 2B - C = A_0 - 2B_0 \tag{8.41}$$

$$B + C + D = B_0 \tag{8.42}$$

where only A and B are present initially.

This situation is quite common, a typical example being saponification of a symmetrical diester such as ethyl succinate. A would represent hydroxide ion, B the diester, and C the half-saponified or monoester. The pair of reactions is series with respect to B and C but parallel with respect to A; hence the name competitive, consecutive reactions. This complex reaction is treated here in some detail, because it illustrates several methods that may be applied to other cases.

If experimental conditions are adjusted so that the reaction is pseudo-first order, for example, by using a large excess of A, the integration becomes equivalent to that discussed earlier in this chapter for two successive first-order reactions.[21] For the general case of conditions where the reaction steps are second order and k_1 and k_2 may take any values, Ingold[22] was the first to devise a successful procedure to determine k_1 from experimental results, but his method was based on approximations and was difficult to apply beyond 30% reaction. Frost and Schwemer[13] have provided a general solution. Assume, for convenience, that equivalent amounts of reactants are present initially, that is, that $A_0 = 2B_0$. Then $C = A - 2B$ and (8.37) becomes

$$\frac{dA}{dt} = (2k_2 - k_1)AB - k_2 A^2 \tag{8.43}$$

Then introduce dimensionless variables α, β, τ, and a dimensionless parameter κ.

$$\alpha = \frac{A}{A_0} \qquad \beta = \frac{B}{B_0} \qquad \tau = B_0 k_1 t \qquad \kappa = \frac{k_2}{k_1} \tag{8.44}$$

Equations (8.43) and (8.38) become

$$\frac{d\alpha}{d\tau} = (2\kappa - 1)\alpha\beta - 2\alpha^2 \tag{8.45}$$

$$\frac{d\beta}{d\tau} = -2\alpha\beta \tag{8.46}$$

Dividing (8.45) by (8.46) yields

$$\frac{d\alpha}{d\beta} = \frac{1 - 2\kappa}{2} + \frac{\kappa\alpha}{\beta} \tag{8.47}$$

which is a linear differential equation with the solution

$$\alpha = \frac{1 - 2\kappa}{2(1 - \kappa)}\beta + \frac{1}{2(1 - \kappa)}\beta^\kappa \tag{8.48}$$

where the constant of integration is determined by the initial condition that when $\alpha = 1, \beta = 1$. Substituting (8.48) in (8.46) and integrating gives

$$\tau = \frac{1 - \kappa}{1 - 2\kappa} \int_\beta^1 \frac{d\beta}{\beta^2 \{1 + [1/(1 - 2\kappa)]\beta^{\kappa - 1}\}} \tag{8.49}$$

The integral in (8.49) can be evaluated in closed form for any κ that is a rational number, that is, a ratio of integers. In practice only ratios of small integers are conveniently handled, but with κ values sufficiently close interpolation can be used for others.

Certain special values of κ give simple integrals and simple second-order kinetic expressions, at least as far as α is concerned. These correspond to limiting cases that are easily solved without recourse to Frost and Schwemer's treatment.

For $\kappa = 0, k_1 \gg k_2$

$$\tau = \ln\left(\frac{\beta + 1}{2\beta}\right) \quad \text{and} \quad \alpha = \frac{\beta + 1}{2} \tag{8.50}$$

Here the first step is much faster than the second and is essentially complete before the second one starts.[23] Each phase of the reaction can be treated separately as a simple second-order reaction. For example, in the saponification of oxalic acid esters the first step is so fast with excess alkali that it is more convenient to hold the hydroxide ion concentration constant and small with a buffer.[24]

For $\kappa = \infty, k_1 \ll k_2$

$$\tau = \frac{1}{2}\left(\frac{1}{\beta} - 1\right) \quad \text{and} \quad \alpha = \beta \tag{8.51}$$

Here the second step is a rapid follow-up to the first. The observed rate is that of a single second-order process, but 2 mol of A react for each mole of B. An example is the hydrolysis of an acetal, the hemiacetal reacting much more rapidly. However, in this example A is water, which is in large excess, causing the kinetics to be pseudo-first order.

For $\kappa = \frac{1}{2}, k_1 = 2k_2$

$$\tau = \frac{1}{\alpha} - 1 \quad \text{and} \quad \alpha^2 = \beta \tag{8.52}$$

This is approximately true for the saponification of glycerol diacetate.[25] For this combination of k's the reaction appears to be simple second order as far as the concentration of A is concerned.

When none of these special cases obtains, a Powell plot or the time-ratio method may be used to obtain rate constants. Both of these are described earlier in this chapter for series first-order reactions. Frost and Schwemer give appropriate plots of α versus $\log \tau$ and of β versus $\log \tau$.[13] Some results of calculating τ as a function of α for various κ's are shown in Table 8.2. From these can be calculated the time

Table 8.2. τ as a function of κ and α

$1/\kappa$	$\alpha = 0.8$ 20% reaction	$\alpha = 0.7$ 30% reaction	$\alpha = 0.6$ 40% reaction	$\alpha = 0.5$ 50% reaction	$\alpha = 0.4$ 60% reaction
2.0	0.2500	0.4286	0.6667	1.000	1.500
3.0	0.2599	0.4564	0.7305	1.133	1.770
4.0	0.2656	0.4741	0.7756	1.239	2.011
5.0	0.2693	0.4865	0.8098	1.327	2.235
6.0	0.2720	0.4957	0.8368	1.404	2.449
7.0	0.2740	0.5028	0.8589	1.471	2.657
8.0	0.2755	0.5085	0.8773	1.531	2.862
9.0	0.2768	0.5131	0.8929	1.586	3.066
10.0	0.2778	0.5170	0.9064	1.637	3.270

Table 8.3.   Time ratios as a function of κ[a]

$1/\kappa$	t_{60}/t_{20}	t_{60}/t_{30}	t_{60}/t_{40}	t_{60}/t_{50}	t_{50}/t_{20}	t_{50}/t_{30}
2.0	6.000	3.500	2.250	1.500	4.000	2.333
3.0	6.812	3.878	2.423	1.562	4.362	2.483
4.0	7.571	4.241	2.592	1.623	4.666	2.614
5.0	8.297	4.593	2.760	1.684	4.928	2.728
6.0	9.003	4.940	2.927	1.745	5.161	2.832
7.0	9.698	5.285	3.094	1.806	5.369	2.925
8.0	10.388	5.629	3.263	1.869	5.558	3.012
9.0	11.078	5.975	3.434	1.933	5.731	3.091
10.0	11.772	6.325	3.607	1.998	5.892	3.166

[a] The ratio t_{60}/t_{20} is the ratio of time for 60% of A reacting to time for 20% reacting, or where $\alpha = 0.4$ and 0.8, respectively.

ratios of Table 8.3, which can then be used for accurate determination of κ by finding various time ratios in a single run. Once κ is determined, Table 8.2 provides values of τ that can be divided by $B_0 t$ to yield k_1, and multiplying $k_1 \times \kappa$ gives k_2 (8.44). Table 8.4 shows how this method was used to obtain k_1 and k_2 from a single run in the saponification of ethyl adipate. Burkus and Eckert[26] have applied this treatment to data on the reactions of diisocyanates, and Reikhsfeld and Prokhorova[27] have applied it to reactions of monoorganosilanes with alcohols.

Wideqvist[28] has developed a graphical-integration time-variable transformation method[29] so that calculations can be made for nonstoichiometric amounts of reactants. Also, he has shown how to treat the data statistically. McMillan[30] has published a relatively simple method for determining the ratio of rate constants κ if the simultaneous concentrations of two substances are known. This method is not restricted to stoichiometric amounts of reactants. Svirbely and coworkers[31] have

Table 8.4.   Calculations of rate constants for the saponification of ethyl adipate[a]

% reaction	t/s	Percentages compared	t ratio	$1/\kappa$ [b]
20	605	60/20	6.661	2.808
30	1060	60/30	3.802	2.795
40	1690	60/40	2.385	2.777
50	2595	60/50	1.553	2.850
50	4030	50/20	4.289	2.785
		50/30	2.448	2.750

% reaction	τ	$k_1/M^{-1}\,s^{-1}$ [c]
20	0.2582	0.08570
30	0.4516	0.08555
40	0.7189	0.08542
50	1.108	0.08574
60	1.716	0.08550

[a] $B_0 = 0.00996\,M$; temperature $= 25.0°C$.
[b] Average $1/\kappa = 2.79$.
[c] Average $k_1 = 0.0856\,M^{-1}\,s^{-1}$; $k_2 = 0.0856/2.79 = 0.0307\,M^{-1}\,s^{-1}$.

extended the treatment of competitive, consecutive second-order reactions to schemes having many more than two steps. They used a computer program involving a least-squares fitting procedure to obtain rate constants for the seven steps in alkaline hydrolysis of triethyl citrate.

REVERSIBLE REACTIONS

For a number of simple reversible reactions the rate law is easily integrated. The derivations that follow make use of the extent of reaction variable x defined in (2.5). Initial concentrations of species A, B, ... are represented by $a, b, \ldots$, and the equilibrium value of x by x_e.

First order

Let

$$A \underset{k'}{\overset{k}{\rightleftharpoons}} B$$

Then

$$\frac{dx}{dt} = k(a-x) - k'(b+x) \tag{8.53}$$

Rearranging and integrating,

$$\int_0^x \frac{dx}{ka - k'b - (k+k')x} = \int_0^t dt$$

$$\ln \frac{ka - k'b}{ka - k'b - (k+k')x} = (k+k')t \tag{8.54}$$

Equations involving approach to equilibrium usually simplify when the equilibrium concentration is introduced. For this case by setting $dx/dt = 0$

$$k(a - x_e) = k'(b + x_e) \tag{8.55}$$

that is, the rates of forward and reverse reactions are equal. Solving (8.55) for $ka - k'b$ and substituting in (8.54) yields

$$\ln \frac{x_e}{x_e - x} = (k + k')t \tag{8.56}$$

Thus the approach to equilibrium is a first-order process with effective rate constant equal to the sum of the constants for the forward and reverse directions. In practice a plot of $\ln(x_e - x)$ versus time [or of $\ln(\lambda_e - \lambda)$ versus time, where λ represents a physical property proportional to concentration] is linear and has a slope equal to $-(k + k')$. If the equilibrium constant K is known independently, k and k' can be determined since

$$K = \frac{B_e}{A_e} = \frac{b + x_e}{a - x_e} = \frac{k}{k'} \tag{8.57}$$

First and Second Order

Suppose that

$$A \underset{k'}{\overset{k}{\rightleftharpoons}} B + C$$

Then

$$\frac{dx}{dt} = k(a-x) - k'(b+x)(c+x) \tag{8.58}$$

and if only A is present initially so that $b = c = 0$

$$\frac{dx}{dt} = k(a-x) - k'x^2 \tag{8.59}$$

At equilibrium $dx/dt = 0$ and from (8.59)

$$k' = \frac{k(a-x_e)}{x_e^2} \tag{8.60}$$

so that

$$\frac{dx}{dt} = k(a-x) - \frac{k(a-x_e)x^2}{x_e^2} \tag{8.61}$$

This may be integrated by the method of partial fractions (see Chapter 2) to give

$$\ln \frac{ax_e + x(a-x_e)}{a(x_e - x)} = k \frac{2a - x_e}{x_e} t \tag{8.62}$$

The result for the case where the forward reaction is second order and the reverse is first order is given in Problem 8.4.

Second-order

Let

$$A + B \underset{k'}{\overset{k}{\rightleftharpoons}} C + D \tag{8.63}$$

and

$$\frac{dx}{dt} = k(a-x)(b-x) - k'(c+x)(d+x) \tag{8.64}$$

This can be integrated by writing (8.64) as

$$\frac{dx}{\alpha + \beta x + \gamma x^2} = dt$$

and introducing

$$q^2 = \beta^2 - 4\alpha\gamma \tag{8.65}$$

The integrated form is[32]

$$\ln \frac{x + (\beta - q)/2\gamma}{x + (\beta + q)/2\gamma} - \ln \frac{\beta - q}{\beta + q} = qt \tag{8.66}$$

If only A and B are present initially,

$$\frac{dx}{dt} = k(a - x)(b - x) - k'x^2 \tag{8.67}$$

and in terms of the equilibrium extent of reaction x_e,

$$\frac{dx}{dt} = k(a - x)(b - x) - k\frac{(a - x_e)(b - x_e)}{x_e^2}x^2 \tag{8.68}$$

This can be integrated by partial fractions to give

$$\ln \left\{ \frac{x[ab - (a + b)x_e] + ab\, x_e}{ab(x_e - x)} \right\} = k\frac{2ab - (a + b)x_e}{x_e}t \tag{8.69}$$

If the initial concentrations of A and B are equal, (8.67) becomes

$$\frac{dx}{dt} = k(a - x)^2 - k'x^2 \tag{8.70}$$

and (8.69) simplifies to

$$\ln \left[\frac{x(a - 2x_e) + ax_e}{a(x_e - x)} \right] = k\frac{2a(a - x_e)}{x_e}t \tag{8.71}$$

Equations (8.70) and (8.71) also apply to a reversible second-order reaction of the form

$$2W \underset{k_r}{\overset{k_f}{\rightleftharpoons}} Y + Z \tag{8.72}$$

For (8.72) the rate law corresponding to (8.70) is

$$\frac{dx}{dt} = k_f(w - 2x)^2 - k_r x^2$$

$$= k_f(2a - 2x)^2 - k_r x^2$$

$$= 4k_f(a - x)^2 - k_r x^2 \tag{8.73}$$

where x is the amount of Y (or Z) formed and $2x$ is the amount of W reacted per unit volume in time t. The rate constants k_f and k_r are defined in terms of Y (or Z) formed or reacted. The initial concentration w of W can be replaced by $2a$, because W of (8.72) corresponds to both A and B of (8.63). Now (8.73) is identical to (8.70) except for the substitutions $k = 4k_f$ and $k' = k_r$, and so the integrated expression (8.71) can be used. In his classical work on decomposition of hydrogen iodide,

$$2HI \rightleftharpoons H_2 + I_2$$

Bodenstein[33] used a formula equivalent to (8.71) for the special case in which the initial concentration a was unity, the concentration unit being 1 mol per 22.4 liters.

EQUILIBRIUM FROM THE KINETIC VIEWPOINT

In the preceding section we set the net rate of reaction, for example, dx/dt of (8.53), (8.58), and (8.64), equal to zero at equilibrium. This idea, of course originated in the famous work of Guldberg and Waage[34] on the law of mass action and the condition of equilibrium deduced from it. For an elementary reaction, in which each reactant has an order equal to its coefficient in the stoichiometric equation, the equating of forward and reverse rates, therefore zero net rate, gives the usual equilibrium constant expression. For a complex mechanism the order with respect to each reactant, or product, cannot be predicted in such a simple way, but the derivation (based on false kinetic assumptions) still results surprisingly in the correct equilibrium relationship. It is of interest to see how equilibrium in a complex reaction can be treated more accurately.[35]

Consider the general case of a complex homogeneous reaction consisting of a number of simple steps. Since corrections for nonideal behavior can readily be made, assume that the reaction involves ideal gases or an ideal solution. According to the *principle of detailed balancing,* when the overall reaction is at equilibrium each simple step must also be at equilibrium, and for each step the rate of the forward reaction equals the rate of the reverse reaction. The principle of detailed balancing can be derived[36-38] from the fact that the equations of classical or quantum mechanics are invariant to time reversal.[†] Suppose that a mechanical system of specified initial coordinates and momenta is permitted to evolve for time t, and then the momenta are exactly reversed in direction, the coordinates remaining unchanged. During an equal period t the system's trajectory is exactly reversed, the system returns to the inital coordinates, and all momenta are returned to their initial magnitudes but are reversed in direction. For a single bimolecular collision, as might be observed in a molecular beam apparatus, in which two reactant molecules of specified translational and internal states produce two products in specified translational and internal states, the processes corresponding to forward and reverse reactions involve equally acceptable solutions of the equations of motion and therefore have equally probable trajectories. Mahan[38] shows how this leads to a relationship between differential cross sections for forward and reverse reactions, and hence, by integrating over all scattering angles and a Maxwell-Boltzmann distribution of velocities, and by summing over all internal states of reactants and products (assuming the distribution of internal states for thermal equilibrium), to the equation

$$\frac{k_f}{k_r} = \frac{Q_{\text{prod}}}{Q_{\text{react}}} e^{-E_0/kt} = K \qquad (8.74)$$

where k_f and k_r are forward and reverse rate constants and Q_{prod} and Q_{react} are total partition functions.

If the equilibrium constant K in (8.74) is replaced by an appropriate quotient

---

[†] This *principle of microscopic reversibility* is more fundamental than detailed balancing, but the terms are often used interchangeably.

of concentrations, say for the elementary bimolecular reversible reaction

$$A + B \underset{k_r}{\overset{k_f}{\rightleftharpoons}} C + D \tag{8.75}$$

then, using A_e, B_e, C_e, and D_e to represent equilibrium concentrations

$$k_f A_e B_e = k_r C_e D_e \tag{8.76}$$

It is clear that the principle of detailed balancing obtains, because the left and right sides of (8.76) are forward and reverse rates. This is true of (8.75), or any other elementary reaction, even if it is a step in a complex mechanism for which unchanging concentrations of all chemical species could be maintained without detailed balancing. The simplest example of such a mechanism is

$$\tag{8.77}$$

in which forward reaction from A to B occurs directly, but reverse reaction from B to A involves intermediate C. Obviously the rate along the path $B \rightarrow C \rightarrow A$ could equal the rate along $A \rightarrow B$, thereby maintaining constant concentrations of A, B, and C; however, detailed balancing requires that the mechanism be[37]

$$\tag{8.78}$$

with each individual step in equilibrium when the overall reaction is at equilibrium.

Now let us apply detailed balancing to obtain the equilibrium constant expression for a sample complex mechanism. Suppose the mechanism is

$$A \underset{k_1'}{\overset{k_1}{\rightleftharpoons}} B + C \tag{8.79a}$$

$$2B \underset{k_2'}{\overset{k_2}{\rightleftharpoons}} D \tag{8.79b}$$

$$C + D \underset{k_3'}{\overset{k_3}{\rightleftharpoons}} E \tag{8.79c}$$

At equilibrium the forward rate equals the reverse rate for each elementary step. Let A represent the equilibrium concentration of A, and so on. Then

$$k_1 A = k_1' BC \tag{8.80a}$$

$$k_2 B^2 = k_2' D \tag{8.80b}$$

$$k_3 CD = k_3' E \tag{8.80c}$$

Rearranging so that all concentrations are on the same side

$$\frac{k_1}{k_1'} = \frac{BC}{A} = K_1 \tag{8.81a}$$

$$\frac{k_2}{k_2'} = \frac{D}{B^2} = K_2 \tag{8.81b}$$

$$\frac{k_3}{k_3'} = \frac{E}{CD} = K_3 \tag{8.81c}$$

If the overall stoichiometry is

$$2A \rightleftharpoons C + E \tag{8.82}$$

then to obtain (8.82), (8.79a) must be multiplied by 2 before it is added to (8.79b) and (8.79c). The overall equilibrium constant K is

$$K = \frac{CE}{A^2} = \left(\frac{B^2 C^2}{A^2}\right)\left(\frac{D}{B^2}\right)\left(\frac{E}{CD}\right) = K_1^2 K_2 K_3 \tag{8.83}$$

That is, the overall equilibrium constant is the product of the equilibrium constants of the individual steps, each raised to a power equal to the multiplier required to cause catalysts and intermediates to disappear from the overall equation. Substituting into (8.83) from (8.81)

$$K = \left(\frac{k_1}{k_1'}\right)^2 \left(\frac{k_2}{k_2'}\right)\left(\frac{k_3}{k_3'}\right) \tag{8.84}$$

and the overall equilibrium constant can be related to the forward and reverse rate constants of the individual steps.

Since at equilibrium the forward and reverse rates must be equal, no matter how complex the reaction, and since the forward and reverse rate laws must combine to give a relationship among concentrations that is consistent with the equilibrium constant expression, it would appear possible to deduce a reverse rate law from a measured forward rate law. This is not necessarily true, however, since equating forward and reverse rates can lead to a relationship among concentrations that might be a power of the usual equilibrium constant expression and still be consistent. As a very simple example of the difficulty involved, consider the reaction whose stoichiometry is

$$A \rightarrow B$$

and for which two mechanisms are possible

$$2A \rightarrow 2B$$

$$2A \rightarrow A + B$$

In both cases the forward rate would be given by $k A^2$. If the first mechanism predominated, then the equilibrium constant expression would be $K = (k/k')^{0.5}$ and the reverse rate would be $(k/K^2)B^2$. However, if the second mechanism was the one operating, then $K = k/k'$ and the reverse rate would be $(k/K)AB$.

Thus, to state what the reverse rate is, knowledge of mechanism is required. Manes, Hofer, and Weller[39] have considered the problem in some detail and show, in addition to the above conclusion, that it is possible, in principle, to determine the rate expressions of forward and reverse reactions by measurements of the net reaction rate near equilibrium over a range of conditions.

THE APPROACH TO EQUILIBRIUM: RELAXATION METHODS

Near equilibrium all reactions, no matter how complex, become first order with respect to any variable that indicates the distance from equilibrium. This important principle is the basis for a range of *relaxation methods* for determining the rates of rapid reactions.[40] A relaxation method involves a small perturbation of a system at equilibrium, either by a single, sudden change in temperature, pressure, or electric field, or by a periodic disturbance such as an ultrasonic wave or an alternating electric field. Since the system is diplaced from equilibrium by only a small amount, it relaxes toward equilibrium in a first-order process. The relaxation time τ is the reciprocal of the first-order rate constant for the approach to equilibrium; τ represents the time needed for the system to traverse a fraction $1/e$ of its path to equilibrium. The concept of a relaxation time was first used by Maxwell to describe physical phenomena.[41] Eigen and others have applied it to chemical systems with great success.[42]

Consider the ionization in water of a weak acid HA.

$$HA + H_2O \underset{k'}{\overset{k}{\rightleftharpoons}} H_3O^+ + A^- \tag{8.85}$$

Let a be the total concentration of acid and x be the concentration of H_3O^+ or A^-. Let $\Delta x = x - x_e$, where x_e refers to equilibrium. Then, if Δx is small, so that $(\Delta x)^2$ can be ignored

$$\frac{dx}{dt} = k(a-x) - k'x^2 \tag{8.86}$$

$$k(a-x_e) = k'x_e^2$$

$$\frac{d(\Delta x)}{dt} = -\frac{\Delta x}{\tau} = -k[a - (x_e + \Delta x)] + k'(x_e + \Delta x)^2$$

$$\cong -k(a-x_e) + k'x_e^2 - (k + 2k'x_e)\Delta x \tag{8.87}$$

Thus the relaxation time is a function of the forward and reverse rate constants and the equilibrium concentration of H_3O^+ (or of A^-).

$$\tau = \frac{1}{k + 2k'x_e} \tag{8.88}$$

For a different mechanism the expression for τ is different. King[104] has derived a general rate equation for reactions governed by a single relaxation time.

If the perturbation from equilibrium is a single step function, it must be accomplished in a time much less than τ. Temperature jumps of $6°C$ can be accomplished in 10^{-7} s in a solution volume of $1\,cm^3$ by discharging a large capacitor through the solution. Pressure jumps and electic field jumps are less commonly used; these require slightly more and slightly less time. Spectroscopic or conductance measurements coupled to an oscilloscope display and/or a dedicated microcomputer are commonly used to observe the approach to the new equilibrium.

A periodic disturbance produces a periodic change in the position of a chemical equilibrium. If the frequency of the disturbance is appropriate, the alternating response of the chemical system lags behind the perturbation and can be resolved into two components: one in phase and one out of phase. The out-of-phase component produces an absorption of energy in the system from the sound wave or the electric field. At low frequencies the chemical system can keep up with the disturbance and little energy is absorbed; at high frequencies the disturbance reverses so rapidly that the chemical system is hardly perturbed. Maximum absorption usually occurs when the frequency equals $1/\tau$.

A complex system has several relaxation times that can be separated to lead to values of several rate constants if a mechanism is assumed for the reaction of the system.[42f] Temperature-jump and ultrasonic absorption methods have been widely used to study proton transfer reactions, formation of intermediates in enzyme-catalyzed reactions, rotational isomerizations and other conformational changes in organic molecules, hydrogen bond formation, and the association of ions to form ion pairs.

ISOTOPE EXCHANGE REACTIONS

Even though a chemical system is at equilibrium, with no net change in concentrations, it is possible to observe the gross rate of reaction, that is, the forward rate or the reverse rate, by the use of isotopic tracers. Consider the exchange reaction.

$$AX + BX^* = AX^* + BX$$

where X^* is an isotope, most conveniently radioactive, that can exchange with a normal isotope X. Let the concentrations be designated as follows

$$[AX] + [AX^*] = a$$
$$[BX] + [BX^*] = b$$
$$[AX^*] = x \qquad [AX] = a - x$$
$$[BX^*] = y \qquad [BX] = b - y$$

Let R be the gross rate of exchange, that is, the rate of exchange of all atoms X whether like or different isotopes. Assume that there is no isotope effect, that is, that R is independent of the various isotope masses. This is not quite true, but except for hydrogen isotopes the error is small. Now, regardless of the mechanism of the exchange reaction, the rate at which X^* in BX^* can exchange with X in AX is proportional to the fraction of AX and AX^* that are AX. The situation is similar for the reverse exchange. The net rate is then

$$\frac{dx}{dt} = -\frac{dy}{dt} = R\left(\frac{y}{b}\right)\left(\frac{a-x}{a}\right)$$
$$-R\left(\frac{b-y}{b}\right)\left(\frac{x}{a}\right) \qquad (8.89)$$

or

$$\frac{dx}{dt} = R\left(\frac{ay - bx}{ab}\right) \tag{8.90}$$

But since

$$y - y_\infty = x_\infty - x$$

and

$$\frac{x_\infty}{y_\infty} = \frac{a}{b}$$

$$\frac{dx}{dt} = \frac{R}{ab}[(a + b)(x_\infty - x)] \tag{8.91}$$

This integrates into

$$\ln\left(\frac{x_\infty}{x_\infty - x}\right) = \frac{R}{ab}(a + b)t = ct \tag{8.92}$$

where c is a constant since R, a, and b are constant during a run and it is assumed that no AX* was present initially. Similarly,

$$\ln\left(\frac{y_\infty - y_0}{y_\infty - y}\right) = ct \tag{8.93}$$

The result is a first-order rate expression, even though no assumptions have been made about the order of the exchange reaction with respect to the chemical constituents. R may be any function of a and b. Equations (8.92) and (8.93) have been derived by several workers[43-47] in essentially this same form. For radioactive tracers at low concentrations $(a - x)/a$ and $(b - y)/b$ in (8.89) are close to unity and are often omitted, but (8.92) and (8.93) remain the same. Alternatively, (8.92) may be written as

$$-\ln(1 - F) = \frac{R}{ab}(a + b)t \tag{8.94}$$

where F is the fraction of exchange that has occurred. For the more general exchange reaction

$$AX_n + nBX^* = AX_n^* + nBX$$

the equivalent expression is

$$-\ln(1 - F) = \frac{R}{nab}(na + b)t \tag{8.95}$$

The references cited illustrate the use of (8.92) for getting information about the mechanism of exchange. Harris[47] has discussed the influence of isotope effects and concludes that, if the relative concentration of the distinguishable isotope is small enough, the isotope exchange will always be closely first order. For large quantities of the differing isotopes, differences in reactivities of the isotopes must be considered. Deuterium isotope effects are widely used in studying reaction mechanisms, and isotope effects of heavier elements, although small, can often provide valuable information about mechanisms.[48]

THE STEADY-STATE APPROXIMATION AND THE EQUILIBRIUM APPROXIMATION

Consider the complex reaction

$$A \underset{k_{21}}{\overset{k_{12}}{\rightleftharpoons}} B \tag{8.96a}$$

$$B \xrightarrow{k_{23}} C \tag{8.96b}$$

This is a special case of the two-step first-order reversible reaction (8.29) treated earlier in this chapter. Substituting $k_{32} = 0$ into (8.36) results in the exact solution (for $B_0 = C_0 = 0$)

$$\frac{A}{A_0} = \frac{k_{12}(\lambda_2 - k_{23})}{\lambda_2(\lambda_2 - \lambda_3)} e^{-\lambda_2 t} + \frac{k_{12}(k_{23} - \lambda_3)}{\lambda_3(\lambda_2 - \lambda_3)} e^{-\lambda_3 t} \tag{8.97a}$$

$$\frac{B}{A_0} = -\frac{k_{12}}{(\lambda_2 - \lambda_3)} e^{-\lambda_2 t} + \frac{k_{12}}{(\lambda_2 - \lambda_3)} e^{-\lambda_3 t} \tag{8.97b}$$

$$\frac{C}{A_0} = \frac{k_{12} k_{23}}{\lambda_2 \lambda_3} + \frac{k_{12} k_{23}}{\lambda_2(\lambda_2 - \lambda_3)} e^{-\lambda_2 t} - \frac{k_{12} k_{23}}{\lambda_3(\lambda_2 - \lambda_3)} e^{-\lambda_3 t} \tag{8.97c}$$

$$\lambda_2 = \tfrac{1}{2} \{ k_{12} + k_{21} + k_{23} + [(k_{12} + k_{21} + k_{23})^2 - 4k_{12} k_{23}]^{1/2} \} \tag{8.98a}$$

$$\lambda_3 = \tfrac{1}{2} \{ k_{12} + k_{21} + k_{23} - [(k_{12} + k_{21} + k_{23})^2 - 4k_{12} k_{23}]^{1/2} \} \tag{8.98b}$$

These equations are unwieldy and of little use for obtaining rate constants from experimental data. Therefore it is useful to make reasonable approximations to simplify the equations.

The steady-state approximation assumes that the concentration of each intermediate, such as B in the present case, remains constant during the course of the reaction. This is a good approximation when the intermediates are very reactive and therefore are present at very small concentrations. The steady-state approximation is applied to the Lindemann mechanism for unimolecular reactions in Chapter 4. It is especially valuable for the chain reactions discussed in Chapter 10. Comparing the exact result given above with that based on the steady-state approximation reveals conditions under which the latter is satisfactory.

For reactions (8.96a) and (8.96b) the differential equations are

$$\frac{dA}{dt} = -k_{12} A + k_{21} B \tag{8.99a}$$

$$\frac{dB}{dt} = k_{12} A - k_{21} B - k_{23} B \tag{8.99b}$$

$$\frac{dC}{dt} = k_{23} B \tag{8.99c}$$

Setting $dB/dt = 0$ in (8.99b) we have

$$B_{ss} = \frac{k_{12} A}{k_{21} + k_{23}} \tag{8.100}$$

where the subscript ss indicates a concentration obtained using the steady-state assumption. Substituting into (8.99c) yields a much simpler rate law

$$\frac{dC}{dt} = -\frac{dA}{dt} = \frac{k_{23} k_{12} A}{k_{21} + k_{23}} \tag{8.101}$$

The accuracy of the steady-state approximation can be measured by calculating

$$\frac{B_{ss}}{B} = \frac{k_{12} A}{(k_{21} + k_{23}) B} \tag{8.102}$$

where A and B are obtained from (8.97a) and (8.97b) at various times during the course of the reaction.[49] Results for several values of k_{21} and k_{23} with $k_{12} = 1\,s^{-1}$ are shown in Fig. 8.9. Two conditions must be met if the steady-state approximation is to hold in the case of (8.96). First, the time must be larger than t_{ss}, the time required for the intermediate concentration to build up to a steady-state value. This can be obtained from (8.97b) by finding dB/dt, setting it to zero, and solving for t:

$$t_{ss} = \ln \frac{\lambda_2/\lambda_3}{\lambda_2 - \lambda_3} \tag{8.103}$$

The second condition is that $(k_{21} + k_{23}) \gg k_{12}$. From (8.100) this requires that $B_{ss} \ll A$, that is, that the concentration of intermediate be small relative to initial reactant.

For a different mechanism the conditions under which the steady-state approximation is valid are different. In each case it must be ascertained that the appropriate conditions are met before the steady-state approximation is used to obtain rate parameters from experimental data. For example, in the case of a series first-order reaction $A \rightarrow B \rightarrow C$, that is, (8.96) with $k_{21} = 0$, the conditions for a valid steady-state approximation are $t \gg 1/k_{23}$ and $k_{23} \gg k_{12}$. Frei and Günthard[50] have compared exact and steady-state solutions of the kinetic equations for $S_N 1$, $S_N 2$, E_1, E_2, and $E_1 CB$ mechanisms. They give practical criteria for recognizing when the steady-state approximation holds, and they conclude that neither the condition of relatively low concentration of intermediate nor the condition that $d[\text{intermediate}]/dt = 0$ is necessary or sufficient for a valid steady-state approximation.

Further simplification of the kinetic equations is possible if it can be assumed that there is a facile equilibrium between reactants and an intermediate. Again using (8.96) as an example, if A and B are in equilibrium, then the equilibrium constant K is

$$K = \frac{k_{12}}{k_{21}} = \frac{B_{eq}}{A} \tag{8.104}$$

and

$$B_{eq} = \frac{k_{12} A}{k_{21}} = KA \tag{8.105}$$

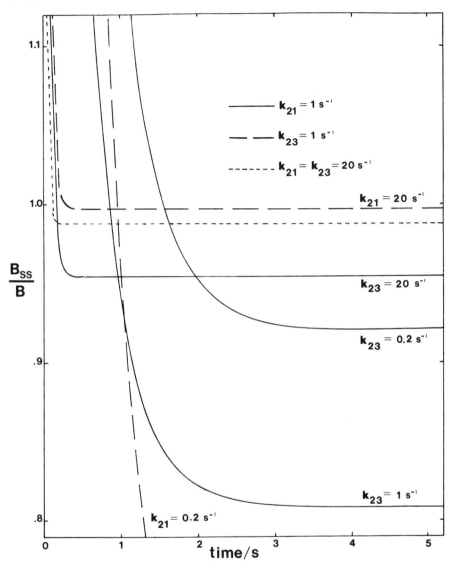

Figure 8.9. Plots of B_{ss}/B [from (8.102)] versus t for complex reaction (8.96) with $k_{12} = 1 \, \text{s}^{-1}$. Values of k_{21} and k_{23} are indicated in the figure.

Comparing (8.105) with (8.100), this corresponds to the steady-state assumption with the further condition that $k_{21} \gg k_{23}$. That is, the reverse step of the process forming the intermediate must be very rapid compared with the rate at which the intermediate reacts to form product. Of course the conditions of the steady-state approximation must also be met. Equation (8.101) now becomes

$$\frac{dC}{dt} = -\frac{dA}{dt} = \frac{k_{23}k_{12}A}{k_{21}} = k_{23}KA \qquad (8.106)$$

This result can be derived simply by assuming that the concentration of B is given by the equilibrium constant expression (8.104) and substituting into (8.99c).

Two examples of situations where the equilibrium approximation comes into play are given in earlier chapters. For the Lindemann mechanism of unimolecular reactions the limiting expression for high pressure, which gives a simple first-order rate, is equivalent to assuming an equilibrium between normal molecules and activated molecules. This also is the situation in which the rate of deactivation is large compared with the rate of reaction. The second example involves reactions of NO that follow a third-order rate expression, such as

$$2NO + O_2 = 2NO_2$$

$$\frac{-d[NO]}{dt} = 2k_{obs}[NO]^2[O_2]$$

From the rate expression[51] the reaction was first interpreted as being trimolecular, but because of its negative activation energy (Table 6.8) an alternative mechanism was proposed. This involves the equilibrium

$$NO + O_2 \underset{}{\overset{K}{\rightleftharpoons}} NO_3 \tag{8.107}$$

preceding a bimolecular step

$$NO_3 + NO \xrightarrow{k} 2NO_2 \tag{8.108}$$

For the alternative mechanism the rate law is similar to (8.106),

$$\frac{-d[NO]}{dt} = 2kK[NO]^2[O_2] \tag{8.109}$$

but the observed third-order rate constant k_{obs} is the product of a bimolecular second-order constant k and an equilibrium constant K for formation of the NO_3. Because dissociation of the dimer is endothermic, K decreases as temperature rises. To account for the negative activation energy it is only necessary to assume that the normal increase in k with rising temperature is smaller than the decrease in K.[52] However, applying transition-state theory to a trimolecular single-step mechanism can also account for the abnormal temperature effect.[53] According to (5.50), for three diatomic molecules forming a nonlinear activated complex in which vibration is not excited, the temperature dependence is as $T^{-7/2}e^{-LE_0/RT}$, and if E_0 is small the rate constant can decrease as T increases. Several other cases exist where there is strong evidence that a preequilibrium reaction step results in small or negative activation energy because an observed rate constant is the product of an equilibrium constant and a true rate constant.[54]

Up to this point we have assumed that the intermediate in equilibrium with the reactant is at very low concentration and so makes no important contribution to the stoichiometry of the reaction. If, on the other hand, as much as a few percent of intermediate is present, the rate of disappearance of reactant becomes complicated. In such a case the intermediate often is present in an observable amount, allowing the kinetics to be handled more directly. The work of Taube and Posey[55] on anation of aquopentaamminecobalt(III) ion by sulfate provides an example of a partially displaced equilibrium. The rate is first order in total cobalt and first order in sulfate at low sulfate concentrations, but reaches a limiting value independent of $[SO_4^{2-}]$ at high sulfate. This corresponds to a rate law of the form

$$\frac{d[Co(NH_3)_5 SO_4^+]}{dt} = \frac{A[Co]_T[SO_4^{2-}]}{1 + B[SO_4^{2-}]} \tag{8.110}$$

where $[Co]_T$ represents the total unreacted cobalt complex. Taube and Posey found that the leveling of rate at high $[SO_4^{2-}]$ resulted from saturation of an equilibrium involving ion-pair formation, the mechanism being

$$RH_2O^{3+} + SO_4^{2-} \overset{K}{\rightleftharpoons} RH_2O^{3+}, SO_4^{2-} \tag{8.111a}$$

$$RH_2O^{3+}, SO_4^{2-} \overset{k}{\longrightarrow} RSO_4^+ + H_2O \tag{8.111b}$$

where $R = Co(NH_3)_5^{3+}$. The total concentration of unreacted cobalt must be apportioned between two forms on the basis of the equilibrium constant K, that is,

$$[RH_2O^{3+}] = \frac{[Co]_T}{1 + K[SO_4^{2-}]}$$

$$[RH_2O^{3+}, SO_4^{2-}] = \frac{[Co]_T K[SO_4^{2-}]}{1 + K[SO_4^{2-}]}$$

Then the rate law becomes

$$\frac{d[RSO_4^+]}{dt} = k[RH_2O^{3+}, SO_4^{2-}]$$

$$= \frac{kK[Co]_T[SO_4^{2-}]}{1 + K[SO_4^{2-}]} \tag{8.112}$$

This corresponds to (8.110) with $B = K$ and $A = kK$. Had Taube and Posey not recognized the presence of (8.111a), an entirely different mechanism would have been proposed. In the limit of high $[SO_4^{2-}]$ equilibrium (8.111a) lies entirely to the right and rate law (8.112) becomes first-order since $K[SO_4^{2-}] \gg 1$. However, the first-order rate law refers to the actual species present in solution, that is, to the ion pair, not to the $Co(NH_3)_5 H_2O^{3+}$ that was initially dissolved. When an intermediate forms almost completely, the kinetics often become simple, but incorrect conclusions may easily be drawn if the existence of the equilibrium is not known.[56]

The possibility of an equilibrium following a rate process also exists. For example, decomposition of dinitrogen pentoxide gives the overall result

$$2N_2O_5 = 2N_2O_4 + O_2$$

but N_2O_4 is in equilibrium with NO_2

$$N_2O_4 \rightleftharpoons 2NO_2$$

Such an equilibrium does not affect the rate in a constant-volume system; however, it can easily affect the method of measurement. In this case the reaction is conveniently followed by observing the increase in pressure of the gaseous reaction mixture. As the products increase in partial pressure the equilibrium shifts, thus necessitating a correction to the pressure increase before the rate of decrease of partial pressure of the reactant can be determined.[57]

PSEUDO-n th-ORDER CONDITIONS

For complex mechanisms in which concentrations of several species appear in the rate expression considerable simplification can be attained by adjusting the reaction conditions so that all but one of the concentrations remains constant. For a reaction whose rate expression is

$$\frac{dx}{dt} = k[A]^u[B]^v[C]^w$$

for example, [B] might be maintained high enough relative to [A] that [B] would not change appreciably during the reaction, and [C] might be buffered by some facile equilibrium. Then the reaction would be pseudo-uth order in A, and concentration–time data for a single run could be treated as for a uth-order reaction. The exponents v and w can then be obtained by varying the constant concentrations of B and C, since

$$\frac{dx}{dt} = k_{obs}[A]^u$$

where

$$k_{obs} = k[B]^v[C]^w \tag{8.113}$$

Birk[58] gives a number of examples in which the logarithmic form of (8.113), that is,

$$\log k_{obs} = \log k + v \log [B] + w \log [C] \tag{8.114}$$

is useful in elucidating a complex rate law. When $A = HCrO_4^-$, $B = Fe(CN)_6^{4-}$, and $C = H^+$, a plot of log [B] versus log k_{obs} (under pseudo-first-order conditions where [B] was varied but [C] was not) gave a straight line of unit slope, implying that $v = 1$. Then $\log k_{obs} - v \log$ [B] was plotted against log [C] for runs in which [H$^+$] was varied. This gave a straight line of unit slope up to [H$^+$] = 0.03M, but the line then leveled off and eventually reached a slope of -2. Downward curvature in the plot implies that the denominator of the rate law contains several terms, each involving a different power of [H$^+$]. Similarly, upward curvature would imply a sum of terms in the numerator. The overall rate law was found to be

$$\frac{dx}{dt} = \frac{a[HCrO_4^-][Fe(CN)_6^{4-}][H^+]}{b + c[H^+] + d[H^+]^2 + [H^+]^3}$$

where values $a = 6.3 \times 10^3 \, M s^{-1}$, $b = 1.04 \times 10^{-2} \, M^3$, $c = 4.4 \times 10^{-2} M^2$, and $d \cong 0$ were obtained by nonlinear least-squares analysis of the data.[59]

COMPUTER SIMULATION IN CHEMICAL KINETICS

The most important evidence for the mechanism of any reaction is identification of reactants, products, and any detectable intermediates. Next in importance is correct prediction of observed concentration–time profiles for reactants, products, and intermediates, followed by extra-kinetic evidence of the type discussed in Chapter 1. Predicting concentration–time profiles usually involves substituting experimentally determined rate constants into an algebraic rate expression. However, examples given earlier in this chapter show that even a relatively simple mechanism of two or three steps can lead to a rate expression complicated enough to make analysis of observed concentration—time data difficult if not impossible. Sometimes complex rate expressions can be made more tractable by using the steady-state or equilibrium approximations, or by adjusting experimental conditions so that the reaction is pseudo-nth order. Even for a fairly simple mechanism, however, the rate constants may be such that these simplifications are not justifiable, and for reactions of moderate or greater complexity approximate methods are usually inadequate and may produce erroneous results.[60]

Computer simulation provides an alternative means of calculating the concentration—time profiles predicted by an assumed mechanism. Most commonly this is done in the following way. For each elementary step in a complex mechanism, assume that the law of mass action correctly describes the dependence of rate on concentrations and that the rate constant at a given temperature is independent of the composition of the reaction mixture. Then, following the methods used earlier in this chapter, write differential equations that relate the rate of change of concentration of each chemical species to appropriate rate constants and concentrations. For example, in the case of a two-step irreversible series first-order reaction, $A \rightarrow B \rightarrow C$, the differential equations (8.9) are appropriate. Finally, solve this set of ordinary differential equations to obtain the concentration of each species as a function of time. Solution can be effected by analog simulation or by numerical integration using a digital computer. It is also possible to carry out a stochastic simulation in which reaction steps and time intervals are selected by a Monte Carlo method.

Analog simulation[61] makes use of electronic circuit components that sum, multiply, integrate, or perform other operations on voltages. An analog program consists of a number of such components arranged in a circuit that is the analog of the particular mechanism under study. A separate circuit is usually required for each different mechanism. Rate constants and other parameters are represented by potentiometers that can be varied to achieve the best match between the analog computer's output and observed experimental data. The work of Williams and Bruice[62] provides an excellent example of analog simulation of a moderately complex chemical system, the biochemically important reduction of carbonyl compounds such as pyruvic acid by 1,5-dihydroflavines. A simplified scheme for such reactions is

$$F1H_2 + \rangle C{=}O \underset{k_{-1}}{\overset{k_1}{\rightleftharpoons}} CA \underset{k_{-2}}{\overset{k_2}{\rightleftharpoons}} Im \qquad (8.115a)$$

$$F1H_2 + \rangle C{=}O \overset{k_3}{\longrightarrow} Fl_{ox} + \overset{H}{\underset{}{\rangle}}C{-}OH \qquad (8.115b)$$

$$Fl_{ox} + CA \overset{k_4}{\longrightarrow} CT + \rangle C{=}O \qquad (8.115c)$$

$$CT \underset{k_{-5}}{\overset{k_5}{\rightleftharpoons}} F1H_2 + Fl_{ox} \qquad (8.115d)$$

in which the reduced flavin $F1H_2$ reacts with the pyruvic substrate to form a carbinolamine CA. This dehydrates to an imine Im in a reversible step. The carbinolamine can revert directly to reduced flavin and substrate or it can combine with the oxidized flavin Fl_{ox} to form a complex CT that dissociates to form reduced and oxidized flavin. The absorbance at 443 nm, where Fl_{ox} has a maximum and other components are transparent, shows an initial burst of formation of Fl_{ox} followed by a slow further increase, as seen in Fig. 8.10.

The eight rate constants of (8.115a)–(8.115d) were varied by adjusting potentiometers to give the best fit to the data of Fig. 8.10. Varying potentiometer settings by ± 5% produced discernably poorer agreement. Once the best fit was obtained, spot checks were made to determine whether the concentrations of F1H₂ and Im were as predicted by the computer and shown in Fig. 8.10. Also, attempts were made, unsuccessfully, to simulate the time course of appearance of Fl_{ox} using only (8.115a) and (8.115b). Many other observations, not specifically related to the computer simulation, lend further support to the proposed mechanism.[62] Such observations are essential, since cases are known where two or more schemes provide essentially the same goodness of fit to experimental data.[63]

Numerical integration by digital computer is discussed in connection with trajectory calculations in Chapter 5. It provides a general method that can be applied to any reaction scheme without the necessity of constructing a new program or circuit, and currently it is by far the most widely used technique for predicting concentration–time profiles. Program packages are available that allow chemical equations and rate constants to be typed into a digital computer in nearly the same form as they would be typewritten in a research paper,[64] following which integration proceeds automatically. The primary limitation on numerical integration methods is the cost of computer time to carry out the calculations. If the form and parameters of the set of differential equations to be solved are such that the equations are *stiff*, that is, if the reaction steps have widely separated time constants and there are one or more species that are maintained at an almost constant concentration by two or more rapid opposing or compensating reactions, then many methods of numerical integration will require excessively long times for

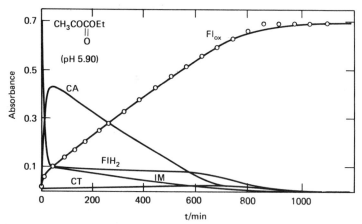

Figure 8.10. Example computer fitting of (8.115) to absorbance–time data for reduction of ethyl pyruvate by 1-5-dihydrolumiflavin-3-acetic acid. (○) Experimental values for Fl_{ox} [Reprinted with permission from R. F. Williams and T. C. Bruice, *J. Am. Chem. Soc.*, **98**, 7752 (1976).]

computation. This happens because the time increment for each step of the integration [δt of (5.14), or a similar parameter for more sophisticated integration methods] must be made small enough so that integration of the equations describing rapid concentration changes will be accurate. Consequently a great many time increments (and hence calculations) are necessary before slow reaction steps with long time constants approach equilibrium.

The problem of stiffness was first recognized in 1952,[65] but nearly two decades passed before real progress was made in overcoming it.[66] Currently the most widely used algorithm for integrating stiff systems of differential equations is that of Gear,[67] an implicit, multistep, predictor–corrector method in which the step size δt is automatically varied to attain a user-specified degree of accuracy at each integration step. Many other algorithms have been constructed and tested,[68] and in some cases programs have made use of the fact that differential equations derived from the law of mass action have a special structure that enables easy and rapid computation of higher derivatives needed for certain numerical methods.[69] Using the Gear algorithm or another of comparable efficiency a digital computer can perform numerical integration for a homogeneous chemical system consisting of as many as 100 elementary steps and 30 or more species within a few seconds. Numerical integration has been very useful in studies of free-radical chain reactions, oscillating reactions, and enzyme kinetics. It also finds applications of great practical importance in studies of hydrocarbon pyrolysis and combustion, of environmental problems such as photochemical smog and stratospheric ozone depletion by chlorofluorocarbons, and even in food chemistry.[70] Examples of several such applications are given subsequently.

Schaad[71g] first described a Monte Carlo method for predicting concentration–time profiles, but his method requires a great deal of computer memory and has not been used extensively. More recently several workers have described algorithms in which the probability of a molecular event (for example a reactive bimolecular collision) is assumed to be proportional to the rate of that process as given by a macroscopic rate expression.[71] For example, Bunker et al.[71e] assumed ideal gas behavior and species concentrations c_i in volume V. They let N_i be the numbers of the various species in the simulation. Then

$$\frac{N_i}{\Sigma N_i} = \frac{c_i}{\Sigma c_i} \tag{8.116a}$$

$$V = \frac{\Sigma N_i}{L \Sigma c_i} \tag{8.116b}$$

where L is Avogadro's constant. For the jth reaction step the instantaneous number of events per second R_j is given by the rate law

$$R_j = \frac{k_j N_i N_1 \cdots}{(LV)^{s_j - 1}} \tag{8.117}$$

where s_j is the overall order of the jth step. The probability p_j that the next molecular event will be reaction j is

$$p_j = \frac{R_j}{\Sigma R_j} \tag{8.118}$$

To select the next reaction event the computer selects a random number r_1 in the range $0 < r_1 \leqslant 1$ and chooses m such that

$$\sum_{j=1}^{m-1} p_j < r_1 \leqslant \sum_{j=1}^{m} p_j \tag{8.119}$$

That is, the probabilities p_j are summed in order until adding p_m exceeds the random number. The mth reaction is then assumed to occur and the N_i are adjusted by subtracting the stoichiometric coefficients for reactants and adding the coefficients for products of the mth step.

Bunker et al. suggested heuristically that the elapsed time for each process described in the preceding paragraph would be $1/\Sigma R_j$. However, there is no requirement that each step in the simulation take the same increment of time. Gillespie[72] has analyzed the situation in terms of the stochastic formulation of chemical kinetics[73] and found that the correct expression for the time increment τ is

$$\tau = \frac{1}{\Sigma R_j} \ln \frac{1}{r_2} \tag{8.120}$$

where r_2 is a second random number in the same range as r_1. The mean of the τ values calculated from (8.120) is $1/\Sigma R_j$, so that the suggestion of Bunker et al. is quite reasonable. The complete algorithm for stochastic simulation involves selecting a molecular event according to (8.119), determining a time increment τ from (8.120), and updating the species numbers and elapsed time. These operations are easily coded for a digital computer and may be repeated as often as necessary to simulate a particular kinetic system. As in the case of numerical integration methods, the principal limitation is the cost of computer time, which here is roughly proportional to the number of reaction steps in the mechanism.

The heuristic discussion of the preceding two paragraphs disguises the fact that stochastic simulation as set forth by Gillespie[72] is based conceptually on a microscopic view of kinetics, while all the methods discussed above, including that of Bunker et al., start with macroscopic, deterministic, differential rate equations. The stochastic simulation method takes full account of molecular-level fluctuations about macroscopic average values, and it does not approximate infinitesimal time increments dt by finite time steps δt. Stiffness instability cannot occur, and so stochastic simulation is especially advantageous for systems in which molecular population levels can change almost instantaneously. Stochastic simulation is a reasonable and consistent simplification of the method of reactive molecular dynamics,[74] in which the classical dynamics of a collection of atoms or molecules is followed as they undergo elastic, inelastic, and reactive collisions according to the laws of mechanics. Stochastic simulation provides reasonable computational efficiency at the expense of the molecular-level detail of a molecular-dynamics calculation. Both of these methods can handle situations in which deterministic, macroscopic, differential rate equations may have multiple solutions.[75] In such cases fluctuations about macroscopic average values become important in determining which of several possible paths a system will take.

Computer simulation of concentration–time profiles can be used in a variety of ways. For relatively simple mechanisms, such as the flavin oxidation example used to illustrate analog simulation, it may be possible both to choose an appropriate mechanism and to fit appropriate rate parameters to the data. For engineering applications, where details of mechanism are much less important than a set of parameters that adequately characterize a system, one may be satisfied simply to find a set of reaction steps that permits accurate modeling, whether or not those reactions are plausible on other grounds. For such a purpose a nonlinear least-squares method such as that described in Chapter 2[76] can be used to obtain the parameters that give the best fit. However, unless initial estimates of the rate parameters are rather close to the best values, standard nonlinear least-squares techniques may diverge, producing poorer rather than better values. This difficulty can be alleviated but not eliminated by using Marquardt's algorithm.[77] If one is primarily concerned with supporting a mechanism for a relatively complicated reaction, then the best approach is usually to obtain rate parameters from independent studies of the individual elementary steps, and vary only the form of the multistep model to obtain the best fit to experimental observations. Sources of rate constants for gas-phase and solution reactions are given in Chapters 6 and 7. In some cases one or two rate parameters may not be known independently. These can often be determined by using theory or analogy with similar reactions to estimate values and then adjusting the rough estimates to achieve good agreement with experiment.

An important adjunct to computer simulation is sensitivity analysis. This involves determining the sensitivity of each dependent variable of the model to changes in the parameters. For example, the effect of variation in the rate constant k_j for the jth reaction step can be expressed as a series of coefficients $\partial c_i/\partial k_j$, where the c_i are concentrations of the various species. Each sensitivity coefficient depends on time and the array $\partial c_i/\partial k_j$ can be evaluated at various times t by solving $N_c \times N_k$ differential equations, where N_c is the number of dependent variables and N_k is the number of rate constants.[78] An alternative method is to superimpose a characteristic oscillation on each of the k_j during the simulation. Fourier analysis of the simulation output determines the extent to which each of the k_j affects each of the c_i.[79] Sensitivity analysis can indicate which concentration profiles are most important for obtaining values of particular rate constants. Conversely, there may be some rate parameters that have little or no effect on concentration profiles over the range covered by the experimental data. These rate parameters need not be varied, then, in seeking a best fit, and attention can be focused on those reaction steps that are more likely to account for remaining discrepancies between model and reality. Gardiner[80] has shown how sensitivity analysis can be used to obtain approximate error limits on rate constants obtained by modeling. Further information on estimating the precision of parameters obtained by computer simulation can be found in treatises on regression analysis.[81] A closely related problem is that of discriminating among several competing models. Systematic procedures[82] are available for deciding which of several models best reproduces experimental observations, and it is important to include all reaction steps whose absence would affect model predictions by magnitudes greater than the errors in observed values. However, extrakinetic considerations remain important, and the statistically best model may still be rejected as not being the most plausible mechanism.[63a]

A model that has been fitted to experimental data over a particular range may be entirely inadequate outside that range. Consequently the range of experimental independent variables should be as broad as possible when a mechanism is being evaluated,[83] or, even better, data should be obtained from several kinds of experiments. Westbrook and Dryer,[84] for example, have fitted a detailed kinetic mechanism involving 84 elementary reactions and 26 chemical species to data on oxidation of methanol from shock-tube and turbulent-flow-reactor studies. The data cover the ranges of 1000–2180 K, 1–5 atm, and fuel–air equivalency ratios 0.05–3.0, broad enough to include temperatures, pressures, and stoichiometries encountered in real combustion systems. Rate constants for 74 of the 84 elementary reactions remained constant at their literature values throughout the study. Reverse rates were calculated from forward rates and equilibrium constants, thermochemical data being obtained from the *JANAF* tables.[85] Temperature dependence of rate constants was calculated from $k = BT^m e^{-Eb/RT}$, with m often zero. The best fit to the experimental data yielded improved rate constants for 5 of the 10 reactions whose parameters were varied. Under different conditions different reactions were important. For example, at a low fuel/air ratio the principal methanol-consuming reaction was

$$CH_3 OH + OH \rightarrow CH_2 OH + H_2 O$$

while under rich conditions

$$CH_3 OH + H \rightarrow CH_2 OH + H_2$$

and

$$CH_3 OH + H \rightarrow CH_3 + H_2 O$$

were most important. The reaction

$$CH_3 OH + M \rightarrow CH_3 + OH + M$$

was important in accounting for shock-tube results but not for turbulent-flow results. For a very rich mixture, with fuel/air equivalency ratio of 6.0, a good fit could not be obtained, apparently because of insufficient detail in the oxidation mechanism for ethene and acetylene.

To test their detailed oxidation mechanism Westbrook and Dryer predicted the laminar flame velocity for a premixed, stoichiometric methanol–air mixture at 298 K and 1 atm, using the model of Lund[86] for one-dimensional transport. The predicted value was $44 \pm 2 \, cm \, s^{-1}$, in good agreement with measured values ranging from 44 to 46 $cm \, s^{-1}$. It should be pointed out that the mathematics of modeling systems that are not uniform throughout the volume under consideration are far less tractable than for homogeneous systems. A set of partial differential equations must be solved to simulate a nonuniform system, and the computer time needed to solve problems of reasonable size is currently excessive, although some means of alleviating this problem can be identified.[87] Since many real situations, such as global atmospheric reactions, chemical processes, and combustion, are nonuniform, there is currently much effort in this area.

CHEMICAL OSCILLATORS

In 1828 a report of an electrochemical cell that produced an oscillating current appeared in the literature,[88] and subsequently other oscillating reactions were noted, including Liesegang's discovery of the periodic precipitation patterns named after him.[89] These phenomena all involved diffusion gradients, and until the 1960s there was a consensus that oscillations could not occur in a homogeneous reaction mixture. This consensus ignored Lotka's theoretical description of a mechanism that could produce oscillations[90] and Bray's observation of oscillation of the concentration of I_2 and of the rate of release of O_2 during the reaction of H_2O_2 with KIO_3 in dilute sulfuric acid.[91] The second law of thermodynamics requires that all spontaneous chemical changes in closed systems must be accompanied by a decrease in Gibbs free energy, and by corollary that a system at constant temperature and pressure must approach equilibrium monotonically, without overshooting and coming back. Consequently it was thought that concentrations of intermediates must either pass through a single maximum or minimum or rapidly reach some steady-state value during the course of a reaction. But this latter behavior is not required by thermodynamics, provided that the reaction mechanism is sufficiently complex and the system is sufficiently far from equilibrium. Oscillations of concentrations of intermediates can occur when one substance in a sequence of reactions is either an activator or an inhibitor for a reaction step that occurs earlier in the sequence.[92] (If the reaction steps are reversible the activated or inhibited step may also occur later in the sequence.) Such activation or inhibition constitutes a chemical feedback loop by means of which intermediate-forming reaction steps can be turned on or off, producing oscillations. When the system is far enough from chemical equilibrium the overall decrease in Gibbs energy provides the driving force for oscillations.[93] As equilibrium is approached this driving force approaches zero, the oscillations are damped out, and all concentrations approach their equilibrium values monotonically.

Chemical oscillators can exhibit periodic behavior in space as well as over time. Oscillations have been observed in homogeneous gas-phase reactions,[94] but are more often encountered in solution. Biochemical oscillators are also known, for example in the glycolytic enzyme system that causes breakdown of glucose to pyruvate in most living cells,[95] and analogies can be drawn with far more complex processes, such as cellular regulation and differentiation in biology, chemical and biological evolution, and ecological cycles.[96] Several reviews of theoretical and experimental aspects of chemical oscillators are available.[97]

The most thoroughly studied chemical oscillator is the Belousov-Zhabotinskii reaction, first observed by Belousov[98] in 1959 and later studied more thoroughly by Zhabotinskii.[99] The net chemical reaction is oxidation of malonic acid by bromate.

$$2H^+ + 2BrO_3^- + 3CH_2(COOH)_2 \rightarrow 2BrCH(COOH)_2 + 3CO_2 + 4H_2O$$

$$(8.121)$$

The reaction is catalyzed by cerium ions, or by some other redox couples, such as Fe^{III}/Fe^{II} and Mn^{III}/Mn^{II}, that involve strong one-electron oxidants. Field, Körös, and Noyes[100] followed the concentration of intermediate Br^- by means of a bromide-sensitive electrode and observed the ratio $[Ce^{IV}]/[Ce^{III}]$ with a tungsten electrode. Their results are shown in Fig. 8.11. There is an initial period of rapid concentration change, a long induction period, and finally a series of oscillations in both $[Br^-]$ and $[Ce^{IV}]/[Ce^{III}]$.

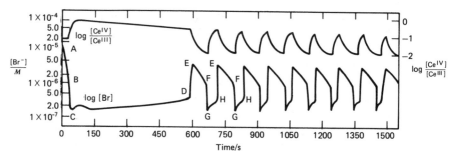

Figure 8.11. Potentiometric traces at room temperature of a bromide-sensitive electrode (lower curve) and of a tungsten electrode (upper curve) that responded to cerium(IV). The $0.8\,M$ sulfuric acid solution contained malonic acid, potassium bromate with a small amount of bromide, and cerium(III) catalyst [Adapted from R. J. Field, E. Körös, and R. M. Noyes, *J. Am. Chem. Soc.*, **94**, 8649 (1972).]

A qualitative rationalization of the observed oscillations is provided by (8.122a)–(8.122c).

$$BrO_3^- + 2Br^- + 3CH_2(COOH)_2 + 3H^+ \rightarrow 3BrCH(COOH)_2 + 3H_2O$$

$$(8.122a)$$

$$BrO_3^- + 4Ce^{3+} + 5H^+ \rightarrow HOBr + 4Ce^{4+} + 2H_2O \qquad (8.122b)$$

$$BrCH(COOH)_2 + 4Ce^{4+} + H_2O + HOBr \rightarrow 2Br^- + 3CO_2 + 4Ce^{3+} + 6H^+$$

$$(8.122c)$$

Clearly none of these is an elementary reaction, the actual mechanism being quite complex. Field and Noyes[101] have proposed a simplified model known as the Oregonator from which (8.122a)–(8.122c) may be obtained, and computer simulation has been applied to more complex models involving up to 20 elementary steps.[102] Nevertheless the general nature of the oscillations is explicable in terms of (8.122a) to (8.122c) and Fig. 8.11.

Initially (region A in Fig. 8.11) cerium ion is present as Ce^{III} and $[Br^-] \cong 10^{-5}\,M$. Reaction (8.122b) is autocatalytic and is also inhibited by Br^-, and so only (8.122a) occurs at first. When region B in Fig. 8.11 is reached, $[Br^-]$ has fallen enough so that (8.122b) can occur at an appreciable rate, and autocatalysis causes a rapid buildup of Ce^{IV}. Also $[Br^-]$ drops even more rapidly because Br^- combines with HOBr and with $HBrO_2$, the latter being an intermediate in (8.122b). There follows an induction period during which [HOBr] and $[Ce^{IV}]$ remain high, $[Br^-]$ remains low, and bromomalonic acid is being produced by the following reaction

$$HOBr + CH_2(COOH)_2 \rightarrow BrCH(COOH)_2 + H_2O \qquad (8.123)$$

Reaction (8.123) is also a step in (8.122a) and may take place directly or through intermediate Br_2 formed from HOBr and Br^-.

As the concentration of bromomalonic acid increases a continually larger concentration of bromide is formed as a result of (8.112c), and eventually, at point D in Fig. 8.11, (8.122b) is again inhibited. There is then a rapid increase in $[Br^-]$, since (8.122b) is no longer producing HOBr or $HBrO_2$ to consume Br^-.

Also the ratio $[Ce^{IV}]/[Ce^{III}]$ begins to fall as (8.122c) proceeds without being balanced by (8.122b). Reaction (8.122a) then consumes Br^- faster than Br^- is produced by (8.122c), and at point F $[Br^-]$ has dropped enough that (8.122b) is turned on again and $[Br^-]$ falls rapidly until point G is reached. In region GH the conditions are similar to those of the induction period CD, except that the concentration of bromomalonic acid is higher and less time is required for $[Br^-]$ to increase sufficiently to stop (8.122b) again. Note from Fig. 8.11 that at longer times the regions corresponding to GH are of shorter duration because of the steeper increase in $[Br^-]$ afforded by the greater concentration of product bromomalonic acid. For large initial concentrations of malonic acid region GH does not even appear.

The essential feature of the Belousov-Zhabotinskii reaction is coupling of the autocatalytic reaction (8.122b) to the other processes by way of Br^- and other intermediates. The sudden bursts of reaction produced by autocatalysis cause intermediate concentrations to overshoot the values that would obtain in a steady state. Otherwise a nonoscillatory steady state could be maintained.

PROBLEMS

8.1. Apply the time-ratio method using Table 8.1 to evaluate k_1 and k_2 from the following data on the hydrolysis of 2,7-dicyanonaphthalene (Kaufler[11b]; see also Swain[11a]).

% reaction	15	35	70
Time/10^2 s	13.2	38.4	151.2

8.2. Consider the case of consecutive, competitive first- and second-order reactions:

$$A \xrightarrow{k_1} B$$

$$A + B \xrightarrow{k_2} C$$

(a) Set up the differential rate equations.

(b) Obtain the steady-state solution. Under what conditions will this be a good approximation?

(c) Derive a more accurate solution for concentrations A and B that will apply if the relative concentration B/A is very small throughout the reaction.

8.3. Use the secular equation method to obtain the integrated solution for two successive first-order reactions.

8.4. Analyze the second- and first-order reversible reaction scheme

$$A + B \underset{k'}{\overset{k}{\rightleftharpoons}} C$$

Assuming equal initial concentrations of reactants $A_0 = B_0 = a$ and that $C_0 = 0$, show that in terms of the extent of reaction variable x the integrated rate expression is

$$\ln \frac{x_e(a^2 - xx_e)}{(x_e - x)a^2} = k'\frac{a^2 - x_e^2}{x_e}t$$

8.5. A mixture of ethyl acetate (E), water (W), and ethanol (A), acidified with hydrochloric acid to provide a constant amount of catalyst, undergoes acid-catalyzed hydrolysis of the ethyl acetate, approaching an equilibrium mixture. The rate may be expressed by

$$\frac{dx}{dt} = k(E - x)(W - x) - k'(A + x)x$$

For an original mixture where $E = 1M$, $W = A = 12.215\,M$, and where 0.06116 N $Ba(OH)_2$ is used to titrate 1-cm³ samples, the following results were observed [O. Knoblauch, *Z. Phys. Chem.*, 22, 268 (1897)]:

Time/min	$V_{Ba(OH)_2}/cm^3$
0	(7.68)
78	8.95
94	9.20
138	9.72
169	9.99
348	11.10
415	11.35
464	11.48
∞	12.00

Obtain a suitable integrated form of the rate equation and evaluate the constants k and k'. (k/k' can be calculated directly from the titer after infinite time.)

8.7. Use (8.36a)–(8.36c) to derive expressions for the concentrations of A, B, and C as a function of time in the mechanism

$$A \underset{k_2}{\overset{k_1}{\rightleftharpoons}} B \overset{k_3}{\longrightarrow} C$$

assuming that B and C are not present initially. That is, derive (8.97a)–(8.97c).

8.8. Use the method of partial fractions to integrate (8.61), hence verifying (8.62).

8.9. Using the notation of (2.1), show that for the general multistep mechanism

$$0 = \nu_{A1}A + \nu_{B1}B + \nu_{C1}C + \dots$$

$$0 = \nu_{A2}A + \nu_{B2}B + \nu_{C2}C + \dots$$

$$0 = \nu_{A3}A + \nu_{B3}B + \nu_{C3}C + \dots$$

$$\begin{array}{cccc} \cdot & \cdot & \cdot & \cdot \\ \cdot & \cdot & \cdot & \cdot \\ \cdot & \cdot & \cdot & \cdot \end{array}$$

the overall equilibrium constant K is given by

$$K = K_1^{n_1}K_2^{n_2}K_3^{n_3} \dots = [A]^{\nu_{At}}[B]^{\nu_{Bt}}[C]^{\nu_{Ct}} \dots$$

where $n_1, n_2, n_3, \dots$ are the factors by which steps 1, 2, 3, $\dots$ must be multiplied before summing to give the overall equation

$$0 = \nu_{At} A + \nu_{Bt} B + \nu_{Ct} C + \ldots$$

(Hint: See A. A. Frost and R. G. Pearson, *Kinetics and Mechanism*, 2nd ed., Wiley-Interscience, New York, 1961, pp. 189–191.)

8.10. For the reaction

$$H_3 AsO_4 + 3I^- + 2H^+ = H_3 AsO_3 + I_3^- + H_2 O$$

the rate law in the forward direction is

$$\text{rate}_f = k_f [H_3 AsO_4] [I^-] [H^+]$$

(a) Would the rate law

$$\text{rate}_r = \frac{k_r [H_3 AsO_3] [I_3^-]}{[I^-]^2 [H^+]}$$

for the reverse reaction be consistent with that of the forward reaction?

(b) Assuming the reverse rate law given in part **a** show that the equilibrium constant $K = k_f/k_r$.

(c) According to the principle of detailed balancing, forward and reverse mechanisms must involve the same species and reactions, and the transition state must be the same for both directions. Are the forward and reverse rate laws given consistent with the same transition state? [Hint: See J. O. Edwards, E. F. Greene, and J. Ross, *J. Chem. Educ.*, **45**, 381 (1968).]

REFERENCES

1. Z. G. Szabó, "Kinetic Characterization of Complex Reaction Systems," in *Comprehensive Chemical Kinetics*, Vol. 2, C. H. Bamford and C. F. H. Tipper, Eds., Elsevier, Amsterdam, 1969.

2. C. Capellos and B. H. J. Bielski, *Kinetic Systems*, Wiley-Interscience, New York, 1972.

3. D. K. Trumpy, O. A. Uyehara, and P. S. Myers, S.A.E. Paper No. 6905.18, Chicago, Illinois, 1969; D. K. Trumpy, Ph.D. Thesis, University of Wisconsin, 1969.

4. H. C. Brown and R. S. Fletcher, *J. Am. Chem. Soc.*, **71**, 1845 (1949).

5. E. D. Hughes, C. K. Ingold, and U. G. Shapiro, *J. Chem. Soc.*, **1936**, 225.

6. W. G. Young and L. J. Andrews, *J. Am. Chem. Soc.*, **66**, 421 (1944).

7. P. Beronius, A.-M. Nilsson, and A. Holmgren, *Acta Chem. Scand.*, **26**, 3173 (1972); A. R. Stein, *J. Org. Chem.*, **38**, 4022 (1973); A. R. Stein, *J. Chem. Educ.*, **52**, 303 (1975).

8. E. D. Hughes, F. Juliusburger, A. D. Scott, B. Topley, and J. Weiss, *J. Chem. Soc.*, **1936**, 1173.

9. W. Esson, *Philos. Trans. R. Soc. (London)*, **156**, 220 (1866).

10. J. L. Jensen and P. A. Lenz, *J. Am. Chem. Soc.*, **100**, 1291 (1978).

11. (a) C. G. Swain, *J. Am. Chem. Soc.*, **66**, 1696 (1944); (b) F. Kaufler, *Z. Phys. Chem. (Leipz.)*, **55**, 502 (1906).

12. (a) R. E. Powell, private communication; D. French, *J. Am. Chem. Soc.*, **72**, 4806 (1950). (b) D. Margerison, "The Treatment of Experimental Data," in *Comprehensive Chemical Kinetics*, Vol. 1, C. H. Bamford and C. F. H. Tipper, Eds., Elsevier, Amsterdam, 1969.

13. W. C. Schwemer, Ph.D. thesis, Northwestern University, 1950; W. C. Schwemer and A. A. Frost, *J. Am. Chem. Soc.*, **73**, 4541 (1951); A. A. Frost and W. C. Schwemer, *J. Am. Chem. Soc.*, **74**, 1268 (1952); C. A. Burkhard, *Ind. Eng. Chem.*, **52**, 678 (1960).

14. N. W. Alcock, D. J. Benton, and P. Moore, *Trans. Faraday Soc.*, **66**, 2210 (1970).

15. A. Skrabal, *Homogenkinetik*, Steinkopf, Dresden and Leipzig, 1941, pp. 190–199; A. Rakowski, *Z. Phys. Chem. (Leipz.)*, **57** 321 (1907); B. J. Zwolinski and H. Eyring, *J. Am. Chem. Soc.*, **69**, 2702 (1947); F. A. Matsen and J. L. Franklin, *J. Am. Chem. Soc.*, **72**, 3337 (1950).

16. T. M. Lowry and W. T. John, *J. Chem. Soc. Trans.*, **1910**, 2634.

17. (a) See reference 1, pp. 29–31, and references therein; (b) see reference 2, pp. 35–38, and references therein; (c) for the solution to a similar problem using matrix methods see E. S. Lewis and M. D. Johnson, *J. Am. Chem. Soc.*, **82**, 5399 (1960).

18. T. S. Peterson, *Chem. Eng. Process Symp. Ser.*, No. 31, 56 (1960).

19. J. Wei and C. D. Prater, in *Advances in Catalysis*, Vol. 13, Academic, New York, 1962, pp. 203–392.

20. E. McLaughlin and R. W. Rozett, *Chem. Technol.*, **1**, 120 (1971); E. McLaughlin and R. W. Rozett, *J. Chem. Educ.*, **49**, 482 (1972).

21. E. Abel, *Z. Phys. Chem. (Leipz.)*, **56**, 558 (1906).

22. C. K. Ingold, *J. Chem. Soc.*, **1931**, 2170.

23. O. Knoblauch, *Z. Phys. Chem., (Leipz.)*, **26**, 96 (1898).

24. A. Skrabal, *Monatsh. Chem.*, **38**, 29, 159 (1917).

25. J. Meyer, *Z. Phys. Chem. (Leipz.)*, **67**, 272 (1909).

26. J. Burkus and C. F. Eckert, *J. Am. Chem. Soc.*, **80**, 5948 (1958).

27. V. O. Reikhsfeld and V. A. Prokhorova, *Kinet. Katal.*, **4**, 483 (1963).

28. S. Wideqvist, *Ark. Kemi*, **8**, 545 (1956); S. Wideqvist, *Acta Chem. Scand.*, **16**, 1119 (1962).

29. S. Wideqvist, *Acta Chem. Scand.*, **4**, 1216 (1950).

30. W. G. McMillan, *J. Am. Chem. Soc.*, **79**, 4838 (1957).

31. F. A. Kundell, D. J. Robinson, and W. J. Svirbely, *J. Phys. Chem.*, **77**, 1552 (1973).

32. S. W. Benson, *The Foundations of Chemical Kinetics*, McGraw-Hill, New York, 1960.

33. M. Bodenstein, *Z. Phys. Chem. (Leipz.)*, **13**, 56 (1894); **22**, 1 (1897).

34. C. M. Guldberg and P. Waage, *J. Prakt. Chem.*, **19**, 71 (1879); collected papers in Ostwald, *Klassiker der exakten Wissenschaften*, W. Englemann, Leipzig, 1899, No. 104.

35. A. A. Frost, *J. Chem. Educ.*, **18**, 272 (1941).

36. R. C. Tolman, *Phys. Rev.*, **23**, 699 (1924); R. C. Tolman, *The Principles of Statistical Mechanics*, Oxford University Press, London, 1938, pp. 152–165.

37. L. Onsager, *Phys. Rev.*, **37**, 405 (1931).

38. B. W. Morrissey, *J. Chem. Educ.*, **52**, 296 (1975); B. H. Mahan, *J. Chem. Educ.*, **52**, 299 (1975).

39. M. Manes, L. J. E. Hofer, and S. Weller, *J. Chem. Phys.*, **18**, 1355 (1950); see also C. A. Hollingsworth, *J. Chem. Phys.*, **27**, 1346 (1957); J. Horiuti, *Z. Phys. Chem. (Neue Folge)*, **12**, 321 (1957).

40. For more detailed treatments than can be given here see G. G. Hammes, Ed., *Investigation of Rates and Mechanisms of Reactions*, Part II, in Techniques of Chemistry, Vol. 6, A. Weissberger, Ed., Wiley-Interscience, New York, 1974; J. N. Bradley, *Fast*

Reactions, Clarendon Press, Oxford, 1975; J. E. Cross, in *Chemical Kinetics* Vol. 9, J. C. Polanyi, Ed., in MTP International Review of Science, Physical Chemistry Series One, Butterworths, London, 1972.

41. J. C. Maxwell, *Philos. Mag. (IV),* **35,** 134 (1868).

42. (a) M. Eigen, G. Kurtze, and K. Tamm, *Z. Elektrochem.,* **57,** 103 (1953); (b) M. Eigen, *Disc. Faraday Soc.,* **17,** 194 (1954); (c) R. G. Pearson, *Disc. Faraday Soc.,* **17,** 187 (1954); (d) M. Eigen, *Z. Elektrochem.,* **64,** 115 (1960); (e) M. Eigen and K. Tamm, *Z. Elektrochem.,* **66,** 93 (1962); (f) R. A. Alberty and G. G. Hammes, *Z. Elektrochem.,* **64,** 124 (1960).

43. H. A. C. McKay, *Nature,* **142,** 997 (1938); H. A. C. McKay, *J. Am. Chem. Soc.,* **65,** 702 (1943).

44. S. Z. Roginsky, *Acta Physiochim.,* **14,** 1 (1941).

45. R. B. Duffield and M. Calvin, *J. Am. Chem. Soc.,* **68,** 557 (1946).

46. T. H. Norris, *J. Phys. Colloid Chem.,* **54,** 777 (1950).

47. G. M. Harris, *Trans. Faraday Soc.,* **47,** 716 (1951).

48. For recent reviews see W. H. Saunders, Jr., in *Investigation of Rates and Mechanisms of Reactions,* E. S. Lewis, Ed., in Techniques of Chemistry, Vol. 6, A. Weissberger, Ed., Wiley-Interscience, New York, 1974; J. Bigeleisen, M. W. Lee, and F. Mandel, *Ann. Rev. Phys. Chem.,* **24,** 407 (1973); M. Wolfsberg, *Ann. Rev. Phys. Chem.,* **20,** 449 (1969).

49. L. Volk, W. Richardson, K. H. Lau, M. Hall, and S. H. Lin, *J. Chem. Educ.,* **54,** 95 (1977); see also H. S. Johnston, *Gas Phase Reaction Rate Theory,* Ronald Press, New York, 1966, pp. 329–332.

50. K. Frei and H. Günthard, *Helv. Chim. Acta,* **50,** 1294 (1967).

51. M. Bodenstein, *Z. Elektrochem.,* **24,** 183 (1918); M. Bodenstein, *Z. Phys. Chem.,* **100,** 68 (1922).

52. M. Bodenstein, *Helv. Chim. Acta,* **18,** 743 (1935); O. K. Rice, *J. Chem. Phys.,* **4,** 53 (1936).

53. H. Gershinowitz and H. Eyring, *J. Am. Chem. Soc.,* **57,** 985 (1935).

54. A. B. Hoffman and H. Taube, *Inorg. Chem.,* **7,** 1971 (1968); R. C. Patel, R. E. Ball, J. F. Endicott, and R. G. Hughes, *Inorg. Chem.,* **9,** 23 (1970).

55. H. Taube and F. A. Posey, *J. Am. Chem. Soc.,* **75,** 1463 (1953).

56. For several other examples see J. H. Espenson, "Homogeneous Inorganic Reactions," in *Investigation of Rates and Mechanisms of Reactions,* Part I, E. S. Lewis, Ed., Techniques of Chemistry, Vol. 6, A. Weissberger, Ed., Wiley-Interscience, New York, 1974.

57. F. Daniels and E. H. Johnston, *J. Am. Chem. Soc.,* **43,** 53 (1921).

58. J. P. Birk, *J. Chem. Educ.,* **53,** 704 (1976) and references therein.

59. J. P. Birk, *J. Am. Chem. Soc.,* **91,** 3189 (1969).

60. L. A. Farrow and D. Edelson, *Int. J. Chem. Kinet.,* **6,** 787 (1974); D. Edelson, *J. Comp. Phys.,* **11,** 455 (1973); D. Edelson and D. L. Allara, *AIChE J.,* **19,** 638 (1973).

61. J. Janata, "Analog Computer Simulation of Kinetic Models," in *Computers in Chemistry and Instrumentation,* Vol. 3, J. S. Mattson, H. B. Mark, Jr., and H. C. MacDonald, Jr., Eds., Dekker, New York, 1973, and references therein; T. R. Crossley and M. A. Slifkin, "Computers in Reaction Kinetics," in *Progress in Reaction Kinetics,* Vol. 5, G. Porter, Ed., Pergamon Press, Oxford, 1970, p. 409.

62. R. F. Williams and T. C. Bruice, *J. Am. Chem. Soc.,* **98,** 7752 (1976).

63. (a) G. Buzzi Ferraris, G. Donati, F. Rejna, and S. Carra, *Chem. Eng. Sci.,* **29,** 1621 (1974); (b) S. J. Wajc, A. H. Mansour, and R. Jottrand, *Rev. Inst. Fr. Pet. Ann. Combust. Liq.,* **20,** 849 (1965).

64. R. S. Butter and P. A. D. DeMaine, *Top. Curr. Chem.*, **58**, 39 (1975); D. Edelson, *Comput. Chem.*, **1**, 29 (1976); R. N. Stabler and J. P. Chesick, *Int. J. Chem. Kinet.*, **10**, 461 (1978); B. C. Finzel and J. W. Moore, *J. Chem. Educ.*, **57**, 250 (1980); see also references 66 and 68.

65. C. F. Curtiss and J. O. Hirschfelder, *Proc. Natl. Acad. Sci. US*, **38**, 235 (1952).

66. L. Lapidus and J. H. Seinfeld, *Numerical Solution of Ordinary Differential Equations*, Academic, New York, 1971; C. W. Gear, *Numerical Initial-Value Problems in Ordinary Differential Equations*, Prentice-Hall, Englewood Cliffs, NJ, 1971; R. A. Willoughby, *Stiff Differential Systems*, Plenum, New York, 1974.

67. C. W. Gear, *Commun. ACM*, **14**, 176 (1971); *ibid.*, 185 (1971).

68. L. Lapidus and W. E. Schiesser, *Numerical Methods for Differential Systems*, Academic, New York, 1976; D. D. Warner, *J. Phys. Chem.*, **81**, 2329 (1977) and references therein; A. Jones, "Recent Advances in the Analysis of Kinetic Data," in *Reaction Kinetics*, Vol. 1, P. G. Ashmore, Ed., Specialist Periodical Report, The Chemical Society, London, 1975; M. L. Michelsen, *Chem. Eng. J.*, **14**, 107 (1977); A. W. Weimer and D. E. Clough, *AIChE J.*, **25**, 730 (1979); J. P. Kennealy, Ph.D. Thesis, Utah State University, 1978.

69. D. D. Warner, "A Partial Derivative Generator," Computer Science Technical Report No. 28, Bell Laboratories, Murray Hill, NJ, 1975; J. P. Kennealy and W. M. Moore, *J. Phys. Chem.*, **81**, 2413 (1977).

70. "Food Quality Improvement Through Kinetic Studies and Modeling," *Food Technol.*, **34**, (1980).

71. (a) M. L. Kibby, *Nature*, **222**, 298 (1969); (b) T. Nakanishi, *J. Phys. Soc. Jap.* **32**, 1313 (1972); (c) D. A. Dixon and R. H. Shafer, *J. Chem. Educ.*, **50**, 648 (1973); (d) W. D. C. Moebs, *Math. Biosci.*, **22**, 113 (1974); (e) D. L. Bunker, B. Garrett, T. Kleindienst, and G. S. Long, III, *Combust. Flame*, **23**, 373 (1974); (f) J. Weber and F. Celardin, *Chimia*, **30**, 236 (1976); (g) L. J. Schaad, *J. Am. Chem. Soc.*, **85**, 3588 (1963).

72. D. T. Gillespie, *J. Phys. Chem.*, **81**, 2340 (1977); D. T. Gillespie, *J. Comput. Phys.*, **22**, 403 (1976).

73. D. A. McQuarrie, *J. Appl. Prob.*, **4**, 413 (1967); D. A. McQuarrie, *Stochastic Approach to Chemical Kinetics*, Vol. 8, in Supplemental Review Series in Applied Probability, Methuen, London, 1967.

74. J. S. Turner, *J. Phys. Chem.*, **81**, 2379 (1977); W. W. Wood in *Fundamental Problems in Statistical Mechanics*, Vol. 3, E. D. G. Cohen, Ed., North-Holland, Amsterdam, 1975, p. 331.

75. I. Prigogine and G. Nicolis, *J. Chem. Phys.*, **46**, 3542 (1967); G. Nicolis and I. Prigogine, *Self-Organization in Nonequilibrium Systems*, Wiley-Interscience, New York, 1977.

76. W. E. Deming, *Statistical Adjustment of Data*, Wiley, New York, 1943, p. 148.

77. D. W. Marquardt, *J. Soc. Ind. Appl. Math.*, **11**, 431 (1963); R. Fletcher, AERE Harwell Report No. R6799, United Kingdom Atomic Energy Authority, 1971.

78. R. W. Atherton, R. B. Schainker, and E. R. Ducot, *AIChE J.*, **21**, 441 (1975); R. P. Dickinson and R. J. Gelinas, *J. Comp. Phys.*, **21**, 123 (1976); J.-T. Hwang, E. P. Dougherty, S. Rabitz, and H. Rabitz, *J. Chem. Phys.*, **69**, 5180 (1978); E. Dougherty, J.-T. Hwang, and H. Rabitz. *J. Chem. Phys.*, **71**, 1794 (1979).

79. R. I. Cukier, H. B. Levine, and K. E. Shuler, *J. Comp. Phys.*, **26**, 1 (1978) and references therein; M. Koda, G. J. McRae, and J. H. Seinfeld, *Int. J. Chem. Kinet.*, **11**, 427 (1979).

80. W. C. Gardiner, Jr., *J. Phys. Chem.*, **83**, 37 (1979).

81. N. R. Draper and H. Smith, *Applied Regression Analysis*, Wiley, New York, 1967.

82. G. E. P. Box and W. J. Hill, *Technometrics*, **9**, 57 (1967).

83. D. B. Olson and W. C. Gardiner, Jr., *J. Phys. Chem.*, **81**, 2514 (1977).

84. C. K. Westbrook and F. L. Dryer, *Combust. Sci. Technol.*, **20**, 125 (1979).

85. D. K. Stull and H. Prophet, Eds., *JANAF Thermochemical Tables*, Dow Chemical Co., Midland, Michigan, 1971.

86. C. Lund, *HCT: A General Computer Program for Calculating Time-Dependent Phenomena Involving One-Dimensional Hydrodynamics, Transport, and Detailed Chemical Kinetics*, University of California Lawrence Livermore Laboratory Report UCRL-52504, 1978.

87. D. Edelson and N. L. Schryer, *Comput. Chem.*, **2**, 71 (1978).

88. A. Th. Fechner, *Schweigg.*, **53**, 141 (1828).

89. W. Ostwald, *Phys. Z.*, **8**, 87 (1899); J. Liesegang, *Naturwiss. Wochschr.*, **11**, 353 (1896).

90. A. J. Lotka, *J. Phys. Chem.*, **14**, 271 (1910); A. J. Lotka, *J. Am. Chem. Soc.*, **42**, 1595 (1920).

91. W. C. Bray, *J. Am. Chem. Soc.*, **43**, 1262 (1921).

92. H. Degn, *J. Chem. Educ.*, **49**, 302 (1972); J. J. Tyson, *J. Chem. Phys.*, **62**, 1010 (1975).

93. I. Prigogine and G. Nicolis, *J. Chem. Phys.*, **46**, 3542 (1967); P. Glaunsdorff and I. Prigogine, *Thermodynamic Theory of Structure, Stability, and Fluctuations*, Wiley-Interscience, New York, 1971.

94. P. Gray, J. F. Griffiths, and R. J. Moule, "Physical Chemistry of Oscillatory Phenomena," *Faraday Symp. Chem. Soc.*, **9**, 103 (1974).

95. B. Chance, A. K. Ghosh, E. K. Pye, and B. Hess, Eds., *Biological and Biochemical Oscillators*, Academic, New York, 1973.

96. G. Nicolis and I. Prigogine, *Self-Organization in Nonequilibrium Systems*, Wiley-Interscience, New York, 1977; T. Ganti, *A Theory of Biochemical Supersystems*, University Park Press, Baltimore, 1979.

97. (a) R. M. Noyes and R. J. Field, *Ann. Rev. Phys. Chem.*, **25**, 95 (1974); (b) B. F. Gray, "Kinetics of Oscillatory Reactions," in *Reaction Kinetics*, Vol. 1, P. G. Ashmore, Ed., Specialist Periodical Report, The Chemical Society, London, 1975; (c) A. Pacault, P. Hanusse, P. DeKepper, C. Vidal, and J. Boissonade, *Acc. Chem. Res.*, **9**, 438 (1976); (d) R. J. Field and R. M. Noyes, *Acc. Chem. Res.*, **10**, 214 (1977); (e) R. M. Noyes and R. J. Field, *Acc. Chem. Res.*, **10**, 273 (1977); (f) D. O. Cooke, *Prog. React. Kinet.*, **8**, 185 (1977); (g) H. Eyring and D. Henderson, Eds., *Theoretical Chemistry, Periodicities in Chemistry and Biology*, Vol. 4, Academic, New York, 1978.

98. B. P. Belousov, *Sb. Ref. Radiat. Med.*, **1**, 145 (1959).

99. A. M. Zhabotinskii, *Dokl. Akad. Nauk SSSR*, **157**, 392 (1964); A. M. Zhabotinskii, *Biofizika*, **9**, 306 (1964); V. A. Vavilin, A. M. Zhabotinskii, and A. N. Zaikin, *Russ. J. Phys. Chem.*, **42**, 1649 (1968).

100. R. J. Field, E. Körös, and R. M. Noyes, *J. Am. Chem. Soc.*, **94**, 8649 (1972).

101. R. J. Field and R. M. Noyes, *J. Chem. Phys.*, **60**, 1877 (1974).

102. K. Showalter, R. M. Noyes, and K. Bar-Eli, *J. Chem. Phys.*, **69**, 2514 (1978); D. Edelson, R. J. Field, and R. M. Noyes, *Int. J. Chem. Kinet.*, **7**, 417 (1975); see reference 97e, and R. M. Noyes, *J. Phys. Chem.*, **81**, 2315 (1977) for comments regarding the latter paper.

103. T. M. Barbara and P. L. Corio, *J. Chem. Educ.*, **57**, 243 (1980).

104. E. L. King, *J. Chem. Educ.*, **56**, 580 (1979).

NINE

HOMOGENEOUS CATALYSIS

Although a great many catalytic reactions of industrial importance are heterogeneous in that the catalyst is present as a distinct solid phase, we restrict ourselves here to the subject of homogeneous catalysis; this is partly because a study of homogeneous catalysis is of more general importance in understanding the mechanism of all chemical reactions and partly because heterogeneous catalysis is an extensive, largely self-contained subject in itself. Consequently its discussion would lead us too far afield.[1]

The accepted definition of a catalyst, due to Ostwald, is that it is a substance that changes the speed of a chemical reaction without undergoing any chemical change itself. Since a reactant or a product may also be a catalyst as well, Bell[2] suggests the definition, "a catalyst is a substance which appears in the rate expression to a power higher than that to which it appears in the stoichiometric equation."

Actually many substances classified as catalysts are destroyed either as a result of the process that gives them their catalytic activity or because of subsequent combination with the products. From a practical point of view, a catalyst is a substance that changes the rate of a desired reaction, regardless of the fate of the catalyst itself. This viewpoint is expressed when we speak of the base-catalyzed halogenation of ketones and similar compounds in which the catalyzing base is converted into its conjugate acid and hence destroyed as a catalyst.

$$CH_3-\overset{O}{\underset{\|}{C}}-CH_3 + B + Br_2 \rightarrow CH_3-\overset{O}{\underset{\|}{C}}-CH_2Br + BH^+ + Br^- \qquad (9.1)$$

The catalyst is used up, but it is recognized again that we have a possible reaction that is very slow in the absence of the base and that the chief function of the base is to speed up the reaction and not to remove the resulting acid.

A somewhat less clearly defined case appears in a number of free-radical chain reactions that are initiated by the addition of small amounts of substances that readily decompose to yield free radicals. For example, benzoyl peroxide is used to polymerize styrene, and the addition of dimethyl mercury enables the pyrolysis of propane to occur at a lower temperature. The essential feature here is that the added substance must be completely destroyed to have any effect, for example, dimethyl mercury, breaking down into mercury atoms and methyl radicals. The methyl radicals then cause a decomposition of the propane by a chain process. It is true that the rate of the desired reaction is greatly increased and that only small or "catalytic" quantities of dimethyl mercury are needed. It is quite common to speak of such added substances as catalysts though the correct term is "initiator."

Completely indefensible is the common practice of saying "the reaction is catalyzed by light" or "the reaction is catalyzed by heat." Such situations where the rate of a reaction is changed by an agency other than a substance should never be classed as catalytic. Also, when the rate is changed by switching to another solvent, it is desirable to discuss the change in rate as being due to a change in medium without speaking of the "catalytic effect" of the solvent. This term has a proper usage as, for example, in acid–base catalysis.

As we see in the discussion that follows, an important criterion of a catalyst is that it changes the mechanism of the parent reaction. Indeed without this change in mechanism, the observed change in rate could not occur. Since most catalysts increase the rate of reaction, it follows that the mechanism must change to one that is easier for the system to follow, involving in general a lower energy barrier. It is frequently said that the function of a catalyst is to lower the activation energy of a given reaction. More correctly, the catalyst changes the mechanism to one having a lower activation energy.

It is true that there are also negative catalysts that slow the rate of a reaction. Actually examples of this behavior do not conform to the strict definition of catalysis, since the negative catalyst is used up or permanently altered. Such substances are properly called inhibitors, and the mechanism of their action is discussed in Chapter 10.

HOMOGENEOUS CATALYSIS IN THE GAS PHASE

A familiar example of catalysis in reactions involving gases is the oxidation of sulfur dioxide to sulfur trioxide by the catalytic action of oxides of nitrogen. There are several probable steps that would be quite typical of catalyzed reactions. The direct oxidation of sulfur dioxide with oxygen is a slow process.

$$2SO_2 + O_2 \xrightarrow{\text{slow}} 2SO_3 \tag{9.2}$$

Because it must occur either by a trimolecular collision or by a very high-energy bimolecular or unimolecular process. But the two reactions

$$2NO + O_2 \rightarrow 2NO_2 \tag{9.3}$$

$$NO_2 + SO_2 \rightarrow NO + SO_3 \tag{9.4}$$

can both occur with reasonable speeds, since trimolecular reactions with nitric oxide are well known, and since nitrogen dioxide can oxidize sulfur dioxide in a low-energy bimolecular process. Since the sum of (9.3) and (9.4), after (9.4) is doubled, is equal to (9.2), it is seen that one slow reaction is replaced by two faster ones to give the same chemical result. Nitrogen dioxide (which is in equilibrium with the tetroxide, N_2O_4) is formed as an intermediate product from which nitric oxide is regenerated again. Schematically many catalytic reactions can be represented as

$$\text{(uncatalyzed) A} + \text{B} \xrightarrow{\text{slow}} \text{products} \tag{9.5}$$

$$\text{(catalyzed) A} + \text{C} \xrightarrow{\text{fast}} \text{X} \tag{9.6}$$

$$\text{X} + \text{B} \xrightarrow{\text{fast}} \text{products} + \text{C} \tag{9.7}$$

where C is the catalyst and the X is the intermediate. The rate of the uncatalyzed reaction is given by

$$\text{rate} = k_1 [\text{A}] [\text{B}] \tag{9.8}$$

and for the catalyzed reaction

$$\text{rate} = k_2 [\text{A}] [\text{C}] \tag{9.9}$$

if (9.7) is much more rapid than (9.6). Equation (9.9) predicts that the rate of the catalyzed reaction is first-order in the catalyst concentration; this is found to be true experimentally in the majority of catalyzed reactions. Actually it need not be, for, depending on the complexity of the mechanism, the rate expression may contain terms independent of the catalyst, terms involving various positive powers of the catalyst concentration, and even terms inversely proportional to the catalyst concentration.

Molecular iodine is an effective catalyst for many thermal decompositions in the gas phase. Since most pyrolytic decompositions of organic compounds occur by a free-radical mechanism, the most reasonable explanation for the catalysis is that iodine molecules dissociate into iodine atoms, which then attack the organic molecule. In this way a chain reaction is started. Consider the decomposition of acetaldehyde as a more specific example. The kinetics are complicated, the order of the reaction being between 1 and 2, depending on the pressure. A free-radical chain mechanism is indicated, for example:

$$CH_3CHO \rightarrow CH_3\cdot + H\cdot + CO \tag{9.10}$$

$$CH_3\cdot \text{ (or H)} + CH_3CHO \rightarrow CH_4 \text{ (or } H_2\text{)} + CH_3\cdot + CO \tag{9.11}$$

The methyl radicals produced in (9.11) then carry the chain, the net result being that a mole of acetaldehyde gives essentially a mole of carbon monoxide and a mole of methane. Iodine probably enters into the reaction as follows[3]:

$$I_2 \rightleftharpoons 2I\cdot \tag{9.12}$$

$$I\cdot + CH_3CHO \rightarrow HI + CH_3\cdot + CO \tag{9.13}$$

$$CH_3 \cdot + I_2 \rightarrow CH_3I + I \cdot$$
$$CH_3 \cdot + HI \rightarrow CH_4 + I \cdot \tag{9.14}$$

The iodine is eventually regenerated, since the final reaction is

$$CH_3I + HI \rightarrow CH_4 + I_2 \tag{9.15}$$

Consequently the net result in the presence of iodine is the same as in its absence, acetaldehyde being decomposed into methane and carbon monoxide.

The acceleration in rate observed when even small amounts of iodine are added is several thousandfold. This is due to a lowering of the activation energy from about $210\,kJ\,mol^{-1}$ for the uncatalyzed reaction (depending on the pressure) to $136\,kJ\,mol^{-1}$ for the catalyzed reaction. This lowering reflects the greater ease of breaking an iodine–iodine bond ($153\,kJ\,mol^{-1}$) than a carbon–carbon bond ($335\,kJ\,mol^{-1}$) or a carbon–hydrogen bond ($420\,kJ\,mol^{-1}$). The rate of decomposition varies directly with the initial iodine concentration, being given by the expression

$$\text{rate} = k[CH_3CHO][I_2]_0 \tag{9.16}$$

for acetaldehyde. Actually the iodine concentration varies during the reaction, but a steady state is reached during the period of experimental measurements. Other reactions catalyzed by iodine have mechanisms leading to a different dependence on the iodine concentration.

An example of catalysis in the gas phase not involving free radicals is the dehydration of tertiary alcohols catalyzed by hydrogen halides.[4] The reaction

$$(CH_3)_3COH \rightarrow (CH_3)_2C=CH_2 + H_2O \tag{9.17}$$

has a rate constant equal to $4.8 \times 10^{14}\, e^{-274\,kJ\,mol^{-1}/RT}\,s^{-1}$ at high temperatures in the vapor. If hydrogen bromide is added, a great increase in rate occurs with no change in the hydrogen halide, which thus acts as a true catalyst. The rate is directly proportional to the concentration of added HBr. The rate constant becomes second order for the catalyzed reaction and is given by $k = 9.2 \times 10^9\, e^{-127\,kJ\,mol^{-1}/RT}\,M^{-1}\,s^{-1}$. Maccoll and Stimson[4] suggest that HBr adds to the alcohol to give a polar complex resembling an ion pair. Hydrogen halides and other acids catalyze a number of other elimination reactions of organic compounds in the vapor state.

Among the most important gas-phase reactions is the decomposition of ozone catalyzed by oxides of nitrogen and by various chlorine-containing compounds such as the freons.

$$2O_3 \xrightarrow{\text{catalyst}} 3O_2 \tag{9.18}$$

These reactions, occurring in the upper atmosphere, reduce the steady-state concentration of ozone appreciably. Since ozone plays an important role in moderating the quantity of ultraviolet light that reaches the earth's surface, this is a serious consequence.

HOMOGENEOUS CATALYSIS IN SOLUTION – METAL IONS

The earliest observed cases of catalysis in solution and still the most frequently encountered are acid- or base-catalyzed reactions. Ester hydrolysis may be mentioned as a familiar example. A more general example of acid catalysis occurs in the decarboxylation of dimethyloxaloacetic acid.[5] The dianion of this acid loses carbon dioxide, forming an enolate dianion:

$$^-O_2C-\overset{\overset{\displaystyle O}{\|}}{C}-C(CH_3)_2-CO_2^- \rightarrow {}^-O_2C-\overset{\overset{\displaystyle O^-}{|}}{C}=C(CH_3)_2 + CO_2 \qquad (9.19)$$

Placing a proton on the carboxylate group or esterifying the carboxyl gives a monoanion, which decarboxylates more rapidly:

$$RO_2C-\overset{\overset{\displaystyle O}{\|}}{C}-C(CH_3)_2-CO_2^- \rightarrow RO_2C-\overset{\overset{\displaystyle O^-}{|}}{C}=C(CH_3)_2 + CO_2 \qquad (9.20)$$

where $R = H$ or C_2H_5. A reasonable explanation is that in losing carbon dioxide, the rest of the molecule must absorb the pair of electrons initially bonding the carboxylate group. Reducing the negative charge on the rest of the molecule must facilitate this transfer of electrons.[5] In agreement with this explanation a number of multiply charged cations act as catalysts for the decarboxylation of the dianion of dimethyloxaloacetic acid. Presumably a complex is formed that reduces the negative charge and increases the ease of accepting a pair of electrons from the carboxyl group, as shown in (9.21).

$$O=\overset{\displaystyle C}{\underset{\displaystyle O}{\diagup}}-\overset{\displaystyle C}{\underset{\displaystyle O^+}{\diagdown\!\!\!\|}}-C(CH_3)_2-CO_2^- \rightarrow O=\overset{\displaystyle C}{\underset{\displaystyle O}{\diagup}}-\overset{\displaystyle C}{\underset{\displaystyle O}{\diagdown}}=C(CH_3)_2 + CO_2$$

$$\underset{\diagup\ \diagdown}{\overset{\diagdown\ \diagup}{Cu}} \qquad\qquad\qquad \underset{\diagup\ \diagdown}{\overset{\diagdown\ \diagup}{Cu}} \qquad (9.21)$$

The catalytic efficiency of a metal ion depends both on its positive charge and on its ability to form a stable complex of the chelate type. The metal ions may be considered to be acting as generalized acids.

Metal ions have some important advantages over the proton. They can have greater charges, leading to greater polarization of the reactant molecule. They can be stabilized by other ligands so that, unlike the proton, they can exist in neutral or even basic solutions. Their high coordination numbers permit the binding of a substrate at more than one site, as shown in (9.21). This advantage helps to make metal ions very efficient catalysts for the hydrolysis of α-amino esters.[6]

$$(9.22)$$

Even more important is the ability of a metal ion to simultaneously bind both a substrate and a reagent. This can have the effect of a template, in which the two reactants are assembled prior to combination.[7]

Consider the following two equilibria in aqueous solution at $25°C$.

$$H_3O^+ \rightleftharpoons H_2O + H^+ \qquad pK_a = -1.74 \qquad (9.23)$$

$$Zn(H_2O)_6^{2+} \rightleftharpoons Zn(H_2O)_5OH^+ + H^+ \qquad pK_a = 8.8 \qquad (9.24)$$

Coordination of a water molecule to a proton obviously creates a much stronger Brønsted acid than coordination to a zinc ion, in spite of the greater positive charge of the latter. More important a hydroxide ion bound to a zinc ion is much more basic (by $10.5\,pK$ units) than a hydroxide ion that has combined with a proton to form water.

The significance of this is that a ligand bound to a metal ion may still have very appreciable basic and nucleophilic character. Indeed it is often found that such ligands are quite reactive.[8] Table 9.1 shows some rate data for the reaction of carbon dioxide in water with various bases. While the products may be varied, the rate step is attack of the base at the carbon atom in all cases.

$$B + CO_2 \xrightarrow{k} B^+\!-\!C\!\!\begin{array}{c} {}^{\displaystyle O} \\[-2pt] {}^{\displaystyle \diagdown O} \end{array}\!\!- \qquad (9.25)$$

From the pK_a values, it can be seen that the basicity is important in determining nucleophilic reactivity. However, it is not the only factor, since secondary amines are less basic than OH^-, but react more rapidly. It can be seen that metal-bound hydroxide ion has quite an appreciable reactivity, consistent with the pK_a values.

The important point is that, unlike OH^- and the amines, a catalyst such as $Co(NH_3)_5OH^{2+}$ can exist at physiological pH values. In living systems the hydration of carbon dioxide by water is too slow to keep up with metabolic demands. A series of enzymes called carbonic anhydrases is needed. These are zinc-bearing enzymes and it is believed that the activity results from nucleophilic attack on CO_2 of a zinc bound hydroxide.[9]

$$E\!-\!ZnOH + CO_2 \rightarrow E\!-\!Zn\!-\!OCO_2H \qquad (9.26)$$

The complexes of zinc(II) are usually very labile and the bicarbonate ligand can quickly be exchanged for water. From the pH dependence of the enzyme activity, the pK_a must be about 7.5. This is a reasonable value. The enzyme has an extraordinarily large rate constant as shown in Table 9.1. We return to this point later.

Many metal ions, particularly of the transition series have several stable oxidation states. This enables them to act as catalysts in certain redox reactions. For example, the direct oxidation of thallous ion with ceric ion is a very slow process in spite of a favorable value for the oxidation potential.

Table 9.1. Second-order rate constants and pK_a's for the reaction of bases with carbon dioxide in water at 25°C

Base	pK_a	$k/M^{-1}\,s^{-1}$
OH^{-a}	15.7	8.5×10^3
$NH_3{}^b$	9.2	3.5×10^2
$CH_3NH_2{}^b$	10.6	2.0×10^4
$(CH_3)_2NH^b$	10.8	5.0×10^4
Piperidine^b	11.0	5.9×10^4
Aniline^b	3.5	5.3×10
$Co(NH_3)_5OH^{2+a}$	6.6	2.2×10^2
$Rh(NH_3)_5OH^{2+a}$	6.8	4.2×10^2
$Ir(NH_3)_5OH^{2+a}$	6.7	5.9×10^2
Glygly $CuOH^{-a}$	9.4	5.8×10^2
H_2O^c	-1.7	6.7×10^{-4}
Human carbonic anhydrase C^d	~ 7.5	1.2×10^8

^{a} Reference 8
^{b} Reference 66. Corrected to 25°C from 18°C.
^{c} R. G. Khalifah, *J. Biol. Chem.*, **246**, 2561 (1971).
^{d} Reference 96.

$$2Ce^{4+} + Tl^+ \rightarrow 2Ce^{3+} + Tl^{3+} \tag{9.27}$$

The reason for the slowness is the requirement for a three-body collision. A stepwise mechanism is also slow because Ce^{2+} is unknown and Tl^{2+} is an unstable oxidation state. Manganous ion, however, catalyzes the oxidation by the following sequence:

$$Ce^{4+} + Mn^{2+} \rightarrow Ce^{3+} + Mn^{3+} \tag{9.28}$$

$$Mn^{3+} + Ce^{4+} \rightarrow Ce^{3+} + Mn^{4+} \tag{9.29}$$

$$Mn^{4+} + Tl^+ \rightarrow Mn^{2+} + Tl^{3+} \tag{9.30}$$

each two-body reaction occurring with normal velocity. Given these special properties of metal ions, it is not surprising to find them catalyzing a wide variety of reactions in organic, inorganic, and biological systems.[10]

THE BASIS OF CATALYTIC ACTION

The preceding examples show some of the ways in which catalysis can occur. A question that quickly arises is when should a reaction that goes through several steps be faster than one that can occur in a single step. If the end products are the same, there can be no difference in thermodynamic driving force. Part of the answer lies in the well-known fact that there is no necessary correlation between the rate of a reaction and the overall free energy change. Nevertheless, as we see below, the number of cases where rates and equilibria are related are sufficiently numerous so that they can be considered the rule rather than the exception. Hence catalytic reactions must either be abnormally fast or the uncatalyzed reaction must be abnormally slow. Both situations exist.

In addition to the slow three-body reactions mentioned above, a second class of slow reactions is those forbidden by orbital symmetry. All that this phrase means, as discussed in Chapter 5, is that a large energy barrier exists, even when the reaction is thermodynamically favorable. Such reactions are prime candidates for catalysis. Since it turns out that very many reactions of homonuclear diatomic molecules are symmetry-forbidden, we find many catalytic reactions of hydrogen, oxygen, and nitrogen. Molecules such as ethylene and acetylene also behave like diatomic molecules because of their high symmetries.[11]

Transition metals turn out to be the best catalysts, in most cases, for catalyzing reactions that are slow for symmetry reasons.[11] This is chiefly due to the partly filled d orbitals of such metals. The d orbitals have symmetry properties that are different from those of s and p orbitals.

To illustrate, consider the reaction of nitric oxide with carbon monoxide

$$2NO + CO \rightarrow N_2O + CO_2 \qquad \Delta G^{\ominus} = -327 \, kJ \, mol^{-1} \qquad (9.31)$$

In spite of the very favorable free energy change, this reaction does not occur to any appreciable extent, even in automobile exhaust gases at high temperatures. However, reaction (9.31) occurs quickly at room temperature with a number of transition metal complexes as catalysts.[12] The reason for the slow, direct reaction is most easily seen if it is assumed that the unstable dimer N_2O_2 forms first.

$$\underset{\text{Four } \pi e}{\overset{N-N}{\underset{O}{\diagdown}}} + \underset{\text{Two } \pi e}{:C\equiv O} \rightarrow \underset{\text{Four } \pi e}{O=N=N} + \underset{\text{Four } \pi e}{O=C=O} \qquad (9.32)$$

The number of π electrons (in orbitals above and below the plane of the page), does not match up in products and reactants. The reducing electrons of CO [shown as dots in (9.32)] are in a σ orbital. This is the source of the symmetry barrier that makes the direct reaction slow since the σ-electrons must transfer to a π-type orbital with which they have zero overlap.

The role of the metal catalyst is to coordinate the two NO molecules in adjacent positions, and then to reduce them to $N_2O_2^{2-}$, the anion of hyponitrous acid. Protonation of this to $HN_2O_2^-$ leads to a rapid decomposition to N_2O and OH^-.[13]

$$\underset{\text{Six } \pi e}{\overset{N-N}{\underset{O}{\diagdown}}} + 2e \xrightarrow{\;H_2O\;} \overset{N=N}{\underset{OH}{\diagdown}} \longrightarrow \underset{\text{Four } \pi e}{O=N=N} + \underset{\text{Two } \pi e}{OH^-} \qquad (9.33)$$

Since the lowest unoccupied molecular orbital of N_2O_2 is of π type, the ion $HN_2O_2^-$ has six π electrons. There are also six π electrons in the products of (9.33), since OH^- has two electrons in a p orbital of this symmetry.

The catalyst must next be regenerated by a two-electron reduction. This is easily effected by carbon monoxide, in a series of reactions. If M^{2+} represents the spent catalyst, and M^0 the regenerated catalyst,

$$M^{2+} + CO \longrightarrow M^{2+}-CO \xrightarrow{2OH^-} M-CO_2 + H_2O$$
$$\longrightarrow M^0 + CO_2 \qquad (9.34)$$

The overall catalytic sequence in (9.33) and (9.34) has many steps, but each of them occurs readily, as is typical of catalytic reactions.

PROTON-TRANSFER REACTIONS

Brønsted acid–base catalysis is effective because proton transfers are generally rapid compared to the making and breaking of other chemical bonds. Hence reactions involving proton transfers as in typical acid or base catalysis are rapid compared to similar reactions of comparable free-energy change. Consider the two reactions

$$CH_3OCH_3 + H_2O \rightleftharpoons CH_3O^- + CH_3OH_2^+ \qquad (9.35)$$

$$CH_3OH + H_2O \rightleftharpoons CH_3O^- + H_3O^+ \qquad (9.36)$$

which are not much different in free energy, (9.36) being favored by about $12\,kJ\,mol^{-1}$. The equilibrium constants thus differ by a factor of about 100. The variations in the rate, however, are infinitely greater: the spontaneous hydrolysis of an ether is so slow that it is never observed, whereas the second reaction is so fast that it cannot be measured by ordinary means. The difference in rates arises from the lack of steric hindrance involved in a nucleophilic displacement on hydrogen as in (9.36) compared to a nucleophilic displacement on carbon as in (9.35). Steric hindrance, that is, the repulsion of nonbonded atoms in the activated complex, is a most important factor in determining activation energies. Since the proton lacks the filled inner electron shells usually responsible for the repulsion and is not surrounded by other groups, it is quite free from steric effects.

Proton transfers involving oxygen–hydrogen bonds are generally rapid, but they are not instantaneous. In the ionization of water, for example,

$$2H_2O \rightleftharpoons H_3O^+ + OH^- \qquad (9.37)$$

it can easily be calculated that the activation energy is at least as great as $57\,kJ\,mol^{-1}$ (the heat of the reverse reaction) and that the entropy of activation is negative. The inference is that the rate constant must be small, and indeed a direct measurement gives $5 \times 10^{-7}\,M^{-1}\,s^{-1}$ for (9.37) reckoned as a bimolecular reaction.[14]

This example makes the important point that an unfavorable equilibrium constant must necessarily make a reaction slow, even if other factors are quite favorable. Given a favorable equilibrium, proton transfers involving oxygen and nitrogen bonds to hydrogen are almost always very fast, approaching diffusion control in many cases. Exceptions can occur if the proton is in a well-shielded position.[15]

Proton transfers to and from carbon atoms are usually quite slow, even for favorable equilibria. For example, for the reaction in water at $0°C$,[16]

$$C_2H_5NO_2 + OH^- \xrightarrow{k} CH_3CHNO_2^- + H_2O \qquad (9.38)$$

$$K_{eq} = \frac{K_a}{K_w} = \frac{2.67 \times 10^{-9}}{1.14 \times 10^{-15}} = 2.34 \times 10^6$$

the rate constant, k, is only $0.65\,M^{-1}\,s^{-1}$. The slowness of such reactions is usually attributed to the extensive structural and electronic rearrangement that occurs as the carbanion is formed. Also, unlike N-H or O-H bonds, the C-H bond is quite nonpolar so that prior association of bases by hydrogen bonding does not occur. In some cases proton transfer to and from carbon can be fast, for example, in the case of acetylene.[15]

$$H-C{\equiv}C-H + B \rightleftharpoons H-C{\equiv}C^- + BH^+ \qquad (9.39)$$

The explanation given is that, unlike other carbanions, the negative charge has no place to go within the molecule.

Another broad class of slow proton transfers is found for the transition metal hydrides.[17] These compounds are often quite good Brønsted acids in spite of the hydride name, for example,

$$H_2Fe(CO)_4 \rightleftharpoons H^+ + HFe(CO)_4^- \qquad pK_a = 4.44 \qquad (9.40)$$

in water at $0°C$. The slow proton transfers again result from nonpolar M-H bonds, and from extensive rearrangement accompanying proton loss or gain.

Rapid proton transfers are often conveniently followed by nuclear magnetic resonance techniques. It is necessary that the magnetically active nucleus (the proton) have two different environments available such that quite different spectra would be observed in the absence of exchange. The exchange occurs under equilibrium conditions, so that no net chemical change occurs. The exchange between sites leads to an averaging of the spectrum in a way that is predictable by the solution of the Bloch equations for magnetic resonance as modified by the inclusion of the exchange phenomenon.[18]

A simple example is provided by the system hydrogen peroxide–water in which two major environments exist for the proton.[19] Figure 9.1 shows how the NMR spectrum of such a system, approximately equimolar, would appear for various circumstances. In this case the important variables are τ, the mean lifetime of a proton in each environment, and δ, the frequency difference in hertz between the two signals in the absence of exchange. In general, values of τ of the order of $1/\delta$ can be found. For τ much larger than $1/\delta$, two separate signals are observed, and for τ much less than $1/\delta$ one averaged signal is observed. Since δ for protons is generally of the order of $100\,Hz$, τ is of the order of $0.01\,s$ for easy measurement. Acid–base catalysis or change in temperature can be used to bring about changes in τ. Table 9.2 shows some proton-transfer rate constants for amines obtained by NMR methods.

The rate constant k_1 is for the direct transfer of a proton from the ammonium ion to the amine. The rate constant k_2 is for the interesting process in which a

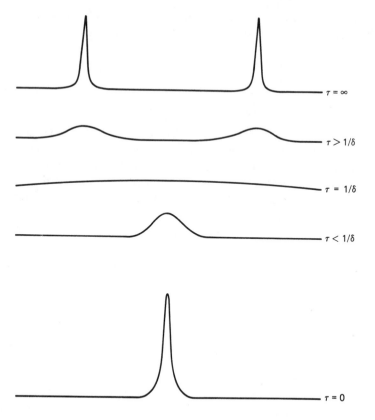

Figure 9.1. The effect of chemical exchange on NMR line width. Resonance frequency increasing from left to right.

water molecule mediates between the two reactants.

$$R_3NH^+ + \overset{H}{\underset{|}{O}}{-}H + NR_3 \xrightarrow{k_2} R_3N + H{-}\overset{H}{\underset{|}{O}} + HNR_3^+ \qquad (9.41)$$

Such an indirect proton transfer is supported not only by the NMR spectra, but also by other evidence.[20] It is quite reasonable since the water molecule would be hydrogen bonded to both reactants both before and after the exchange. Furthermore, from Table 9.2 it can be seen how the sterically hindered amines take more and more advantage of this path compared to sterically free ammonia.

A study of the exchange reaction between benzoic acid and methanol by NMR shows that two methanol molecules are required.[21] This has been explained by a cyclic transition state such as

$$\text{(9.42)}$$

Table 9.2. Rates for amine-ammonium ion proton exchange at $22°C$

System	$B + BH^+ \rightleftharpoons BH^+ + B$ $k_1/M^{-1} s^{-1}$	$k_2/M^{-1} s^{-1}$
$NH_3 + NH_4^+$	10.6×10^8	0.9×10^8
$CH_3 NH_2 + CH_3 NH_3^+$	2.5×10^8	3.4×10^8
$(CH_3)_2 NH + (CH_3)_2 NH_2^+$	0.4×10^8	5.6×10^8
$(CH_3)_3 N + (CH_3)_3 NH^+$	0.0×10^8	3.1×10^8

Source. Data from A. Loewenstein and S. Meiboom, *J. Chem. Phys.,* **27**, 1067 (1957); and S. Meiboom, A. Loewenstein, and S. Alexander, *ibid.,* **29**, 969 (1958).

Two methanol molecules have exchanged protons for every one molecule of acid. Notice that this mechanism avoids the formation of ions, a thermodynamically less favored path.

Concerted processes in which several reactant molecules are arranged in a precise fashion often appear attractive energetically. It must be remembered, however, that statistically they are rather improbable. The entropy of activation would be very unfavorable in most cases. Bordwell has shown that there is a great deal of evidence against highly concerted reactions in which simultaneous making and breaking of many bonds is required.[22] In the cases mentioned above, the fact that the extra molecules are solvent molecules, which would be hydrogen bonded in any case, makes the suggested mechanisms much more plausible.

While a proton-transfer reaction is often represented as a single step as in (9.38) or (9.39), a consideration of how bimolecular reactions occur in solution leads to a more complex mechanism.

$$AH + B^- \underset{k_{-a}}{\overset{k_a}{\rightleftharpoons}} AH\text{---}B^- \underset{k_{-b}}{\overset{k_b}{\rightleftharpoons}} A^{\doteq}\text{---}HB \underset{k_{-c}}{\overset{k_c}{\rightleftharpoons}} A^- + HB$$

$$\text{Encounter} \qquad \begin{array}{c}\text{Proton}\\\text{transfer}\end{array} \qquad \text{Separation} \qquad (9.43)$$

A special feature is the hydrogen bonding that stabilizes the encounter complex. This general mechanism was first discussed by Eigen.[23] If we apply the steady-state treatment to the concentrations of the two hydrogen-bonded species, the result is

$$k_{obs} = \frac{k_a k_b k_c}{k_{-a} k_{-b} + k_{-a} k_c + k_b k_c} \qquad (9.44)$$

for the forward reaction only.

Three limiting cases are often found, depending on the relative values of the various rate constants.

1. $k_{obs} = k_a$, encounter rate limiting $\qquad\qquad (9.45)$
2. $k_{obs} = k_b k_a / k_{-a}$, proton transfer rate limiting $\qquad\qquad (9.46)$
3. $k_{obs} = k_c k_a k_b / k_{-a} k_{-b}$, separation rate limiting $\qquad\qquad (9.47)$

Examples are known of all these cases,[24] case **1** often being diffusion controlled and leading to very large rate constants. In the case of carbon acids and other non-hydrogen-bonding acids, k_a may be much less than in the diffusion-controlled case.

The values of k_b and k_{-b} can be as large as a vibrational frequency ($\sim RT/Lh$), but there can also be a substantial energy barrier as the proton passes through a position half-way between A^- and B^-. Figure 9.2 shows a typical one-dimensional potential-energy diagram for a hydrogen-bonded system. The hydrogen atom has two equilibrium positions, close to A^- or to B^-. A few rare cases exist, such as HF_2^-, in which the hydrogen atom lies equidistant from each heavy atom, and a single potential minimum exists. The potential barrier in Fig. 9.2 can be as large as $20-40\,kJ\,mol^{-1}$. Each $5.7\,kJ\,mol^{-1}$ would lower the proton transfer rate by a factor of 10.

Case **3** involves the phenomenon of internal return and is well documented in isotope exchange studies of weak carbon acids.[25]

$$B^- + D-R \rightleftharpoons BD + R^- \xrightarrow{\text{BH}} R-H + B^- \qquad (9.48)$$

The slow step is escape of the carbanion, R^-, into the solvent pool. Reaction (9.39) is an example of extensive internal return.[15]

In spite of the complex mechanism, proton transfers still are rapid. It remains to be shown how this property can lead to acid—base catalysis. By considering the addition reactions of nucleophiles to the carbonyl group, Jencks has shown how acid—base catalysis must occur in some situations.[26] A good illustration is given by the aminolysis of simple esters.[27]

$$RNH_2 + CH_3COOR' \rightarrow CH_3CONHR + R'OH \qquad (9.49)$$

In such reactions, also called acyl transfers, an important feature is the formation

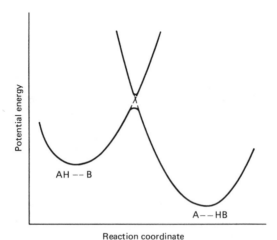

Figure 9.2. Potential energy diagram for proton-transfer process. The curve is generated by superimposing two diatomic potential curves (HA and HB) with allowance for an interaction that lowers the energy.

of the so-called tetrahedral intermediate.[28]

$$RNH_2 + CH_3COOR' \underset{k_{-a}}{\overset{k_a}{\rightleftharpoons}} \overset{+}{R}NH_2 - \underset{\underset{CH_3}{|}}{\overset{\overset{O^-}{|}}{C}} - OR'$$
$$T^{\pm} \tag{9.50}$$

The intermediate is labeled $T^{\pm}$ since it is a zwitterion. Depending on the nature of R', $T^{\pm}$ may break down to products by expelling $R'O^-$ and a proton with the help of solvent molecules. In this case k_a is rate determining. More commonly, RNH_2 is expelled more readily than $R'O^-$, and the reaction becomes very slow. Catalysis then occurs by either acids or bases.

$$T^{\pm} + B \rightleftharpoons RNH - \underset{\underset{CH_3}{|}}{\overset{\overset{O^-}{|}}{C}} - OR'$$
$$T^- \tag{9.51}$$

$$T^{\pm} + HA \rightleftharpoons \overset{+}{R}NH_2 - \underset{\underset{CH_3}{|}}{\overset{\overset{OH}{|}}{C}} - OR' + A^-$$
$$T^+ \tag{9.52}$$

The tetrahedral intermediate T^- is much less likely to eject RNH^- than $R'O^-$, and product is formed. The intermediate T^+ has an acidic proton on nitrogen, which it can lose to A^- or to the solvent, forming T^0.

$$\overset{+}{R}NH_2 - \underset{\underset{CH_3}{|}}{\overset{\overset{OH}{|}}{C}} - OR' + H_2O \rightarrow RNH - \underset{\underset{CH_3}{|}}{\overset{\overset{OH}{|}}{C}} - OR' + H_3O^+$$
$$T^0 \tag{9.53}$$

The same T^0 can also be formed by an intramolecular proton switch in $T^{\pm}$ without catalysis. Just as for T^-, the intermediate T^0 loses $R'O^-$ more readily than RNH^-, and product results.

The above reactions constitute rather complicated paths for the apparently simple reaction (9.49). Nevertheless the various steps are well documented, and because of the rapidity of proton transfers, the mechanisms are efficient. There are also other ways in which acids and bases can catalyze acyl transfers and similar addition–elimination reactions.[29]

ACID–BASE CATALYSIS

Since catalysis by acids and bases is by far the most common in homogeneous reactions, a more detailed analysis of the kinetic situations that are usually encountered is given here. There are several distinct types of acid–base catalysis, almost all of which involve a proton transfer in at least one step. These are treated in general in terms of a substrate S or SH on which the acids or bases work, S^- the conjugate base of the substrate SH, B a general base, BH^+ its conjugate acid, HA a general acid, and A^- its conjugate base. B and A^- may or may not be the same as may BH^+ and HA. R is some other reactañt not acting as a base or acid.

Case I

Prior equilibrium between substrate and hydrogen ion followed by a rate-determining reaction with another reagent:

$$S + H^+ \rightleftharpoons SH^+ \tag{9.54}$$

$$SH^+ + R \xrightarrow{\text{slow}} \text{products} \tag{9.55}$$

$$\text{rate} = k[SH^+][R] = kK_{eq}[S][H^+][R] \tag{9.56}$$

where K_{eq} is the equilibrium constant for (9.54) and k is the rate constant for (9.55).

An important feature of this type of reaction is that the concentration of the complex SH^+ depends only on the hydrogen ion concentration in solution, regardless of the source of the hydrogen ion. Hence the rate also is dependent on the hydrogen ion only and not on the concentration of any other acids in the solution. This is an example of specific hydrogen ion catalysis. The observed kinetics depend a great deal on the value of the equilibrium constant of (9.54) and the relative concentrations of H^+ and S. If K_{eq} is large enough, S or H^+, whichever is present in least amount, is completely converted to SH^+. The observed kinetics are second order, first order in SH^+, and first order in R. The more usual case is for K_{eq} to be small so that S and H^+ are present in amounts initially added minus amounts that have reacted. The observed kinetics are third order, first order in H^+, S, and R. Since H^+ is frequently constant, being regenerated in step (9.55), the kinetics may be pseudo-second order or pseudo-first order, depending on the change in R. The pseudo-order constants, however, are dependent on the hydrogen ion concentration in solution.

Examples of case I (all with K_{eq} small) are found in the hydrolysis and alcoholysis of esters, the inversion of sucrose, the hydrolysis of ethyl orthoformate, and the hydrolysis of acetals and ethers. The acid hydrolysis of amides provides examples of K_{eq} large enough to cause the rate to level off at about $3 M$ hydrogen ion concentrations.

Case II

Prior equilibrium between substrate and hydrogen ion followed by the rate-determining proton transfer:

$$HS + H^+ \rightleftharpoons HSH^+ \tag{9.57}$$

$$HSH^+ + B \xrightarrow{\text{slow}} BH^+ + SH \tag{9.58}$$

$$\text{rate} = k[HSH^+][B] = kK_{eq}[H^+][HS][B] = kK_{eq}K_a[BH^+][HS] \tag{9.59}$$

The rate of the reaction is proportional to the concentrations of hydrogen ion and of base, but their product in turn is proportional to the concentration of the conjugate acid BH^+. Thus the kinetics satisfy the conditions for general acid catalysis. The rate increases not with the hydrogen ion concentration but with the concentration of acid BH^+. Also, if a number of acids are present,

$$\text{rate} = K_{eq}[HS] \sum_i k_i K_{ai}[BH_i^+] \qquad (9.60)$$

All acids contribute to the rate to an extent determined by their concentrations, acid ionization constants, and the specific rate constants of their conjugate bases.

An example of case II is the hydrolysis of alkyl benzimidates and diarylformadimium ions.[30] Some keto–enol changes, the depolymerization of dihydroxyacetone, and the acid-catalyzed mutarotation of glucose are probably also in this category.

Case III

A prior equilibrium involving hydrogen bonding of the substrate with an acid followed by a rate-determining proton transfer:

$$HS + HA \rightleftharpoons HS \cdot HA \qquad (9.61)$$

$$HS \cdot HA + B \xrightarrow{\text{slow}} \text{products} \qquad (9.62)$$

$$\text{rate} = k[HS \cdot HA][B] = kK_{eq}[HS][HA][B] \qquad (9.63)$$

Alternatively, this system could operate by complex formation between B and HS followed by reaction with HA, or HA and B can attack the substrate simultaneously in a trimolecular step. The trimolecular hypothesis seems less likely, unless either HS or B is a solvent molecule, on the basis that trimolecular solute collisions are uncommon. The possibilities are kinetically indistinguishable if K_{eq} is small. Both an acid and a base are needed for this mechanism, and, if a number of acids and bases are present, there is a term in the rate equation for all possible pairs.

$$\text{rate} = [HS] \sum_{ij} k_{ij} K_i [HA_i][B_j] \qquad (9.64)$$

Since a hydroxylic solvent can play the role of either an acid or base, the kinetics frequently reduce to

$$\text{rate} = kK_{eq}[HS][HA][H_2O] \qquad (9.65)$$

in the presence of a single acid, and to

$$\text{rate} = kK_{eq}[HS][B][H_2O] \qquad (9.66)$$

in the presence of a single base. However, in an inert solvent both acid and base must be added.[31]

Examples of Case III are certain keto–enol changes such as that of acetone, the methyleneazomethine rearrangement, and the mutarotation of tetramethylglucose and similar substances in inert solvents. In nonpolar solvents case III becomes more probable than case II, because ionization of the acid is not required. The kinetics in such media are, however, frequently complicated, simple orders not being observed. There are other reactions in which the rate law contains terms in which the product of $[HA][B]$ occurs. Equation (9.69) can be of this form if the reagent R is a base, such as an amine.

Case IV

A prior equilibrium involving hydrogen bonding of a substrate with an acid followed by a slow step not involving a proton transfer:

$$S + HA \rightleftharpoons S \cdot HA \tag{9.67}$$

$$S \cdot HA + R \xrightarrow{\text{slow}} \text{products} \tag{9.68}$$

$$\text{rate} = k[S \cdot HA][R] = kK_{eq}[S][HA][R] \tag{9.69}$$

This is an example of general acid catalysis. The mechanism seems to apply to the hydrolysis of ethylorthoacetate and ethylorthocarbonate in which the reagent R is a water molecule, and also to the hydration of aldehydes. It is sometimes called electrophilic catalysis since HA increases the electrophilicity of S.

Case V

Reaction of an anion with an acid to give either the final product or an intermediate that rapidly gives the final product:

$$S^- + HA \xrightarrow{\text{slow}} SH + A^- \tag{9.70}$$

$$SH \xrightarrow{\text{fast}} \text{products} \tag{9.71}$$

$$\text{rate} = k[S^-][HA] \tag{9.72}$$

This is an example of general acid catalysis. For a number of acids

$$\text{rate} = [S^-] \sum_i k_i[HA_i] \tag{9.73}$$

This behavior is found in the conversion of the anions of the nitroparaffins into the nitro forms. Also the decomposition of the diazoacetate ion and the azodicarbonate ion seems to involve such a process.

Case VI

A prior ionization of the substrate that comes to equilibrium followed by a slow reaction of the resulting anion with another reagent:

$$HS + B \rightleftharpoons S^- + BH^+ \tag{9.74}$$

$$S^- + R \xrightarrow{\text{slow}} \text{products} \tag{9.75}$$

$$\text{rate} = k[S^-][R] = kK_{eq} \frac{[SH][B][R]}{[BH^+]}$$

$$= \frac{kK_{eq}}{K_b}[SH][R][OH]^- \tag{9.76}$$

The rate depends on the ratio $[B]/[BH^+]$, which, however, is proportional to the hydroxide ion concentration in aqueous solution (or in general, the lyate anion in other protic solvents). Consequently the rate is a linear function of the hydroxide ion concentration, and the reaction is subject to specific hydroxide ion catalysis.

Examples of case VI include the condensation of acetone to diacetone alcohol and many similar condensations in organic chemistry, such as the Claisen, Michael, Perkin, and Aldol condensations.

In alcohol solution where many such condensations are carried out, the ethoxide ion takes the place of the hydroxide ion in the rate expression. Also obeying the kinetics of (9.76) are the brominations of β-disulfones and β-dinitriles,[32] which have the unusual feature that the rate of reaction of the carbanion formed in (9.74) with bromine is slow compared to the reverse of (9.74).

Case VII

A slow ionization of the substrate followed by a rapid reaction of the anion to give the products:

$$HS + B \xrightarrow{\text{slow}} S^- + BH^+ \tag{9.77}$$

$$S^- + R \xrightarrow{\text{fast}} \text{products} \tag{9.78}$$

$$\text{rate} = k[HS][B] \tag{9.79}$$

Such a reaction shows general base catalysis. For a number of bases in solution each contributes to the rate

$$\text{rate} = [HS] \sum_i k_i [B_i] \tag{9.80}$$

The kinetics indicated for case VII are shown in the decomposition of nitramide, the halogenation, isomerization, and deuterium exchange of many organic substances containing an acidic hydrogen, and the racemization of such substances when the acidic hydrogen is on an asymmetric atom. Also general base catalyzed is the aldol condensation of acetaldehyde, which has the feature of the condensation step being faster than the reversal of (9.77). A similar situation occurs in the condensation of glycerinaldehyde or dihydroxyacetone.

Case VIII

A reversible reaction of R and S to give an intermediate that can be trapped by either an acid or a base. After trapping, the product is formed.

$$R + S \rightleftharpoons T \tag{9.81}$$

$$\text{rate} = [T] \left\{ \sum_i k_i [B_i] + \sum_j k_j [HA_j] \right\} \tag{9.82}$$

This includes the example of acyl transfers discussed earlier. Also some aromatic substitution reactions, some condensations, and some reactions of carbonium ions show trapping catalysis.[26]

Case IX

A prior equilibrium involving addition of a base to the substrate followed by a slow reaction of the complex with or without another reagent:

$$HS + B \rightleftharpoons B \cdot HS \tag{9.83}$$

$$B \cdot HS + (R) \xrightarrow{\text{slow}} \text{products} \tag{9.84}$$

$$\text{rate} = k[B \cdot HS][R] = kK_{eq}[B][HS][R] \tag{9.85}$$

This again is an example of general base catalysis; actually most reactions of this type show considerable specificity in the base. The common examples involve addition of one base to a carbonyl group followed by the elimination of another base as in ester hydrolysis. Different products are formed, depending on the base, and the base is generally used up. An exception is in the hydrolysis of certain activated esters such as p-nitrophenyl acetate. Here true catalysis by substances such as pyridine and trimethylamine occurs.[33]

$$CH_3\overset{\overset{\displaystyle O}{\|}}{C}-OC_6H_4NO_2 + B \xrightarrow{\text{slow}} CH_3-\overset{\overset{\displaystyle O}{\|}}{C}-B^+ + {}^-O-C_6H_4-NO_2$$

$$CH_3-\overset{\overset{\displaystyle O}{\|}}{C}-B^+ + H_2O \xrightarrow{\text{fast}} CH_3-\overset{\overset{\displaystyle O}{\|}}{C}-OH + BH^+ \tag{9.86}$$

$$BH^+ + NO_2-C_6H_4-O^- \xrightarrow{\text{fast}} B + NO_2-C_6H_4-OH$$

Since this kind of mechanism does not involve a proton transfer prior to the rate-determining step, it is usual to refer to it as nucleophilic catalysis.[29]

Another example of case IX is the alcoholysis of esters, diketones, ketoesters, and the like. Here the base catalyst is regenerated:

$$R'-\overset{\overset{\displaystyle O}{\|}}{C}-B + OR^- \rightleftharpoons R'-\overset{\overset{\displaystyle O^-}{|}}{\underset{\underset{\displaystyle R}{|}}{\overset{\displaystyle |}{C}}}-B \rightarrow R'-\overset{\overset{\displaystyle O}{\|}}{C}-OR + B^-$$

$$B^- + ROH \rightarrow BH + RO^- \tag{9.87}$$

However, this is possible only if the catalyst base is the anion of the solvent and it corresponds to a case of specific rather than general base catalysis. A similar situation is met in the exchange of O^{18} in water with the oxygen of carbonyl compounds, a reaction that is specifically catalyzed by hydroxide ion.

In the base-catalyzed hydrolysis of these compounds, a new feature emerges. The rate is often dependent on the square of the hydroxide ion concentration.[34] An example is the cleavage of acetylmanganese pentacarbonyl[35]

$$CH_3COMn(CO)_5 + 2OH^- \rightarrow CH_3COO^- + Mn(CO)_5^-$$

$$k_{obs} = k_2[OH^-] + k_3[OH^-]^2 \tag{9.88}$$

The mechanism proposed is as follows:

$$CH_3-\overset{\overset{O}{\|}}{C}-Mn(CO)_5 + OH^- \rightleftharpoons CH_3-\overset{\overset{O^-}{|}}{\underset{\underset{A}{OH}}{C}}-Mn(CO)_5 \tag{9.89}$$

$$A + OH^- \rightleftharpoons CH_3-\overset{\overset{O^-}{|}}{\underset{\underset{B}{O_-}}{C}}-Mn(CO)_5 \tag{9.90}$$

$$\begin{aligned} A &\rightarrow CH_3COOH + Mn(CO)_5^- \\ B &\rightarrow CH_3COO^- + Mn(CO)_5^- \end{aligned} \tag{9.91}$$

Removal of the second proton provides an additional driving force for carbon–metal bond cleavage. It is possible that general base catalysis exists for (9.89), if (9.91) is sufficiently rapid.

The rate laws for the preceding cases are written for certain limiting situations. It must be appreciated that intermediate situations can occur. As an illustration, take the base hydrolysis of haloammines of cobalt(III). The mechanism is well established.[36]

$$Co(NH_3)_5Cl^{2+} + B \underset{k_{-1}}{\overset{k_1}{\rightleftharpoons}} Co(NH_3)_4NH_2Cl^+ + BH^+ \tag{9.92}$$

$$Co(NH_3)_4NH_2Cl^+ \xrightarrow{k_2} Co(NH_3)_4NH_2^{2+} + Cl^- \tag{9.93}$$

$$Co(NH_3)_4NH_2^{2+} + H_2O \xrightarrow{fast} Co(NH_3)_5OH^{2+} \tag{9.94}$$

It is called an S_N1CB mechanism, for unimolecular dissociation of the conjugate base of the reactant. Applying the steady-state treatment to the conjugate base, leads to the rate law

$$\text{rate} = \frac{k_1k_2[B][\text{complex}]}{k_2 + k_{-1}[BH^+]} \tag{9.95}$$

If $k_{-1}[BH^+] \gg k_2$, we obtain the rate law (9.76). However, if the reverse is true, we obtain (9.79). The usual situation is for (9.76) to be followed, but a few exceptions are known.[36]

THE BRØNSTED CATALYSIS LAW

For those reactions that are subject to general acid or general base catalysis an expected relationship exists between the strength of the acid or base, as determined by its ionization constant, and its efficiency as a catalyst, determined by the observed rate constant. If we denote this constant (also called the catalytic constant) by k_a or k_b for acid or base catalysis, respectively, then the relationship is best shown by the Brønsted catalysis law,[37]

$$k_a = C_A K_a^\alpha \qquad k_b = C_B K_b^\beta \tag{9.96}$$

where K_a and K_b are the acid and base ionization constants, and $C_A, C_B, \alpha,$ and β are constants characteristic of the reaction, the solvent, and the temperature. Normally α and β are positive and have values between zero and one.

To illustrate (9.96) consider the rates and equilibrium of the following reversible reaction:

$$HS + B \underset{k_a}{\overset{k_b}{\rightleftharpoons}} S^- + HB^+ \tag{9.97}$$

where HS is a constant substrate and B is a base that may be varied. The equilibrium constant is proportional to K_b in the particular solvent used, or to K_a if the reverse reaction is considered. If a change is made from B to a similar base B′ that is a stronger base by 2 powers of 10, the equilibrium constant is increased by the same amount. Since $K_{eq} = k_b/k_a$, only one rate constant need be affected, but generally both are. For example, k_b may be increased by a power of 10, since B′ is a better proton acceptor, and k_a may be decreased by a power of 10, since B′H⁺ is a poorer proton donor or acid. This corresponds to the case

$$k_a = C_A K_a^{0.5} \qquad k_b = C_B K_b^{0.5} \tag{9.98}$$

Now as to whether other bases and acids will conform to the same equation depends chiefly on how similar they are to the first two bases. A group of amines may all fit a single equation and a group of carboxylic acid anions may fit a similar equation with slightly different constants. Since most acid–base-catalyzed reactions involve a proton transfer somewhere, the Brønsted relationship frequently holds even for those cases involving a more complex series of reactions.

For exactness, a statistical correction should be made. If for a given catalyst p is the number of equivalent protons on the acid and q is the number of equivalent positions where a proton can be accepted in the conjugate base, (9.96) should be written as

$$\frac{k_a}{p} = C_A \left[\left(\frac{q}{p} \right) K_a \right]^\alpha \qquad \frac{k_b}{q} = C_B \left[\left(\frac{p}{q} \right) K_b \right]^\beta \tag{9.99}$$

These corrections were suggested by Brønsted.[37] If transition-state theory is used, then the symmetry numbers for the reactants and activated complex should be used to correct k_a.[38] The symmetry numbers of HA and A⁻ should be used to correct K_a. The relationship becomes

$$\frac{k_a \sigma^{\ddagger}}{\sigma_{HA} \sigma_R} = C_A \left(\frac{K_a \sigma_{A^-}}{\sigma_{HA}} \right)^\alpha \tag{9.100}$$

where R is the molecule that reacts with HA. Unfortunately, the use of (9.100) requires an assumption about the transition-state structure. Nevertheless, the use of symmetry numbers is probably the best way to make statistical corrections in reactions of all kinds.[39]

Figure 9.3 shows data obtained by Bell and Higginson on the dehydration of acetaldehyde hydrate in acetone solution, using phenols and carboxylic acids as catalysts.[40]

$$CH_3 CH(OH)_2 \xrightarrow[k_a]{HA} CH_3 CHO + H_2O \tag{9.101}$$

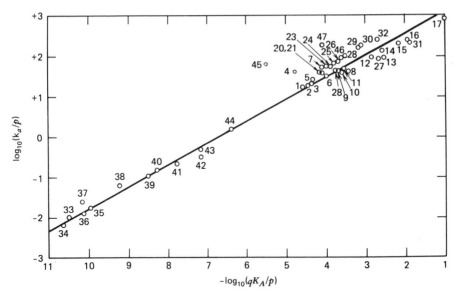

Figure 9.3. Brønsted plot of catalytic constants of various acids in the dehydration of acetaldehyde hydrate: (*1*) propionic, (*2*) acetic, (*3*) β-phenylpropionic, (*4*) trimethylacetic, (*5*) crotonic, (*6*) phenylacetic, (*7*) cinnamic, (*8*) formic, (*9*) β-chloropropionic, (*10*) diphenylacetic, (*11*) glycollic, (*12*) phenoxyacetic, (*13*) bromacetic, (*14*) chloracetic, (*15*) cyanacetic, (*16*) phenylpropiolic, (*17*) dichloracetic, (*18*) o-methoxybenzoic, (*19*) o-toluic, (*20*) p-hydroxybenzoic, (*21*) p-methoxybenzoic, (*22*) m-hydroxybenzoic, (*23*) p-toluic, (*24*) m-toluic, (*25*) benzoic, (*26*) p-chlorobenzoic, (*27*) o-chlorobenzoic, (*28*) m-chlorobenzoic, (*29*) m-nitrobenzoic, (*30*) p-nitrobenzoic, (*31*) o-nitrobenzoic, (*32*) salicylic, (*33*) thymol, (*34*) hydroquinone, (*35*) phenol, (*36*) resorcinol, (*37*) pyrocatechol, (*38*) p-chlorophenol, (*39*) o-chlorophenol, (*40*) m-nitrophenol, (*41*) 2,4-dichlorophenol, (*42*) o-nitrophenol, (*43*) p-nitrophenol, (*44*) 2,4,6-trichlorophenol, (*45*) pentachlorophenol, (*46*) 2,6-dinitrophenol, (*47*) 2,4-dinitrophenol. (Data from Bell and Higginson.)[40]

This reaction can be followed dilatometrically by the increase in volume that occurs. The data are plotted in the form

$$\log\left(\frac{k_a}{p}\right) = \log\left(\frac{K_a q}{p}\right) + \log C_A \qquad (9.102)$$

where K_a is the value in water and $\alpha = 0.54$. Good agreement is seen for acids varying over 10 powers of 10 in acid strength. Water also is a Brønsted acid, with a pK_a of 15.74 at 25°C ($10^{-14}/55.5 = K_a$). Its rate constant is not shown on the figure, but is 40 times as large as predicted. Abnormal values for H_2O, OH^-, and H_3O^+ are often found in Brønsted relationships. Reaction (9.101) is also subject to general base catalysis,[41] and the water rate is in better agreement with its expected value as a base.

The hydration reactions of acetaldehyde and formaldehyde have been extensively studied because they are models for the more general formation of acetals and ketals.[42] They are also models for the tetrahedral intermediates involved in acyl transfer reactions. The mechanism for the acid-catalyzed hydration is probably given by case IV.

$$H_2O + {\textstyle >}C{=}O\text{---}HA \rightleftharpoons H_2O^+\text{---}\overset{|}{\underset{|}{C}}{-}OH\text{---}\bar{A}$$

(9.103)

$$A^- + H_2\overset{+}{O}{-}\overset{|}{\underset{|}{C}}{-}OH \rightleftharpoons AH + HO{-}\overset{|}{\underset{|}{C}}{-}OH$$

with the first step being rate determining.

The base catalysis probably goes by a modification of case IX, in which the base associates with the water molecule, increasing its nucleophilicity in attacking the carbonyl group.

$$B\text{---}H_2O + {\textstyle >}C{=}O \rightleftharpoons B^+\text{---}\text{---}H\text{---}\overset{\overset{\displaystyle H}{|}}{O}{-}\overset{|}{\underset{|}{C}}{-}O^- \rightleftharpoons$$

(9.104)

$$BH^+ + HO{-}\overset{|}{\underset{|}{C}}{-}O^- \rightleftharpoons HO{-}\overset{|}{\underset{|}{C}}{-}OH + B$$

(9.105)

with (9.104) the rate-determining step. This is an example of nucleophilic catalysis, just as (9.103) is electrophilic catalysis.

GENERAL AND SPECIFIC CATALYSIS

In the Brønsted equation a low value of α signifies a low sensitivity of the catalytic constant to the strength of the catalyzing acid, whereas a large value of α indicates a high sensitivity. Since the solvated hydrogen ion is usually the strongest acid available in ordinary solvents, reactions with large values of α show a dependence on the hydrogen ion concentration so large that catalysis by other acids is difficult to detect. If the reaction is sufficiently fast, buffered solutions may be prepared that are basic enough so that the hydrogen ion reaction is slow but that contain sufficient quantities of other acids to test their catalytic activities. The best procedure for detecting general acid catalysis is to measure the rate of the reaction in a series of such buffers with a constant ratio of acid to conjugate base, but of different total concentrations. If the reaction is general acid catalyzed, the observed rate constant increases with total concentration. If the reaction is subject to specific hydrogen ion catalysis, the rate constant varies only slightly, since the hydrogen ion concentration is determined chiefly by the ratio, which remains constant. To exclude variations due to secondary salt effects on the ionization of the acid, some care should be used to hold the total ionic strength constant. Although it is true that there are some mechanisms that demand specific hydrogen or hydroxide ion catalysis, the experimental finding of specific catalysis does not prove that such a mechanism is operating, since it may be a case of a large value of α or β that makes catalysis by other acids or bases small.

A value of α or β close to zero may in some cases cause the general acid or basic catalysis of a reaction not to appear experimentally, since the solvent molecules, which are present in large excess, may act as acid or base. An example of this is found in the acid-catalyzed enolization of ketones, where as the acidity of the ketone increases, catalysis by acids becomes less important, finally disappearing. Thus the enolization of acetone is general acid catalyzed, but the enolization of acetylacetone is independent of all acids, including hydrogen ion. The place of the acid in the latter case is presumably taken by water or other solvent molecules.

Table 9.3 shows some data on the hydrolysis of acetal in formate ion–formic acid buffers.[43] The ratio of acid to ion is held constant at 2.96 (some corrections are needed because of the ionization of formic acid). The last four values show the constancy of the experimental rate constant when the ionic strength is maintained at 0.1 by the addition of sodium chloride while the total concentration of the buffer is changed. This indicates a specific hydrogen ion catalysis. The salt effect on the ionization constant of formic acid would lead to an increase in the rate constant and an apparent catalysis by formic acid molecules as the first four entries show. This is an example of the secondary salt effect.

LINEAR FREE-ENERGY CHANGES

The Brønsted equation is only one example of a more general phenomenon that may be called the principle of linear free-energy changes in a series of related reactions.[44] This principle may be expressed in one of several ways relating equilibrium constants and rate constants:

$$K_i = C_1 K_j^\alpha \qquad k_i = C_2 K_i^\alpha$$
$$k_i = C_3 K_j^\alpha \qquad k_i = C_4 k_j^\alpha \tag{9.106}$$

where the K_i's and K_j's are a series of equilibrium constants for two related reactions involving a group of similar reactants and the k_i's and k_j's are the corresponding rate constants. The equations introduced by Hammett,[44]

$$\log k = \log k_0 + \sigma\rho$$
$$\log K = \log K_0 + \sigma\rho \tag{9.107}$$

are alternative ways of expressing the same relationships shown in (9.106). Here ρ is identical with α and is a measure of the susceptibility of the reaction to the

Table 9.3. Hydrolysis of acetal

Formic acid concentration/M	Sodium formate concentration/M	Sodium chloride concentration/M	Ionic strength	$k_c \times 10^3/\text{min}^{-1}$
0.0296	0.0100	–	0.011	12.5
0.0592	0.0200	–	0.020	13.4
0.1480	0.0500	–	0.050	15.1
0.2220	0.0750	–	0.075	16.4
0.2960	0.1000	–	0.100	17.8
0.1776	0.0600	0.0400	0.100	17.6
0.0987	0.0333	0.0667	0.100	17.7
0.0222	0.0075	0.0925	0.100	18.2

changes caused by *meta* and *para* substitution in the benzene ring, and σ is the log of the ratio of the ionization constant of the substituted benzoic acid to that of benzoic acid itself and is a measure of the effect of the substituent. Expressed in the more general ways of (9.106)–(9.108) the ρ's and α's may be either positive or negative and greater or less than unity.

A very large number of examples of (9.106) have appeared in the literature.[45] The first theoretical interpretations were based on potential energy curves, showing a linear relationship between activation energies and overall heats of reaction.[46]

$$\delta \Delta E^{\ddagger} = \alpha \delta \Delta E \tag{9.108}$$

The symbol δ is used to indicate a variation in the quantity due to a substituent change in one reactant.

Unfortunately it is not true that the variations in rates and equilibria for a series of related reactions are due only to changes in energies. Quite often entropy changes are as important, or more important, particularly for reactions in solution. This illustrates the point made in Chapter 7 that potential energy curves in solution can be dramatically changed from those in the gas phase, but only at the expense of large entropy effects. The two factors tend to compensate in the free energy, $\Delta G = \Delta H - T \Delta S$, so that frequently the free energy changes in a predictable manner, while the enthalpy and entropy do not.

An intuitive derivation of the equation $k_i = C_2 K_i^{\alpha}$ may be given.[47]

$$\text{reactants} \underset{}{\overset{k_i}{\rightleftharpoons}} \text{products} \qquad K_i \tag{9.109}$$

As small changes in the reactants are made by substituents, it is reasonable to assume that the free energy of the transition state can be represented as a linear combination of changes in the reactants and in the products.

$$\delta G^{\ddagger} = \alpha \delta G_P^{\ominus} + (1 - \alpha) \delta G_R^{\ominus} \tag{9.110}$$

The parameter α lies between 0 and 1. It measures the extent to which the transition state resembles the product. A value near unity means close resemblance to the products, and a value near zero means close resemblance to the reactants. Therefore α is a measure of how far along on the reaction coordinate the transition state appears. Equation (9.110) is consistent with the Hammond postulate,[48] which proposes that if two states occur consecutively during a reaction process and have nearly the same energy, their interconversion requires only a small amount of nuclear movement. Thus in a highly endothermic reaction the transition state resembles the products, and in an exothermic reaction the transition state more closely resembles the reactants. Using transition-state theory, (9.110) leads to

$$\delta \Delta G^{\ddagger} = \alpha \delta \Delta G^{\ominus} \tag{9.111}$$

or

$$\ln k_i = \alpha \ln K_i + C \tag{9.112}$$

which is the desired result.

A more mathematical derivation of the relationship between k_i and K_i has been given by Marcus.[49] The theory was originally developed for electron-transfer reactions, but can also be applied to proton transfers, or indeed the transfer of any atom or group between two sites. It corresponds to the situation given by (9.46), where there is a rate-determining proton transfer following formation of an encounter complex. A model of two identical parabolic potential energy curves, as in Fig. 9.2, is used. Changes in $\Delta G^{\ominus}$ are made by simply moving the right-hand parabola up and down. The results are

$$k_r = Z e^{-\Delta G^{\ddagger}/RT} \tag{9.113}$$

$$\Delta G^{\ddagger} = w^r + \Delta G_0^{\ddagger}\left(1 + \frac{\Delta G^{\ominus}}{4\Delta G_0^{\ddagger}}\right)^2 \tag{9.114}$$

where Z is the collision frequency in solution (about $10^{11} \, M^{-1} \, s^{-1}$), and w^r is the work required to bring the reactants close enough together for proton transfer to occur. The term $\Delta G_0^{\ddagger}$ represents the intrinsic energy barrier, that is, the barrier when $\Delta G^{\ominus}$ is equal to zero. It corresponds to the exchange reaction .

$$\text{HA} + \text{A}^- \rightleftharpoons \text{A}^- + \text{HA} \tag{9.115}$$

When $|\Delta G^{\ominus}| \geqslant 4\Delta G_0^{\ddagger}$, $\Delta G^{\ddagger}$ has limiting values of 0 and $\Delta G^{\ominus}$, for $\Delta G^{\ominus}$ negative and positive, respectively. Equation (9.114) leads to

$$\alpha = \frac{\delta \Delta G^{\ddagger}}{\delta \Delta G^{\ominus}} = \frac{1}{2}\left(1 + \frac{\Delta G^{\ominus}}{4\Delta G_0^{\ddagger}}\right) + \frac{\delta w^r}{\delta \Delta G^{\ominus}} \tag{9.116}$$

so that α is not constant, but varies slowly about the value of 0.5 as $\Delta G^{\ominus}$ changes.

Equations (9.113) and (9.114) fit a number of data on proton-transfer reactions, as well as some data on methyl-transfer reactions.[50] Of course they contain two adjustable constants, w^r and $\Delta G_0^{\ddagger}$ and are rather flexible. These are assumed to be constant for a given series, but one can readily imagine that w^r will change. A number of other theoretical treatments have been given that yield results very similar to those of Marcus.[51]

Figure 9.4 shows a plot of $\log k$ against pK_a for a series of nucleophilic displacement reactions on benzyl chloride in DMSO solution.[52]

$$\text{(9.117)}$$

$$G = \text{CN}, \text{CO}_2\text{CH}_3, \text{C}_6\text{H}_5\text{SO}_2, \text{C}_6\text{H}_5\text{S}, \text{C}_6\text{H}_5, \text{C}_6\text{H}_5\text{O}, \text{C}_6\text{H}_5\text{CH}_2, \text{CH}_3, (\text{CH}_3)_3\text{C}$$

The nucleophiles are the 9-substituted fluorenyl anions indicated, and further family members substituted at the 2- and 2,7-positions. The pK_a values were measured for the conjugate acids in DMSO.

It is evident that substituents at the 9-position exert a steric hindrance that affects the rates. Within a family, however, a good linear relationship is found, the

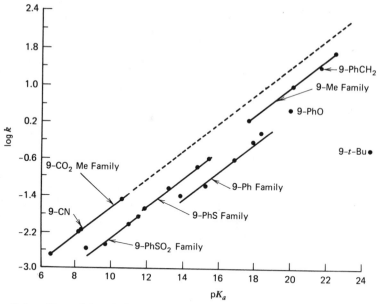

Figure 9.4.   Plot of log k for reactions of 9-substituted florenyl carbanions reacting with benzyl chloride in DMSO at 25°C. [Reprinted with permission from F. G. Bordwell and D. L. Hughes, *J. Org. Chem.*, **45**, 3314 (1980). Copyright 1980, American Chemical Society.]

family members having a constant steric effect. All the data can be represented by the equation

$$\log k = \beta pK_a + C \qquad (9.118)$$

where C is different for each family, but β is constant at 0.31 for all families. The pK_a values cover a range of 18 units, which is unusually large.

Similar results have been found for other nucleophilic displacements on carbon.[53] Good Brønsted plots are obtained as long as the bases used are very similar and have a constant donor atom. However, this is not the case if the bases are allowed to vary more widely. The rate constants for the reaction of a large number of nucleophiles with methyl iodide have been measured in methyl alcohol.[54] A plot of $\log k$ against pK_a gives a shotgun pattern, similar to that shown in Fig. 9.5. A characteristic feature is that bases in which the donor atom is of low electronegativity and high polarizability, (P compared to N, S compared to O, and I compared to F) are much more reactive than expected.

The reactions of planar complexes of platinum(II) show this enhanced reactivity to an even greater degree. Figure 9.5 shows a plot of $\log k$ against pK_a for various bases reacting with *trans*-Pt(pyridine)$_2$Cl$_2$ in methanol.[54a]

$$Pt(py)_2Cl_2 + B \rightarrow Pt(py)_2ClB^+ + Cl^- \qquad (9.119)$$

It should be noted that the strongest base possible in methanol, the methoxide ion, has no measurable reaction rate with the platinum complex; also $(C_2H_5)_3P$ is more than 10^6 times more reactive than $(C_2H_5)_3N$, though the latter is a much stronger base.

Methanol is often a convenient solvent for kinetic studies since it resembles water, but is a much better solvent than water for covalent organic and inorganic

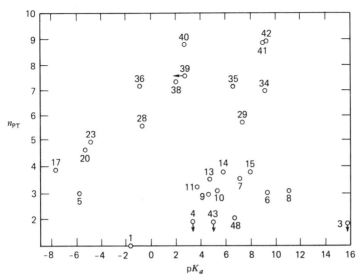

Figure 9.5.   Plot of $n_{Pt} = \log k_b/k_{CH_3OH}$ against pK_a for various nucleophiles listed by number: (*1*) CH_3OH, (*3*) CH_3O^-, (*4*) F^- (*5*) Cl^- (*6*) NH_3 (*7*) imidazole, (*8*) piperidine, (*9*) aniline, (*10*) pyridine, (*11*) NO_2^-, (*13*) N_3^-, (*14*) NH_2OH, (*15*) NH_2NH_2, (*17*) Br^-, (*20*) $(CH_3)_2S$, (*23*) $(CH_2)_4S$, (*28*) SCN^-, (*29*) SO_3^{2-}, (*34*) CN^-, (*35*) $C_6H_5S^-$, (*36*) $SC(NH_2)_2$, (*38*) $S_2O_3^{2-}$, (*39*) $(C_2H_5)_3As$, (*40*) $(C_6H_5)_3P$, (*41*) $(n$-$C_4H_9)_3P$, (*42*) $(C_2H_5)_3P$, (*43*) CH_3COO^-, (*48*) α-picoline. [Reprinted by permission from R. G. Pearson, H. Sobel and J. Songstad, *J. Am. Chem. Soc.*, **90**, 319 (1968). Copyright 1968, American Chemical Society.]

substances. For acid–base reactions, we need to know the ion product of methanol, which is 1.96×10^{-17} at $25°C$.[55] A number of pK_a values have been measured in methanol.[56] Because of its lower dielectric constant, ion pairing is more prevalent in methanol than in water.[57] In mixtures of water and methanol, it is important to know the following (at $25°C$):

$$K_2 = \frac{a_{H_3O^+}a_{CH_3OH}}{a_{CH_3OH_2^+}a_{H_2O}} = 137 \tag{9.120}$$

$$K_3 = \frac{a_{OH^-}a_{CH_3OH}}{a_{OCH_3^-}a_{H_2O}} = 0.13 \tag{9.121}$$

The ratio of activities of methanol and water may be replaced by their mole fraction ratio, with no great error.[55]

We see from (9.120) and (9.121) that water is a considerably stronger base than methanol, and water is also a weaker acid than methanol. Thus a solution of base in a mixture of water and methanol should contain relatively more CH_3O^- than OH^-, but the latter would be expected to be more nucleophilic because it is a stronger base. The experimental results for the reaction with methyl iodide in water–methanol mixtures are surprising.[58] Except, possibly, for very water–rich mixtures, methoxide ion is much more reactive than hydroxide ion, as shown in Fig. 9.6. This is another example of increasing base polarizability and illustrates how difficult it is to predict relative reactivities in even the simplest cases.

While the rate constants for reactions of nucleophiles with Pt(II) complexes

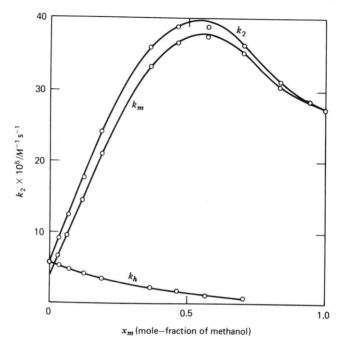

Figure 9.6. Rate constants for reactions of methyl iodide with base in methanol—water mixtures; k_2 is the observed second-order rate constant and k_h and k_m are the rate constants for OH^- and CH_3O^-, respectively. (Data from Murto.[58])

cannot be predicted from the basicities of the nucleophiles (Fig. 9.5), the rate constants for reaction with other Pt(II) complexes can be predicted from the data for *trans*-Pt(pyridine)$_2$Cl$_2$.[59] This is an application of the linear free energy relationship

$$\log k_i = \alpha \log k_j + C \qquad (9.122)$$

Alternative forms of this equation are often used. For example, the rate constants for reactions of the standard, k_j, may be set equal to nucleophilic reactivity parameters.[60]

In a somewhat more complex application, the rate constants for solvolysis of a given substrate in a series of solvents might be represented by the equation[61]

$$\log\left(\frac{k}{k_0}\right) = lN + mY \qquad (9.123)$$

where N is the nucleophilic power of a solvent and Y is its ionizing power. The rate constant, k_0, is for a standard solvent, and l and m measure the sensitivity of the substrate to nucleophilic ability and ionizing ability, respectively. By selecting model compounds, such as t-butyl chloride, where only Y should be important, and methyl chloride, where both Y and N would play a role, various solvents and solvent mixtures can be characterized by values of N and Y. The rate behavior of the substrate in these solvents can then indicate to what extent it solvolyzes by S_N1

and S_N2 mechanisms. The success of equations such as (9.123) depends on the validity of linear free energy relationships, including variation of the solvent as a perturbation.

THE POSITION OF THE TRANSITION STATE

A question that has long intrigued chemists interested in reaction mechanisms is "just where on the potential energy surface does the transition state occur?" This is equivalent to asking whether the activated complex resembles the products, the reactants, or neither of these. There are a number of reasons for this interest. For example, in reactions forming ions from neutral molecules the effect of solvents depends largely on the quantity of charge developed in the transition state. In three-body reactions in the gas phase the position of the activated complex determines what kind of energy is effective at causing reaction (see p. 204). Most important, perhaps, is the relation to the selectivity–reactivity principle.

This principle can be derived qualitatively from the Hammond postulate, or from (9.111).[62] The reactivity of a reagent depends on its rate constant for reaction with a standard substrate. Its selectivity depends on its difference in reactivity with any two substrates. The conclusion is that the selectivity is small for very reactive reagents and large for unreactive reagents. The transition state should resemble the reactants in the former case, and the products in the latter case.

It should be remembered that there is no unambiguous experimental method for locating the transition state on the potential energy surface. The best we could do would be a purely theoretical *ab initio* calculation of the surface, a procedure that is not practical at the present time, except for the simplest cases. The next best that we can do is to construct a more or less detailed empirical surface that correctly predicts the available kinetic information about the reaction. While this has been done in some detail for a few gas-phase reactions (Chapters 5 and 6), reactions in solution usually call for drastic simplification. Usually only a two-dimensional curve of potential energy versus reaction coordinate is considered.

The reaction coordinate is actually a very complex mixture of nuclear motions. It is customary to simplify it to a single parameter, λ, ranging from $\lambda = 0$ for the reactants to $\lambda = 1$ for the products. If the making or breaking of one bond characterizes the reaction, then λ, or $1 - \lambda$, is the bond order of that bond. Generally at least two bonds are involved in a reaction, for example,

$$A + B{-}C \rightarrow A{-}{-}{-}B{-}{-}{-}C \rightarrow A{-}B + C \qquad (9.124)$$

Usually the assumption is made that the sum of the bond orders of A-B and of B-C is constant. In this case either bond can be taken as a measure of λ.

This assumption may be reasonably good in the case of gas-phase reactions. In the case of reactions in solution, it is much more questionable. If ions are involved, either as reactants or products, then interactions with the solvent can play a leading role. That is, as the bond B-C breaks, solvation can take the place of the new bond A-B.[63] In the limit, an ion-pair would be formed that reacts with A in a separate step.

Ideally one would like to relate $\lambda^{\ddagger}$, the value of λ corresponding to the activated complex, to the quantities most readily available from experiment, E_a and ΔE.

Just such a relationship has been derived.[64]

$$\lambda^{\ddagger} = \frac{E_a^{\bullet}}{2E_a - \Delta E} \tag{9.125}$$

The derivation requires that the path followed has a minimum arc length in the $\lambda - E$ plane. If we define the reaction coordinate so that the potential energy is a linear function of λ, we can obtain $\lambda^{\ddagger}$, as shown in Fig. 9.7. The path shown is the shortest distance that connects the energies of reactants, products, and activated complex. However, $\lambda^{\ddagger}$ defined in this way does not have an obvious meaning in terms of bond distances and bond angles in the activated complex. The limiting values, $\lambda^{\ddagger} = 1/2$ when $\Delta E = 0$, $\lambda^{\ddagger} = 1$ when $\Delta E = E_a$, and $\lambda^{\ddagger} = 0$ when $\Delta E = -\infty$, are all reasonable.

The values of α and β in the Brønsted equations (9.96) are usually interpreted in terms of the degree of bond breaking or formation. That is, in proton transfers α is the fraction of the original HA bond that has been broken, and β is the fraction of the new BH bond that has been made. In nucleophilic substitutions, such as (9.117) and (9.119), β is considered the fraction of the bond made by the nucleophile to the substrate in the activated complex. Nothing is inferred about the degree of breaking of the C-Cl or Pt-Cl bonds, however.

To illustrate the use of β as a mechanistic probe, the aminolysis of phenyl acetates and carbonates serves admirably.[65] The mechanism is essentially (no catalysis)

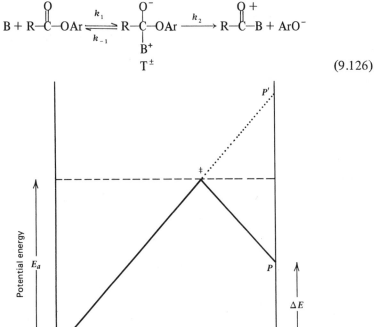

$$\tag{9.126}$$

Figure 9.7. Plot of potential energy versus reaction coordinate. P' is the mirror image of P, if the dashed line is the mirror.

When a Brønsted plot of the observed rate constants and the base strengths of B is made, a sharp break in the value of β is seen. For basic amines, $\beta \cong 0.30$, and for less basic amines $\beta \cong 1.0$. The interpretation is that k_1 is rate determining for the basic amines, and that the new N-C bond is 30% formed at the transition state. Expulsion of the aryl oxide ion is rate determining for the less basic amines, and the N-C bond is 100% formed in the intermediate $T^{\pm}$. This is also an example of the reactivity–selectivity principle.

In Table 9.1 the rate constants are given for the reaction of a number of bases with carbon dioxide. An interesting comparison of the rate constants for some of these same bases reacting with cyanic acid has been made.[66]

$$RNH_2 + HCNO \rightarrow RNHCONH_2 \tag{9.127}$$

A linear free energy relationship is found,

$$\log k_{HCNO} = \log k_{CO_2} - 1.08 \tag{9.128}$$

in which the value of β is unity. This does not mean complete formation of the new bond in the activated complex, but simply that the degree of bond formation is the same in (9.127) and (9.25) at the transition state. Incidentally, this study is also the best evidence that the urea reaction (p. 265) proceeds by way of molecular, rather than ionic, reactants.

The reactivity–selectivity principle requires that α and β do not remain constant for a sufficiently large variation in the reactants. To see this in a simple way, consider the proton-transfer reaction again

$$HS + B \underset{k_a}{\overset{k_b}{\rightleftharpoons}} S^- + HB^+ \tag{9.97}$$

For a given substrate, HS, variation of B produces a Brønsted relationship with a certain value of β. Now as HS is varied so that it is more reactive, the value of β should decrease. This effect has been demonstrated by Bell[67] for a series of carbon acids. In the limit, the rate of (9.97) becomes diffusion controlled and β drops to zero.[68] Similarly, for the reverse reaction, when HS is a very weak acid (unreactive), S^- is very reactive and k_a is diffusion controlled so that $\alpha = 0$. When HS is a strong acid (reactive), α must go toward unity. The sum of α and β must always equal unity since we have the equilibrium relationship

$$K_{eq} = \frac{k_b}{k_a} = \frac{K_b K_{HS}}{K_w} = \frac{K_w}{K_{S^-} K_a} \tag{9.129}$$

if the solvent is water. K_{HS} is the acid ionization constant of the substrate, and K_{S^-} is the base constant of its conjugate base.

It is not necessary to go to the limit of diffusion control to observe changes in α or β even with a single substrate. Figure 9.8 shows a Brønsted plot for the elimination reaction in water[69]:

$$CH_3-\overset{\overset{\displaystyle O}{\|}}{C}-CH_2-CH_2-OC_6H_4NO_2 + B^- \xrightarrow{k_b}$$

$$CH_3-\overset{\overset{\displaystyle O}{\|}}{C}-CH=CH_2 + BH + O_2NC_6H_4O^- \tag{9.130}$$

A variety of criteria show that proton removal is the rate-determining step (case VII). The bases used were oxyanions, including strongly basic alkoxide ions. The bases also included some oximate anions that show enhanced nucleophilic reactivity toward attack at carbon. This is the so-called α-effect, since it occurs for nucleophiles with an unshared pair of electrons on the atom α to the nucleophilic atom.[70] As previously shown by Bruice, there is no α-enhancement in proton removals.[71]

The Brϕnsted slope, β, changes from 0.75 for weak bases to about 0.3 for the strongest bases. Such a change is predicted by the Marcus equations (9.114 and 9.116). The curved line passing through the experimental points in Fig. 9.8 is calculated from these equations with $\Delta G_0^{\ddagger} = 10.45 \text{ kJ mol}^{-1}$ and $w^r = 63.1 \text{ kJ mol}^{-1}$. While the fit is excellent, the value of $\Delta G_0^{\ddagger}$ requires that the pK_a of the substrate be 11.1 ($\alpha = 0.50$ when $\Delta G^{\ominus} = 0$). The expected value of pK_a for this ketone is 16.5. Also Bell's data[67] imply that $\Delta G_0^{\ddagger} = 41.8 \text{ kJ}$ and $w^r = 16.7 \text{ kJ}$ would be much more reasonable values. A better interpretation of Fig. 9.8 is that the curvature is due to a change in w^r.[69] The strongly basic anions, like OH^- and

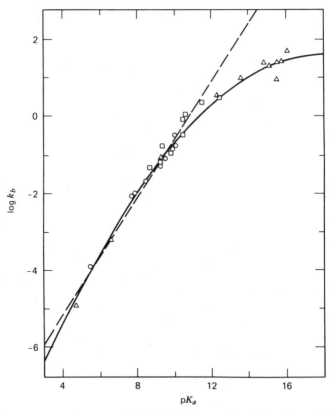

Figure 9.8. A plot of log k_b for (9.130) versus the pK_a values of the catalysts including ($\circ$) phenoxides, ($\triangle$) alkoxides, hydroxide, and acetate ions, and ($\square$) oximate anions. The dashed line has a slope of 0.75. The curve is computed from the Marcus equation (see text). [Reprinted with permission from D. J. Hupe and D. Wu, *J. Am. Chem. Soc.*, **99**, 4219 (1977). Copyright 1977, American Chemical Society.]

$C_2H_5O^-$, are most tightly solvated, and more work would be needed to desolvate them before reaction could occur (see p. 271). Thus the term $(dw^r/d\Delta G^\ominus)$ in (9.116) cannot be ignored.

There are numerous examples of free-energy relationships that cover a wide range of reactivities with no apparent curvature. Reaction (9.117) is one instance. These represent cases of failure of the reactivity–selectivity relationship,[72] though sometimes special circumstances are responsible. In proton-transfer reactions, because of the complex mechanism shown in (9.43), α and β may appear constant over a wide range.[51] While it seems inherently reasonable that a highly reactive molecule would show little selectivity in its reactions, the facts often indicate the opposite. Thus reactive methyl fluorosulfonate shows the same range of relative reactivities toward substituted pyridines as does the much less reactive ethyl iodide.[63] The hydrogen atom is surely a most reactive entity, but its reaction rates with various substrates vary by many powers of 10.

ISOTOPE EFFECTS AND TUNNELING

Another method that has been widely used to study transition-state structure is the measurement of kinetic isotope effects, especially for hydrogen-atom, or proton-transfer reactions.[73] There are primary isotope effects, when the isotopic change is made in a bond that is being formed or broken in the elementary reaction involved, and smaller secondary isotope effects, when the isotope bonds remain intact. There are also solvent isotope effects.[74] Except for those of hydrogen, kinetic isotope effects are quite small.

The major factor in determining isotope effects is the zero point energy, $h\nu/2$, that exists for all vibrations of molecules. Since the frequency, ν, depends inversely on the reduced mass of a vibration, it changes upon isotopic substitution even though the force constant remains the same. The situation is demonstrated in Fig. 9.9 for a diatomic hydride, H—X and D—X. When X is heavy the frequencies are different by nearly the square root of 2. It can be seen that the dissociation into X and H atoms requires less energy than that into X and D atoms. The frequencies also enter into the vibrational partition function so there is also some entropy difference in the equilibrium constants for dissociation. These are usually of little importance except at high temperatures. A safe rule to use is that the heavier isotope prefers to stay where the bonding is strongest.

In a rate constant, of course, it is the difference in bonding between the reactants and the transition state that must be considered. If the bonding (of the isotopic atom) is weaker in the transition state than in the reactants, the heavier isotope reacts more slowly. For the primary isotope effect, this is usually the case. Using transition-state theory, and considering specifically hydrogen and deuterium, we can write (Chapter 5)

$$\frac{k_H}{k_D} = \frac{q_D \cdot q_H^\ddagger}{q_H \cdot q_D^\ddagger} \tag{9.131}$$

where q_D refers to the total partition function for the deuterium-containing reactant, and so on. It is assumed that no statistical corrections due to symmetry

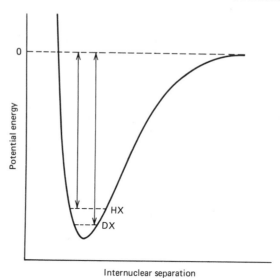

Figure 9.9. Potential energy diagram for diatomic molecules HX and DX. The lengths of the arrows represent the dissociation energies of the two isotopic molecules.

numbers are needed.[75] At room temperature (9.131) can be simplified to

$$\frac{k_H}{k_D} = \prod_i \exp \frac{-h(\nu_{Hi}^{\ddagger} - \nu_{Di}^{\ddagger} - \nu_{Hi} + \nu_{Di})}{2kT} \qquad (9.132)$$

The isotope effect for tritium can be found if that for deuterium is known.[76]

To estimate the kinetic isotope effect in a proton transfer reaction, the procedure of Westheimer is followed.[77] Two limiting cases of a symmetric transition state and an early (or late) transition state are taken.

A---H---B   A—H---B

Symmetric TS   Early TS

A and B are polyatomic molecules containing heavy atoms that bond to hydrogen. They are treated as single heavy atoms for simplicity. The pseudo-triatomic molecule has a symmetric stretch and an asymmetric stretch, similar to those for CO_2 and other linear, triatomic molecules. One stretch corresponds to motion along the reaction coordinate and has an imaginary frequency. It is factored out from the partition functions $q_H^{\ddagger}$ and $q_D^{\ddagger}$ of (9.131) and contributes nothing to (9.132). The other motion is a real vibration and is included in (9.132).

In the case of the symmetric transition state, the asymmetric stretch must be the reaction coordinate (r.c.). Furthermore, the symmetric stretch involves no motion of the central H atom and therefore $\nu_H^{\ddagger} = \nu_D^{\ddagger}$, with no isotope effect.

$\overset{\leftarrow}{A}$---$\vec{H}$---$\overset{\leftarrow}{B}$ $\overset{\leftarrow}{A}$---$\overset{\leftarrow}{H}$---$\vec{B}$

r.c. = ν_{as} ν_s

But with an early (or late) transition state, it must be the symmetric stretch that is the reaction coordinate, since HA approaches B as a unit with little change in the HA bond length.

$$\underset{\nu_{as}}{\overset{\leftarrow\quad\rightarrow}{A-H---B}} \qquad \underset{r.c. \ = \ \nu_s}{\overset{\rightarrow\ \ \rightarrow\quad\leftarrow}{A-H---B}}$$

We know from studies on hydrogen bonding that the asymmetric stretch in the early transition state is only a few hundred wave numbers lower than that in HA. Also the two bending modes of the pseudo-triatomic molecule are of much lower frequency than the stretching modes, and are roughly matched by the two hydrogen bending modes of the real HA molecules, since hydrogen atom motions are primarily involved in both cases. The conclusions, finally, are that an early (or late) transition state gives a very small isotope effect, whereas a symmetric transition state gives a maximum isotope effect.

$$\frac{k_H}{k_D} \cong e^{-h(\nu_H - \nu_D)/2kT} \tag{9.133}$$

Table 9.4 shows the maximum kinetic isotope effect expected for breaking various bonds to hydrogen. Note that there is no distinction made between transfer of a hydrogen atom, a proton, or a hydride ion. With certain important exceptions discussed later, experimental results are in good agreement in that k_H/k_D ranges between 1 and the maximum value indicated. There are a few cases of an inverse isotope effect where $k_H/k_D < 1$, even though a bond to hydrogen is being broken.[78] These are readily explained by the theory given earlier, for example,

$$R_2C{=}CH_2 + HMn(CO)_5 \rightarrow R_2CCH_3 + \cdot Mn(CO)_5 \tag{9.134}$$

$$\frac{k_H}{k_D} = 0.4$$

The metal–hydrogen stretching frequency is only $1800 \ cm^{-1}$ in the reactant and could easily be greater in the transition state since a bond to carbon is being formed.

Table 9.4.   Maximum kinetic isotope effects, k_H/k_D, for different bonds[a, b]

Bond	ν_H/cm^{-1}	$e^{h(\nu_H - \nu_D)/2kT}$
C–H	2900	7.0
O–H	3600	11.0
$\overset{+}{N}$–H	2700	6.0
S–H	2550	5.4
F–H	4140	14.9
M–H	1500–2200	2.7–4.2

[a] Except for HF ν_D is taken as $\nu_H/1.38$.
[b] M refers to various transition metals. The frequencies are for terminal metal–hydrogen bonds. Bridging bonds have lower frequencies.

According to the Westheimer treatment the study of kinetic isotope effects can be useful in two ways: first in showing that a hydrogen atom transfer is part of the rate-determining step of a mechanism, and, second, in locating the degree of transfer in the activated complex. Fortunately, the conclusions drawn from the second application generally agree with those from other methods. In the base-catalyzed ionization of ketones and nitroalkanes, k_H/k_D changes as β changes in the expected manner.[79] Reaction (9.39) has $\beta = 1$ and $k_H/k_D = 1$, agreeing with the proton transfer being complete in the transition state.[80]

A more convincing demonstration is shown in Fig. 9.10, where k_H/k_D is plotted for a number of ketones and nitro compounds as a function of ΔpK, the pK_a of the substrate minus the pK_a of the base.[73b] A maximum is seen at $\Delta pK = 0$, which is consistent with the expectation that the transition state would be symmetric for such cases. However, Fig. 9.10 is somewhat misleading since it only includes values of the isotope effect that fall roughly within the theoretical maximum limit of Table 9.4. There are a large number of cases where values of k_H/k_D are as large as 50 at room temperature, and even larger at lower temperatures.[81] Such large ratios can only be explained by tunneling, which is discussed in the next section.

QUANTUM-MECHANICAL TUNNELING

It is well known in quantum mechanics that a small particle can exist in regions where classically it is excluded, that is, where its energy is less than the potential

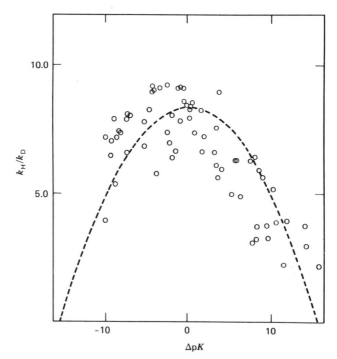

Figure 9.10. Plot of k_H/k_D as a function of ΔpK for the ionization of ketones and nitroalkanes. $\Delta pK = pK$ (substrate) $- pK$ (base).

energy. In kinetics this means that such a particle may leak through, or tunnel through, a potential energy barrier, rather than pass over it. It is a consequence of the Heisenberg uncertainty principle and its importance in any given case may be gauged by calculating the DeBroglie wavelength.

$$\lambda = \frac{h}{mv} \tag{9.135}$$

For a hydrogen atom traveling with a velocity of 2×10^5 cm s^{-1}, the DeBroglie wavelength is 2 Å. The same value would be found for an electron traveling at 3.6×10^8 cm s^{-1}. In a hydrogen bond, A—H---B, a typical value of the H---B distance would be 1.7 Å,[82] so that tunneling is a possibility.

The tunneling correction Q may be defined as

$$Q = \frac{k_{\text{obs}}}{k_{\text{class}}} \tag{9.136}$$

where k_{class} is the expected rate constant in the absence of tunneling. To calculate Q it is necessary to assume some reasonable form for the potential energy function.[83] The one-dimensional Schrödinger equation is then solved to give an expression for G, the probability of penetrating the barrier. This is a function of the energy W of the particle. Assuming a Boltzmann distribution of the energy, the value of Q can be found by averaging

$$Q = \frac{\int_0^\infty G(W) e^{-W/kt} \, dW}{\int_E^\infty e^{-W/kt} \, dW} \tag{9.137}$$

E is the height of the energy barrier. The denominator has the classical value, $e^{-E/kt}$, which results from setting $G = 0$ for $E > W$, and $G = 1$ for $E \leqslant W$.

The results are rather complicated, even for simple model barriers. Q depends on the height of the barrier, its thickness, the mass of the particle, and temperature. For reasonable model potentials, it is possible to account for observed values of Q_H/Q_D as large as 50–100. Large ratios are most likely when there are strong steric factors in forming the activated complex. It is correct to say that tunneling is always a factor in reactions involving transfer of hydrogen atoms, protons, or hydride ions. Even for high, wide barriers, a molecule that has been activated to an energy somewhat less than the critical energy can easily tunnel through the remaining barrier. A theory for tunneling corrections in unimolecular reactions has also been given.[84]

For elements heavier than hydrogen, tunneling can be neglected. For the electron, as mentioned above, tunneling is quite common. In the ground-state hydrogen atom the electron has a 24% probability of being more than a distance of $2a_0$ from the nucleus, that is, in the classically forbidden region. Some redox reactions occur by so-called outer-sphere electron-transfer mechanisms.[85] A collision complex is formed first between the reactants, for example,

$$\text{Fe(CN)}_6^{4-} + \text{Ru(NH}_3)_6^{3+} \overset{K_{os}}{\rightleftharpoons} \text{Ru(NH}_3)_6^{3+}, \text{Fe(CN)}_4^{4-} \tag{9.138}$$

The equilibrium constant, K_{os}, for this outer-sphere complex can be calculated by methods described earlier (p. 263), or in some cases it can be measured directly.[86] Electron transfer then occurs within the outer-sphere complex.

$$Ru(NH_3)_6^{3+}, Fe(CN)_6^{4-} \xrightarrow{k_{et}} Ru(NH_3)_6^{2+}, Fe(CN)_6^{3-} \qquad (9.139)$$

Values of k_{et} fall in the range of 10^{-3} to 10^9 s^{-1}, if thermodynamically allowed. There is a symmetry restriction in the electron-transfer step in that there must be a net positive overlap between the orbital from which the electron originates and that in which it eventually resides in the product. There is probably some degree of tunneling for the electron-transfer step in the usual case. The theory of electron tunneling is similar to that of proton tunneling, except that it is not a function of temperature.[87] Hence at low temperatures tunneling becomes the favored mechanism.

At very low temperatures proton tunneling must also become independent of temperature, since the energy of the reactants becomes the zero-point energy. Thus hydrogen-atom-transfer reactions can occur even at the absolute zero of temperature. Deuterium atom transfer would be completely frozen out under these conditions.[88] At 77 K, ratios of k_H/k_D as large as 3×10^4 have been observed.

THE PRINCIPLE OF MICROSCOPIC REVERSIBILITY — FREE ENERGY PROFILES

It is recognized that the addition of a catalyst, according to the strict definition of the term, merely speeds up the attainment of equilibrium but does not change the point of equilibrium. That is, the maximum yield in a given reaction cannot be changed by adding a catalyst unless it is added in amounts sufficient to constitute a change in medium or unless it combines with one of the products. This consequence follows immediately from thermodynamic arguments, since the change in free energy that controls the equilibrium constant is dependent only on the initial and final states of the system and is independent of the path or mechanism of changing from one state to another.

Another proof, useful for kinetic purposes, of the independence of the point of equilibrium on the reaction mechanism comes from the principle of microscopic reversibility and detailed balancing (p. 307). Taking a chemical example in the transition $A \rightleftharpoons B$, which can go by a number of mechanisms, namely,

$$A \underset{k_2}{\overset{k_1}{\rightleftharpoons}} B \qquad (9.140)$$

$$A + C \underset{k_4}{\overset{k_3}{\rightleftharpoons}} X \underset{k_6}{\overset{k_5}{\rightleftharpoons}} B + C \qquad (9.141)$$

$$A + D \underset{k_8}{\overset{k_7}{\rightleftharpoons}} Y \underset{k_{10}}{\overset{k_9}{\rightleftharpoons}} B + D \text{ etc.} \qquad (9.142)$$

where C and D are catalysts for the reaction and X and Y are the corresponding intermediates, then at equilibrium the number of molecules of B formed from A in unit time by each of the mechanisms shown in (9.140)–(9.142) must be equal to the number of molecules of A formed from B by the reverse of each mechanism. This sets up the condition (as can be readily verified by setting the rate of formation of each component equal to zero and using the proper activity coefficients in the rate expressions)

$$\frac{a_B}{a_A} = K_{eq} = \frac{k_1}{k_2} = \frac{k_3 k_5}{k_4 k_6} = \frac{k_7 k_9}{k_8 k_{10}} \tag{9.143}$$

Accordingly it is ensured that the equilibrium constant is unchanged for the catalyzed paths and also that certain restrictions exist for the various rate constants involved.

The great importance of the principle of microscopic reversibility in mechanism determinations is that it enables the mechanism of the reverse reaction to be known with as much accuracy as the mechanism of the forward reaction is known. For example, if it is known that for a given set of conditions A is converted to B exclusively by the catalyst C as in (9.141), then, under the same conditions, B must be converted to A catalyzed by C and through the same intermediate X. The possibility that the reverse reaction operates through the direct process as in (9.140) or, more disconcertingly, by some unsuspected new mechanism is eliminated. In general, the principle states that if a certain mechanism (series of steps) is followed for the forward reactions, then under the same conditions the reverse reaction consists of the same steps traversed backward. To illustrate, for a given reaction, if a diagram of free energy as a function of the extent of reaction is drawn showing the various intermediates and activated complexes, a diagram such as Fig. 9.11 is obtained. If the energy barriers and nature of the intermediate are known for the forward reaction, the principle of microscopic reversibility enables the mechanism of the reverse reaction to be described by simply reversing the procedure starting with the products and going from right to left on the energy diagram.

Figure 9.11 should not be confused with other figures we use earlier in which the potential energy is shown as a function of the reaction coordinate. It is the free energy that is represented, and, to be precise, it is the standard free energy, $G^{\ominus}$. The diagram is useful in showing the various intermediates and activated complexes in a reaction mechanism. The differences in free energy determine not only the

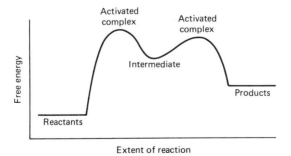

Figure 9.11.   Free-energy profile for reaction involving an intermediate. The same intermediate and activated complexes apply to both forward and reverse reactions.

rate constants (from absolute rate theory), but also equilibrium constants. The rates of going from one extremum to another depend not only on $\Delta G^{\ddagger}$ but on the actual concentrations prevailing.

There has been some discussion about the propriety of such free energy profiles, since activated complexes are defined in terms of maximum potential energies along the reaction coordinate.[89] Certainly for individual molecules, it is the potential energy that is significant. Because of the entropy contribution, the free energy has meaning only for an assembly of molecules. In terms of the reaction coordinate, the maximum in free energy would not necessarily be reached at the same point as the maximum potential energy. In spite of this objection, free energy profiles are commonly shown with the reaction coordinate as the abscissa, but the scale is now arbitrary.

As an illustration of the use of free energy profiles and the principle of microscopic reversibility, Fig. 9.12 represents the mechanism of acid-catalyzed hydration of formaldehyde discussed in (9.103) and (9.104). The mechanism of acid-catalyzed dehydration can be obtained by reading from right to left. The hydrated aldehyde is protonated in a fast, reversible step. The rate-determining step is proton removal from the OH group by A^-, with a synchronous elimination of water.

The condensation of acetaldehyde to aldol and of acetone to diacetone alcohol are two reactions that may be discussed together since they have almost identical mechanisms as far as the steps involved are concerned. Small differences, however, are sufficient to make the observed kinetics of the two reactions quite dissimilar. The reaction is

$$R\text{—}\overset{\overset{\displaystyle O}{\|}}{C}\text{—}CH_3 + R\text{—}\overset{\overset{\displaystyle O}{\|}}{C}\text{—}CH_3 \rightleftharpoons R\text{—}\overset{\overset{\displaystyle CH_3}{|}}{\underset{\underset{\displaystyle H}{\overset{|}{O}}}{C}}\text{—}CH_2\text{—}\overset{\overset{}{}}{\underset{\underset{\displaystyle O}{\|}}{C}}\text{—}R$$

(9.144)

where R is H or CH_3, basic catalysts being used in each case. The reaction is reversible though subsequent reactions in the aldol condensation (dehydration and polycondensation) tend to make it irreversible. The point of equilibrium is well to the

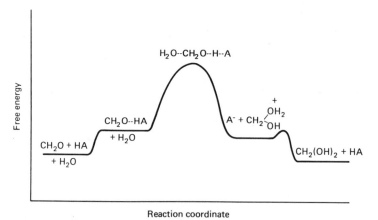

Figure 9.12. Free energy versus reaction coordinate for acid-catalyzed hydration of formaldehyde.

right for acetaldehyde, even in dilute aqueous solution, but it is well to the left for acetone even in concentrated acetone solution. For this reason the aldol condensation is conveniently studied only from the acetaldehyde side and the acetone reaction only from the diacetone alcohol side. The reverse process can in each case be reconstructed from the mechanism of the forward process so that a complete comparison of (9.144) can be made for $R = CH_3$, and $R = H$.

The assumed mechanism is one of great importance in organic chemistry, since it is operative for most condensation reactions. The general sequence can be given as follows

$$HA + B \underset{k_2}{\overset{k_1}{\rightleftharpoons}} BH^+ + A^- \qquad (9.145)$$

$$A^- + C \xrightarrow{k_3} \text{product} \qquad (9.146)$$

where HA is an organic compound with an acidic hydrogen that can be ionized by a base B, forming a carbanion A^-. The carbanion can either regenerate the reactants by pulling a proton from BH^+, or react with a substrate C to form the desired product of the reaction. In the present example, HA and C are the same substance, both being either acetaldehyde or acetone. The kinetics of (9.145) and (9.146) are complicated unless, as is usual, the carbanion is reactive enough to come to a kinetic equilibrium so that the steady-state method may be used for its concentration. This method gives

$$[A^-] = \frac{k_1 [HA] [B]}{k_3 [C] + k_2 [BH^+]} \qquad (9.147)$$

$$\frac{d[\text{product}]}{dt} = \frac{k_1 k_3 [HA] [B] [C]}{k_3 [C] + k_2 [BH^+]} \qquad (9.148)$$

Two limiting cases are of interest in which $k_3 [C]$ is either much greater or much smaller than $k_2 [BH^+]$, assuming only one base present. The first case leads to

$$\text{rate} = k_1 [HA] [B] \qquad (9.149)$$

and the second to

$$\text{rate} = \frac{k_1 k_3 [HA] [C] [B]}{k_2 [BH^+]} = \frac{k_1 k_3 [HA] [C] [S^-]}{k_2 K_B} \qquad (9.150)$$

where K_B is the ionization constant of the base and S^- is the anion of the solvent SH. Equation (9.149) corresponds to general base catalysis, and (9.150) to a specific catalysis by S^-, the usual case being catalysis by hydroxide ion in aqueous solution.

For acetone and acetaldehyde (9.145) and (9.146) are shown in

$$R-\overset{\overset{\displaystyle O}{\|}}{C}-CH_3 + B \underset{k_2}{\overset{k_1}{\rightleftharpoons}} R-\overset{\overset{\displaystyle O}{\|}}{C}-CH_2^- + BH^+ \qquad (9.151)$$

$$R-\overset{\overset{\displaystyle O}{\|}}{C}-CH_2^- + R-\overset{\overset{\displaystyle O}{\|}}{C}-CH_3 \underset{k_4}{\overset{k_3}{\rightleftharpoons}} R-\overset{\overset{\displaystyle O}{\|}}{C}-CH_2-\overset{\overset{\displaystyle R}{|}}{\underset{\underset{\displaystyle CH_3}{|}}{C}}-O^- \qquad (9.152)$$

$$R\text{—}\overset{\overset{\displaystyle O}{\|}}{C}\text{—}CH_2\text{—}\overset{\overset{\displaystyle R}{|}}{\underset{\underset{\displaystyle CH_3}{|}}{C}}\text{—}O^- + BH^+ \underset{k_6}{\overset{k_5}{\rightleftharpoons}} R\text{—}\overset{\overset{\displaystyle O}{\|}}{C}\text{—}CH_2\text{—}\overset{\overset{\displaystyle R}{|}}{\underset{\underset{\displaystyle CH_3}{|}}{C}}\text{—}OH + B \qquad (9.153)$$

the last step being very rapid. The rate equations (9.149) and (9.150) become, in water,

$$\text{rate} = k_1 \left[R\text{—}\overset{\overset{\displaystyle O}{\|}}{C}\text{—}CH_3 \right][B] \qquad (9.154)$$

$$\text{rate} = \frac{k_1 k_3}{k_2 K_B} \left[R\text{—}\overset{\overset{\displaystyle O}{\|}}{C}\text{—}CH_3 \right]^2 [OH^-] \qquad (9.155)$$

The factor that determines whether (9.154) or (9.155) is observed is the relative velocity of the carbanion toward condensing at the carbonyl compared to reaction with BH$^+$. A high rate of condensation leads to (9.154), and a high rate of neutralization or a low rate of condensation to (9.155). The base B is not used up during a reaction so that B, BH$^+$, and hydroxide ion remain constant during a run. The reaction is thus either first or second order in the carbonyl compound, and the rate varies from run to run with either the concentration of the base B or the hydroxide ion. Since in water solution more than one base is usually present, (9.154) is more correctly written as

$$\text{rate} = \left[R\text{—}\overset{\overset{\displaystyle O}{\|}}{C}\text{—}CH_3 \right] \sum_j k_{1j} [B_j] \qquad (9.156)$$

Figure 9.13 represents schematically how the free energy varies along the reaction coordinate. The heights of the free energy barriers that the carbanion A$^-$ must overcome in going backward or forward determine the kinetics found experimentally. For acetaldehyde the rate law is given by (9.154) or (9.156), since condensation occurs easily.[90] For acetone the rate law is given by (9.155), since condensation is difficult (dashed line in Fig. 9.13).[90]

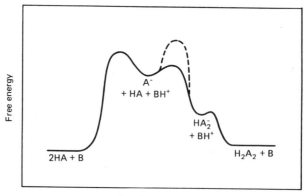

Reaction coordinate

Figure 9.13. Free energy barriers in condensation reactions of carbanions. Full line is for easy condensation step. Dashed line is for difficult condensation step.

Figure 9.13 may also be used to construct the reverse reactions. For acetaldehyde the rate-limiting step is

$$^-CH_2CHO + BH^+ \xrightarrow{k_2} CH_3CHO + B \tag{9.157}$$

The rate equation is then

$$\text{rate} = k_2[BH^+][A^-] = \frac{k_2k_4k_6[H_2A_2][B]}{k_3k_5[HA]} \tag{9.158}$$

where H_2A_2 is the condensed product, that is, aldol. The cleavage reaction is thus general base catalyzed and inhibited by the presence of the monomer. Equating the rates of the forward reaction given by (9.154) and the reverse reaction gives the equilibrium condition:

$$\frac{[H_2A_2]}{[HA]^2} = \frac{k_1k_3k_5}{k_2k_4k_6} = K_{eq} \tag{9.159}$$

For acetone the rate-limiting step is the dissociation of HA_2^-.

$$CH_3-\overset{O}{\underset{}{C}}-CH_2-\overset{CH_3}{\underset{O_-}{\overset{|}{\underset{|}{C}}}}-CH_3 \xrightarrow{k_4} CH_3-\overset{O}{\underset{}{C}}-CH_2^- + CH_3-\overset{O}{\underset{}{C}}-CH_3 \tag{9.160}$$

$$\text{rate} = k_4[HA_2^-] = \frac{k_4k_6[H_2A_2][B]}{k_5[BH^+]} = \frac{k_4k_6[H_2A_2][OH^-]}{k_5K_B} \tag{9.161}$$

Setting the rate of the forward reaction (9.155) equal to the rate to the reverse reaction gives the equilibrium constant

$$\frac{[H_2A_2]}{[HA]^2} = \frac{k_1k_3k_5}{k_2k_4k_6} = K_{eq} \tag{9.162}$$

Note that in all cases, both forward and reverse reactions and both reactants, the observed rate depends only on the difference in free energy between the reactants, 2HA or H_2A_2, and the highest energy activated complex encountered.

The principle of microscopic reversibility applied to isotope exchange reactions is revealing. Obviously the mechanism must be precisely the same for the forward and reverse reactions, and the free energy profile must be symmetric. The existence of only one path for exchange, implies that at some point the exchanging atoms must become equivalent either in a transition state or in an intermediate. As an example, take the iron(II)–iron(III) exchange reaction of their aquo ions.[91]

$$Fe^*(H_2O)_6^{2+} + Fe(H_2O)_6^{3+} \rightleftharpoons Fe^*(H_2O)_6, Fe(H_2O)_6^{5+} \rightleftharpoons Fe^*(H_2O)_6^{3+} + Fe(H_2O)_6^{2+} \tag{9.163}$$

Electron transfer can occur in the ion-pair intermediate only if the bond distances between iron(III) and ligand water are lengthened and those between iron(II) and ligand water are decreased to some equal intermediate value.[91] Otherwise the electron transfer would produce iron(II) and iron(III) in the wrong coordination environments, and the energy would abruptly increase. The two iron atoms must be equivalent to prevent this from happening.

The exchange reaction (9.163) is catalyzed by a number of anions, including chloride ion. One logical path for this catalysis is an inner-sphere atom transfer mechanism.[85]

$$Fe^*(H_2O)_5Cl^{2+} + Fe(H_2O)_6^{2+} \rightleftharpoons (H_2O)_5Fe^*-Cl-Fe(H_2O)_5^{4+}$$

$$\rightleftharpoons Fe^*(H_2O)_6^{2+} + Fe(H_2O)_5Cl^{2+} \qquad (9.164)$$

Iron(II) being labile, $Fe(H_2O)_6^{2+}$ can readily lose a water molecule and bridge to iron(III) by way of the chloro ligand. In this path for exchange, both iron atoms again become indistinguishable before chlorine atom transfer can occur.

A careful study reveals that another mechanism is also operating.[92] This is a simple electron transfer between the ions.

$$Fe^*(H_2O)_5Cl^{2+} + Fe(H_2O)_6^{2+} \rightarrow Fe^*(H_2O)_5Cl^+ + Fe(H_2O)_6^{3+} \quad (9.165)$$

followed by rapid loss of Cl^- from the labile iron(II) chloro complex. Now this mechanism causes concern because if Cl is not bridging the two iron atoms in the activated complex, they cannot be equivalent, and the free energy profile would be asymmetric. However, this does not rule out the mechanism, as is sometimes incorrectly assumed. We need only add the reverse of (9.165) as an additional path for isotope exchange.

$$Fe(H_2O)_6^{2+} + Cl^- \rightleftharpoons Fe(H_2O)_5Cl^+ + H_2O$$

$$Fe(H_2O)_5Cl^+ + Fe^*(H_2O)_6^{3+} \rightarrow Fe(H_2O)_5Cl^{2+} + Fe^*(H_2O)_6^{2+}$$

$$(9.166)$$

A free energy profile that includes both (9.165) and (9.166), going from left to right, is perfectly symmetric upon reflection at the midpoint. A number of other examples of exchange reactions that occur by asymmetric, multiple, paths have been presented.[93]

CATALYSIS BY ENZYMES

The subject of enzymatic catalysis is now so vast that it must be treated in specialized monographs, as is done for heterogeneous catalysis.[94] Still it is worthwhile to give a brief discussion, particularly to try to relate enzymatic catalysis to ordinary catalysis. Enzymes are interesting, not only because of their biological importance, but also because of their remarkable efficiency and specificity. On a molar basis other catalysts rarely come within a factor of a million of matching enzyme reactivity. A major question, obviously, is can this extraordinary behavior be understood on the basis of what we know about simple catalysts, and, if not, what are the additional factors?

Enzymes are macromolecules with molecular weights in the range of 10^4-10^6. A portion of all enzymes is protein, and sometimes they are entirely protein. The protein part is called the apoenzyme, and the nonprotein part is the prosthetic group or coenzyme. Metal ions are also an important part of many enzymes. Enzymes are classified as oxidoreductases, transferases, hydrolases, lyases, isomerases, and ligases, according to the kinds of reaction they catalyze. Some enzymes work only on a single substance, called the substrate. Other enzymes catalyze the reactions of all molecules containing a certain functional group. In the living cell enzyme concentrations are in the range of 10^{-9} to $10^{-5}\,M$.

In studying enzyme kinetics in the laboratory, concentrations are usually in the range of 10^{-8} to $10^{-10}\,M$ to produce easily measured rates. Substrate concentrations are very much higher. For convenience, usually only initial rates, v, are measured as a function of various concentrations, pH, temperature, and so on. The initial step is invariably complex formation between the enzyme and one substrate. The specificity and efficiency of enzyme catalysis is partly due to a complementarity of structures of enzyme and substrate: the substrate molecule fits into some part of the enzyme surface as a key fits a lock (Emil Fischer, 1894). Further reactions then occur within the enzyme–substrate complex. Other necessary reactant molecules may also add to the complex.

A number of different kinetic situations can develop, depending on the system.[95] We discuss only the simplest case, represented by the equations

$$\text{E} + \text{S} \underset{k_2}{\overset{k_1}{\rightleftharpoons}} \text{ES} \overset{k_3}{\longrightarrow} \text{E} + \text{P} \qquad (9.167)$$

where ES is the enzyme–substrate complex, and P is the product. Figure 9.14 shows how the initial rate varies as a function of substrate concentration in such a system, as well as in many other enzymatic reactions. There is a saturation at high values of [S], which is attributed to complete formation of the complex. Assuming a steady state for [ES], the rate equations for (9.167) are as follows:

$$\frac{d[\text{ES}]}{dt} = k_1\,[\text{E}]\,[\text{S}] - k_2\,[\text{ES}] - k_3\,[\text{ES}] = 0 \qquad (9.168)$$

We define the Michaelis-Menten constant K_m and the maximum rate V_m,

$$\frac{k_2 + k_3}{k_1} = K_m \qquad v = V_m = k_3 E_0 \qquad (9.169)$$

where $E_0 = [\text{E}] + [\text{ES}]$, the initial enzyme concentration. In these terms (9.168) can be written as

$$v = \frac{V_m\,[\text{S}]}{K_m + [\text{S}]} \qquad (9.170)$$

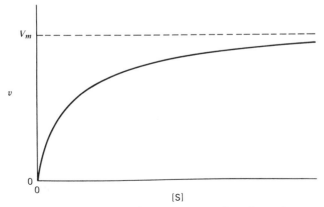

Figure 9.14. The initial velocity of an enzyme-catalyzed reaction as a function of substrate concentration.

which has the properties needed to match Fig. 9.14. The rate is proportional to the enzyme concentration at low substrate, and reaches a limiting value, V_m, at high substrate concentrations.

The rate law (9.170) is a hyperbolic equation, which can be converted to a linear relationship by taking reciprocals.

$$\frac{1}{v} = \frac{K_m}{V_m [S]} + \frac{1}{V_m} \tag{9.171}$$

Figure 9.15 shows a plot of this equation. The slope gives K_m/V_m, and the intercept is $1/V_m$. In enzyme kinetics such a graph is called a Lineweaver-Burk plot. Note the similarity of (9.170) and (9.171) to those equations found in heterogeneous catalysis when the Langmuir adsorption isotherm is used. It may also be noted that (9.167) can be expanded to include any number of intermediates formed from E + S before the product is released.

$$ES_1 \rightleftharpoons ES_2 \rightleftharpoons ES_3 \ldots \tag{9.172}$$

As long as the steady-state assumption is valid for these species, the same form of the rate law is found, with a change in the meaning of K_m and V_m.

The Michaelis-Menten constant is an approximation to the dissociation constant, K_d, for the enzyme–substrate complex. The exact value is given by k_2/k_1. Fortunately, in a number of cases values of k_2 and k_1 have been obtained directly by using rapid reaction techniques such as stopped flow and relaxation methods.[95a] Some typical results are given in Table 9.5. The values of k_1 are in the range of 5×10^6 to $10^9 \, M^{-1} \, s^{-1}$, approaching diffusion control. The values of k_2 are in the range of $10^2 - 10^5 \, s^{-1}$. The dissociation constants are usually about $10^{-4} \, M$. The rate constant k_3 can be found from V_m. It is called the turnover number and gives the number of moles of product per second per mole of binding sites on the

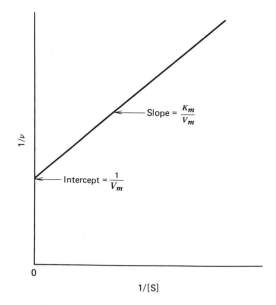

Figure 9.15. Lineweaver-Burk plot of $1/v$ against $1/[S]$.

enzyme. Typical values of k_3 range from 10 to 10^7 s^{-1}, the latter value being for catalase, which catalyzes the decomposition of hydrogen peroxide.

We are now in a position to compare rates of ordinary reactions in solution with those of enzymatic reactions. For a bimolecular reaction, according to Chapter 7,

$$A + B \underset{}{\overset{K'}{\rightleftharpoons}} A,B \xrightarrow{k'_3} \text{product} \tag{9.173}$$

so that the second-order rate constant is normally given by $k' = K'k'_3$. The association constant for spherical molecules is (7.78).

$$K' = \frac{4}{3} \pi d_{AB}^3 L e^{-U/kT} = \frac{1}{K_d} \tag{9.174}$$

which normally is equal to 10^{-1} to 10^2, the latter value corresponding to oppositely charged ions or hydrogen-bonded complexes where U is favorable. If K' were much larger, then A,B would be completely formed at the usual concentrations for kinetics, and we would not be dealing with a second-order reaction. Clearly this does not happen often.

In enzyme reactions the second-order rate constant (low substrate concentrations) is given by $k = k_3/K_m$. If $k_2 \gg k_3$, k is equal to k_3/K_d. Since $1/K_d \cong 10^4$, there is an advantage to the enzyme reaction of roughly this factor. That is the enzyme–substrate complex is at least 10^4 times as stable as the typical collision complex of (9.173). If $k_3 \gg k_2$, then $k = k_1$, the rate of the enzyme-catalyzed reaction is equal to the rate of formation of the complex. This is also a consequence of the increased stability of the complex, since it lasts long enough so that the product-forming step (k_3) can occur.

It appears certain that nature has picked the optimum value for the formation constant of the enzyme–substrate complex in each case. A less stable complex would reduce the rate. A more stable complex would saturate at very low substrate concentrations, and no further increase in rate would occur at fixed enzyme concentration. Also an enzyme that forms very stable complexes might hold the

Table 9.5. Rate data for enzyme–substrate complex formation in water at 25°C

Enzyme	Substrate	k_1/M^{-1} s^{-1}	$k_2/$s^{-1}
Peroxidase[a]	H_2O_2	9×10^6	> 1.4
Aspartate amino-transferase[b]	NH_2OH	3.7×10^6	38
Ribonuclease A[c]	Cytidine 3'-phosphate	4.6×10^7	4.2×10^3
Ribonuclease A[c]	Uridine 3'-phosphate	7.8×10^7	1.1×10^4
Creatine kinease[d]	ADP	2.3×10^7	1.8×10^4
Creatine kinase[d]	MgADP	5.3×10^6	5.1×10^3
Chymotripsin[e]	Furyloyl-L-tryptophanamide	6×10^6	1×10^4

[a] B. Chance, *Arch. Biochem.* **22**, 229 (1949).
[b] P. Fasella and G. G. Hammes, *J. Am. Chem. Soc.,* **85**, 3929 (1963).
[c] G. G. Hammes and F. G. Walz, Jr., *J. Am. Chem. Soc.,* **91**, 7179 (1969).
[d] G. G. Hammes and J. K. Hurst, *Biochemistry,* 8, 1083 (1969).
[e] G. P. Hess in reference 94a, Vol. 3, p. 213.

product tightly. By microscopic reversibility, (9.167) should be written more correctly as

$$E + S \rightleftharpoons ES \rightleftharpoons EP \rightleftharpoons E + P \qquad (9.175)$$

so that dissociation of the product from the enzyme can become rate limiting.

To compare k_3 and k_3' values, it is necessary to select a definite reaction since k_3' can vary over a very wide range for ordinary reactions. A suitable comparison is the hydration of carbon dioxide, catalyzed by carbonic anhydrase and by other bases shown in Table 9.1. Carbonic anhydrase is characterized[96] by a turnover number (k_3) of 1×10^6 s^{-1} and a K_m value of 9×10^{-3} M at 25°C, which leads to a second-order rate constant of 1.1×10^8 M^{-1} s^{-1}. Since k_3 is much larger than the values of k_2 shown in Table 9.5, it is reasonable that this is an example of $k = k_1$, that is, at the low substrate concentrations, the rate of the enzyme reaction is equal to the rate of formation of the enzyme–substrate complex. The complex always lasts long enough to react.

The value of k_3 may now be compared to values of k_3' from the data in Table 9.1, assuming K' is of the order of unity in all cases, except when water is the reactant. For the best reagent, piperidine, the enzyme has an advantage of $1 \times 10^6/6 \times 10^4 = 16$. For metal-ion-bound hydroxide, the factor is about 1.6×10^3. These are not large factors favoring the enzyme, considering its advantages.

The enzyme–CO_2 complex would have the Zn–OH group and CO_2 positioned perfectly for reaction, whereas a $Rh(NH_3)_5OH^{2+}$–CO_2 complex might be quite random. Furthermore, the enzyme would have other groups near the active site that would be favorably placed to stabilize the transition state for (9.26). The enzyme may conveniently be regarded as a microregion of special solvent in which the reaction occurs. The factors mentioned above are only modest examples of solvent effects.

In this respect surfactants in aqueous solution mimic enzymes when they aggregate to form micelles.[97] Surfactants, or detergents, have both hydrophilic and hydrophobic groups. In micelles the hydrophobic groups concentrate in the interior and the hydrophilic groups point outward to the solvent. Rates of many organic and inorganic reactions are accelerated by micelles, sometimes by a factor of several thousand. The reaction occurs in the micellar pseudo-phase. Rate enhancement can occur because of a concentration of the reactants into a small volume, or because of a favorable medium effect on the rate constant.

Another catalytic system that has some resemblance to enzymes is the so-called phase-transfer catalysis. This is a method of achieving reaction between mutually insoluble reagents, such as small ions and large organic molecules.[98] Two immiscible solvents are used, such as water and benzene. The ion is drawn into solution in the benzene by using a large benzene-soluble counterion, such as a quaternary ammonium cation. The reaction then goes readily in the benzene phase since the small anion is unusually reactive in such a medium. We see again a concentration effect and a favorable medium effect, such as exists in enzyme catalysis.

In summary, there appears to be nothing in enzyme action that is not understandable in conventional kinetic terms. Enzymes have had the advantage of an evolutionary development over long periods of time. There is no reason why chemists cannot synthesize tailor-made catalysts that do as well as enzymes for specific reactions. Indeed considerable progress is being made in this direction.[99]

PROBLEMS

9.1. The hydrolysis of the hydroxylamine disulfonate ion proceeds as follows:

$$HON(SO_3)_2^{2-} + H_2O \rightarrow HONHSO_3^- + H^+ + SO_4^{2-}$$

From the data of Naiditch and Yost given below determine the rate constants for the uncatalyzed reaction and for the reaction catalyzed by hydrogen ion (the former can be obtained most readily from the initial readings and the latter from the later readings). The reaction was followed by titrating the acid produced. Original concentrations: $K_2 HON(SO_3)_2$, $0.0473 M$, no acid initially added. Temperature $25°C$.

Time/min	Acid $\times 10^3$/mol liter^{-1}
11	1.12
24	1.34
38	1.67
58	2.03
81	2.68
100	3.29
122	4.32
144	5.32
164	6.65
189	8.93
218	12.08
256	16.17

Source: S. Naiditch and D. M. Yost, *J. Am. Chem. Soc.*, **63**, 2123 (1941).

9.2. The mutarotation of glucose is subject to catalysis by both acids and bases. Evaluate as well as possible the catalytic constants of water, hydrogen ion, mandelic acid, mandelate ion, pyridine, and pyridinium ion from the observed first-order rate constants given. K_a for mandelic acid is 4.3×10^{-4}. K_b for pyridine is 2.9×10^{-9}. Temperature, $18°C$.

Catalyst conc./mol liter^{-1}	$k_{obs} \times 10^4$/s^{-1}
HCl 10^{-4}	0.88
HCl 10^{-5}	0.87
HClO$_4$ 10^{-4} + 0.1 N KNO$_3$	0.90
HClO$_4$ 0.0048	1.00
HClO$_4$ 0.0247	1.49
HClO$_4$ 0.0325 + 0.2 N KNO$_3$	1.67

Sodium mandelate	mandelic acid	
0.050	0.001	0.97
0.100	0.001	1.06
0.125	0.001	1.10
0.100	0.050	1.11
0.100	0.100	1.17
Pyridine	pyridine perchlorate	
0.025	0.025	1.24
0.050	0.050	1.55
0.050	0.100	1.60
0.050	0.200	1.57

9.3. The decomposition of nitramide depends on general base catalysis, the rate-determining step being the removal of a proton.

$$NH_2 NO_2 = N_2O + HO_2$$

Plot the catalytic constants obtained for a series of negatively charged bases given below and obtain the constants of the Brønsted equation. Apply statistical corrections where necessary. Are there any observable deviations with structure or charge? The acid ionization constant of the conjugate acid of each base is given. All the thermodynamic and kinetic data are at $15°C$.

Catalyst	K_a	k_b
Hydroxide ion	1.1×10^{-16}	1×10^6
Secondary phosphate ion	5.8×10^{-8}	86
Succinate ion	2.4×10^{-6}	1.8
Malate ion	7.8×10^{-6}	0.72
Trimethylacetate ion	9.4×10^{-6}	0.822
Propionate ion	1.3×10^{-5}	0.649
Acetate ion	1.8×10^{-5}	0.504
Tartrate ion	4.1×10^{-5}	0.165
Acid succinate ion	6.5×10^{-5}	0.320
Oxalate ion	6.8×10^{-5}	0.104
Phenylacetate ion	5.3×10^{-5}	0.232
Benzoate ion	6.5×10^{-5}	0.189
Formate ion	2.1×10^{-4}	0.082
Acid malate ion	4.0×10^{-4}	0.077
Acid tartrate ion	9.7×10^{-4}	0.036
Acid phthalate ion	1.2×10^{-3}	0.029
Salicylate ion	1.0×10^{-3}	0.021
Monochloracetate ion	1.4×10^{-3}	0.016
Primary phosphate ion	7.6×10^{-3}	0.0079
o-Nitrobenzoate ion	7.3×10^{-3}	0.0042
Dichloracetate ion	5.0×10^{-2}	0.0007

9.4. In a buffered solution containing $0.05 M$ acetic acid and $0.05 M$ sodium acetate, calculate the percentage reaction due to hydrogen ion, acetic acid, and water in acid-catalyzed reactions where the coefficient α in the Brønsted equation is equal to $0.1, 0.5$, and 0.9, respectively.

9.5. Consider (9.130) and assume $pK_a = 16.5$ for this ketone. Using the Marcus equation, calculate α for the range of bases shown in Fig. 9.8. Assume $\Delta G_0^{\ddagger} = 10.45 \, kJ \, mol^{-1}$ and $w^r = 63.1 \, kJ$, for one set of calculations, and $\Delta G_0^{\ddagger} = 41.8 \, kJ$ and $w^r = 16.7 \, kJ$ for a second set. Compare your calculated values with those given for Fig. 9.8.

9.6. The enzyme fumarase catalyzes the hydration of fumarate ion to malate.

$$\begin{array}{c} \text{}^-O_2C \\ \diagdown \\ CH{=}CH \\ \diagdown \\ CO_2^- \end{array} + H_2O \xrightarrow{\text{fum.}} HO{-}\underset{\underset{CO_2^-}{|}}{CH}{-}CH_2{-}CO_2^-$$

Initial rates are obtained at $pH = 6.5$ and $25°C$ for various fumarate concentrations:

Fumarate Concentration $\times 10^5 / M$	v(arbitrary units)
5	1.90
10	2.86
20	4.0
50	5.0

Find V_m and K_m by plotting these data.

REFERENCES

1. For a short but readable account see R. L. Burwell, Jr., in *Surv. Prog. Chem.*, **8**, 2 (1977); a standard treatise is M. Boudart, *Kinetics of Chemical Processes*, Prentice-Hall, Englewood Cliffs, NJ, 1968.

2. R. P. Bell, *Acid-Base Catalysis*, Oxford University Press, Oxford, 1941.

3. G. K. Rollefson and R. F. Faull, *J. Am. Chem. Soc.*, **59**, 625 (1937).

4. A. Maccoll and V. R. Stimson, *J. Chem. Soc.*, **1960**, 2836.

5. R. Steinberger and F. H. Westheimer, *J. Am. Chem. Soc.*, **71**, 4158 (1949); **73**, 429 (1951).

6. H. Kroll, *J. Am. Chem. Soc.*, **74**, 2036 (1952).

7. For examples, see F. Basolo and R. G. Pearson, *Mechanisms of Inorganic Reactions*, 2nd ed., Wiley, New York, 1968.

8. R. B. Martin, *J. Inorg. Nucl. Chem.*, **38**, 511 (1976).

9. P. Wooley, *Nature (Lond.)*, **258**, 677 (1975); M. F. Dunn, *Struct. Bonding (Berl.)*, **1975**, 23.

10. For examples see, among others, H. Sigel, Ed., *Metal Ions in Biological Systems*, Dekker, New York, 1976; J. Kochi, *Organometallic Mechanisms and Catalysis*, Academic, New York, 1978; M. M. Taqui Khan and A. E. Martell, *Homogeneous Catalysis by Metal Complexes*, Academic, New York, 1974.

11. R. G. Pearson, *Symmetry Rules for Chemical Reactions,* Wiley-Interscience, New York 1976.

12. B. F. G. Johnson and S. Bhaduri, *Chem. Commun.,* **1973,** 650; B. Haymore and J. A. Ibers, *J. Am. Chem. Soc.,* **96,** 3325 (1974); D. E. Hendriksen and R. Eisenberg, *ibid.,* **98,** 4662 (1976).

13. J. R. Buchholz and R. E. Powell, *J. Am. Chem. Soc.,* **85,** 509 (1963).

14. M. Eigen and L. DeMaeyer *Z. Elektrochem.,* **59,** 986 (1955).

15. A. J. Kresge, *Acc. Chem. Res.,* **8,** 354 (1975).

16. S. H. Maron and V. K. LaMer, *J. Am. Chem. Soc.,* **60,** 2588 (1938).

17. C. A. Tolman, *J. Am. Chem. Soc.,* **92,** 4217 (1970); H. W. Walker, C. T. Kresge, P. C. Ford, and R. G. Pearson, *ibid.,* **101,** 7428 (1979).

18. H. S. Gutowsky, D. W. McCall, and C. P. Schlichter, *J. Chem. Phys.,* **21,** 279 (1953); H. M. McConnell, *ibid.,* **28,** 430 (1958).

19. M. Anbar, A. Loewenstein, and S. Meiboom, *J. Am. Chem. Soc.,* **80,** 2630 (1958).

20. C. G. Swain and M. M. Labes, *J. Am. Chem. Soc.,* **79,** 1084 (1957); C. G. Swain, J. T. McKnight, and V. P. Kreiter, *ibid.,* 1088.

21. E. Grunwald, C. F. Jumper, and S. Meiboom, *J. Am. Chem. Soc.,* **85,** 522 (1963).

22. F. G. Bordwell, *Acc. Chem. Res.,* **3,** 281 (1970); **5,** 374 (1972).

23. M. Eigen, *Angew. Chem., Int. Ed.,* **3,** 1 (1964).

24. For a discussion see J. R. Murdoch, *J. Am. Chem. Soc.,* **94,** 4410 (1972).

25. D. J. Cram, D. A. Scott, and W. D. Nielsen, *J. Am. Chem. Soc.,* **83,** 3696 (1961); A. Streitwieser, Jr., W. C. Langworthy, and D. E. Van Sickle, *ibid.,* **84,** 251 (1962).

26. W. P. Jencks, *Acc. Chem. Res.,* **9,** 425 (1976).

27. A. C. Satterthwaite and W. P. Jencks, *J. Am. Chem. Soc.,* **96,** 7018 (1974).

28. M. L. Bender, *J. Am. Chem. Soc.,* **73,** 1626 (1951).

29. For discussions see E. M. Kosower, *Molecular Biochemistry,* McGraw-Hill, New York, 1962; T. C. Bruice and S. J. Benkovic, *Bio-Organic Mechanisms,* Benjamin, New York, 1966; W. P. Jencks, *Catalysis in Chemistry and Enzymology,* McGraw-Hill, New York, 1969; M. L. Bender, *Mechanisms of Homogeneous Catalysis,* Wiley-Interscience, New York, 1971. These treatises cover many examples of acid–base catalysis important in organic and biochemistry.

30. R. H. DeWolfe, *J. Am. Chem. Soc.,* **82,** 1585 (1960); R. H. DeWolfe and F. B. Augustine, *J. Org. Chem.,* **30,** 699 (1965).

31. C. G. Swain and J. F. Brown, *J. Am. Chem. Soc.,* **74,** 2534 (1952).

32. S. Wideqvist, *Ark. Kemi,* **2,** 303 (1950).

33. W. P. Jencks and J. Carriulo, *J. Am. Chem. Soc.,* **82,** 1778 (1960).

34. R. G. Pearson, D. H. Anderson, and L. L. Alt, *J. Am. Chem. Soc.,* **77,** 527 (1955).

35. R. W. Johnson and R. G. Pearson, *Inorg. Chem.,* **10,** 2091 (1971).

36. E. Ahmed, M. L. Tucker, and M. L. Tobe, *Inorg. Chem.,* **14,** 1 (1975).

37. J. N. Brønsted and K. Pedersen, *Z. Phys. Chem.,* **A108,** 185 (1923); J. N. Brønsted, *Chem. Rev.,* **5,** 322 (1928).

38. S. W. Benson, *J. Am. Chem. Soc.,* **80,** 5151 (1958).

39. E. Pollak and P. Pechukos, *J. Am. Chem. Soc.,* **100,** 2984 (1978); D. R. Coulson, *ibid.,* 2992.

40. R. P. Bell and W. C. E. Higginson, *Proc. R. Soc. (London)*, **A197**, 141 (1949).

41. R. P. Bell, M. H. Rand, and K. M. A. Wynne-Jones, *Trans. Faraday Soc.*, **52**, 1093 (1956).

42. For an extensive discussion see R. P. Bell, *Adv. Phys. Org. Chem.*, **4**, 1 (1966); L. H. Funderburk, L. Aldwin, and W. P. Jencks, *J. Am. Chem. Soc.*, **100**, 5444 (1978).

43. J. N. Brønsted and W. F. K. Wynne-Jones, *Trans. Faraday Soc.*, **25**, 59 (1929).

44. L. P. Hammett, *Chem. Rev.*, **17**, 125 (1935); *Physical Organic Chemistry*, 2nd ed., McGraw-Hill, New York, 1970.

45. See (a) J. E. Leffler and E. Grunwald, *Rates and Equilibria of Organic Reactions*, Wiley, New York, 1963; (b) *Advances in Linear Free Energy Relationships*, N. B. Chapman and J. Shorter, Plenum, London, 1972.

46. R. P. Bell, *Proc. R. Soc. (London)*, **A154**, 414 (1936); M. G. Evans and M. Polanyi, *Trans. Faraday Soc.*, **32**, 1333 (1936).

47. J. E. Leffler, *Science*, **117**, 340 (1953).

48. G. S. Hammond, *J. Am. Chem. Soc.*, **77**, 334 (1955).

49. R. A. Marcus, *J. Phys. Chem.* **72**, 891 (1968); *J. Am. Chem. Soc.*, **91**, 7224 (1969).

50. A. O. Cohen and R. A. Marcus, *J. Phys. Chem.*, **72**, 4249 (1968); W. J. Albery and M. M. Kreevay, *Adv. Phys. Org. Chem.*, **16**, 87 (1978).

51. For references see J. R. Murdoch, *J. Am. Chem. Soc.*, **102**, 71 (1980).

52. F. G. Bordwell and D. L. Hughes, *J. Org. Chem.*, **45**, 3314 (1980).

53. R. F. Hudson, in *Chemical Reactivity and Reaction Paths*, G. Klopman, Ed., Wiley-Interscience, New York, 1974.

54. (a) R. G. Pearson, H. Sobel, and J. Songstad, *J. Am. Chem. Soc.*, **90**, 319 (1968); (b) R. G. Pearson and P. E. Figdore, *ibid.*, **102**, 1541 (1980).

55. J. Koskikallio, *Suom. Kemistil.*, **30B**, 38, 111, 155 (1957).

56. R. P. Bell, *The Proton in Chemistry*, 2nd ed., Cornell University Press, Ithaca, NY, 1973.

57. R. D. Guthrie and N. S. Cho, *J. Am. Chem. Soc.*, **97**, 2280 (1975).

58. J. Murto, *Ann. Acad. Sci. Fenn.*, Series **AII**, **1962**, 117; see also M. L. Bender and W. A. Glasson, *J. Am. Chem. Soc.*, **81**, 1590 (1959).

59. U. Belluco, L. Cattalini, F. Basolo, R. G. Pearson, and A. Turco, *J. Am. Chem. Soc.*, **87**, 241 (1965).

60. C. G. Swain and C. B. Scott, *J. Am. Chem. Soc.*, **75**, 141 (1953).

61. T. W. Bentley and P. von R. Schleyer, *Adv. Phys. Org. Chem.*, **14**, 1 (1977).

62. See reference 45a, p. 162 ff.

63. For an excellent example, see E. M. Arnett and R. Reich, *J. Am. Chem. Soc.*, **102**, 5892 (1980).

64. N. Agmon, *J. Chem. Soc., Faraday Trans. II*, **74**, 388 (1978); A. R. Miller, *J. Am. Chem. Soc.*, **100**, 1984 (1978).

65. W. P. Jencks and M. Gilchrist, *J. Am. Chem. Soc.*, **90**, 2622 (1968); M. J. Gresser and W. P. Jencks, *ibid*, **99**, 6963 (1977).

66. M. B. Jensen, *Acta Chem. Scand.*, **13**, 289 (1959).

67. Reference 56, p. 203.

68. Diffusion control has been shown by Eigen, reference 23.

69. D. J. Hupe and D. Wu, *J. Am. Chem. Soc.*, **99**, 7653 (1977).

70. J. O. Edwards and R. G. Pearson, *J. Am. Chem. Soc.*, **84**, 16 (1962).

71.   R. F. Pratt and T. C. Bruice, *J. Org. Chem.*, **37**, 3563 (1972).

72.   See A. Pross, *Adv. Phys. Org. Chem.*, **14**, 69 (1977); C. D. Johnson, *Chem. Rev.*, **75**, 755 (1975), for reviews of selectivity-reactivity.

73.   For reviews see (a) C. J. Collins and N. S. Bowman, *Isotope Effects in Chemical Reactions*, Van Nostrand-Reinhold Co., New York, 1970; (b) R. A. More O'Ferrall in *Proton Transfer Reactions*, V. Gold and E. F. Caldin, Eds., Chapman and Hall, London, 1975, p. 201.

74.   For a review see W. J. Albery, in *Proton Transfer Reactions*, E. F. Caldin and V. Gold, Eds., Chapman and Hall, London, 1975, p. 263.

75.   For a more detailed treatment see J. Bigeleisen, *J. Chem. Phys.*, **17**, 675 (1949).

76.   C. G. Swain, E. C. Stivers, J. F. Reuwer, and L. J. Schaad, *J. Am. Chem. Soc.*, **80**, 5885 (1958).

77.   F. H. Westheimer, *Chem. Rev.* **61**, 265 (1961). See also L. Melander, *Isotope Effects on Reaction Rates*, Ronald Press, New York, 1960.

78.   R. L. Sweany and J. Halpern, *J. Am. Chem. Soc.*, **99**, 8335 (1977).

79.   R. P. Bell and J. E. Crooks, *Proc. R. Soc. (Lond.)*, **A286**, 285 (1965).

80.   A. J. Kresge and A. C. Lin, *J. Chem. Soc., Chem. Commun.*, **1973**, 761.

81.   The first such examples were by E. S. Lewis and J. D. Allen, *J. Am. Chem. Soc.*, **86**, 2022 (1964); L. Funderburk and E. S. Lewis, *ibid.*, 2531; for a review see R. P. Bell, *Chem. Soc. Rev.*, **3**, 513 (1974).

82.   W. C. Hamilton and J. A. Ibers, *Hydrogen Bonding in Solids*, Benjamin, New York, 1968, Chap. 2.

83.   See E. F. Caldin, *Chem. Rev.*, **69**, 135 (1969).

84.   W. H. Miller, *J. Am. Chem. Soc.*, **101**, 6810 (1979).

85.   For a historical review see H. Taube, *Electron Transfer Reactions of Complex Ions in Solution*, Academic, New York, 1970.

86.   P. L. Gaus and J. L. Villanueva, *J. Am. Chem. Soc.*, **102**, 1934 (1980).

87.   R. J. Marcus, B. J. Zwolinski, and H. Eyring, *J. Phys. Chem.*, **58**, 432 (1954).

88.   R. J. LeRoy, H. Murai, and F. Williams, *J. Am. Chem. Soc.*, **102**, 2325 (1980); E. D. Sprague, *J. Phys. Chem.*, **81**, 516 (1977).

89.   K. J. Laidler, *Theories of Chemical Reaction Rates*, McGraw-Hill, New York, 1969, p. 76 ff.; F. R. Cruickshank, A. J. Hyde, and D. Pugh, *J. Chem. Educ.*, **54**, 288 (1977); R. K. Boyd, *J. Chem. Educ.*, **55**, 84 (1978).

90.   For a detailed discussion see A. A. Frost and R. G. Pearson, *Kinetics and Mechanism*, 2nd ed., Wiley, New York, 1961, p. 335 ff.

91.   W. F. Libby, *J. Phys. Chem.*, **56**, 863 (1952).

92.   R. J. Campion, T. J. Conocchioli, and N. Sutin, *J. Am. Chem. Soc.*, **86**, 4591 (1964).

93.   R. L. Burwell, Jr. and R. G. Pearson, *J. Phys. Chem.*, **70**, 300 (1966).

94.   (a) P. D. Boyer, Ed., *The Enzymes*, Academic, New York, Vol. II, 1971; (b) "Enzyme Kinetics and Mechanism," in *Methods in Enzymology*, Vols. 63, 1979 and 64, 1980, D. L. Purich, Ed., Academic, New York.

95.   For shorter accounts than reference 94, see (a) G. G. Hammes, *Principles of Chemical Kinetics*, Academic, New York, 1978, Chap. 9; (b) D. Piszkiewicz, *Kinetics of Chemical and Enzyme-Catalyzed Reactions*, Oxford University Press, New York, 1977.

96.   R. G. Khalifah, *Proc. Natl. Acad. Sci., US*, **70**, 1986 (1973).

97. C. A. Bunton, Micellar Catalysis and Inhibition, in *Solution Chemistry of Surfactants,* Vol. 2, K. L. Mittal, Ed., Plenum Press, New York, 1979, p. 519; J. H. Fendler, *Acc. Chem. Res.,* **9,** 153 (1976).

98. M. Makosza and M. Wawrzyniewicz, *Tetrahedron Lett.* **53,** 4659 (1969); C. M. Starks and C. Liotta, *Phase Transfer Catalysis,* Academic, New York, 1978.

99. J. Boger and J. R. Knowles, *J. Am. Chem. Soc.,* **101,** 7631 (1979); R. Breslow, M. F. Czarniecki, J. Emert, and H. Hamaguchi, *ibid.,* **102,** 762 (1980).

CHAIN REACTIONS, PHOTOCHEMISTRY

A chain reaction is one whose mechanism involves low concentrations of reactive intermediates, called *chain carriers*, that participate in a cycle of reaction steps such that the chain carriers are reformed after each cycle. The chain carriers are originally formed in a *chain-initiation* step. They then participate in the *chain-propagation* steps until they are removed from the reaction system by a *chain-termination* or chain-breaking step. The commonest chain reactions involve free radicals as chain carriers, although cations or anions can also serve as reactive intermediates.

Chain reactions are important in many industrial processes, such as polymerization, halogenation, and combustion, and they form the basis for atmospheric reactions, including formation of photochemical smog and maintenance of the stratospheric ozone layer.

THE REACTION H$_2$ + Br$_2$

This reaction, the first for which a chain mechanism was proposed to account for experimental results, provides examples of several techniques used to study chain reactions. In 1906 Bodenstein and Lind[1] reported rate measurements in the gas phase over the temperature range of 200–300°C. Their results agreed with the empirical equation

$$\frac{d\,[\text{HBr}]}{dt} = \frac{k\,[\text{H}_2]\,[\text{Br}_2]^{1/2}}{1 + k'\,[\text{HBr}]\,/\,[\text{Br}_2]} \tag{10.1}$$

where $k' = 0.10$ and is temperature independent and $k \propto e^{E_a/RT}$ with $E_a = 175\,\text{kJ}\,\text{mol}^{-1}$. Despite the complexity of the rate equation (10.1), they were able to integrate it for various initial conditions and so compare their results with the integrated form.

Thirteen years later the form of (10.1) was explained by Christiansen, Herzfeld, and Polanyi[2] in terms of the chain mechanism

Initiation: $Br_2 \xrightarrow{k_1} 2Br$ step 1

Propagation: $Br + H_2 \xrightarrow{k_2} HBr + H$ step 2

                    $H + Br_2 \xrightarrow{k_3} HBr + Br$ step 3

Inhibition: $H + HBr \xrightarrow{k_4} H_2 + Br$ step 4

Termination: $2Br \xrightarrow{k_5} Br_2$ step 5

Here the chain carriers are Br and H atoms; the inhibition step is necessary to account for the terms in the denominator of (10.1). More recent experimental work[3, 4] has extended the conditions under which the reaction has been studied over a much wider range than that of Bodenstein and Lind.

To derive (10.1) from the proposed mechanism, write the rate of formation of HBr as

$$\frac{d[HBr]}{dt} = k_2[Br][H_2] + k_3[H][Br_2] - k_4[H][HBr] \qquad (10.2)$$

This expression contains the not easily measured concentrations of H and Br. These could be eliminated by solving (10.2) simultaneously with other differential equations, including:

$$\frac{d[Br]}{dt} = 2k_1[Br] - k_2[Br][H_2] + k_3[H][Br_2] + k_4[H][HBr] - 2k_5[Br]^2$$
$$(10.3)$$

$$\frac{d[H]}{dt} = k_2[Br][H_2] - k_3[H][Br_2] - k_4[H][HBr] \qquad (10.4)$$

However, such a solution is difficult and it is convenient to apply the steady-state approximation to both chain carriers, that is, to set the right-hand sides of (10.3) and (10.4) equal to zero. The requirements for invoking the steady-state approximation are discussed in Chapter 8. Comparisons of steady-state results with accurate numerical integration have been made for some chain mechanisms, with varying degrees of agreement.[5] For the $H_2 + Br_2$ reaction the chain carriers are certainly at a very low concentration, and Matsen and Franklin[6] estimate the induction period to be about 10^{-9} s. Consequently it is reasonable to subtract the right-hand side of (10.4), which equals zero, from (10.2) to yield

$$\frac{d[HBr]}{dt} = 2k_3[H][Br_2] \qquad (10.5)$$

To obain [H], add the right-hand sides of (10.3) and (10.4) to get

$$2k_1[Br_2] - 2k_5[Br]^2 = 0$$

and therefore

$$[Br] = \left(\frac{k_1}{k_5}[Br_2]\right)^{1/2} \tag{10.6}$$

Substitute this result into (10.4) to obtain

$$[H] = \frac{k_2[H_2]}{k_3[Br_2] + k_4[HBr]}[Br] = \frac{k_2(k_1/k_5)^{1/2}[H_2][Br_2]^{1/2}}{k_3[Br_2] + k_4[HBr]} \tag{10.7}$$

and in turn substitute (10.7) into (10.5)

$$\frac{d[HBr]}{dt} = \frac{2k_3k_2(k_1/k_5)^{1/2}[H_2][Br_2]^{1/2}}{k_3[Br_2] + k_4[HBr]}$$

$$= \frac{2k_2(k_1/k_5)^{1/2}[H_2][Br_2]^{1/2}}{1 + (k_4[HBr]/k_3[Br_2])} \tag{10.8}$$

This has the same form as the empirical expression (10.1) and agrees with it quantitatively if

$$k = 2k_2\left(\frac{k_1}{k_5}\right)^{1/2} \tag{10.9}$$

and

$$k' = \frac{k_4}{k_3} = 0.10 \tag{10.10}$$

The constant ratio $k_4/k_3 = 0.10$ can be rationalized by considering bond-dissociation energies for Br_2, HBr, and H_2. The bond dissociation energy D is the standard enthalpy change for a gas-phase reaction

$$AB \rightarrow A + B$$

where A and B are atoms or free radicals that were bonded together in molecule AB. For polyatomic molecules bond-dissociation energies must be distinguished from bond energies, which often are average values for bonds surrounded by a variety of groups. Such average values may be used as a rough guide, but bond-dissociation energies can vary significantly from the average, as in the case of D_{C-C}, which is $356 \, kJ \, mol^{-1}$ in ethane or graphite but less than $40 \, kJ \, mol^{-1}$ in hexaarylethanes. Based on bond-dissociation energies $D_{Br_2} = 194 \, kJ \, mol^{-1}$, $D_{HBr} = 366 \, kJ \, mol^{-1}$, and $D_{H_2} = 436 \, kJ \, mol^{-1}$, the steps governed by k_3 and k_4 in the chain mechanism are exothermic, and since both are atom reactions their activation energies should be small, perhaps less than $15 \, kJ \, mol^{-1}$. The ratio k_3/k_4 could very well be temperature independent, and the difference from unity might be due to different steric factors or entropies of activation. Such a conclusion is confirmed by independent measurements of the rates of steps 3 and 4,[7] which yield the parameters reported in Table 6.3. The observed activation energies are 15 and $9 \, kJ \, mol^{-1}$, and the observed frequency factors are 11.97 and $10.79 \, M^{-1} \, s^{-1}$.

The rate constant k_2 can be evaluated from (10.9), since (10.6) shows that k_1/k_5 is the equilibrium constant, K, for dissociation of Br_2, a value that has been determined independently.[8] This yields the experimental expression[9]

$$k_2 = 10^{11.43}e^{-(82.4 \, kJ \, mol^{-1})/RT}M^{-1} \, s^{-1} \tag{10.11}$$

The reported frequency factor does not require an unusually small steric factor, and the activation energy of 82.4 kJ mol^{-1} is also reasonable. On the basis of bond-dissociation energies the k_2 step is endothermic by 70 kJ mol^{-1}, and so its activation energy must be at least this great. Adding to this the 9 kJ mol^{-1} activation energy for step 4 (the reverse of step 2) leaves only a slight discrepancy. It is interesting to note that the overall activation energy of the reaction, 175 kJ mol^{-1}, is less than that for the initiation step, which must be at least 194 kJ mol^{-1}. This is because the overall activation energy reflects the temperature dependence of $k_2 K^{1/2}$, and $K \propto e^{-D_{Br_2}/RT}$. Therefore E_a (overall) $= E_a(2) + D_{Br_2}/2$. For chain reactions where the bond broken in the initiation step is stronger than the Br—Br bond this difference in activation energies is even more pronounced. The efficiency of chain mechanisms is due to this lowering of the overall activation energy.

The proposed mechanism has now been shown to reproduce the form of the empirical rate expression, with reasonable activation energies and steric factors for the several steps. However, for the mechanism to be entirely satisfactory, it must be shown why certain other possible reaction steps are not included, for example:

$$\text{Initiation:} \qquad H_2 \rightarrow 2H$$

$$\text{Inhibition:} \qquad Br + HBr \rightarrow Br_2 + H$$

$$\text{Termination:} \qquad \begin{cases} H + Br \rightarrow HBr \\ H + H \rightarrow H_2 \end{cases}$$

$$HBr \rightarrow H + Br$$

If any of these were important the rate expression would probably be different from (10.1), but all can be shown to be negligibly slow as compared with other competing reactions in the mechanism. Dissociation of H_2 and HBr should be slow compared with dissociation of Br_2 because Br_2 has a much smaller bond-dissociation energy. The Br + HBr reaction should be unimportant compared to the Br + H_2 reaction because, although the Br + H_2 reaction has an activation energy of 73.8 kJ mol^{-1}, the Br + HBr reaction is endothermic by 172 kJ mol^{-1} and requires at least that much activation energy. From independent studies of Br + HBr, $E_a = 174 \text{ kJ mol}^{-1}$.[10] Possible competition of H + Br or H + H as chain terminating steps would be determined by relative concentrations of the atoms, the activation energies being zero for these atom recombination reactions. Equation (10.7) can be rearranged to give

$$\frac{[H]}{[Br]} = \frac{k_2 [H]}{k_3 [Br_2] + k_4 [HBr]} \cong \frac{k_2}{k_3}$$

at least in the early stages of the reaction and where the starting concentrations of H_2 and Br_2 are about equal. Using parameters from Table 6.3,

$$k_2/k_3 = [10^{11.43} e^{-(82.4 \text{ kJ mol}^{-1})/RT}]/[10^{11.97} e^{-(15 \text{ kJ mol}^{-1})/RT}]$$

$$= 10^{-0.54} e^{-(67.4 \text{ kJ mol}^{-1})/RT}.$$

At a typical temperature, say 573 K, $k_2/k_3 = 2.1 \times 10^{-7}$. Therefore a recombination of H with Br would be less than a millionth as likely as a recombination of

Br with Br, and of H with H less than 1 in 10^{13} as likely. Although the H-atom concentration is exceedingly low, it is about 10^4 times that resulting from the equilibrium $H_2 \rightleftharpoons H + H$. Whereas in this reaction there is an equilibrium of Br with Br_2, the H-atom concentration is determined by the steady state involving the second, third, and fourth steps of the mechanism.

Up to this point it is assumed that the first and last steps of the mechanism are unimolecular and bimolecular, respectively. Because the steady-state assumption leads to equilibrium relationship (10.6) between Br and Br_2, it does not matter what mechanism is assumed for attainment of this equilibrium. In fact, Rabinowitch and Lehmann[11] have shown that the recombination of atoms is trimolecular, with rate $= k'_5 [M] [Br]^2$, where M is a third body. The first step, dissociation, must then also involve M in a bimolecular process with rate $= k'_1 [M] [Br_2]$. The resulting expression for the rate of HBr formation has $k'_1 [M]$ in place of k_1 and $k'_5 [M]$, replacing k_5. But only the ratio appears in (10.8); therefore [M] cancels and does not make itself apparent.

Direct proof that the fifth step must be trimolecular under the conditions of the chain reaction can be obtained by initiating the chain photochemically rather than by thermal dissociation. Bodenstein and Lütkemeyer,[12] Jost and Jung,[13] and Ritchie[14] have studied the photochemical reaction at temperatures near $200°C$. Over a wide pressure range Jost and Jung found that the rate could be expressed as

$$\frac{d [HBr]}{dt} = \frac{k [H_2] I_{abs}^{1/2}}{p^{1/2} [1 + (k' [HBr] / [Br_2])]} \qquad (10.12)$$

where I_{abs} is the intensity of light absorbed and p is the total pressure. This form may be derived by substituting photochemical dissociation of Br_2 for thermal dissociation as the initiation step

$$Br_2 + h\nu \rightarrow 2 Br$$

with rate $= 2 k''_1 I_{abs}$. Then

$$\frac{d [Br]}{dt} = 2 k''_1 I_{abs} - 2 k'_5 [M] [Br]^2 \qquad (10.13)$$

and in the steady state, when $d [Br] / dt = 0$, the equation corresponding to (10.6) is

$$[Br] = \left(\frac{k''_1 I_{abs}}{k'_5 [M]} \right)^{1/2} \qquad (10.14)$$

Using (10.14) instead of (10.6) in the previous derivation yields

$$\frac{d [HBr]}{dt} = \frac{2 k_2 (k''_1 / k'_5)^{1/2} [H_2] I_{abs}^{1/2}}{[M]^{1/2} [1 + (k_4 [HBr] / k_3 [Br_2])]} \qquad (10.15)$$

This has the same form as (10.12) if [M] is proportional to p. This is a fair assumption, but the constant of proportionality differs for different gases whose molecules have different efficiencies in facilitating recombination of atoms. It is the presence of $p^{1/2}$ in the denominator of (10.12) that confirms the trimolecular character of the recombination process.

THE ROTATING-SECTOR METHOD

The photochemical reaction of $H_2 + Br_2$ has also been studied over a broader range of conditions,[15-17] Briers and Chapman[16] employed the rotating sector method,[18] in which the beam of light impinging on the reaction mixture is interrupted periodically. Because the steady state assumed in the derivation of (10.14) is not attained instantaneously, use of light intermittent in time can lead to evaluation of the rate constant for the chain termination process. The time-dependent solution of (10.13) is

$$[Br]_{td} = (k_1'' I_{abs} / k_5' [M])^{1/2} \tanh \{ t (4 k_1'' I_{abs} k_5' [M])^{1/2} \}$$

$$= [Br]_{ss} \tanh \left(\frac{t}{\tau} \right) \tag{10.16}$$

where the subscripts td and ss refer to time-dependent and steady-state concentrations, the latter being given by. (10.14), and $\tau = (4 k_1'' I_{abs} k_5' [M])^{-1/2}$. The interpretation of (10.16) is that illumination for a time greater than τ is required to establish the steady state. If the light source is interrupted with a period several times greater than τ, then the steady state will be reached early in each cycle. The overall rate of reaction will be given by the usual treatment for chain reactions with light intensity I_{abs}, but only operating $1/(r + 1)$ of the time, where r is the ratio (time dark)/(time illuminated). If the flashing period is much less than τ, the full steady state described by (10.14) is never reached. Instead, a concentration of chain carriers corresponding to a uniform intensity of $I_{abs}/(r + 1)$ forms. In other words, (10.14) must be replaced by

$$[Br]_{ss} = \left(\frac{k_1'' I_{abs}}{k_5' [M] (r + 1)} \right)^{1/2} \tag{10.17}$$

For values of the flashing period of the order of τ, a transition between the two limiting values occurs. From this behavior τ, and hence the rate constant of the termination step, can be evaluated.

These effects of intermittent light are only significant if the chain termination step is other than first order in chain carrier. [If the last term in (10.13) contained [Br] instead of $[Br]^2$, then (10.14) and (10.17) would not involve a square root, and it would not matter whether I_{abs} or the overall rate were divided by $r + 1$.] Briers and Chapman found that for low intensities of illumination the rate became proportional to intensity, while for high intensities the rate was proportional to the square root of the intensity. From this they concluded that bromine atoms were destroyed both by the three-body process described above and by a wall-recombination process that is first-order in [Br]. Ritchie[14] compared the importance of these two recombination processes and studied the relative efficiency of chaperone molecules for three-body recombination. He found that CO_2, O_2, N_2, and Ar were highly efficient, while H_2 and He fell into a low-efficiency category. Sullivan[17] has reported that the atomic chain mechanism is inadequate to explain the results of photolysis of $H_2 + Br_2$ at room temperature, since the rates of HBr formation are faster by a factor of more than 200 than can be accounted for by the

Br + H$_2$ → HBr + H reaction. He proposes several trimolecular reactions involving bromine atoms to rationalize the experimental results.

SHOCK TUBE STUDIES

The initiation step of the H$_2$ + Br$_2$ reaction has been studied independently in shock tubes, as has the overall reaction.[4] The shock method is similar to a single-impulse relaxation method, such as the temperature-jump method described in Chapter 8, except that the perturbation applied to a premixed equilibrium system is extremely violent and the system is driven far from the equilibrium state. Usually the applied disturbance is a shock wave that passes through the system with super-sonic velocity. In a shock tube an inert gas at high pressure and a reactive gas at low pressure are separated by a diaphragm. When the diaphragm is ruptured, a pressure wave caused by the high-pressure gas compresses the low-pressure gas adiabatically, producing temperatures as high as 2000°C. The increase in temperature can be calculated from the thermodynamic properties of the gases and from measurement of the shock velocity. Chemical change subsequent to the rapid temperature increase can be followed spectrophotometrically. A number of general treatments of the shock method are available.[19]

Britton and Cole[4c] investigated the H$_2$ + Br$_2$ and D$_2$ + Br$_2$ systems in a shock tube over the temperature range 1300–1700 K. The reaction was followed spectro-photometrically by observing Br$_2$ at a wavelength of 500 nm. For the second step of the chain mechanism they found $k = 10^{11.36} e^{-(82.8 \text{ kJ mol}^{-1})/RT}$, in reasonable agreement with (10.11), which was obtained at much lower temperatures. They also found $k_4/k_3 = 0.099$, and they observed an isotope effect in the second reaction step that has also been observed at lower temperatures.[3b] Shock-tube studies of dissociation of Br$_2$ span the interval 1200–2300 K.[20] The expression for k_1, when argon is the third body, is $k_1 = 1.39 \times 10^8 T^{1/2} (\Delta E_0/RT)^{1.97} e^{-\Delta E_0/RT}$ $K^{-1/2} M^{-1} s^{-1}$. Here $\Delta E_0 = D_{Br_2}$ is the bond-dissociation energy for Br$_2$. This equation reproduces the rate constant from room temperature to 2300 K. It has the same form as (4.30), the activation-in-many-degrees-of-freedom rate constant. Since Br$_2$ has only a single vibrational mode, the power of $\Delta E_0/RT$ equal to nearly 2 must mean that rotational energy can be drawn upon for dissociation. Using noble-gas diluents other than argon results in little variation in dissociation.

CHAIN LENGTH

The concept of chain length is of great importance but somewhat indefinite. Chain length may be defined theoretically as the number of successful chain-propagation steps resulting from a single original chain carrier. Or it may be considered the number of propagation cycles that occur before the chain is interrupted. This number would be half that calculated on the basis of the number of steps if two chain carriers (such as Br and H for the mechanism discussed above) were involved. The chain length might also be taken as the number of molecules of the desired product produced from each initial chain carrier. This definition would yield a lower value for the chain length of the $H_2 + Br_2$ reaction than would the first definition because the reverse of one of the propagation steps inhibits product formation.

Experimental definitions of chain length are often used. For a thermal reaction the ratio of the overall rate of reaction to the rate of the first step may be used. The rate of the first step may be taken as the rate of the completely inhibited reaction, that is, with chains cut short by rapid removal of chain carriers by NO or other inhibitor. Or, if the reaction is initiated by decomposition of some other substance, the independently determined rate of decomposition of the initiator may be used. This would include substances such as aliphatic azo compounds, peroxides, and hydroperoxides.

For photochemical initiation, the experimental chain length may be taken as the quantum yield. The quantum yield Φ is the number of molecules of reactant consumed (or of product formed) per photon of light absorbed. Depending on the reaction mechanism, the quantum yield may or may not coincide with any of the theoretical definitions of chain length given above. For example, in the case of the $H_2 + Br_2$ reaction, each photon absorbed produces two Br atoms and so

$$\Phi = \frac{\text{number of HBr molecules produced}}{\text{number of quanta absorbed}}$$

$$= \frac{\text{number of HBr molecules produced}}{0.5 \text{ (number of Br atoms produced)}}$$

If we take the third theoretical definition of chain length given above, then

$$\text{chain length} = \frac{\text{number of HBr molecules}}{\text{number of Br atoms}}$$

$$= 0.5 \Phi$$

For the other theoretical definitions the relationship with quantum yield is more complex.

To have a long chain the rate of propagation should be large and the rate of termination small. Also, if the termination step is second order in the chain carrier, a small rate of initiation favors long chains because carriers in different chains are less likely to interact and combine, owing to their low concentration. In general the relative rates of propagation and termination depend on reaction conditions, and so chain length does also. At 500 K and 0.1 atm the chain length for the $H_2 + Br_2$ reaction is about 100. For the similar reaction of H_2 with Cl_2 the chain-propagation steps are very much faster,[21] and the chain length approaches 10^6. This reaction is much more difficult to study because the effects of wall recombination and impurities are large and hard to eliminate. The reaction

$$Cl + H_2 \xrightarrow{k_2} HCl + H$$

is responsible for rapid chain propagation. It has been studied independently[9] and obeys the equation $k_2 = 10^{10.92} e^{-(23 \text{ kJ mol}^{-1})/RT} M^{-1} \text{ s}^{-1}$, the activation energy of 23 kJ mol^{-1} being much less than the 82.4 kJ mol^{-1} for the corresponding bromine reaction (10.11). In fact the propagation steps are too fast to be balanced by gas-phase recombination of Cl atoms and so a steady state is not reached in the homogeneous reaction unless the temperature is so low ($< 200°C$) that the rate of thermal reaction becomes negligible. A steady state can be achieved in the photochemically initiated reaction below 172°C. The similarity of mechanism between the Br_2 and the Cl_2 reaction is supported by the fact that between 1100 and 1600 K the rate of propagation of the bromine reaction also becomes too fast for the chain carriers to achieve a steady-state concentration.[4a, 22]

At low temperatures the reaction of H_2 and I_2 involves a nonchain mechanism, as described in Chapter 6, but above 600 K a chain mechanism becomes increasingly important.[23] The chain propagation step

$$I + H_2 \rightarrow HI + H$$

has an activation energy of $140 \, kJ \, mol^{-1}$ and hence is slow at low temperatures. The reaction of H_2 with F_2 is by far the most exothermic of the hydrogen–halogen series, owing to the very strong H—F bond and the very weak F—F bond. This reaction is very fast, even at liquid hydrogen temperatures, and is strongly influenced by surface effects. The propagation step

$$F + H_2 \rightarrow HF + H$$

has been extensively studied, as reported in Chapter 6.

INHIBITION

The slowing down of a reaction upon addition of a constituent to the reaction mixture is called inhibition. The constituent causing the effect is an inhibitor. Occasionally such an effect is referred to as negative catalysis, but, since there usually is net consumption of the inhibitor, inhibition is the preferred term. Inhibition always occurs in reversible reactions where, upon approaching an equilibrium, addition of a product decreases the net rate of reaction. An example of inhibition by a reversible step is provided by the $H_2 + Br_2$ chain mechanism. However, chain reactions and catalytic reactions are commonly susceptible to much more spectacular inhibition in which a mere trace of inhibitor can cause a marked decrease in rate. Such an effect can be explained in at least two ways: the inhibitor may combine with a catalyst and prevent it from operating, as is the case for many enzyme-catalyzed reactions; or the inhibitor may interrupt the propagation cycle in a chain reaction.[24] This latter event may occur because the chain carrier combines with the inhibitor to form a radical that cannot attack reactant molecules to propagate the chain. Or the chain carrier may abstract hydrogen from the inhibitor with the same effect. Some inhibitors are relatively stable free radicals, such as NO, that combine with chain carriers to form nonradical species. Inhibition resulting from addition of a small amount of an appropriate substance provides strong evidence for the presence of a chain mechanism.

Inhibition plays an important role in controlling autoxidation, the slow reaction with molecular oxygen of most organic and some inorganic substances at temperatures below about 150°C and usually in the liquid state. Autoxidation is sometimes an undesirable reaction that should be inhibited, as in spoilage of food or deterioration of lubricating oils, in which case easily oxidized substances such as phenols or amines can be used to inhibit the reaction. In other cases autoxidation is the desired reaction in large-scale commercial processes. It is one of the most important reactions known to chemistry and has been studied very extensively.[25]

Usually autoxidation can be represented by the following mechanism:

$$\text{initiator} \xrightarrow{k_i} 2R' \cdot \tag{10.18a}$$

$$R' \cdot + O_2 \to R'O_2 \cdot \tag{10.18b}$$

$$R'O_2 \cdot + RH \to R'O_2H + R \cdot \tag{10.18c}$$

$$R \cdot + O_2 \xrightarrow{\text{fast}} RO_2 \cdot \tag{10.18d}$$

$$RO_2 \cdot + RH \xrightarrow{k_p} RO_2H + R \cdot \tag{10.18e}$$

$$2RO_2 \cdot \xrightarrow{k_t} \text{inactive products} \tag{10.18f}$$

It is a chain reaction with a hydroperoxide as the first product. This may be oxidized further or may decompose thermally to start new chains. Assuming that (10.18d) is fast but (10.18e) is slow, because of lesser reactivity of $RO_2 \cdot$ compared to $R \cdot$, and assuming long chains, the mechanism (10.18) leads to the rate expression

$$\text{rate} = \left(\frac{k_i}{k_t}\right)^{1/2} k_p \, [\text{initiator}]^{1/2} \, [\text{RH}] \tag{10.19}$$

The termination step involves two $RO_2 \cdot$ rather than $R \cdot$ because of the higher steady-state concentration of the former.

Experimentally the rate of oxygen consumption obeys (10.19) at oxygen pressures above 50 mm Hg or so. At lower pressures there is a dependence on $[O_2]$. This is expected since, if less oxygen is dissolved in the liquid phase, (10.18d) is no longer faster than (10.18e). Studies of photooxidations in the liquid phase[26] indicate that (10.18d) is as much as 10^6 times as fast as step (10.18e), the former occurring at nearly every collision. Initiation can be produced by thermal decomposition of substances naturally present or added to the reaction mixture. Photoinitiation is also possible. The square-root dependence on initiator concentration, or on light intensity, is generally found, at least under well-defined experimental conditions. An example of photoinitiation in the gas phase is the work of McDowell and Sharpless,[27] who found that the rate of oxidation of acetaldehyde to peroxyacetic acid followed the form of (10.19) with [initiator] replaced by I_{abs}.

Metal-ion catalysis (see Chapter 9) is often used to increase the rate of initiation, if it is due to the decomposition of peroxides.[28] These are usually formed as a result of reaction in any case. The mechanism is illustrated by a study of autoxidation of cumene initiated by 2,2'-azodiisobutyronitrile (ABN).[29] The effect of cobalt(II) acetate catalysis and of inhibition by 2,6-di-t-butyl-p-cresol is shown in Fig. 10.1.

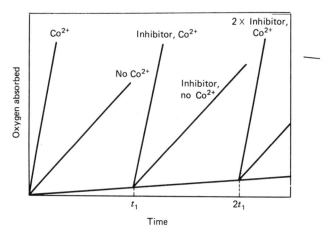

Figure 10.1. Effect of catalyst and inhibitor on oxidation of cumene in glacial acetic acid (Data from Blanchard.)[29]

The inhibitor produces an induction period of very low rate, with the length of the induction period proportional to the amount of inhibitor. Inhibition ceases when the inhibitor is all consumed. Cobalt(II) ion greatly increases the rate but, at low concentrations, has no effect on the length of the induction period. This must mean that the inhibitor is used up only by the radicals produced from ABN and that the catalyst metal ion does not produce acceleration by increasing the rate of decomposition of ABN.

The reactivity of a carbon–hydrogen bond to the atom-transfer reaction (10.18e) is greatly affected by the presence of a double bond or aromatic ring or by allyl substituents on the carbon atom. This is because the resulting free radical $R \cdot$ is greatly stabilized. Addition of $RO_2 \cdot$ to the double bond can also occur.

Szwarc and coworkers[30] have completely inhibited the chains in certain organic vapor decompositions by using toluene as a carrier and an inhibitor. This is important as a means for getting precise information about the chain-initiation reactions and hence obtaining bond-dissociation energies. For the elementary act of decomposition of a molecule into unstable atoms or radicals the bond-dissociation energy is probably equal to the activation energy. This results because experimental evidence suggests that the reverse reaction has zero activation energy, though usually not a zero temperature dependence. A typical example is dissociation of benzyl bromide, which takes place as shown below when carried through a heated tube at low pressure by a large excess of toluene:

$$C_6H_5CH_2Br \rightarrow C_6H_5CH_2 \cdot + Br \tag{10.20a}$$

$$Br + C_6H_5CH_3 \rightarrow C_6H_5CH_2 \cdot + HBr \tag{10.20b}$$

$$2C_6H_5CH_2 \cdot \rightarrow (C_6H_5CH_2)_2 \tag{10.20c}$$

Since the toluene is present in great excess, it effectively removes Br atoms according to (10.20b) even though it is intrinsically less reactive than benzyl bromide. Toluene is a suitable carrier because it is reactive enough to remove radicals such as Br but still thermally stable at the high temperatures necessary to obtain appreciable rates for fission reactions such as (10.20a). Aniline is also useful in this technique,[31] especially when one of the products of bond breaking is a methyl radical, since $CH_3 \cdot$ might be formed by pyrolysis of toluene as well as by the step corresponding to (10.20a). For dissociation of benzyl bromide Szwarc and coworkers[30] obtained

$$k = 10^{13.0} \exp \left[\frac{-(211 \pm 8)\,kJ}{RT\,mol} \right] s^{-1}$$

To obtain the bond-dissociation energy in benzyl bromide a term RT should be added to the observed activation energy, since the bond-dissociation energy is an enthalpy change. However, at $298\,K$ $RT = 2.5\,kJ\,mol^{-1}$, well within the error limits of E_a, and so this correction is commonly ignored. Bond-dissociation energies can be obtained by other kinetic methods[32] as well as by spectroscopic measurements, thermodyamic calculations, and other techniques.[33]

CHAIN-TRANSFER REACTIONS: POLYMERIZATION

A process in which a chain carrier reacts with a molecule to give a new radical species that is not the normal chain carrier is called a chain-transfer reaction.[34] An example would be reaction of a radical with the solvent for a chain reaction in solution

$$R \cdot + SH \rightarrow RH + S \cdot$$

Since a new radical $S \cdot$ is formed, the chain is not necessarily broken; rather the original chain is stopped and a new one potentially started. The reactivity, or lack of reactivity, of $S \cdot$ may play a part in the observed kinetics.

The most important case of chain transfer occurs in polymerization reactions.[35] Formation of an addition polymer is a chain process carried by free radicals, anions, or cations. There is an initiation act that produces an active radical or ion. This species adds to the double bond of an olefinic compound to produce a new, larger radical or ion that can add to another monomer. A chain reaction occurs in which the chain is the growing polymer molecule. The chain is broken by some termination reaction, such as recombination or disproportionation of two free radicals, and neutralization of the ionic charge.

Take as a specific example the radical polymerization of a styrene initiated by thermal decomposition of ABN:

$$(CH_3)_2 \underset{\underset{CN}{|}}{C} - N = N - \underset{\underset{CN}{|}}{C}(CH_3)_2 \xrightarrow{k_i} 2(CH_3)_2 \underset{\underset{CN}{|}}{C} \cdot + N_2 \qquad (10.21)$$

Representing this radical by $R' \cdot$

$$R' \cdot + CH_2 = CHAr \xrightarrow{k_p} R' - CH_2 - CHAr \cdot \qquad (10.22)$$

$$R'CH_2 CHAr \cdot + CH_2 CHAr \xrightarrow{k_p} R'CH_2 CHAr - CH_2 - CHAr \cdot \qquad (10.23)$$

$$2R'(CH_2 CHAr)_n CH_2 CHAr \cdot \xrightarrow{k_t} \text{recombination or disproportionation} \qquad (10.24)$$

Experiments reveal that the overall rate of polymerization, or disappearance of monomer, is first order in monomer concentration and half-order in initiator. This result can be deduced theoretically from the above mechanism if it is assumed that the rate constant for propagation k_p is independent of the size of the growing polymer chain. The usual steady-state theory for the concentration of all free radicals then leads to

$$\text{rate} = \left(\frac{k_i}{k_t}\right)^{1/2} k_p [ABN]^{1/2} [\text{monomer}] \tag{10.25}$$

In the absence of chain transfer, the molar mass, or degree of polymerization of the polymer, is directly proportional to the chain length. If disproportionation of the two polymer radicals is the means of termination, that is,

$$2R - CH_2 - CHAr \cdot \rightarrow R - CH = CHAr + R - CH_2 - CH_2 Ar \tag{10.26}$$

then the chain length is the degree of polymerization. The chain length ν is given by

$$\nu = \frac{k_p [\text{monomer}]}{(k_i k_t [ABN])^{1/2}} \tag{10.27}$$

This equation indicates that chains started late in the reaction have a different length from those initiated earlier when the monomer and catalyst concentrations are high. Actually there is a statistical distribution of chain lengths since (10.27) only gives the average for any instantaneous set of concentrations.[36] Typical values of k_p are of the order of $10^3 M^{-1} s^{-1}$ at 80°C.[37] For k_t a typical figure is $10^7 M^{-1} s^{-1}$. Since k_i for ABN is $1.5 \times 10^{-4} s^{-1}$ at this temperature, the degree of polymerization is about 1000 for $10^{-3} M$ initiator and $1 M$ monomer. Activation energies for propagation are of the order of 20–40 kJ mol^{-1} and for termination about zero.[37] The reaction of two large radicals is considerably slower than that for two methyl or ethyl radicals, in agreement with the transition-state theory.

If a transfer reaction occurs with the solvent, for example,

$$R-CH_2-CHAr \cdot + CCl_4 \xrightarrow{k_s} R-CH_2-CHClAr + Cl_3 C \cdot \tag{10.28}$$

the average chain length becomes

$$\nu = \frac{k_p [\text{monomer}]}{(k_i k_t [ABN])^{1/2} + k_s [S]} \tag{10.29}$$

Equation (10.29) follows from the definition of chain length as rate of propagation divided by rate of formation of chain carriers, since a new chain is started by the tricholormethyl radical.

Chain transfer in which a hydrogen atom is abstracted from the middle of a polymer chain is also of great importance in producing cross-linking and branches in polymers.

Cross-linking and branching are much less common in the case of ionic polymerization, an excellent example of a chain reaction where the chain carriers are not free radicals. Cationic chain carriers can be produced by using initiators that are acids. Anionic polymerization, which is of greater commerical importance, is initiated using bases such as lithium amide, n-butyllithium and other organometallic compounds, or active metals. For example, polymerization of styrenes in ether solution can be initiated by sodium and naphthalene

$$Na + naphthalene \rightleftharpoons Na^+, naphthalene^{\overline{\cdot}}$$

$$Na^+, naphthalene^{\overline{\cdot}} + CHAr=CH_2 \rightarrow naphthalene + CHAr=CH_2^{\overline{\cdot}}, Na^+$$

$$2CHAr=CH_2^{\overline{\cdot}}, Na^+ \rightarrow Na^+, {}^-CH_2=CH-CH=CH_2^-, Na^+$$
$$\qquad\qquad\qquad\qquad\qquad\quad \underset{Ar}{|}\quad \underset{Ar}{|}$$

The naphthalene radical anion formed by transfer of an electron from sodium in turn transfers an electron to the styrene. The styrene radical anion dimerizes to produce a dianion that is the true chain initiator and begins to grow at both ends. By contrast with radical polymerization, termination reactions between polymer anions are slow, often requiring days or weeks. The process starts with hydride elimination

$$R-CH_2-CHAr^-, Na^+ \rightarrow R-CH=CHAr + Na^+, H^-$$

but this is really a chain-transfer step since H^- is a strong nucleophile that can initiate another chain. If another polymer anion abstracts the allylic hydrogen of the unsaturated end group, then termination does occur, since the diaryl allyl anion formed in (10.30) is unreactive. Because normal chain termination is so slow,

$$R-CH_2-CHAr^- + R'-CH-CH=CH \rightarrow R-CH_2-CH_2 Ar$$
$$\qquad\qquad\qquad\quad \underset{Ar}{|}\qquad \underset{Ar}{|}$$

$$+ R'-C\cdots\cdots CH\cdots\cdots CH \qquad\qquad (10.30)$$
$$\qquad \underset{Ar}{|}\qquad\qquad \underset{Ar}{|}$$

impurities or molecules not otherwise involved in the chain reaction may provide the chief route for disappearance of polymer anions. Under carefully controlled conditions there may be no termination steps at all.[38] Once rapid polymerization has used up all the monomer present, the polymer remains "living" — further addition of the same monomer results in further polymerization, forming longer chains of greater molar mass. Addition of a different monomer after the initial polymerization results in block copolymers. Each chain consists of one long section containing only the first monomer and another that contains only the second monomer. Living polymers can be "killed" by adding compounds like water that react with carbanions.

In anionic or cationic polymerization the charge of the growing polymer chain must be balanced by some counterion. Ion pairing of the counterion with the charged end of the chain often affects the stereochemistry of the polymer that results. When an anionic polymer chain is coordinate-covalently bound to a transition-metal complex, a high degree of stereospecificity is possible. Reactions of this type are called coordination polymerizations and are exemplified by the heterogeneous Ziegler-Natta system and its homogeneous analogues.[39] The Ziegler-Natta system employs a fibrous form of $TiCl_3$ produced by reacting $TiCl_4$ and $Al(C_2H_5)_3$. The accepted mechanism was originally proposed by Cossee.[40] It involves binding of ethylene or propylene to a vacant coordinate site on a titanium atom that also contains an alkyl ligand. The alkyl group migrates to the olefin, forming a larger alkyl ligand, as shown below:

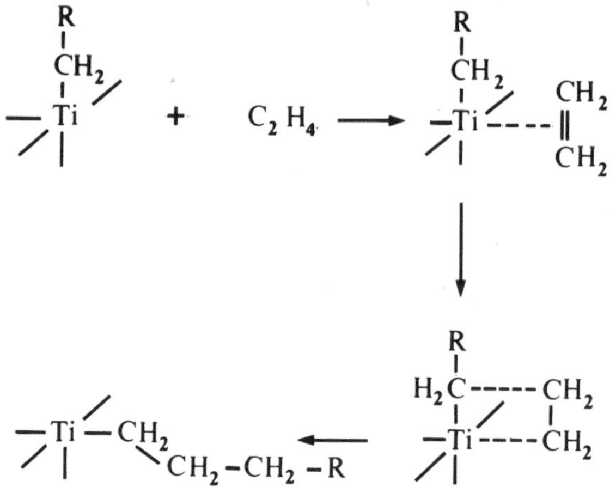

This mechanism is in accord with the rules of orbital symmetry developed in Chapter 5,[41] and the direction of electron flow is that expected if the polymer chain migrates as an anion. Stereochemically, *cis* addition to the olefin is required during each cycle of the chain, and migration is predicted to occur with retention of configuration if the growing alkyl group contains an asymmetric carbon adjacent to the metal. Recently an alternative mechanism has been proposed[42] that involves a carbene species generated by transferring a hydrogen from the α carbon of the polymer chain to the metal. The carbene then bonds to the cordinated olefin and the hydrogen is transferred back from the metal.

RICE-HERZFELD MECHANISMS OF ORGANIC MOLECULE DECOMPOSITION

Thermal decomposition of gaseous hydrocarbons is important both in the development of gas-phase kinetics and in industry, where cracking of petroleum provides higher quality gasoline and raw materials for the petrochemical industry. Pyrolyses of hydrocarbons, ketones, aldehydes, alcohols, ethers, and so on were originally thought to be unimolecular since the kinetics were generally first order. However, Rice and Rice[43] detected free radicals in these decompositions by means of the Paneth mirror technique. This method, which provided the first experimental confirmation that free radicals could exist,[44] is based on the fact that alkyl and other radicals can remove a mirror of Pb (or some other metal) by forming volatile organometallic species. The presence of radicals suggested that pyrolysis occurred by way of a chain mechanism. Rice and Herzfeld[45] provided a general discussion of possible mechanisms and showed that first-order kinetics were possible despite the complexity of a chain mechanism.

Rice and Herzfeld obtained first-order kinetics by proposing the mechanism

$$M_1 \xrightarrow{k_1} R_1 + M_2$$

$$R_1 + M_1 \xrightarrow{k_2} R_1H + R_2$$

$$R_2 \xrightarrow{k_3} R_1 + M_3$$

$$R_1 + R_2 \xrightarrow{k_4} M_4$$

where R represents a radical and M a molecule. Here two unlike radicals react in the chain termination step. Applying the steady-state approximation to the radical species gives

$$\frac{d[R_1]}{dt} = 0 = k_1[M_1] - k_2[R_1][M_1] + k_3[R_2] - k_4[R_1][R_2] \quad (10.31)$$

$$\frac{d[R_2]}{dt} = 0 = k_2[R_1][M_1] - k_3[R_2] - k_4[R_1][R_2] \quad (10.32)$$

Now add (10.31) to (10.32), solve for $[R_2]$, substitute this result into (10.31), and solve for $[R_1]$ to obtain

$$[R_1] = \frac{k_1/4 \pm \sqrt{(k_1/4)^2 + k_1k_2k_3/2k_4}}{k_2}$$

If the chains are long k_2 and k_3 must be much greater than k_1 and k_4 so that

$$k_2[R_1] \cong \sqrt{\frac{k_1k_2k_3}{2k_4}}$$

Now substitute this into the expression for the rate of disappearance of M_1

$$\frac{-d[M_1]}{dt} = k_1[M_1] + k_2[R_1][M_1]$$

$$= k_1[M_1] + \left(\frac{k_1k_2k_3}{2k_4}\right)^{1/2}[M_1]$$

This is the desired first-order result. However, this is not the only way to get a first-order rate. Various combinations of unimolecular or bimolecular initiation with bimolecular or trimolecular termination can result in overall orders ranging from 0 to 2.[46]

As an example of a Rice-Herzfeld mechanism consider pyrolysis of ethane, a major means by which the chemical industry produces ethylene. The industrial process employs a catalyst, but we limit attention here to the homogenous, uncatalyzed reaction. In the temperature range 700–900 K and at pressures above 100 mm Hg, 99% of the reaction in its early stages follows the stoichiometry

$$C_2H_6 \rightarrow C_2H_4 + H_2$$

In addition small yields of methane and butane are observed. As the reaction proceeds, the methane product becomes more important and increased yields of propylene are observed. Focusing on the early stages where stoichiometry is relatively straightforward, the reaction is first order in ethane

$$\frac{-d[C_2H_6]}{dt} = k[C_2H_6]$$

Pyrolysis of ethane is inhibited by NO and other radical scavengers. For a long time it was thought that the residual reaction that occurs in the presence of high concentrations of inhibitor occurred by way of a molecular mechanism, not involving free radicals. However, Rice and Varnerin's isotopic mixing studies[47] showed that there is no significant molecular component. When C_2D_6 decomposes in the presence of CH_4, mixed products, like CH_3D, form by way of radical reactions such as

$$C_2D_6 \rightarrow 2CD_3$$

$$CD_3 + C_2D_6 \rightarrow CD_4 + C_2D_5$$

$$C_2D_5 + CH_4 \rightarrow C_2D_5H + CH_3$$

$$CH_3 + C_2D_6 \rightarrow CH_3D + C_2D_5$$

Rice and Varnerin found that the ratio $[CH_3D]/[CH_4]$ was proportional to the extent of reaction, and that the concentration of NO had no effect on the constant of proportionality; that is, addition of NO reduces mixing rates, which are known to involve radicals, by the same proportion that the pyrolysis rate is reduced. Consequently the residual pyrolysis must also involve radicals.

For decomposition of ethane the Rice-Herzfeld mechanism is

$$C_2H_6 \xrightarrow{k_1} 2CH_3 \cdot \qquad (10.33)$$

$$CH_3 \cdot + C_2H_6 \xrightarrow{k_2} C_2H_5 \cdot + CH_4 \qquad (10.34)$$

$$C_2H_5 \cdot \xrightarrow{k_3} C_2H_4 + H \cdot \qquad (10.35)$$

$$H \cdot + C_2H_6 \xrightarrow{k_4} C_2H_5 \cdot + H_2 \qquad (10.36)$$

$$2C_2H_5 \cdot \xrightarrow{k_5} n-C_4H_{10} \qquad (10.37)$$

$$2C_2H_5 \cdot \xrightarrow{k_6} C_2H_4 + C_2H_6 \qquad (10.38)$$

Initiation involves fission of the weakest bond in ethane, and the methyl radicals so produced react with ethane to form ethyl radicals that, along with hydrogen atoms, are the chain carriers. Termination involves combination or disproportionation of ethyl radicals. The overall rate is

$$\frac{-d[C_2H_6]}{dt} = k_1[C_2H_6] + k_2[CH_3][C_2H_6] + k_4[H][C_2H_6] - k_6[C_2H_5]^2$$

$$(10.39)$$

Concentrations of radical species can be eliminated from (10.39) by applying the steady-state approximation to $[CH_3]$, $[H]$, and $[C_2H_5]$:

$$\frac{d[CH_3]}{dt} = 0 = 2k_1[C_2H_6] - k_2[CH_3][C_2H_6]$$

$$\frac{d[C_2H_5]}{dt} = 0 = k_2[CH_3][C_2H_6] - k_3[C_2H_5] + k_4[H][C_2H_6]$$

$$- 2k_5[C_2H_5]^2 - 2k_6[C_2H_5]^2$$

$$\frac{d[H]}{dt} = 0 = k_3[C_2H_5] - k_4[H][C_2H_6]$$

Solving these simultaneous equations for the concentrations of radicals and substituting into (10.39) yields

$$\frac{-d[C_2H_6]}{dt} = \left(\frac{3k_1 - k_6k_1}{k_5 + k_6}\right)[C_2H_6] + k_3\left(\frac{k_1}{k_5 + k_6}\right)^{1/2}[C_2H_6]^{1/2} \quad (10.40)$$

The first term on the right-hand side of (10.40) involves rate constants for initiation and termination only, while the second term involves the rate constant for a propagation step as well. That the chain is long can be deduced from the fact that 99% of the product is C_2H_4, which implies a chain length on the order of 100 at least. Consequently the propagation steps must be faster than initiation and termination, and to a good approximation the first term on the right of (10.40) may be omitted. This leads to a prediction that the reaction should have order 0.5,

$$\frac{-d[C_2H_6]}{dt} = k_3\left[\frac{k_1}{k_5 + k_6}\right]^{1/2}[C_2H_6]^{1/2} \quad (10.41)$$

which does not agree with experiment. However, reaction (10.35) is unimolecular and at very low pressures its rate is given by $k_3'[M][C_2H_5]$ instead of $k_3[C_2H_5]$, which was used to obtain (10.41). For small extent of reaction $[M] \cong [C_2H_6]$ since the total pressure is almost entirely due to ethane. The rate of (10.35) becomes $k_3'[C_2H_6][C_2H_5]$, and k_3 in (10.41) should be replaced by $k_3'[C_2H_6]$. This predicts $\frac{3}{2}$-order dependence on the concentration of ethane, a limiting value that has been observed at low pressures.[48] Under the usual conditions of pyrolysis (10.35) is in the fall-off region, its rate is $k_3'[C_2H_6]^{1/2}[C_2H_5]$, and the overall reaction follows first-order kinetics.[49]

Since methane is produced only by (10.34), the rate of methane formation is a measure of the rate of initiation. The rate of methane formation is first order in ethane at high pressures, and the rate constant falls off as expected, but only at pressures considerably lower than that for (10.35).[49, 50] Since for any chain mechanism the rate of termination must equal the rate of initiation, and since the ratio k_5/k_6 is known independently,[51] the rate of butane formation by way of (10.37) can also provide information about the rate of initiation. Conclusions based on butane formation agree with those from methane production.[50]

The increased yields of methane and propylene observed in later stages of ethane pyrolysis can be attributed to reaction of $C_2H_5\cdot$ with the increasingly concentrated ethylene product.

$$C_2H_5 \cdot + C_2H_4 \rightleftharpoons 1-C_4H_9 \cdot$$

$$1-C_4H_9 \cdot + C_2H_4 \rightleftharpoons 1-C_6H_{13} \cdot$$

The 1-hexyl radical can isomerize by way of a 1,5 intramolecular hydrogen transfer to produce 2-hexyl radical. This then decomposes as follows:

$$2-C_6H_{13} \cdot \rightarrow C_3H_7 \cdot + C_3H_6$$

$$C_3H_7 \cdot \rightarrow C_2H_4 + CH_3 \cdot$$

$$CH_3 \cdot + C_2H_6 \rightarrow CH_4 + C_2H_5 \cdot$$

The result is formation of equal amounts of methane and propylene.

Rice-Herzfeld mechanisms can explain the kinetics of pyrolysis of heavier hydrocarbons,[52] but these reactions are more complicated because any one of several carbon—carbon bonds may be broken in the initiation steps. Thus a greater variety of radicals is produced and there is a correspondingly greater number of possible propagation steps.

BRANCHING CHAIN REACTIONS

A branching chain reaction involves at least one step in which two or more chain carriers are produced, instead of the customary one. A familiar example is the nuclear fission chain reaction, in which neutrons are the chain carriers. Fission of a single $^{235}_{92}U$ nucleus produces, on average, three neutrons, each of which can cause another fission if it strikes a $^{235}_{92}U$ nucleus before escaping from the uranium-containing substance. Such a reaction can be controlled by adjusting the rate of absorption of neutrons by inert materials and/or the rate of escape of neutrons from the sample so that two of the three neutrons produced in the chain-branching step do not contribute to the chain reaction. Such a situation corresponds to an infinitely long linear chain of the type discussed above. In nuclear-fission terminology the situation is referred to as critical. In a subcritical sample the chain carriers escape or are absorbed faster than they are produced and little or no chain reaction can occur. In a supercritical situation the reaction rate increases exponentially and, unless the rate of escape or absorption of neutrons increases, an explosion results.

Branched-chain chemical reactions are especially interesting, both because of the extraordinary consequences that can be produced by a branching elementary process and because of the commerical importance of combustion reactions, which usually involve branched-chain mechanisms. Branched-chain combustion reactions sometimes proceed at a steady rate, but beyond certain critical experimental conditions these reactions accelerate rapidly, resulting in explosion. Such conditions are called explosion limits, and the onset of explosive reaction is called ignition. Two types of explosion may be distinguished. A thermal explosion occurs when the rate of evolution of heat by an exothermic reaction exceeds the rate of dissipation of heat by conduction, convection, and radiation. If the reaction has an Arrhenius temperature dependence and significant activation energy, increasing

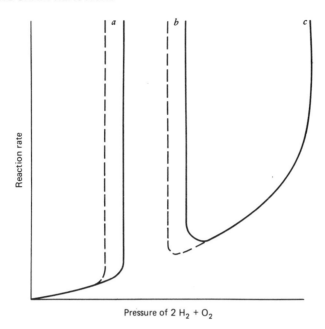

Pressure of 2 H_2 + O_2

Figure 10.2.   Variation of reaction rate with reactant pressure for $2H_2 + O_2$ at constant temperature. First, second, and third explosion limits are labeled *a, b,* and *c*, respectively. Displacement of the first limit upon adding inert gas or increasing size of reaction vessel and displacement of the second limit upon adding inert gas are shown by dashed lines.

temperature in the reaction vessel increases the rate dramatically. Semenov[53] has treated this situation quantitatively. An isothermal explosion is one that does not require such an acceleration by increasing temperature, though since the reaction is usually exothermic there is almost always a thermal contribution. In an isothermal explosion the fundamental reason for accelerating rate is an increase in the concentration of chain carriers as a result of a branching step in the mechanism.[53]

As an example of a branched-chain mechanism consider the extensively studied reaction of hydrogen with oxygen,[54] which serves as a prototype for combustion reactions in general.[55] Experimentally the $2H_2 + O_2$ reaction is found to be very dependent on the composition of the vessel surface. For a stoichiometric mixture at 550°C in a silica vessel, the reaction proceeds very slowly at pressures below about 4 torr. Between this pressure and about 100 torr, however, explosion occurs. Above about 100 torr there is a normal reaction whose rate increases with increasing pressure until finally a third explosion limit is reached at quite high pressures. Because the normal reaction is already quite rapid at these pressures the third limit is difficult to study quantitatively. The high-pressure explosion appears to involve both thermal and isothermal components. The behavior of explosion limits in the $2H_2 + O_2$ system is shown diagrammatically in Figs. 10.2 and 10.3. Notice that increasing the diameter of the reaction vessel or adding an inert gas lowers the pressure at which the first explosion limit occurs, and adding an inert gas also lowers the pressure of the second explosion limit, though vessel diameter

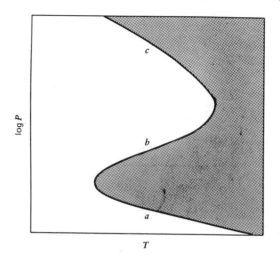

Figure 10.3.   Explosion limits as a function of temperature and pressure for $2H_2 + O_2$. First, second, and third explosion limits are labeled *a, b,* and *c*. Shaded area is region of explosion.

has no effect. No explosion occurs below about 400°C, and the mixture always explodes above about 600°C.

As many as 40 elementary steps have been proposed for the $2H_2 + O_2$ reaction, and at least 15 are required to account for the slow reaction under controlled conditions.[54] However, the following simple mechanism provides a reasonable qualitative and semiquantitative account of the observations:

$$\text{Initiation:} \qquad H_2 + O_2 \xrightarrow{\text{wall}} 2OH \cdot \qquad\qquad (10.42)$$

$$\text{Propagation:} \qquad OH \cdot + H_2 \xrightarrow{k_2} H_2O + H \cdot \qquad\qquad (10.43)$$

$$\text{Branching:} \qquad H \cdot + O_2 \xrightarrow{k_3} OH \cdot + O: \qquad\qquad (10.44)$$

$$O: + H_2 \xrightarrow{k_4} OH \cdot + H \cdot \qquad\qquad (10.45)$$

$$\text{Termination:} \qquad H \cdot \xrightarrow{k_5} \text{wall} \qquad\qquad (10.46)$$

$$H \cdot + O_2 + M \xrightarrow{k_6} HO_2 \cdot + M \qquad\qquad (10.47)$$

Initiation might also involve dissociation of H_2 or of O_2, which then could carry the chain by way of (10.44) or (10.45). At low pressures termination occurs by way of the heterogeneous reaction (10.46), while at higher pressures the homogeneous step (10.47) becomes important. The relatively unreactive hydroperoxide radical probably diffuses to the wall without reacting in the gas phase. The propagation step (10.43) is fast and exothermic, but the branching step (10.44) is endothermic, has a large activation energy, and only becomes important above about 400°C. Reaction (10.45) is much faster than (10.44).

At pressures below the first explosion limit the rate is governed by competition for H· atoms by diffusion to the wall (10.46) and the slow branching step (10.44). At pressures above the second explosion limit the homogeneous termination step (10.47) is the principal competitor with (10.44). Considering both termination reactions and applying the steady-state approximation, which is reasonable outside the explosion region, yields for the concentration of hydrogen atoms

$$[H]_{ss} = \frac{2r_i}{k_5 + k_6 [O_2][M] - 2k_3 [O_2]} \qquad (10.48)$$

where r_i is the rate of the initiation step (10.42). Below the first limit $k_6 [O_2]$ $[M] \ll k_5$. As the pressure increases toward the limit, $2k_3 [O_2]$ approaches k_5 and $[H]_{ss}$ tends to infinity, causing the overall rate also to become infinite. Increasing the diameter of the vessel decreases k_5 because more time is required for hydrogen atoms to diffuse the longer distance to the wall. Adding inert gas at constant partial pressure of $2H_2 + O_2$ also decreases k_5 because the rate of diffusion is inversely proportional to total pressure. Thus (10.48) can account for lowering of the explosion limit in a larger vessel or with added inert gas. The former effect is analogous to the critical mass requirement for explosive nuclear fission, where, owing to escape of neutrons from a reactor, there is a critical size below which only very slow reaction occurs.

Above the second explosion limit $k_6 [O_2][M] \gg k_5$ and the denominator of (10.48) becomes approximately $k_6 [O_2][M] - 2k_3 [O_2]$. In the stoichiometric reaction mixture M may be either H_2 or O_2, which have different efficiencies in reaction (10.47). Nevertheless, to a good approximation $[M] = k'P$, and so as pressure decreases $k_6 [O_2][M]$ decreases more rapidly than $2k_3 [O_2]$. The explosion limit is reached when these two terms are equal, that is, when $[M] = 2k_3/k_6$, which again causes $[H]_{ss}$ and the overall rate to go to infinity. Adding inert gas increases $[M]$ without affecting the rate of branching, and so lowers the explosion limit in terms of partial pressure of $2H_2 + O_2$. Changing the diameter of the vessel, however, has no effect on either of the terms remaining in the denominator of (10.48), since both refer to homogeneous reaction steps.

The third explosion limit can be explained qualitatively by assuming that at very high pressures some hydroperoxide radicals react before reaching the wall:

$$HO_2 \cdot + H_2 \rightarrow H_2O + OH \cdot$$

This diminishes the rate of termination, allowing chain branching to become dominant again.

PHOTOCHEMISTRY

Photochemical initiation of chain reactions, the concept of quantum yield, and the rotating sector method are discussed in preceding sections of this chapter, and some photochemical methods are described earlier in this book. Here we consider molecular absorption and emission of electromagnetic energy, modes and rates of redistribution of that energy within molecules, properties of excited states and their influence on reaction rates, and photochemical techniques for determining rates of very rapid processes.

There are a great many important photochemical reactions. Photosynthesis of carbohydrate from CO_2 and H_2O in green plants provides the most obvious example, but atmospheric reactions such as maintenance of earth's stratospheric ozone layer and formation of photochemical smog also have tremendous impact on human activities. Human vision, formation of vitamin D and skin cancer by sunlight, and photography all depend on photochemical reactions, and a great deal of recent research has been devoted to finding photochemical processes by which solar energy can be stored in chemical fuels or converted to more manageable electrical form. From a theoretical standpoint photochemical reactions can be studied much more selectively than thermal reactions. Monochromatic radiation can be used to produce excited molecules distributed over a very narrow range of energies, as opposed to the broad distribution that results from increased temperatures. Laser pulses on a picosecond time scale provide extremely high temporal resolution for investigation of very rapid reactions.

Usually one photon excites one molecule and only one molecule. This is known as the Stark-Einstein law. It results because normal light intensities are low enough and lifetimes of typical excited states are short enough ($\sim 10^{-8}$ s) so that an excited molecule undergoes deactivation before a second photon can strike it. With lasers much higher intensities can be achieved and multiple-photon absorptions can occur. Also, there are unusual circumstances where one photon can excite two adjacent molecules.[56] The energy of 1 mol of photons is often referred to as one *einstein*. It is given by $E = Lh\nu$, where L is Avogadro's constant, h is Planck's constant, and ν is the frequency of the radiation. For visible wavelengths of 400–700 nm 1 einstein ranges from 300 to 171 kJ mol^{-1}. The former value is comparable to many bond-dissociation energies, while the latter is sufficient to break only the weakest bonds. Under conditions where the Stark-Einstein law applies, ultraviolet, and perhaps visible, light is capable of initiating reactions by dissociating molecules.

The first law of photochemistry was stated by Grotthuss and Draper in 1818 and seems so obvious now as hardly to need mention: only radiation that is absorbed by a sample is effective in producing a photochemical change. However, not all molecules absorb light in an energetically adequate and experimentally convenient region of the spectrum. In such cases photochemical reactions can still occur if an appropriate *photosensitizer* absorbs radiant energy and transfers that energy to a reactant molecule. Mercury, for example, is a photosensitizer for decomposition of hydrogen:

$$Hg + h\nu \rightarrow Hg^*$$

$$Hg^* + H_2 \rightarrow Hg + H\cdot + H\cdot$$

Photosensitization may also involve reaction of the photosensitizer itself to form species that initiate the desired reaction. An example is photosensitization of the $2H_2 + O_2$ reaction by NO_2, which photodissociates to produce oxygen atoms that are chain carriers in the mechanism (10.42)–(10.47).

At normal temperatures and pressures nearly all molecules are in their ground electronic states, and for most molecules that state is a singlet. Furthermore, about

99.99% of these molecules are in the ground vibrational state. Since rotational quanta are generally of very low energy and have little effect on the kinetics of photochemical processes, we do not include them here. Thus in most cases absorption of a photon in a primary photochemical process involves an electronic transition that originates in the ground vibrational level of a singlet electronic state. Electronic selection rules require with a high degree of probability that the excited state also be a singlet, but usually it involves considerable vibrational excitation — often enough so that bond fission occurs. If the excited molecule does not dissociate, then a variety of photophysical processes can occur. These are summarized schematically in Fig. 10.4. Excess vibrational energy can be transferred to other molecules by intermolecular collisions or it can be lost by vibrational quantum jumps that result in emission of infrared photons. For polyatomic molecules in the gas phase at normal temperatures and pressures the former energy-transfer process has a lifetime of about 10^{-6} s while the latter requires on the order of 10^{-3} s, so in general collisional deactivation is the mechanism for vibrational

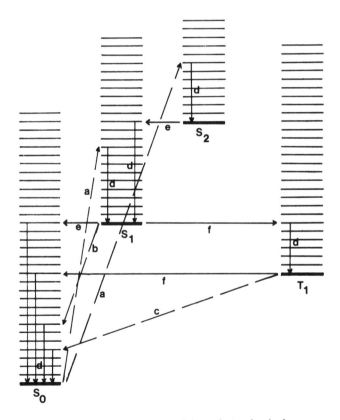

Figure 10.4. Schematic diagram summarizing photophysical processes that may occur in an excited molecule. Dashed lines represent absorption or emission of radiation. Solid lines are radiationless transitions. (*a*) Absorption; (*b*) fluorescence; (*c*) phosphorescence; (*d*) collisional deactivation; (*e*) internal conversion; (*f*) inter-system crossing. S_0 is the ground singlet electronic state, S_1 and S_2 are excited singlets, and T_1 is the lowest triplet state.

relaxation to an equilibrium population of vibrational levels. In the liquid phase, where collisions occur more often, collisional deactivation is invariably favored over infrared emission. The latter is only observed in the gas phase at very low pressures.

If the ground-state molecule is excited to other than the lowest singlet, for example, to S_2 in Fig. 10.4, there is usually vibrational relaxation to the ground vibrational state followed by internal conversion to an excited vibrational level of the lowest excited singlet, S_1. Then collisional deactivation can occur in S_1 as described above. Thus no matter which excited singlet is attained initially, the molecule usually relaxes rapidly to the ground vibrational state of the lowest excited singlet (Kasha's rule). There are several routes by which the excited state S_1 may return to the ground state. Radiationless deactivation involves a transition from S_1 to an excited vibrational level of S_0 that has the same energy. Then vibrational relaxation of S_0 by collisional deactivation takes place, the overall effect being a transition to the ground electronic and vibrational state without emission of radiation. A second route is fluorescence, that is, emission of a photon as a result of an electronic transition between two states of the same multiplicity. The energy of the photon emitted in fluorescence is generally less than the energy of the exciting photon since vibrational relaxation in the excited state and transitions to excited vibrational levels of S_0 usually occur. Typical lifetimes for fluorescence are of the order of 10^{-8} s. It is also possible to have intersystem crossing from S_1 to T_1, the lowest triplet state of the molecule. Such a transition, from one excited state to another of different multiplicity, is forbidden, but electronic selection rules are less stringent in polyatomic molecules because vibrations of appropriate symmetry can couple with the electronic change. Heavy atoms in the molecule, or in the medium, can also facilitate spin changes. Following intersystem crossing there is usually rapid vibrational relaxation in T_1. This may be followed by the rather slow process of reverse intersystem crossing (to S_0) or by phosphorescence, emission of radiation that involves a forbidden transition between states of different multiplicity. Lifetimes for phosphorescence range from 10^{-3} to 1 s for typical polyatomic molecules in the absence of collisions. Whether it disappears as a result of intersystem crossing or phosphorescence emission, the triplet state is metastable, that is, relatively long lived, and consequently can play important roles in molecular photochemistry.

The preceding generalizations are not absolute. Some photochemical reactions do involve higher excited states than S_1 or T_1,[57] and there is good reason to believe that there are many crossings of potential energy surfaces for excited states. As nuclear arrangements vary the system is likely to move from one surface to another, eventually returning to the S_0 surface. Since the same ground-state hypersurface applies to products as well as reactants, such a return may correspond to chemical reaction. In general products of photochemical reactions are formed in their ground electronic states; they do not form in excited states and then fluoresce or phosphoresce. Energy conservation requires, however, that return to S_0 involve highly excited vibrational states, which result in reaction, provided that they are not immediately deactivated by intermolecular collisions. For reaction to compete successfully with deactivation, return to the S_0 surface should occur at a nuclear configuration close to that of the activated complex of the thermal reaction. In many cases minima on excited-state surfaces have the same configurations as maxima on ground-state surfaces, thus providing regions in configuration space through which reactions can occur.[58]

EXCITED-STATE PROPERTIES

An electronically excited molecule has its own structure, physical properties, and chemical properties; it is not just a ground-state molecule with excess energy, and it may behave quite differently. For example, acid–base strengths of excited states can be remarkably different,[59] as shown by the following values for protonation of 2-naphthylamine and deprotonation of 2-naphthol:

	pK_{S_0}	pK_{S_1}	pK_{T_1}
2-Naphthylamine	4.1	-1.5 to -2.0	3.1–3.3
2-Naphthol	9.5	2.5 to 3.0	7.7–8.1

The large decrease in pK_{S_1} can be attributed to the formation of the excited state by an n, π^* transition in which one electron from a lone pair is transferred to an antibonding π orbital on the aromatic ring. In the case of 2-napthol, S_1 corresponds to

Since there is less electron density on the oxygen, acidity should be much greater. Acidity of the triplet state is much less affected, perhaps because T_1 results from a π, π^* transition.[60]

Excited-state molecules also have unusual redox properties. The vacancy created in the bonding orbitals when an electron is promoted more readily accepts an electron from an external source. Also, an electron raised to an antibonding orbital as a result of absorption of a photon is more likely to be lost in an electron-transfer process.

Redistribution of electron density in the excited state can also affect the position in a molecule at which reaction will occur. For example, metal carbonyl complexes readily undergo photochemical substitution reactions in which a ligand is expelled and then another ligand is picked up. For a 6-coordinate complex containing a single noncarbonyl ligand L the two possible ligand expulsion processes are

$$M(CO)_5 L \xrightarrow{h\nu} M(CO)_5 + L \qquad (10.49)$$

$$M(CO)_5 L \xrightarrow{h\nu} M(CO)_4 L + CO \qquad (10.50)$$

Reaction (10.49) leads to ligand exchange if additional L is present while (10.50) leads to *cis* and *trans* isomers of $M(CO)_4 L_2$. Studies of quantum yields for (10.49) and (10.50) as a function of wavelength reveal that higher-energy radiation favors expulsion of CO and decreases expulsion of L. This can be explained if two reactive ligand-field excited states are involved.[61] Low-energy excitation populates the d_{z^2} metal orbital, whose sigma antibonding character is directed primarily along the axis containing L. Higher-energy radiation populates $d_{x^2-y^2}$, thus labilizing the four CO ligands in the xy plane. Reaction (10.49) also occurs in the presence of short-wavelength irradiation, apparently as a result of internal conversion from the state in which $d_{x^2-y^2}$ is populated to the one involving d_{z^2}. The rate of internal conversion relative to the rate of reaction from the higher-energy excited state determines whether the reaction is wavelength dependent or not.

Often the minimum-energy geometry of an excited-state molecule is significantly different from that of the ground state. Structural changes usually occur on the same time scale as molecular vibrations, whereas excited states have lifetimes of 10^{-9} to 10^{-6} s for photochemical reactions or deexcitation — some 10^3–10^6 times as long. Consequently conformational and other structural changes can occur prior to photochemical reaction, and molecules having different shapes are expected to react differently. For example, thermal pyrolysis of nitro compounds usually results in C—N bond fission:

$$C_2 H_5 NO_2 \xrightarrow{\Delta} C_2 H_5 \cdot + NO_2 \cdot$$

while photolysis involves hydrogen abstraction as well.

$$C_2 H_5 NO_2 \xrightarrow{h\nu} C_2 H_4 + HONO$$

The latter reaction takes place because in the excited state N is no longer surrounded by coplanar bonds. This brings an oxygen close enough to a β hydrogen atom so that abstraction can occur. Changes in excited-state structure that alter the molecular point group can affect predictions of reactivity that are based on orbital topology arguments of the type presented in Chapter 5.

An important example where geometry of an excited molecule plays a significant role is *cis*–*trans* isomerization of olefins. Figure 10.5 shows the energy of each of the three lowest states of ethene as a function of twist angle θ of one CH_2 group relative to the other.[62] The height of the ground-state curve at $\theta = 90°$ comes from the activation energy for thermal *cis*–*trans* isomerization. Based on Fig. 10.5, excitation to either S_1 or T_1 should result in isomerization, and the quantum yield should be 0.50 unless deactivation can occur before the torsional angle changes to $90°$. The expected quantum yield is observed, within reasonable error limits, when *cis*- or *trans*-1,2-dideuterioethene vapor is irradiated so as to populate the T_1 state.[63] However, the behavior of other substituted olefins is not so simple. Photo-isomerization of *cis*- and *trans*-stilbene provides an example of the problems involved.

Either *cis*- or *trans*-stilbene gives a photostationary state when irradiated in hydrocarbon solvents. A photostationary state is a steady state that is set up when an equilibrium system is irradiated so as to affect the forward rate differently from

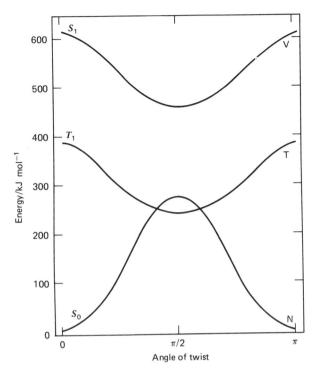

Figure 10.5. Energy of ground state S_0, lowest excited triplet T_1, and lowest excited singlet S_1 as a function of twist angle θ in ethene. The labels, N, T, and V are commonly used for the three states. (Reprinted with permission from R. G. Pearson, *Symmetry Rules for Chemical Reactions,* Wiley-Interscience, 1976, p. 990, based on A. J. Merer and R. S. Mulliken, *Chem. Rev.,* **69**, 639 (1969).

the reverse rate. For stilbene, the *cis* isomer is favored in the photostationary state, although the *trans* isomer predominates at equilibrium. Without attempting to identify precisely the photochemically excited species, this situation can be represented by

$$cis \xrightarrow{\ k_c\ } trans \qquad\qquad r_1 \ =\ k_c\,[cis]$$

$$trans \xrightarrow{\ k_t\ } cis \qquad\qquad r_2 = k_t\,[trans]$$

$$trans + h\nu \rightarrow trans^* \qquad r_3 \ =\ I_a$$

$$trans^* \xrightarrow{\ k_f\ } trans + h\nu \qquad r_4 \ =\ k_f[trans^*]$$

$$trans^* \xrightarrow{\ k_i\ } cis \qquad\qquad r_5 \ =\ k_i[trans^*]$$

where $r_1 - r_5$ are rates of individual steps and I_a is the intensity of light absorbed. The rate of conversion of *trans* to *cis* isomer is

$$\frac{d\,[cis]}{dt} \ =\ k_t[trans] + k_i[trans^*] - k_c\,[cis]$$

and of *cis* to *trans* is

$$\frac{d\,[trans]}{dt} = k_c\,[cis] + k_f[trans^*] - k_t\,[trans] - I_a$$

Applying the steady-state approximation to obtain $[trans^*]$, setting $d\,[cis]/dt = d\,[trans]/dt$, and rearranging yields

$$\frac{[cis]}{[trans]} = \frac{k_t}{k_c} + \frac{I_a}{k_c\,[trans]}\left(\frac{k_i}{k_f + k_i}\right)$$

Thus the ratio $[cis]/[trans]$ in the photostationary state differs from k_t/k_c, the value of the equilibrium constant for the thermal reaction. Actually the photo-stationary state for stilbenes is more complex than this, involving photosensitization and more than one excited state.[64]

The singlet excited state from *trans*-stilbene fluoresces while that from the *cis* isomer does not. At low temperatures only fluorescence occurs ($k_f \gg k_i$ in the simple scheme above), while at high temperatures only isomerization takes place, with $\Phi = 0.50$. This competition between fluorescence and isomerization indicates that the latter process involves a potential barrier on the order of $8\,\text{kJ mol}^{-1}$. This barrier has been suggested to exist between *cis* and *trans* forms of the triplet state or between the singlet and triplet states, but there is considerable evidence that isomerization occurs directly from the singlet, without intersystem crossing.[64] Thus there apparently are different *cis* and *trans* singlets, S_{1c} and S_{1t}, rather than the single twisted state suggested by Fig. 10.5.

One explanation of the difference between S_{1c} and S_{1t} is that, because of π conjugation to the phenyl rings, there is a barrier to twisting, even though the twisted form is more stable. The barrier must be larger for the *trans* isomer than for the *cis*. Such a situation is illustrated by the dashed line in Fig. 10.6. Another possi-bility is to recognize that the structure of excited stilbene should be similar to that of hydrazobenzene, which is definitely not all coplanar. In this case two stereo-isomeric forms arise, as shown by the following Newman projections.

$$S_{1c} \qquad\qquad S_{1t}$$

Interconversion requires inversion at one carbon atom, which might have the life-time of 10^{-8} to 10^{-9} s needed to explain the results. A plausible energy diagram is shown by the solid curves in Fig. 10.6.

Another example where an excited state behaves quite differently from the ground state is photocycloaddition of ketones to olefins, a well-established

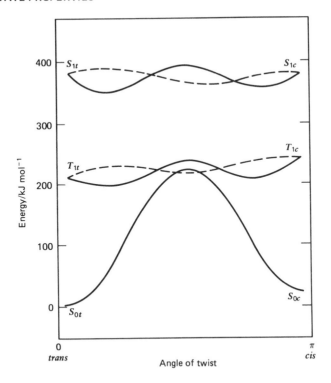

Figure 10.6. Potential energy of three lowest states of stilbene as a function of twist angle θ. Solid lines assume that excited states involve pyramidal structures at the two carbon atoms; dashed lines assume barriers due to π conjugation.

synthetic reaction leading to oxetanes. The reaction is initiated by an n,π^* transition in the ketone. An electron goes from a lone-pair orbital on oxygen to the π antibonding orbital (concentrated on carbon) of the carbonyl group. The excited, diradical species may be represented by the Lewis diagram

$$\begin{array}{c} R \\ \diagdown \\ \overset{\textstyle .}{C}-\ddot{O}: \\ \diagup \cup \\ R \end{array}$$

where the unpaired electron on oxygen is in a lone-pair orbital in the plane of the page and the odd electron on carbon is in a π orbital perpendicular to the plane of the page. The electron density in the excited diradical is such that oxygen is quite electron deficient and can attack, for example, electron-rich olefins. Conversely, the carbon atom is a nucleophilic free radical and can attack electron-poor centers, such as olefins that have electron-withdrawing substitutents. For some combinations of ketone and olefin a π complex is formed initially,[65] which then rearranges to the more stable diradical. Such a molecular complex that exists only in the excited state is called an exciplex. It appears that prior exciplex formation is quite common in cycloaddition reactions of olefins, ketones, dienes, and enones.

QUENCHING OF FLUORESCENCE

As is described earlier, an electronically excited molecule may emit a quantum of energy by fluorescence or may transfer its excitation energy by collision with other molecules. The latter process is called quenching. Presumably quenching does not require an activation energy, other than that of diffusion, and the chance of reaction occurring on the first collision is high. However, different molecules do have different quenching efficiencies, more complex molecules being more efficient. For example, Myers, Silver, and Kaufman[66] have reported rate constants for quenching of NO_2 fluorescence by various gases. The number of gas collisions required, on average, to bring about quenching ranges from 25 for He to 5 for NO to 2 for H_2O.

A simple scheme for excitation–deactivation is as follows, where F is the fluorescing molecule and Q is the quenching molecule:

$$F + h\nu \rightarrow F^* \tag{10.51}$$

$$F^* \xrightarrow{k_f} F + h\nu' \tag{10.52}$$

$$Q + F^* \xrightarrow{k_q} F + Q^* \tag{10.53}$$

(More complex schemes that include internal conversion, intersystem crossing, etc. as individual steps are necessary to provide a framework for detailed understanding of excitation–deactivation.) For this simple scheme the fluorescent yield, or ratio of intensities of light emitted to light absorbed, is given by

$$\frac{I_f}{I_a} = \frac{k_f[F^*]}{k_f[F^*] + k_q[F^*][Q]} \tag{10.54}$$

Rearranging we obtain the Stern-Volmer equation.[67]

$$\frac{I_a}{I_f} = 1 + \frac{k_q[Q]}{k_f} \tag{10.55}$$

A Stern-Volmer plot of the reciprocal of the fluorescent yield against the concentration of the quencher gives the ratio k_q/k_f. Since k_f is the Einstein coefficient for spontaneous emission, it is often known or can be determined from the absorption spectrum, in which case k_q can be found. Table 10.1 shows some results on quenching of fluorescence of β-naphthylamine by carbon tetrachloride in the gas phase and in solution obtained by Rollefson and Curme.[68] These results depend on

Table 10.1. Quenching of fluorescence of β-napthylamine by CCl_4

Medium	$k_q/M^{-1}\,s^{-1}$
Gas	5.9×10^{10}
Isooctane	$2.0 \times 10^{11} \exp(-6694\,J\,mol^{-1}/RT)$
Cyclohexane	$4.5 \times 10^{11} \exp(-10\,334\,J\,mol^{-1}/RT)$

Source. Data from Rollefson and Curme.[68]

Table 10.2. Acid—base reactions of ROH, β-Naphthol, $25°C$

Reaction	K_{eq}	$k/M^{-1} s^{-1}$
$H_3O^+ + RO^{-*}$	650	4.8×10^{10}
$HCOOH + RO^{-*}$	0.11	2.8×10^{8}
$CH_3COOH + RO^{-*}$	0.011	3.3×10^{7}
$CH_3COO^- + ROH^*$	88	2.9×10^{9}
$HCOO^- + ROH^*$	8.8	2.4×10^{9}
$H_2PO_4^- + ROH^*$	0.20	6.0×10^{8}

Source. Data from A. Weller.[69]

a value of k_f based on quenching experiments using oxygen (an efficient quencher) with the assumption that every collision leads to deactivation. This assumption is probably good only to a factor of 2 or 3. It is of interest, however, to see that the gas-phase value agrees with the collision theory and is temperature independent. The solution rate constants are smaller and have apparent activation energies similar to, but not identical with, the activation energies for diffusion.

When the excited molecule is capable of chemical reaction without deactivation, a situation that is most common for acid—base reactions, rate constants may be obtained for reactions of the excited molecules by means of fluorescence measurements.[69] For example, β-naphthol is a weak acid and may be converted to its anion when in the excited state.

$$ROH^* + B \rightleftharpoons RO^{-*} + BH^+ \qquad (10.56)$$

Since ROH^* and RO^{-*} have different fluorescence spectra, the extent to which (10.56) has occurred can be detected experimentally as a function of various concentrations of B and BH^+. The results depend on the interplay of the rate constants for fluorescence and of the proton transfers of (10.56). Table 10.2 shows some rate constants obtained by Weller using the above approach. It should be remembered that both rate and equilibrium constants are for excited-state molecules and would not be the same for ground-state molecules.

FLASH PHOTOLYSIS

Around 1950 Norrish and Porter[70] developed this technique, which employs an intense flash of light to produce transient species, radicals or excited molecules, in concentrations many thousands of times greater than in conventional systems. Up to 10^5 J may be absorbed in a few microseconds, the light being produced by discharging a condenser through a quartz tube containing an inert gas. Wavelengths range from the lowest transmitted by quartz (about 200 nm) through the UV and visible regions. A weaker flash, timed to follow the initiating flash by a known delay period, can be used to measure the spectra of transient species. By varying the delay time a series of spectra of the reaction mixture can be obtained that provide a record of transient species concentrations over microsecond—millisecond time scales. This method is called flash spectroscopy. Alternatively, reaction can be detected by kinetic spectrophotometry. Here a continuous trace of absorbance as a function of time can be made at a particular wavelength using a steady light source, monochromator, and photoelectric detector. First-order rate constants as large as 10^5 s^{-1} and second-order rate constants as large as 10^{11} M^{-1} s^{-1} can be measured, and the method is equally applicable in the gas phase or in solution.

If the contents of a flash-photolysis vessel have low heat capacity, as for a gas at low pressure, there is an almost instantaneous temperature increase of as much as several thousand kelvins. Such conditions may be taken advantage of to study high-temperature pyrolyses and explosions, for example, of hydrides.[71] On the other hand, if the reaction vessel contains an inert gas or solvent to increase its thermal capacity, the temperature rise can be kept below 10 K and reactions can be studied under reasonably isothermal conditions.

An example of a reaction that has been studied by flash photolysis is the recombination of methyl radicals,

$$CH_3 \cdot + CH_3 \cdot \xrightarrow{\ k_r\ } C_2H_6 \qquad (10.57)$$

Basco, James, and Suart[72] generated methyl radicals by photolyzing azomethane, while van den Berg, Callear, and Norstrom[73] used both azomethane and dimethyl-mercury as a source of radicals. Reactions were carried out at room temperature in a large excess of purified nitrogen, and methyl-radical concentrations were monitored by following the absorbance A at 215 nm. Plots of $1/A$ versus time were linear, showing that the rate law for (10.57) is

$$-\frac{1}{2} \frac{d[CH_3 \cdot]}{dt} = k_r[CH_3 \cdot]^2$$

Since $A = \epsilon l [CH_3 \cdot]$, where ϵ is the molar absorption coefficient and l is the cell length, the slope of the second-order plot is $2k_r/\epsilon l$.

A recurrent problem in flash photolysis studies is determination of molar absorption coefficients. Usually these are not available independently because the species under study are transient intermediates, and in most cases the concentrations of such transient species are not measurable by any nonspectrophotometric method. In this case ϵ was determined by measuring the rise and fall of absorbance due to $CH_3 \cdot$ during, as well as after, the photoflash. It was assumed that the total number of methyl radicals generated was twice the number of ethane molecules present at the end of the experiment. Then the absorbance at the end of the flash was corrected to take account of methyl-radical recombination during the flash, and ϵ was calculated from this corrected absorbance, twice the final concentration of ethane, and the path length. Both groups of investigators obtained $k_r = 10^{10.4}$ M^{-1} s^{-1}, in excellent agreement with results obtained at higher temperatures by rotating sector and other methods (see Table 6.1). Since methyl-radical recombination has no activation energy, direct comparison of these various experiments is permissible.

LASER METHODS

Recent advances in laser technology have extended the flash photolysis technique from microsecond to nanosecond and even picosecond time scales.[74] Pulsed lasers can provide single pulses of a few hundred kilowatts power lasting a few nanoseconds. Trains of pulses of a few picoseconds duration and separated by about 10 ns can be obtained (at much lower power levels) from mode-locked lasers. For flash spectroscopy studies a beam splitter transmits part of the laser pulse to a scintillation solution adjacent to the photolysis cell. The scintillation solution fluoresces rapidly, emitting a continuum of wavelengths that provide a secondary analyzing flash of the same duration as the primary laser pulse. This background flash can be delayed for a known time by varying the path length of the secondary beam, 3 m of extra path length correspondong to a 10 ns delay, for example. For kinetic spectrophotometry a conventional flash discharge of several hundred microseconds duration provides a steady background illumination so that changes in absorbance of transient species can be observed over the several nanoseconds to a few microseconds of experimental interest. By contrast with the usual flash photolysis light source, a laser provides monochromatic radiation and hence a finer probe of molecular excited states. On the other hand, unless the laser is tunable over a fairly broad range, a variety of laser sources must be used to span the UV and visible regions.

   Chuang, Hoffman, and Eisenthal's study[75] of photodissociation and recombination of I_2 in CCl_4 solution provides an example of the application of picosecond spectroscopy. Figure 10.7 shows transmittance of a probe beam of wavelength 530 nm at various times before and after excitation of the sample by a 5 ps laser pulse whose energy ranged from 50 to 100 mJ and whose wavelength was

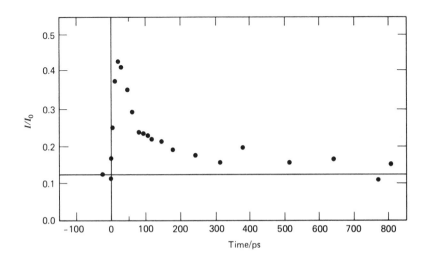

Figure 10.7.   Probe transmittance I/I_0 as a function of time after excitation for I_2 in carbon tetrachloride. [(Reprinted with permission from T. J. Chuang, G. W. Hoffman, and K. B. Eisenthal, *Chem. Phys. Lett.* **25**, 201 (1974).].

also 530 nm. The peak in transmittance of probe light occurs about 20 ps after the pulse. This continuing increase after the excitation can be explained by assuming that energy can be absorbed not only by ground-state I_2 molecules, but also by those in the excited bound state $^3\Pi_{0+u}$. After the excitation pulse depletes ground-state I_2, there is a further increase in transmittance because of collision-induced dissociation of $^3\Pi_{0+u}$ molecules. A rate constant of 10^{11} s^{-1} is obtained for this process.

Betweeen 20 and about 800 ps after the excitation pulse there is a decrease in transmittance to an essentially constant value that is larger than before excitation. This residual transmittance can be attributed to escape of about 25% of the I atoms from the solvent cage in which they were formed by photodissociation. That is, 75% of the I atoms are trapped in the solvent cage and recombine with their original partners. The time constant for this "immediate" recombination can be obtained from the falling transmittance curve and is found to be 140 ps in carbon tetrachloride. The residual 25% of I atoms escape the solvent cage and recombine with different partners in a diffusion-controlled process. Based on the concentration and diffusion coefficient of I atoms, this must take at least 10^{-8} s and hence does not occur to a significant extent on the picosecond time scale. Thus the effect of the solvent cage on recombination rate has been observed directly in this picosecond spectroscopic experiment.

Another important and interesting result of laser development is dissociation of polyatomic molecules as a result of multiple-photon absorption by a single molecule in an intense infrared laser beam.[76] Absorption of enough infrared photons to dissociate many ground-electronic-state molecules — as many as 20–30 photons in some cases — can occur on a time scale much shorter than the collisional lifetime for gases at low pressure. Molecules can be selectively activated by irradiating specific bands, and true unimolecular reactions can occur without the bimolecular, energy-exchanging collisions necessary in a thermal unimolecular process. In some cases laser-induced dissociation is isotope selective;[77] separation of isotopes can be achieved by choosing appropriate wavelengths to activate molecules containing one isotope but not another. Success in laser isotope separation has led to the hope that specific radicals could be synthesized by energizing and breaking particular chemical bonds, but this has been more difficult to achieve. A recent report[83] indicates that molecules pumped to a narrow range of vibrational energy by an infrared laser can more readily undergo absorption of two visible photons (from a second laser) to cause dissociation. This provides a means for studying dynamic behavior of species having large, uniform vibrational excitation.

Multiple-photon excitation at first seems unlikely because all photons in a laser beam have almost identical energies, while the energy difference between vibrational levels decreases with increasing vibrational quantum number as a result of vibrational anharmonicity. Thus a photon that could undergo resonant absorption in raising a molecule from its ground to its first vibrational level would not match the energy difference between, say, the ninth and tenth levels. However, for polyatomic molecules above $v = 3$ or so the density of states is very large, corresponding to a quasicontinuum, and exact coupling between laser photons and a particular vibrational mode becomes less important. Near the dissociation limit there is a true continuum, and absorption becomes essentially nonselective. Thus once the bottleneck of the first few vibrational levels has been passed additional photons are easily absorbed.

Infrared multiple-photon dissociation has been applied extensively to the problem of statistical versus nonstatistical energy distribution in unimolecular reactions. For example, Hicks et al.[78] have studied methyl isocyanide, a reactant whose isomerization to acetonitrile is unimolecular and has been thoroughly studied. Photolysis of CH_3NC was initiated by the 10.696-μm frequency of a CO_2 TEA laser, which pumps the C—NC stretching frequency. In addition to isomerization there was dissociation to form CN, CH, and C_2. These fragments were identified by scanning appropriate wavelengths with a dye laser and observing fluorescence spectra about 300 ns after the excitation pulse. From the laser-induced fluorescence spectrum it was ascertained that CN obtained from CH_3NC was at a significantly higher vibrational and rotational temperature than CN obtained by photolysis of CH_3CN. This might imply a nonstatistical (non RRKM) energy distribution. Other than this difference, however, a remarkable similarity was noted between the reactions of CH_3NC and CH_3CN, the same product fragments being observed in the same sequence and at essentially the same onset time for both reactants. Hence Hicks et al. concluded that the most likely mechanism for photolysis was isomerization of CH_3NC to CH_3CN followed by photolysis of the latter.

Infrared laser absorption and mass spectrometry have also been used to probe fragments obtained from infrared multiple-photon dissociations. Y. T. Lee and coworkers,[79] for example, have carried out elegant experiments using crossed laser and molecular beams. This eliminates any possibility that molecular collisions subsequent to IR excitation may influence the results. These experiments do not provide any evidence that contradicts statistical theory. RRKM calculations satisfactorily predict both the excess energy above the dissociation threshold and the translational energy distribution of the products.

Other applications of lasers include generation of reactive species by visible and ultraviolet photodissociation and monitoring of reactive intermediates by a variety of methods. Further details on the techniques involved may be obtained from references 74 and 76. An example is the multiple-photon photolysis of allene, which produces the C_3 radical.[84] C_3 concentration can be followed by laser-induced fluorescence and rate constants for the reactions of C_3 with C_3H_4, NO, and O_2 can be obtained.

CHEMILUMINESCENCE

Some chemical reactions are accompanied by emission of electromagnetic radiation that does not arise simply from thermal excitation. Examples of such chemiluminescent phenomena are the light of fireflies and certain deep-sea fish, and commercially available safety lights that can be illuminated by mixing two solutions. A typical reaction is oxidation of luminol by ferricyanide in alkaline solution, which produces a greenish-blue emission.

Luminol

Although chemiluminescence in solution is of greatest practical importance, detailed understanding of the underlying excitation mechanisms is restricted essentially to gas-phase systems. In the gas phase recombination reactions and exchange reactions are the main types of processes that produce chemiluminescence. An example of recombination chemiluminescence is the yellow afterglow of nitrogen that has been subjected to an electric discharge. The chemiluminescent reaction is a three-body recombination of ground-state (^4S) nitrogen atoms to form N_2:

$$2N(^4S) + M \rightarrow N_2(^3\Pi_g) + M \qquad (10.58)$$

That the N_2 is in the $^3\Pi_g$ state is inferred from the fact that most of the visible emission corresponds to the electronic transition $^3\Sigma_u^+ \leftarrow ^3\Pi_g$. However, the $^3\Pi_g$ state dissociates to one ground-state and one excited-state (^2D) nitrogen atom, rather than the two ^4S nitrogens. Consequently (10.58) cannot occur simply as written, but must involve some intermediate electronic state that does correlate with $2N(^4S)$. Apparently the $^3\Pi_g$ molecular state is populated by a highly efficient collision-induced radiationless transition from the $^3\Sigma_u^+$ state.[82] Also, the molecular ground state, $^1\Sigma_g^+$, does correlate with $2N(^4S)$, and therefore a considerable fraction of ground-state N_2 is expected to form without electronic excitation. Measurements of absolute emission efficiency indicate that roughly one-third to one-half of total recombination proceeds according to (10.58).

An example of an exchange process that produces chemiluminescence is the much-studied reaction of atomic fluorine with molecular hydrogen.[80] In this case the product HF molecule is not formed in an excited electronic state, but rather 74% of the available energy $(\Delta H + E_0)$ is channeled into vibrational and rotational excitation. For those vibrations that are infrared active this excess energy can be dissipated by infrared chemiluminescence, observation of which makes possible determination of nascent vibrational, rotational, and (by difference) translational energies of reaction products.

The apparatus for infrared chemiluminescence measurements consists of a flow reaction vessel with walls at liquid-nitrogen temperature and a total pressure of 10^{-4} torr. Partial pressures of molecular species are maintained at 10^{-6} to 10^{-7} torr to ensure that product molecules undergo almost no collisions before being condensed out on the cold walls. Infrared emission can be monitored either by conventional grating spectroscopy or by Fourier-transform spectroscopy. Under the conditions of the experiment vibrational relaxation is arrested and the even more rapid process of rotational relaxation occurs only partially. Polanyi and

Woodall[80] found that rate constants for formation of product HF in the $v = 1$, $v = 2$, and $v = 3$ vibrational states were in the ratio 0.31:1:0.47. For DF formed by reacting F with D_2 the relative rates corresponding to the $v = 1$ to $v = 4$ product states were 0.28:0.65:1:0.71.

These rate constants imply that HF and DF are produced with population inversions in their vibrational levels. That is, the $v = 2$, 3, and 4 levels are more highly populated than would be expected on the basis of the Boltzmann distribution law. Such a population inversion implies that it should be possible to observe laser emission of the infrared chemiluminescence, and indeed the $F + H_2$ reaction served as the basis for one of the first chemical lasers.[81] The energy necessary to maintain a population inversion is obtained from the exoergic reaction rather than from optical pumping. Since the $F + H_2$ reaction is an elementary step in the $H_2 + F_2$ chain reaction, which can provide a continuous source of energy, the HF chemical laser has received more study than any other. Chemical lasers can also be used to obtain information about the kinetics of the reactions on which they are based. By monitoring emissions from product molecules produced by pulsed chemical lasers, the time dependence of vibrational-energy-level populations can be studied.

PROBLEMS

10.1. By applying the steady-state approximation to the mechanism (10.42)–(10.47), verify (10.48).

10.2. Consider the chain mechanism for autoxidation of a hydrocarbon RH.

(a) Using the steady-state approximation derive an expression for $-d[O_2]/dt$ under conditions of very low oxygen pressure. Use the initiation reaction

$$2ROOH \xrightarrow{k_i} R\cdot + RO_2\cdot + H_2O_2$$

and the termination reactions

$$
\begin{array}{l}
2R\cdot \\
\text{or} \qquad\qquad\qquad \longrightarrow \text{ inactive products} \\
R\cdot + RO_2\cdot
\end{array}
$$

(b) What relationships among the various activation energies would be necessary for long chains?

10.3. The homogeneous gas-phase reaction of iodine with formaldehyde produces exclusively CO and HI under most conditions [R. Walsh and S. W. Benson, *J. Am. Chem. Soc.*, **88**, 4570 (1966)]. The rates fit the expression

$$\frac{d[HI]}{dt} = \frac{2kK^{1/2}[I_2]^{1/2}[CH_2O]}{1 + k'[HI]/[I_2]}$$

where K is the equilibrium constant for dissociation of I_2. The parameter k follows the expression

$$k = 10^{10.92}e^{-72.9\,\text{kJ mol}^{-1}/RT}$$

over the temperature range $180{-}300°C$.

(a) Devise a plausible mechanism and show that it is in accord with the rate expression given above.

(b) Given an estimated activation energy of $6.3\,kJ\,mol^{-1}$ for the process

$$\cdot CHO + HI \rightarrow CH_2O + I\cdot$$

that $\Delta H_f^{\ominus}(CH_2O) = -115.9\,kJ\,mol^{-1}$, and that $\Delta H_f^{\ominus}(CO) = -110.5\,kJ\,mol^{-1}$, use bond-dissociation energies from Table 6.3 to estimate the bond-dissociation energy for a C—H bond in CH_2O and for the C—H bond in $\cdot CHO$. (Assume negligible temperature dependence for these quantities.)

(c) Is your result in part **b** consistent with the work of R. Klein and L. J. Schoen [*J. Chem. Phys.*, **24**, 1094 (1956)], who observed initiation of a free-radical reaction in photolysis of CH_2O with 365-nm radiation?

10.4.   Dickenson and Carrico [*J. Am. Chem. Soc.*, **56**, 1473 (1934)] have reported the photochlorination of tetrachloroethylene

$$Cl_2 + C_2Cl_4 \rightarrow C_2Cl_6$$

The data of four experiments involving radiation of wavelength 435.8 nm are reported below.

p_{Cl_2} /torr	$p_{C_2Cl_4}$ /torr	$I_0/10^{-4}\,J\,s^{-1}$	$(-d[Cl_2]/dt)/10^{-8}M^{-1}\,s^{-1}$
22.4	11.2	15.8	3.19
13.2	1.9	15.8	1.29
46.7	5.8	14.9	8.79
43.2	2.3	4.21	3.83

I_0 is the total radiant energy flux incident on a reaction cell whose path length was 12 cm. The molar absorption coefficient of Cl_2 is $1.64\,M^{-1}\,cm^{-1}$

(a) Use the Beer-Lambert law to calculate the intensity of radiation transmitted by the reaction mixture and hence I_{abs} (in einsteins s^{-1}).

(b) Calculate the quantum yield for consumption of Cl_2 in each experiment. What can be inferred from the magnitudes of these results?

(c) Assuming that the rate law is of the form

$$\frac{-d[Cl_2]}{dt} = kI_{abs}^a[Cl_2]^b$$

determine the exponents a and b.

(d) Suggest at least one mechanism that is in accord with these experimental observations.

REFERENCES

1.   M. Bodenstein and S. C. Lind, *Z. Phys. Chem.*, **57**, 168 (1907).

2.   J. A. Christiansen, *K. Dan. Vidensk. Selsk., Mat.-Fys. Medd.*, **1**, 14 (1919); K. F. Herzfeld, *Ann. Phys.*, **59**, 635 (1919); M. Polanyi, *Z. Elektrochem.*, **26**, 49 (1920).

3. (a) A Levy, *J. Phys. Chem.,* **62,** 570 (1958); (b) R. B. Timmons and R. E. Weston, Jr., *J. Chem. Phys.,* **41,** 1654 (1964).

4. (a) D. Britton and N. Davidson, *J. Chem. Phys.,* **23,** 2461 (1955); (b) M. N. Plooster and D. Garvin, *J. Am. Chem. Soc.,* **78,** 6003 (1956); (c) D. Britton and R. M. Cole, *J. Phys. Chem.,* **65,** 1302 (1961).

5. M. D. Porter and G. B. Skinner, *J. Chem. Educ.,* **53,** 366 (1976); see also reference 60 in Chapter 8.

6. F. A. Matsen and J. L. Franklin, *J. Am. Chem. Soc.,* **72,** 3337 (1950).

7. S. D. Cooley and R. C. Anderson, *Ind. Eng. Chem.,* **44,** 1402 (1952); G. B. Skinner and G. H. Ringrose, *J. Chem. Phys.,* **43,** 4129 (1965).

8. M. Bodenstein, *Z. Elektrochem.,* **22,** 327 (1916).

9. G. C. Fettis and J. H. Knox, *Prog. React. Kinet.,* **2,** 1 (1964); see also A. F. Trotman-Dickenson, *Adv. Free Radical Chem.,* **1,** 1 (1965).

10. S. W. Benson and J. H. Buss, *J. Chem. Phys.,* **28,** 301 (1958).

11. E. Rabinowitch and H. L. Lehmann, *Trans. Faraday Soc.,* **31,** 689 (1935).

12. M. Bodenstein and H. Lütkemeyer, *Z. Phys. Chem.,* **114,** 208 (1925).

13. W. Jost and G. Jung, *Z. Phys. Chem.,* **B3,** 83 (1929); see discussion in G. K. Rollefson and M. Burton, *Photochemistry and the Mechanism of Chemical Reactions,* Prentice-Hall, New York, 1939, Chapter 11.

14. M. Ritchie, *Proc. R. Soc. (Lond.),* **A146,** 828 (1934).

15. M. Bodenstein and G. Jung, *Z. Phys. Chem.,* **121,** 127 (1926); M. Bodenstein, W. Jost, and G. Jung, *J. Chem. Soc.,* **1929,** 1153.

16. F. Briers and D. L. Chapman, *J. Chem. Soc.,* **1928,** 1802.

17. J. H. Sullivan, *J. Chem. Phys.,* **49,** 1155 (1968).

18. A. Berthoud and H. Bellenot, *Helv. Chim. Acta,* **7,** 307 (1924); D. L. Chapman, F. Briers, and E. Walters, *J. Chem. Soc.,* **1926,** 562; G. M. Burnett and H. W. Melville, *Chem. Rev.,* **54,** 225 (1954); G. M. Burnett and H. W. Melville, in *Investigation of Rates and Mechanisms of Reactions,* Technique of Organic Chemistry, Vol. 8, S. L. Friess, E. S. Lewis, and A. Weissberger, Eds., Interscience, New York, 1963, Chap. XX; R. M. Noyes in *Investigation of Rates and Mechanisms of Reactions,* Part II, Techniques of Chemistry, Vol. 6, A. Weissberger and G. G. Hammes, Eds.., Wiley-Interscience, New York, 1974, Chap. IX.

19. S. H. Bauer, *Science,* **141,** 867 (1963); J. N. Bradley, *Shock Waves in Chemistry and Physics,* Methuen, London, 1962; E. F. Greene and J. P. Toennies, *Chemical Reactions in Shock Waves,* Academic, New York, 1964; A. G. Gaydon and I. R. Hurle, *The Shock Tube in High Temperature Chemical Physics,* Reinhold, New York, 1963.

20. D. Britton and N. Davidson, *J. Chem. Phys.,* **25,** 810 (1956); H. B. Palmer and D. F. Hornig, *J. Chem. Phys.,* **26,** 98 (1957); D. Britton, *J. Phys. Chem.,* **64,** 742 (1960); C. D. Johnson and D. Britton, *J. Chem. Phys.,* **38,** 1455 (1963); R. K. Boyd, G. Burns, T. R. Lawrence, and J. H. Lippiatt, *J. Chem. Phys.,* **49,** 3804 (1968); M. Warshay, *J. Chem. Phys.,* **54,** 4060 (1971); K. Westberg and E. F. Greene, *J. Chem. Phys.,* **56,** 2713 (1972).

21. M. Bodenstein, *Z. Phys. Chem.,* **85,** 329 (1931); J. C. Morris and R. N. Pease, *J. Am. Chem. Soc.,* **61,** 391, 396 (1939); W. J. Kramers and L. A. Moignard, *Trans. Faraday Soc.,* **45,** 903 (1949); P. G. Ashmore and J. Chanmugam, *Trans Faraday Soc.,* **49,** 254 (1953).

22. J. C. Giddings and H. Shin, *Trans. Faraday Soc.,* **57,** 468 (1961); H. Shin, *J. Chem. Phys.,* **39,** 2937 (1963).

23. J. H. Sullivan, *J. Chem. Phys.,* **30,** 1292, 1577 (1959); **36,** 1925 (1962); **39,** 3001 (1963).

24. G. S. Hammond, C. E. Boozer, C. E. Hamilton, and J. N. Sen, *J. Am. Chem. Soc.*, 77, 3238 (1955).

25. (a) G. A. Russell, *J. Chem. Educ.*, 36, 111 (1959); (b) V. N. Kondratiev, in *Comprehensive Chemical Kinetics*, Vol. 2, C. H. Bamford and C. F. H. Tipper, Eds., Elsevier, New York, 1969, pp. 165–173.

26. M. Niclause, J. Lemaire, and M. Letort, *Adv. Photochem.*, 4, 25 (1966).

27. C. A. McDowell and L. K. Sharples, *Can. J. Chem.*, 36, 251, 268 (1958).

28. A. Robertson and W. A. Waters, *Trans. Faraday Soc.*, 42, 201 (1946).

29. H. S. Blanchard, *J. Am. Chem. Soc.*, 82, 2014 (1960); see also reference 25a.

30. M. Szwarc, *Chem. Rev.*, 47, 75 (1950); M. Szwarc, B. N. Ghosh, and A. H. Sehon, *J. Chem. Phys.*, 18, 1142 (1950); M. Szwarc, C. H. Leigh, and A. H. Sehon, *J. Chem. Phys.*, 19, 657 (1951).

31. G. L. Esteban, J. A. Kerr, and A. F. Trotman-Dickenson, *J. Chem. Soc.*, 1963, 3873.

32. J. A. Kerr, *Chem. Rev.*, 66, 465 (1966).

33. T. L. Cottrell, *The Strengths of Chemical Bonds*, 2nd ed., Butterworths, London, 1958; S. W. Benson, *J. Chem. Educ.*, 42, 502 (1965); J. Drowart and P. Goldfinger, *Angew. Chem.*, 6, 581 (1967); D. A. Johnson, *Some Thermodynamic Aspects of Inorganic Chemistry*, Cambridge University Press, London 1968, p. 267.

34. P. J. Flory, *J. Am. Chem. Soc.*, 59, 241 (1937).

35. For general discussions see (a) P. E. M. Allen and C. R. Patrick, *Kinetics and Mechanisms of Polymerization Reactions*, Wiley, New York, 1974; (b) A. M. North, *The Kinetics of Free Radical Polymerization*, Pergamon, Oxford, 1966; (c) C. Walling, *Free Radicals in Solution*, Wiley, New York, 1957; (d) P. J. Flory, *Principles of Polymer Chemistry*, Cornell University Press, Ithaca, 1953.

36. M. Dole, *Introductory Principles of Statistical Thermodynamics*, Prentice-Hall, New York, 1954, Chap. 3.

37. G. M. Burnett and H. W. Melville, *Chem. Rev.*, 54, 225 (1954).

38. M. Szwarc, *Carbanions, Living Polymers, and Electron-Transfer Processes*, Wiley-Interscience, New York, 1968.

39. A. Yamamoto and T. Yamamoto, *J. Polym. Sci. Macromol. Rev.*, 13, 161 (1978); J. A. Moore, Ed., *Macromolecular Synthesis*, Vol. 1, Wiley, New York, 1978; W. M. Saltman, Ed., *The Stereorubbers*, Wiley, New York, 1977; T. Keii, *The Kinetics of Ziegler-Natta Polymerization*, Chapman & Hall, London 1972; J. C. W. Chien, Ed., *Coordination Polymerization*, Academic, New York, 1974; G. N. Schrauzer, Ed., *Transition Metals in Homogeneous Catalysis*, Dekker, New York, 1971; G. Natta and F. Danusso, Eds., *Stereoregular Polymers and Stereospecific Polymerizations*, Vols. 1 and 2, Pergamon, Oxford, 1967.

40. P. J. Cossee, *J. Catalysis*, 3, 80 (1964); E. J. Arlman and P. J. Cossee, *J. Catalysis*, 3, 99 (1964).

41. R. G. Pearson, *Fortschr. Chem. Forsch.*, 41, 75 (1973).

42. M. L. H. Green and R. Mahtab, *J. Chem. Soc., Dalton Trans.*, 1979, 262; K. J. Ivin, J. J. Rooney, and C. D. Stewart, *J. Chem. Soc., Chem. Commun.*, 1978, 603; M. L. H. Green, *Pure Appl. Chem.*, 50, 27 (1978).

43. F. O. Rice and K. K. Rice, *The Aliphatic Free Radicals*, Johns Hopkins Press, Baltimore, 1935.

44. F. A. Paneth and W. Hofeditz, *Chem. Ber.*, 62B, 1335 (1929).

45. F. O. Rice and K. F. Herzfeld, *J. Am. Chem. Soc.*, 56, 284 (1934).

46. P. Goldfinger, M. Letort, and M. Niclause, in *Contrib. étude structure mol.*, Vol. commem. Victor Henri, Desoer, Liège, 1947–48, pp. 283–296; M. F. R. Mulcahy, *Gas Kinetics*, Wiley, New York, 1973, pp. 87–92.

47. F. O. Rice and R. E. Varnerin, *J. Am. Chem. Soc.*, 76, 324 (1954).

48. M. C. Lin and M. H. Back, *Can. J. Chem.*, 45, 3165 (1967).

49. C. P. Quinn, *Proc. R. Soc. (London)*, A275, 190 (1963).

50. M. C. Lin and M. H. Back, *Can. J. Chem.*, 44, 505, 2357 (1966).

51. A. F. Trotman-Dickenson and J. A. Kerr, in G. Porter, Ed., *Prog. React. Kinet.*, 1, Pergamon, Oxford, 1961.

52. J. H. Purnell and D. A. Lethard, *Annu. Rev. Phys. Chem.*, 21, 197 (1970).

53. N. N. Semenov, *Some Problems in Chemical Kinetics and Reactivity*, Vol. 2, Princeton University Press, Princeton, 1959, translated by M. Boudart.

54. R. R. Baldwin and R. W. Walter, *Essays Chem.*, 3, 1 (1972).

55. For detailed treatments of combustion reactions see C. H. Bamford and C. F. H. Tipper, Eds., *Comprehensive Chemical Kinetics*, Vol. 17, Elsevier, Amsterdam, 1977; J. N. Bradley, *Flame and Combustion Phenomena*, Chapman and Hall, London, 1972; R. C. Anderson, *J. Chem. Educ.*, 44, 248 (1967).

56. E. A. Ogryzlo, *J. Chem. Educ.*, 42, 647 (1965).

57. J. Michl, *J. Am. Chem. Soc.*, 93, 523 (1971); E. F. Ullman, *Acc. Chem. Res.*, 1, 353 (1968).

58. R. C. Dougherty, *J. Am. Chem. Soc.*, 93, 7187 (1971); J. Michl, *Mol. Photochem.*, 4, 243 (1972); H. E. Zimmerman, K. S. Kamm, and D. P. Werthemann, *J. Am. Chem. Soc.*, 96, 7821 (1974) and references therein.

59. J. F. Ireland and P. A. H. Wyatt, *Adv. Phys. Org. Chem.*, 12, 131 (1976); T. Förster, *Pure Appl. Chem.*, 34, 225 (1973); T. Förster, *Pure Appl. Chem.*, 24, 443 (1970); A. Weller, *Prog. React. Kinet.*, 1, 189 (1961); A. Weller, *Disc. Faraday Soc.*, 27, 28 (1959).

60. G. Jackson and G. Porter, *Proc. R. Soc. (London)*, A260, 13 (1961).

61. M. Wrighton, *Chem. Rev.*, 74, 401 (1974).

62. A. J. Merer and R. S. Mulliken, *Chem. Rev.*, 69, 639 (1969).

63. A. Bylina and Z. R. Grabowski, *Trans. Faraday Soc.*, 65, 458 (1969); Z. R. Grabowski and A. Bylina, *Trans. Faraday Soc.*, 60, 1131 (1964); D. F. Evans, *J. Chem. Soc.*, 1960, 1735.

64. J. Saltiel, J. D'Agostino, E. D. Megarity, L. Metts, K. R. Neuberger, M. Wrighton, and O. C. Zafiriou, in *Organic Photochemistry*, Vol. 3, O. L. Chapman, Ed., Dekker, New York, 1973 and references therein.

65. R. A. Caldwell, G. W. Sovocool, and R. P. Gajewski, *J. Am. Chem. Soc.*, 95, 2549 (1973); F. D. Lewis, C. E. Hoyle, and D. E. Johnson, *J. Am. Chem. Soc.*, 97, 3267 (1975).

66. G. H. Myers, D. M. Silver, and F. Kaufman, *J. Chem. Phys.*, 44, 718 (1966).

67. O. Stern and M. Volmer, *Phys. Z.*, 20, 183 (1919).

68. H. G. Curme and G. K. Rollefson, *J. Am. Chem. Soc.*, 74, 3766 (1952).

69. T. Förster, *Naturwissenschaften*, 36, 186 (1949); T. Förster, *Z. Elektrochem.*, 54, 42, 531 (1950); A. Weller, *Z. Elektrochem.*, 64, 55 (1960); A. Weller, *Z. Phys. Chem. (Neue Folge)*, 17, 224 (1958).

70. R. G. W. Norrish and G. Porter, *Nature*, 164, 658 (1949); G. Porter, *Proc. R. Soc., (London)*, A200, 284 (1950); R. G. W. Norrish and G. Porter, *Disc. Faraday Soc.*, 17, 40 (1954); G. Porter, *Z. Elektrochem.*, 64, 59 (1960); G. Porter and M. A. West, in *Investi-*

gation of Rates and Mechanisms of Reactions, Part II, G. G. Hammes, Ed., Techniques of Chemistry, Vol. 6, Wiley-Interscience, New York, 1974, Chap. X.

71. R. G. W. Norrish, *Chem. Br.,* **1**, 289 (1965).

72. N. Basco, D. G. L. James, and R. D. Suart, *Int. J. Chem. Kinet.,* **2**, 215 (1970).

73. H. E. van den Berg, A. B. Callear, and R. J. Norstrom, *Chem. Phys. Lett.,* **4**, 101 (1969).

74. M. C. Lin and J. R. McDonald, in D. W. Setser, Ed., *Reactive Intermediates in the Gas Phase,* Academic, New York, 1979, Chap. 4; K. B. Eisenthal, *Annu. Rev. Phys. Chem.,* **28**, 207 (1977); S. Kimel and S. Speiser, *Chem. Rev.,* **77**, 437 (1977); K. J. Kaufmann and P. M. Rentzepis, *Acc. Chem. Res.,* **8**, 407 (1975); P. M. Rentzepis, *Science,* **169**, 239 (1970); G. Porter and M. R. Topp, *Proc. R. Soc. (London),* **A315**, 163 (1970); J. R. Novak and M. W. Windsor, *Proc. R. Soc. (London),* **A308**, 95 (1968).

75. T. J. Chuang, G. W. Hoffman, and K. B. Eisenthal, *Chem. Phys. Lett.,* **25**, 201 (1974).

76. P. A. Schulz, Aa. S. Sudbø, D. J. Kranjovich, H. S. Kwok, Y. R. Shen, and Y. T. Lee, *Ann. Rev. Phys. Chem.,* **30**, 379 (1979); K. V. Reddy, R. G. Bray, and M. J. Berry, in A. H. Zewail, Ed., *Advances in Laser Chemistry,* Springer-Verlag, New York, 1978, p. 48; N. Bloembergen and E. Yablonovitch, *Phys. Today,* May 1978, p. 23; R. V. Ambartsumian and V. S. Letokhov, Chapter 2 in C. B. Moore, Ed., *Chemical and Biochemical Applications of Lasers,* Academic, New York, 1977.

77. R. V. Ambartzumian, V. S. Letokhov, E. A. Ryabov, and N. V. Chekalin, *Pis'ma Zh. Eksp. Teor. Fiz.,* **20**, 597 (1974); J. L. Lyman, R. V. Jensen, J. Rink, C. P. Robinson, and S. D. Rockwood, *Appl. Phys. Lett.,* **27**, 87 (1975).

78. K. W. Hicks, M. L. Lesiecki, S. M. Riseman, and W. A. Guillory, *J. Phys. Chem.,* **83**, 1936 (1979).

79. Aa. Sudbø, P. Schulz, D. Krajnovich, Y. R. Shen, and Y. T. Lee, in *Advances in Laser Chemistry,* A. H. Zewail, Ed., Springer-Verlag, New York, 1978, p. 308, and references therein.

80. J. C. Polanyi and K. B. Woodall, *J. Chem. Phys.,* **57**, 1574 (1972).

81. K. L. Kompa, *Chemical Lasers, Top. Cur. Chem.,* **37** (1973); J. C. Polanyi, *Appl. Opt.,* **10**, 1717 (1971).

82. I. M. Campbell and B. A. Thrush, *Proc. R. Soc. (London),* **A296**, 201 (1967); B. A. Thrush, *Chem. Br.,* **2**, 287 (1966).

83. R. C. Dunbar, J. D. Hays, J. P. Honovich, and N. B. Lev, *J. Am. Chem. Soc.,* **102**, 3950 (1980).

84. M. L. Lesiecki, K. W. Hicks, A. Orenstein, and W. A. Guillory, *Chem. Phys., Lett.,* **71**, 72 (1980).

APPENDIX

```
C          ***** LEAST SQUARES *****
C     PROGRAM FOR FITTING PARAMETERS A, B, C,... ETC. TO OBSERVED
C     VALUES OF VARIABLES X, Y, Z,... ETC., GIVEN A FUNCTION
C          FØ(X,Y,Z,...,A,B,C,...) = Ø
C     PARTIAL DERIVATIVES OF FUNCTION FØ WITH RESPECT TO PARAMETERS
C     A, B, C,... MUST BE DEFINED AS FA, FB, FC,...
C     IF WEIGHTING OF DATA POINTS IS TO BE USED, PARTIAL DERIVATIVES
C     OF FØ WITH RESPECT TO X, Y, Z,... MUST BE DEFINED AS FX, FY, FZ,...
C     PROGRAM AS WRITTEN HANDLES WEIGHTING AND INCLUDES 2 VARIABLES
C     (X,Y) AND 2 PARAMETERS (A,B).
C     INITIAL ESTIMATES OF PARAMETERS A AND B MUST BE SUPPLIED AS INPUT.
C     THESE ARE CORRECTED BY FACTORS AA AND BB.  IF INITIAL ESTIMATES
C     ARE GOOD, FIRST CORRECTION WILL GIVE GOOD RESULTS. IF NOT, SEVERAL
C     ITERATIONS MAY BE NECESSARY.  THIS PROGRAM DOES EIGHT ITERATIONS.
C     CHANGE NITER FOR MORE. IF AA AND BB ARE OF SAME ORDER OF MAGNITUDE
C     AS A AND B, CONVERGENCE MAY NOT OCCUR. A FACTOR FRAC IS MULTIPLIED
C     INTO AA AND BB BEFORE THEY ARE APPLIED TO A AND B.  REDUCING FRAC
C     FROM THE DEFAULT VALUE OF 1.Ø WILL IMPROVE CONVERGENCE WHEN
C     INITIAL ESTIMATES OF A AND B ARE POOR.
C     METHOD USED IS DESCRIBED BY W. E. DEMING, "STATISTICAL ADJUSTMENT
C     OF DATA," WILEY, N.Y., 1943, CHAPTER 9.
C          DIMENSION X(1ØØ),Y(1ØØ),FF(2,2),WTX(1ØØ),WTY(1ØØ),LABEL(4Ø)
C
C     INPUT IS FROM UNIT NUMBER LIN AND OUTPUT IS TO UNIT NUMBER LOUT.
C     DEFINE APPROPRIATE UNIT NUMBERS IN NEXT TWO LINES.
      LIN=2
      LOUT=3
      NITER=8
      FRAC=1.Ø
C
C     PROGRAM CONTINUES UNTIL BLANK LABEL CARD IS ENCOUNTERED.
    1 READ(LIN,1ØØ1)(LABEL(I),I=1,4Ø)
 1ØØ1 FORMAT (4ØA2)
      DO 2  I=1,4Ø
      IF(LABEL(I) .NE. '  ')GO TO 3
    2 CONTINUE
      GO TO 999
    3 WRITE(LOUT,1ØØ2)(LABEL(I),I=1,4Ø)
 1ØØ2 FORMAT (1H1,4ØA2/)
      WRITE(LOUT,1ØØ3)
 1ØØ3 FORMAT(15H       INPUT X(I),3X,12HINPUT WTX(I),3X,1ØHINPUT Y(I),3X,
     1  12HINPUT WTY(I))
C
C     READ EXPERIMENTAL DATA FOR VARIABLES.
C     WTX(I) AND WTY(I) ARE RECIPROCAL SQUARE ROOTS OF WEIGHTS ASSIGNED
C     TO THE ITH DATA PAIR.  THUS WTX(I) CAN BE SET TO THE STANDARD
C     DEVIATION IN X(I), ETC., IF DESIRED.
C     LAST DATA CARD SHOULD HAVE NEGATIVE WEIGHTS TO SIGNAL
C     END OF EXPERIMENTAL VALUES.
      I=Ø
    4 I=I+1
      IF(I .GT. 1ØØ) WRITE(IOUT,1Ø16)
 1Ø16 FORMAT(55H TOO MANY DATA POINTS! PROGRAM DIMENSIONED FOR 1ØØ MAX.)
      IF(I .GT. 1ØØ) GO TO 5
      READ(LIN,1ØØ4) X(I),WTX(I),Y(I),WTY(I)
 1ØØ4 FORMAT (4E12.5)
      IF(WTX(I) .LT. Ø.Ø .OR. WTY(I) .LT. Ø.Ø)GO TO 5
      WRITE(LOUT,1ØØ5) X(I),WTX(I),Y(I),WTY(I)
 1ØØ5 FORMAT (1H ,4(1PE14.5))
      WTX(I)=1.Ø/(WTX(I)**2)
      WTY(I)=1.Ø/(WTY(I)**2)
      GO TO 4
    5 NPTS=I-1
      READ(LIN,1ØØ6)    A,B
 1ØØ6 FORMAT (2E12.5)
```

434

```
      DO 10 KJ=1,NITER
      WRITE (LOUT,1017) KJ
 1017 FORMAT(1H0,14HITERATION NO. ,I2)
      WRITE(LOUT,1007) A
 1007 FORMAT (25H THE INPUT VALUE OF A WAS,1PE14.5)
      WRITE(LOUT,1008) B
 1008 FORMAT (25H THE INPUT VALUE OF B WAS,1PE14.5)
      DO 7 I=1,2
      DO 6 J=1,2
    6 FF(I,J)=0.0
    7 CONTINUE
      FX1=0.0
      FX2=0.0
      FFOO=0.0
      WRITE(LOUT,1014)
 1014 FORMAT(1H ,6X,4HX(I),10X,4HY(I),11X,2HF0)
      DO 8  I=1,NPTS
C
C     MODIFY THIS SECTION TO SUIT FUNCTION TO BE FITTED.  SEE DEFINITIONS
C     OF F0,FA,FB,FX, AND FY IN INITIAL COMMENTS ABOVE.
C     IN STATEMENTS BELOW X REPRESENTS TIME, Y REPRESENTS CONCENTRATION,
C     A REPRESENTS INITIAL CONC., AND B REPRESENTS RATE CONSTANT.
      F0 = A*EXP(-B*X(I)) - Y(I)
      FX = -A*B*EXP(-B*X(I))
      FY = -1.0
      FA = EXP(-B*X(I))
      FB = -A*X(I)*EXP(-B*X(I))
C     END OF FUNCTION SECTION.
C
      ZL=FX*FX/WTX(I)+FY*FY/WTY(I)
      FF(1,1)=FF(1,1)+FA*FA/ZL
      FF(1,2)=FF(1,2)+FA*FB/ZL
      FF(2,2)=FF(2,2)+FB*FB/ZL
      FX1=FX1+F0*FA/ZL
      FX2=FX2+F0*FB/ZL
      FFOO=FFOO+F0*F0/ZL
C
C     PRINT VALUE OF F0 FOR EACH DATA POINT AS A CHECK ON GOODNESS OF FIT
      WRITE(LOUT,1015) X(I),Y(I),F0
 1015 FORMAT(3(1PE14.5))
    8 CONTINUE
      DENOM=FF(1,1)*FF(2,2)-FF(1,2)*FF(1,2)
      AA=(FX1*FF(2,2)-FX2*FF(1,2))/DENOM
      BB=(FX2*FF(1,1)-FX1*FF(1,2))/DENOM
      COFA=(FF(2,2))/DENOM
      COFB=(FF(1,1))/DENOM
      SUMSQ=FFOO-FX1*AA-FX2*BB
      AT=N
      SIGMA=SQRT(ABS(SUMSQ/(AT-2.0)))
      ERRORA=SQRT(ABS(COFA))*SIGMA
      ERRORB=SQRT(ABS(COFB))*SIGMA
      A=A-AA*FRAC
      B=B-BB*FRAC
      WRITE(LOUT,1009) A
 1009 FORMAT (28H THE CORRECTED VALUE OF A IS,1PE15.5)
      WRITE(LOUT,1010) B
 1010 FORMAT (28H THE CORRECTED VALUE OF B IS,1PE15.5)
      WRITE(LOUT,1011) ERRORA
 1011 FORMAT (28H THE STANDARD ERROR IN A IS ,1PE15.5)
      WRITE(LOUT,1012) ERRORB
 1012 FORMAT (28H THE STANDARD ERROR IN B IS ,1PE15.5)
      WRITE(LOUT,1013) SIGMA
 1013 FORMAT (28H THE STANDARD DEVIATION IS  ,1PE15.5)
   10 CONTINUE
      GO TO 1
  999 CONTINUE
      END
```

AUTHOR INDEX

SUBJECT INDEX